Thermal Power Plants - Advanced Applications

Thermal Power Plants - Advanced Applications

Editor:

Harinirina Randrianarisoa, Ph.D.

ArclerPress

www.arclerpress.com

Thermal Power Plants - Advanced Applications

Editor: Harinirina Randrianarisoa, Ph.D.

708 3rd Avenue, 6th Floor
New York
NY 10017
United States of America
www.arclerpress.com

ISBN: 978-1-68094-363-4

Library of Congress Control Number: 2016950942

About the Editor

Harinirina Randrianarisoa, Ph.D.

Hari Randrianarisoa obtained is Ph.D from Institute of Electronics, Microelectronics and Nanotechnology (IEMN) of the University of Science and Technology – Lille (France). His thesis focused for designing and manufacturing of thermal micro sensors. His interests span from on renewable energy especially solar energy. He has a strong knowledge in technology transfer field. He is currently working as a Business advisor for economic development organization in Québec City and as a consultant Scientist advisor.

Contents

List of Contributors

Linas Gelažanskas
Department of Engineering, Lancaster University, Bailrigg, Lancaster LA1 4YR, UK

Audrius Baranauskas
Department of Power Systems, Kaunas University of Technology, Studentu Str. 48, Kaunas LT-51367, Lithuania
Strategy and Research Division, LITGRID AB, A. Juozapaviciaus Str. 13, Vilnius LT-09311, Lithuania

Kelum A.A. Gamage
Department of Engineering, Lancaster University, Bailrigg, Lancaster LA1 4YR, UK

Mindaugas Ažubalis
Department of Power Systems, Kaunas University of Technology, Studentu Str. 48, Kaunas LT-51367, Lithuania

Masoud Rokni
Department of Mechanical Engineering, Thermal Energy Section, Technical University of Denmark, Nils Koppels Allé, Copenhagen 2800, Denmark

Zhongguang Fu
Key Laboratory of Condition Monitoring and Control for Power Plant Equipment, North China Electric Power University, Ministry of Education, Changping District, Beijing 102206, China

Ke Lu
Key Laboratory of Condition Monitoring and Control for Power Plant Equipment, North China Electric Power University, Ministry of Education, Changping District, Beijing 102206, China

Yiming Zhu
Key Laboratory of Condition Monitoring and Control for Power Plant Equipment, North China Electric Power University, Ministry of Education, Changping District, Beijing 102206, China

Rogelio Peón Menéndez
Group TSK, 33203 Gijón, Asturias, Spain

Juan Á. Martínez
Department of Electrical Engineering, Universidad de Oviedo, 33203 Gijón, Asturias, Spain

Miguel J. Prieto
Department of Electrical Engineering, Universidad de Oviedo, 33203 Gijón, Asturias, Spain

Lourdes Á. Barcia
González Soriano S.A., 33420 Llanera, Asturias, Spain

Juan M. Martín Sánchez
ADEX, S.L., 28031 Madrid, Spain

A G Malchik
Yurga Institute of Technology, National Research Tomsk Polytechnic University Affiliate, 652050, Yurga, ul. Leningradskaya 26

S V Litovkin
Yurga Institute of Technology, National Research Tomsk Polytechnic University Affiliate, 652050, Yurga, ul. Leningradskaya 26

P V Rodionov
Yurga Institute of Technology, National Research Tomsk Polytechnic University Affiliate, 652050, Yurga, ul. Leningradskaya 26

V V Kozik
Yurga Institute of Technology, National Research Tomsk Polytechnic University Affiliate, 652050, Yurga, ul. Leningradskaya 26
National Research Tomsk State University, pr. Lenina 36, Tomsk, 634050 Russia

M A Gaydamak
Yurga Institute of Technology, National Research Tomsk Polytechnic University Affiliate, 652050, Yurga, ul. Leningradskaya 26
National Research Tomsk State University, pr. Lenina 36, Tomsk, 634050 Russia

R S Fediuk
Far Eastern Federal University, Vladivostok, Russia

A M Yushin
Far Eastern Federal University, Vladivostok, Russia

Manuel Romero
IMDEA Energy Institute, Ramón de la Sagra 3, 28935 Móstoles, Spain

José González-Aguilar
IMDEA Energy Institute, Ramón de la Sagra 3, 28935 Móstoles, Spain

Xiaogang Xu
School of Energy Power and Mechanical Engineering, North China Electric Power University, 619 Yonghua North Street, Baoding, 071003, China

Songling Wang
School of Energy Power and Mechanical Engineering, North China Electric Power University, 619 Yonghua North Street, Baoding, 071003, China

Jinlian Liu
School of Energy Power and Mechanical Engineering, North China Electric Power University, 619 Yonghua North Street, Baoding, 071003, China

Xinyu Liu
Department of Mechanical Engineering, McGill University, 817 Sherbrooke Street West, Montreal, Quebec H3A 0C3, Canada

Shengli Liao
Institute of Hydropower System & Hydroinformatics, Dalian University of Technology, Dalian 116024, China

Zhifu Li
Institute of Hydropower System & Hydroinformatics, Dalian University of Technology, Dalian 116024, China

Gang Li
Institute of Hydropower System & Hydroinformatics, Dalian University of Technology, Dalian 116024, China

Jiayang Wang
Institute of Hydropower System & Hydroinformatics, Dalian University of Technology, Dalian 116024, China

Xinyu Wu
Institute of Hydropower System & Hydroinformatics, Dalian University of Technology, Dalian 116024, China

Aitor Ibarguren
IK4-Tekniker, Calle Iñaki Goenaga 5, Eibar 20600, Spain

Jorge Molina
IK4-Tekniker, Calle Iñaki Goenaga 5, Eibar 20600, Spain

Loreto Susperregi
IK4-Tekniker, Calle Iñaki Goenaga 5, Eibar 20600, Spain

Iñaki Maurtua
IK4-Tekniker, Calle Iñaki Goenaga 5, Eibar 20600, Spain

Lourdes A. Barcia
[1]Normagrup Technology S.A. Llanera 33420, Asturias, Spain

Rogelio Peón Menéndez
Grupo TSK, Gijón 33203, Asturias, Spain

Juan Á. Martínez Esteban
Department of Electrical Engineering, University of Oviedo, Gijón 33203, Asturias, Spain

Miguel A. José Prieto
Department of Electrical Engineering, University of Oviedo, Gijón 33203, Asturias, Spain

Juan A. Martín Ramos
Department of Electrical Engineering, University of Oviedo, Gijón 33203, Asturias, Spain

F. Javier de Cos Juez
Department of Explotation and Prospection of Mining, University of Oviedo, Asturias 33004, Spain

Antonio Nevado Reviriego
Department of Electrical, Electronics and Control Engineering, National Distance Education University, Madrid 28040, Spain

E. B. Elanchezhian
Department of Electrical Engineering, Annamalai University, Chidambaram, India

S. Subramanian
Department of Electrical Engineering, Annamalai University, Chidambaram, India

S. Ganesan
Department of Electrical Engineering, Annamalai University, Chidambaram, India

Punnaiah Veeraboina
Department of Electrical Engineering, Centre for DNA Fingerprinting and Diagnostics (CDFD), Hyderabad, India

G. Yesuratnam
Department of Electrical Engineering, Osmania University, Hyderabad, India

Shengli Liao
Institute of Hydropower and Hydroinformatics, Dalian University of Technology, Dalian 116024, China

Chuntian Cheng
Institute of Hydropower and Hydroinformatics, Dalian University of Technology, Dalian 116024, China

Jing Wang
Huadian Electric Power Research Institute, Hangzhou 310030, China

Zhongkai Feng
Institute of Hydropower and Hydroinformatics, Dalian University of Technology, Dalian 116024, China

Rajashekar P Mandi
Energy Efficiency and Renewable Energy Division, Power Research Institute, Bangalore, India

S Seetharamu
Energy Efficiency and Renewable Energy Division, Power Research Institute, Bangalore, India

Udaykumar R Yaragatti
Dept. of Electrical and Electronic Engineering Central, National Institute of Technology, Surathakal, India

Zong-Xiao Yang
Institute of Systems Science and Engineering, Henan University of Science and Technology, Luoyang 471003, China

Lei Song
Institute of Systems Science and Engineering, Henan University of Science and Technology, Luoyang 471003, China

Chun-Yang Zhang
Institute of Systems Science and Engineering, Henan University of Science and Technology, Luoyang 471003, China

Chong Li
Institute of Systems Science and Engineering, Henan University of Science and Technology, Luoyang 471003, China

Xiao-Bo Yuan
Institute of Systems Science and Engineering, Henan University of Science and Technology, Luoyang 471003, China

Selvaraj Durai
Department of Electrical Engineering, Annamalai University, Chidambaram, India

Srikrishna Subramanian
Department of Electrical Engineering, Annamalai University, Chidambaram, India

Sivarajan Ganesan
Department of Electrical Engineering, Annamalai University, Chidambaram, India

Paul Tait
University of Edinburgh, Institute for Materials and Processes, School of Engineering, Sanderson Building, The King's Buildings, Mayfield Road, Edinburgh EH9 3JL, United Kingdom

Bill Buschle
University of Edinburgh, Institute for Materials and Processes, School of Engineering, Sanderson Building, The King's Buildings, Mayfield Road, Edinburgh EH9 3JL, United Kingdom

Ilja Ausner
Sulzer Chemtech Ltd, Sulzer-Allee 48, 8404 Winterthur, Switzerland

Prashant Valluri
University of Edinburgh, Institute for Materials and Processes, School of Engineering, Sanderson Building, The King's Buildings, Mayfield Road, Edinburgh EH9 3JL, United Kingdom

Marc Wehrli
Sulzer Chemtech Ltd, Sulzer-Allee 48, 8404 Winterthur, Switzerland

Mathieu Lucquiaud
University of Edinburgh, Institute for Materials and Processes, School of Engineering, Sanderson Building, The King's Buildings, Mayfield Road, Edinburgh EH9 3JL, United Kingdom

Lakhoua M.N
The National Engineering School of Carthage, University of Carthage, Research Unit: Signals & Mechatronic Systems, Tunisia

Ben Hamouda M
The National Engineering School of Carthage, University of Carthage, Research Unit: Signals & Mechatronic Systems, Tunisia

Glaa R
The National Engineering School of Carthage, University of Carthage, Research Unit: Signals & Mechatronic Systems, Tunisia

El Amraoui L
The National Engineering School of Carthage, University of Carthage, Research Unit: Signals & Mechatronic Systems, Tunisia

Svetlana A. Kuznetsova
Institute of Power Engineering, National research Tomsk Polytechnic University, Tomsk 634050, Russia

Vyacheslav I. Maksimov
Institute of Power Engineering, National research Tomsk Polytechnic University, Tomsk 634050, Russia

Vitali Dubrovskiy
National Research Tomsk Polytechnic University, 634050 Tomsk, Russia

Marina Zubova
Institute of Economics and Business Process Management, Department of Economics and Organization of energy and transport sectors, Russia

Nikolai Sedelnikov
National Research Tomsk Polytechnic University, 634050 Tomsk, Russia

Anna Dihnova
Institute of Economics and Business Process Management, Department of Economics and Organization of energy and transport sectors, Russia

Robert A. MIKULANDRIC
Faculty of Mechanical Engineering and Naval Architecture, University of Zagreb, Zagreb, Croatia

Dražen M. LON^AR
Faculty of Mechanical Engineering and Naval Architecture, University of Zagreb, Zagreb, Croatia

Dejan B. CVETINOVI
Vin~a Institute of Nuclear Science, University of Belgrade, Belgrade, Serbia

Gabriel SPIRIDON
R&D Institute for Scientific Research and Technological Engineering for Automation - IPA SA, Bucharest, Romania

Daniel R. SCHNEIDER
Faculty of Mechanical Engineering and Naval Architecture, University of Zagreb, Zagreb, Croatia

Preface

Thermal power plants are one of the most important process industries for engineering professionals. The most fundamental challenge is meeting the growing power demand in sustainable and efficient ways. Therefore, technological advancement is the requirement of the hour in every field of industrialization for improvement in their efficiencies by reduction in operating expenditure, judicious utilization of natural resources and protection damage to our natural environment. There will come a day when much quantity will be limited like charcoal. In addition, the use of these fossil fuels results in greater pollution, despite the measures taken.

The details of the research results and the new development in fields are highlighted in this book. The subject looked after research and development, starting from power generation to environmental aspects of power plants. With regards to the thermal power plants research, the technological advance discussed in this book include experimentation, investigation assessment and analysis new models, new processes and control process concept for improving efficiencies and reducing emissions of CO. Advances of the use of solar energy to realize a new type of thermal power plant are also discussed.

Editor
Harinirina Randrianarisoa, Ph.D.

CHAPTER 1

HYBRID WIND POWER BALANCE CONTROL STRATEGY USING THERMAL POWER, HYDRO POWER AND FLOW BATTERIES

Linas Gelažanskas[a], Audrius Baranauskas[b, c], Kelum A.A. Gamage[a], Mindaugas Ažubalis[b],

[a]Department of Engineering, Lancaster University, Bailrigg, Lancaster LA1 4YR, UK

[b]Department of Power Systems, Kaunas University of Technology, Studentu Str. 48, Kaunas LT-51367, Lithuania

[c]Strategy and Research Division, LITGRID AB, A. Juozapaviciaus Str. 13, Vilnius LT-09311, Lithuania

ABSTRACT

The increased number of renewable power plants pose threat to power system balance. Their intermittent nature makes it very difficult to predict power output, thus either additional reserve power plants or new storage and control technologies are required. Traditional spinning reserve cannot fully compensate sudden changes in renewable energy power generation. Using new storage technologies such as flow batteries, it is feasible to balance the variations in power and voltage within very short period of time. This paper summarises the controlled use of hybrid flow battery, thermal and hydro power plant system, to support wind power plants to reach near perfect balance, i.e. make the total power output as close as possible to the predicted value. It also investigates the possibility of such technology to take part in the balance of the Lithuanian power system. A dynamic model of flow battery is demonstrated where it evaluates the main parameters such as power, energy, reaction time and efficiency. The required battery size is tested based on range of thermal and hydro power plant reaction times. This work suggests that power and energy of a reasonable size flow battery is sufficient to correct the load and wind power imbalance.

Keywords

Flow battery; Power system balance; Wind farms; Wind power generation

INTRODUCTION

During the past decade, the number of renewable energy sources has increased dramatically. It is forecasted that the growth of green energy generation will increase even further. Policy makers in developed countries create many incentives in favour of the development of low-carbon technologies and subsidise green energy generation. This should help to reduce carbon footprint and climate change. On the other hand, most of renewable energy comes from generators that are inherently very hard to control [1], thus it introduces further complexity in system balancing task.

Up to now in many cases wind turbines or solar panels are being connected to the grid with minimal control. Due to hardly predictable natural resources, like wind or solar irradiation, the

Citation: Linas Gelažanskas, Audrius Baranauskas, Kelum A.A. Gamage, Mindaugas Ažubalis, Hybrid wind power balance control strategy using thermal power, hydro power and flow batteries, doi:10.1016/j.ijepes.2015.08.002.

errors between actual energy output and forecasted generation are relatively large. This increases the difficulty of the energy balance problem: corresponding operators need either more tools or new technologies to come in hand [2]. Increasing advanced spinning reserve to back up intermittent generation would require inadequate level of investment considering exponential growth of power generation using green technologies. Also, this type of reserve has limited power variation capabilities (in the order of minutes) whereas solar power output can drop nearly instantly. The alternative is to use new highly responsive storage technologies [3] and [4] that could be incorporated into the system and shave over-generation as well as generate energy when demand overtakes supply.

Lithuanian Power System (PS) and other Baltic States currently operate synchronously with IPS/UPS synchronous zone and are connected to BRELL power ring, which consists of Belarus, Russian, Estonian, Latvian, and Lithuanian power systems. According to BRELL regulations Baltic States are not required to have automatic secondary power control, however Baltic States are planning to synchronically connect to the power grid of Continental Europe in 2020. This means decentralisation of power system control and responsibility to maintain power and energy balance within strict boundaries [5]. Therefore, it is important to investigate the feasibility of Lithuanian PS to automatically maintain power balance.

National Renewable Energy Laboratory in USA focuses on researching economic feasibility of energy storage and clearly states that high penetration of variable generation increases the need for all flexibility options including storage [6]. They also note that the economic value of energy storage devices is at its best when selling to the entire grid, instead of any single source. However the role of storage for wind generation requires continuous analysis and additional studies including new techniques to evaluate more dynamic grid operation.

Droste-Franke et al. analysed German power system balancing options and concluded that technological progress is needed in the following areas. Firstly, grid expansion and inter-regional connectivity compensating regional shortages of supply from renewable sources in Europe. Secondly, load management could become feasible through technologies such as smart metering, and finally, storage capacities need to be extended[7]. They also conclude that interrupted renewable power smoothing using battery storage system [4] and [8] is the cheapest option at present.

Söder and Hamon investigate wind power capabilities to participate in balancing services. They conclude that wind power plants do not usually participate in balancing services because they must be set to produce less than they are capable in order to be used for up regulation [9]. Margins are kept by spilling the wind, which cannot be stored. A method is proposed to select a certain tertiary reserve control in order to minimise the total cost of the system and maintain stability of the power system with larger portions of wind power. This means that they deal with emergency power system operation modes while our proposed method covers secondary control reserve and optimal share of reserve power between different kinds of generation sources.

Lubosny and Bialek proposed wind farm supervisory control scheme which is suitable to control individual wind mill or separate wind farm in two different ways – using additional storage device or power reserve achieved through part-loading one or more turbines in a wind farm. Authors suggest using wind power filter in order to separate the variability of wind power. They also concluded that elimination of larger power variations can be done more effectively using a central or single energy storage [10]. Therefore our proposed control strategy differs due to the fact that it deals with central control of all wind farms instead of individual ones.

Jiang and Wang similarly to [10] suggest to control wind power plant using power filter. Additionally, they proposed the optimisation model of corresponding filter parameters. However, due to the uncertainty inherent in wind power generation, optimal control during long time periods has difficulties predicting wind power and is unpractical in actual real-time operation [11]. Besides it requires additional computational resources and time. Active power losses and state of charge of storage devices depend on wind power generation, therefore it is hard to maintain the proper charge level and mitigate wind power fluctuation. Authors conclude that two-time-scale coordination control method gives controversial results because the required battery power reaches 33% of installed wind power (in our case it reaches 5–25% depending on power system operation mode, discussed later in the paper) while the power fluctuation allowance is up to 10%. Finally, the capacity component of the battery dominates (comparing to power) which means that storage devices are controlled according to wind power trends. The control strategy proposed in this paper controls storage devices according to high frequency component of wind power imbalance and it allows reaching 100% power balance with reasonably lower ratings of storage device.

Abbey et al. suggests using filters and neural networks to control two different types or multiple storage devises [12]. It is novel and interesting method but too complex for wind power balancing purposes on real-time operation. In principle, multiple levels of storage is needed only in new areas such as island household networks with renewable sources or micro-grids while wind power integration to conventional power systems usually require only short-term storage because the trends of wind power imbalance could be compensated by thermal or hydro power plants in more economical way. Our proposed hybrid wind power balance control strategy composes of conventional generation and energy storage control from power system operator point of view, which means central control in more efficient manner.

Wang et al. studied operational reliability of power system with high wind power penetration [13]. They have concluded that energy storage systems dramatically increase reliability of systems with wind farms. Authors also analyse and show the required battery sizing for certain reliability index.

Khalid and Savkin proposed new semi-distributed storage configuration [14] and using model predictive control [15] identified the optimal capacity of battery energy storage system. However the purpose was to identify the optimal capacity only taking into account the system ramp rates while our paper also deals with installed power of energy storage, conventional power plants optimal control and active power reserve optimisation. In addition we have used actual wind data of 10 days with a time step of one second instead of 1 day and 10 min time step. Yuan et al. proposed dual-battery energy storage system[16] which consist of two separate battery storage systems. One of them is suitable for positive error compensation where the other one is suitable for negative ones. They also proposed three indices for the assessment of the performance on wind power dispatchability which could be identified by using sequential Monte Carlo simulation. However the time step is one hour which means that little dynamic behaviour could be represented. The authors also do not introduce any optimisation task.

Ansari and Velusami have been investigating the dynamic stability of hybrid autonomous wind – diesel with battery energy storage system. They proposed dual mode linguistic hedge fuzzy logic controller [17] and have shown its advantages comparing to traditional fuzzy logic and PI controllers. Kalantar and Mousavi G. replaces less effective and high pollution diesel generator to more flexible and reliable microturbine with the addition of solar array system to earlier investigated one. In order to maximise power outputs of wind and solar power plants they proposed a model reference adaptive Lyapunov controller[18] and improve the system behaviour comparing to fuzzy logic and PID controllers. Latter Mousavi G. have adapted the proposed method to offshore wind and tidal hybrid system with microturbine and BESS [19]. The authors provide an in depth investigation/review of the autonomous rural hybrid system in literature [17], [18] and [19], however the proposed methods are not suitable for wind integration to large power systems with conventional generation, therefore our paper deals with this issue. In addition, our paper represents the optimal wind balancing power allocation between conventional power plants and energy storage devices.

A feasibility study of hybrid solar-wind-battery system for remote location can be found in[20]. Although it shows that it is possible to replace diesel generators by 100% renewable energy, about 48.6% of energy is dumped due to lack of storage and energy management.

Traditional Automatic Generation Control (AGC) system calculates error of the control area and allocates the required regulating power plants. Then they participate in the system balance according to participation factors [21] in order to keep power system in balance. The participation factors are usually determined according to power plant's parameters such as rate limits [22] and [23], available spinning reserve [24] or economic (cost) characteristics. There are many methods to determine them: major part of power is allocated to the cheapest power plant, the fastest response power plant [25] or combined[21] method. This paper describes energy management method for increasing the quality of wind energy output using conventional Thermal Power Plant (TPP), Hydro Power Plant (HPP) and Flow Batteries (FB).

More proposed ideas of balancing wind power intermittency using energy storage systems can be found in other publications [26], [27] and [28]. In [3] authors show through simulation how flywheel ESS can be used for wind power smoothening.

THE MAIN FEATURES OF THERMAL AND HYDRO POWER PLANTS

In general the limits of power changing rate of TPP and HPP are quite different. The maximum load rate of TPP is about 2% of the installed unit capacity (per minute) while for HPP the maximum load rate could reach 100% of unit size (per minute). However it is impossible to perfectly match the area control error continuously in relation to variations in plants characteristics such as the system frequency, load or wind power plant output. The energy storage devices, such as flow batteries, could significantly improve the flexibility of the system control and reduce the power systems imbalances.

The following traditional transfer function of a classical hydraulic turbine could be expressed:

$$W_{H(s)} = \frac{\Delta \bar{P}_m}{\Delta G} = \frac{1 - T_w s}{1 + 0.5 T_w s} \quad (1)$$

where $\Delta \bar{P}_m$ is the change in turbine mechanical power, ΔG is the change in gate position, and T_w is the water time constant.

Equation shows how the turbine output power reacts to a change in position of gate. Fig. 1clearly shows that the initial power output of hydraulic turbine is two times opposite the value when the gate is opened immediately. This is due to water inertia which is represented by T_w. The complete response of hydraulic turbine in respect to gate step change can be seen in Fig. 1.

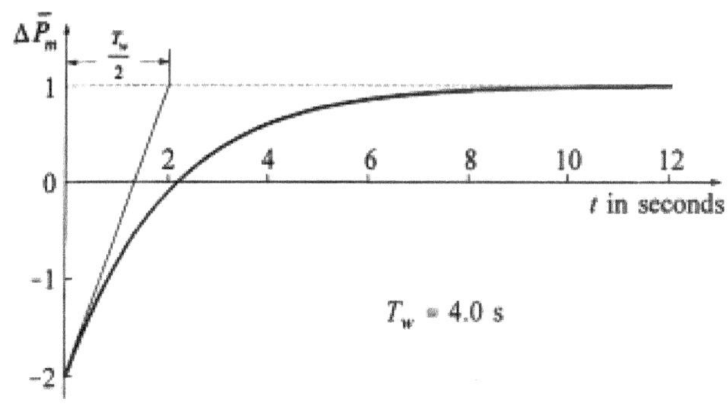

Figure 1. Hydraulic turbine mechanical power output response to gate position step change [29].

Fig. 2 is included in order to demonstrate the behaviour of main variables of HPP, when the gate position change is a ramp function during one second. It represents the main HPP parameters – head, power output and water velocity.

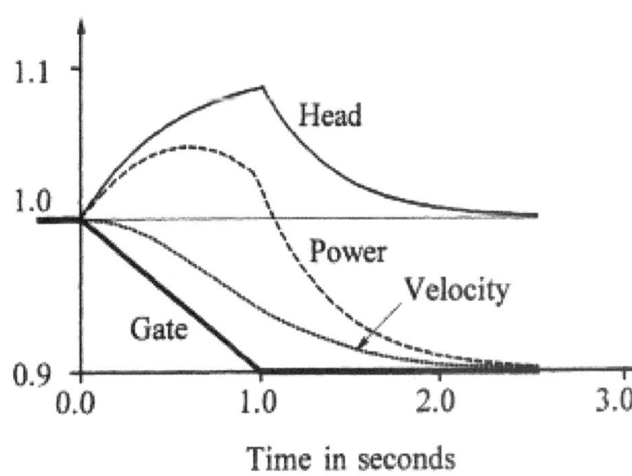

Figure 2. HPP main parameters respect to reduction in gate opening [29].

In order to get stable operation of HPP it is necessary to have permanent and temporary droop compensation when implementing the governors of the hydraulic turbine. The result is a governor with a high droop for fast speed deviations and a low droop for steady state [30].

The simplified transfer function of the steam turbine with only high pressure section and disclaimer of crossover piping in comparison to reheater could be expressed:

$$W_{Ts}(s) = \frac{\Delta T_m}{\Delta V_{CV}} = \frac{1 + T_{RH}F_{HP}s}{(1 + T_{CH}s)(1 + T_{RH}s)} \quad (2)$$

It shows how turbine mechanical torque reacts to change of control valve position. T_{RH} and T_{CH} represent inertia time constant of the reheater and inlet steam chest while F_{HP} shows the fraction of high steam pressure section. Also, T_m is turbine mechanical torque and V_{CV} is control valve position.

The turbine control function is similar to hydraulic turbine. However in order to get stable operation of thermal power plant, it is enough to implement governor with a 4–5% speed droop. So there is no need of two types of droop compensation compare to hydro power plant. On the other hand thermal power plant output highly depends on primary fuel system and boiler operation and control. Fig. 3 illustrates the power output change of TPP in respect to control mode.

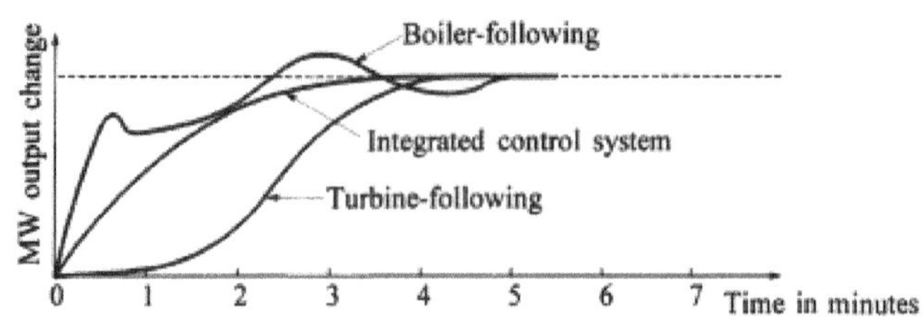

Figure 3. TPP power output in respect to control mode [29].

In the boiler following mode the turbine control valves initiates the changes in generation, while in turbine following mode the changes in generation are implemented by combustion controller [31].

The typical power outputs of steam and hydraulic turbines are shown in Fig. 4. Depending on the boiler type, the control mode and the size of the load change, the power output of thermal power plant night change significantly slower than illustrated. However hydraulic power plant output with a low head could be significantly faster than considered here.

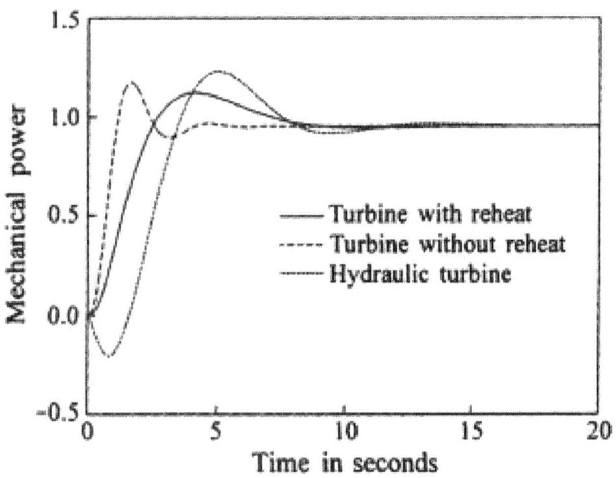

Figure 4. Typical responses of steam and hydraulic units [29].

Flow batteries, sometimes called redox batteries (i.e. oxidation and reduction reactions), are electrochemical systems, which are an alternative between the usual batteries and fuel cells [32]. Flow batteries could be charged as ordinary batteries and provide energy as long as charged electrolyte is supplied. The charging and discharging cycles are possible due to reversible electrochemical reaction between two electrolytes [33]. Conversely charged electrolyte is pumped through separate contours and reaction takes place in special ionic membrane as shown in Fig. 5[34].

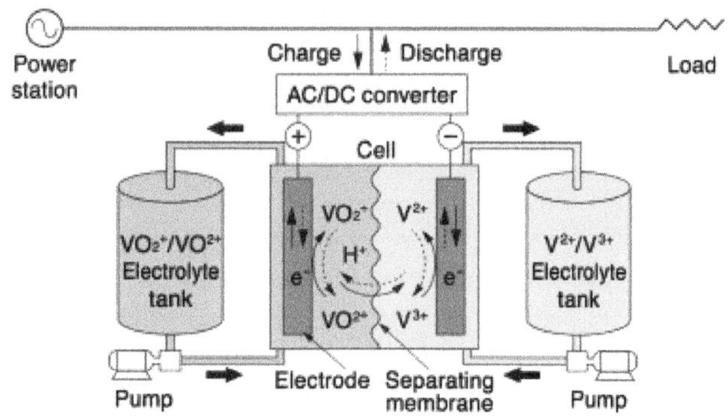

Figure 5. Principle and configuration of a flow battery [34].

The power of flow batteries depends on the surface area of the electrodes. It also depends on the quantity of the bi-pole electrodes. The capacity of flow battery could be increased by expanding the volume of the electrolyte reservoirs, thus increasing the amount of the electrolyte [35] and [36]. The modules of flow batteries are connected into groups in series to ensure the required voltage, but hydraulic circuits are connected in parallel in order to share electrolyte between groups, therefore the same charging level is ensured [37].

Some advantages of flow batteries are that they can operate in low temperature and pressure conditions. Also, the electrolyte could be discharged completely [38]. The electrochemical processes are very fast so the reaction time of the batteries is rather small (0.04–0.06 s) and mostly depends on the operation of power electronics. Thus, flow batteries with power ratings in megawatt range could be very useful for power system balance. As it will be seen from the results of investigations the necessary power ratings of FB reaches tens of MW.

DYNAMIC MODELS OF HYBRID POWER SYSTEMS

MATLAB Simulink software environment was chosen to model and simulate the system. The goal of this research is to investigate the potential of flow battery technology to serve as energy balancing tool. TPP and HPP are used to cover low and mid-frequency imbalances respectively, whereas FB acts on high frequencies thus a trade-off between conventional power plant equipment wear and required battery size can be observed. It is important to determine the required flow battery parameters (power and capacity ratings) in order to maintain balance in the power system. During this research a model of TPP, HPP and FB was proposed with a control strategy.

Control Strategy of Hybrid Power System

The proposed energy balancing method aims to reduce the imbalance of a virtually isolated electric power system. It consists of thermal power plant, hydro power plant, flow battery and a PI controller that mainly deals with the compensation of energy losses related to the flow battery charge/discharge cycle (Fig. 6). The model is designed to offer a tradeoff between equipment wear and the required size of the battery. As discussed later in the paper, bigger time constant in the low-pass filter (LPF) requires larger battery size and higher power ratings. The initial error is the difference between the actual power and the forecasted power:

$$P_{error}=P_{act}-P_f \quad (3)$$

where P_{error} is the initial error, P_{act} is actual power and P_f is forecasted power.

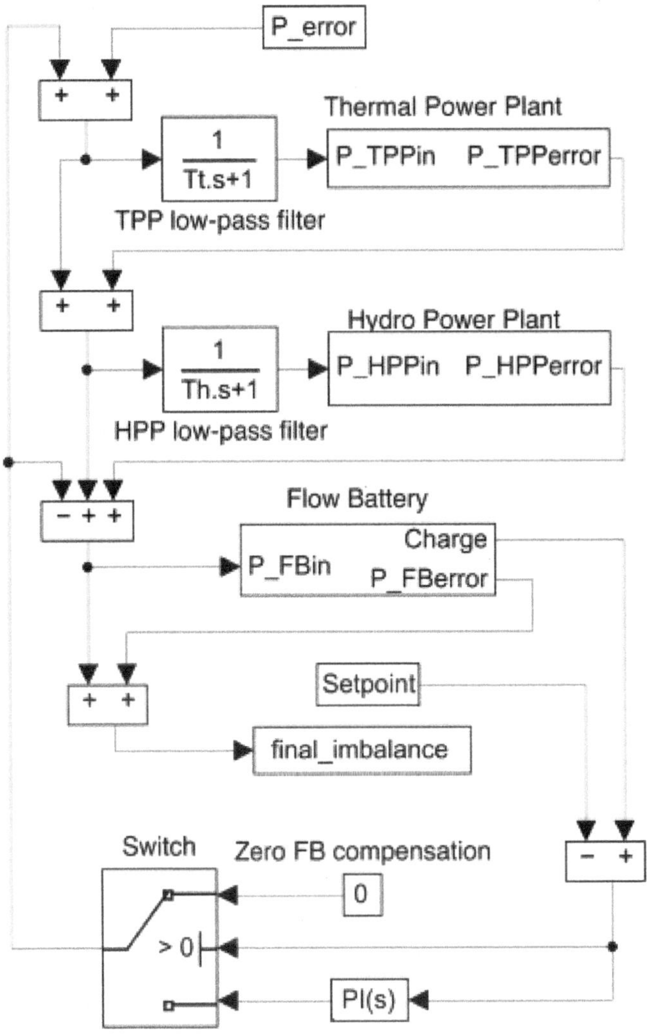

Figure 6. Model of the system.

The error between the actual and forecasted generation is first fed through rate limiting low-pass filter. It cuts off mid and high frequencies and reduces TPP depreciation costs. Next, the error left after TPP, is fed through another low-pass filter with a slightly lower time constant. This separates the mid-frequency band, which then goes to the HPP as a control input. It is clear that HPP changed its power output according to mid-frequency variations. This technique extends the lifetime of the HPP servo equipment. Generally HPP cannot change its power output very fast due to physical limitations, such as slew rate of the servomechanism and water inertia, which might cause water hammer. Thus, HPP responds by adjusting its power output to compensate only mid-frequency component.

On the other hand, the error that is left (mainly consisting of high frequencies) is then fed to a storage device – a flow battery. The small time constant and high charge/discharge power handles high frequency power fluctuations and smoothens the total power output from the system. Due to the fact that flow battery has a cycle efficiency of about 85% [39], additional energy to compensate energy losses is required. This is done by the feedback loop that signals TPP and HPP to overcome the losses. The signal adds nearly constant power compensation.

The proposed model is designed so that the average charge in the battery stays around half of its capacity. This deviation in battery charge from half of the capacity is fed to the PI controller as the error i.e. the desired battery charge set point is half the battery size. The controller responds by signalling TPP and HPP to adjust its generation to maintain the charge of FB at the desired level. Again, the charge in FB fluctuates but on average battery charge is kept constant. This level is proposed to be half the total capacity of the battery to be able to equally compensate both energy shortage and surplus. Also, a switch is added to compensate battery losses only when the battery's state of charge is below 50%. The operation of the battery is discussed in greater detail in Section 'The results of proposed balancing technique'.

To sum up, the frequency spectrum of the initial imbalance is divided into three bands – low, mid and high. The lowest frequencies are handled by TPP, mid-frequencies are cancelled by HPP and what is left – high frequencies – using a flow battery storage device.

Flow Battery Model

The main characteristics of flow batteries were estimated during the process of modelling the flow battery. The model does not take into account any electrochemical processes inside the cell nor the kinetic energy of the electrolyte itself. The main parameters considered were the limits of the power and energy ($P_{min,} P_{max,} E_{min,} E_{max}$), losses (efficiency) and reaction time. The main objective of the flow battery model was to simulate the response to power imbalance. When balancing wind power, the power to be balanced by the FB is determined as the difference between the initial error and power generated by TPP and HPP:

$$P_{FBin}=P_{error}-P_{TPP}-P_{HPP} \quad (4)$$

where P_{FBin} is the power to be balanced by the flow battery, P_{error} is the initial power imbalance, P_{TPP} and P_{HPP} is power generated by thermal and hydro power plants correspondingly.

The main principle is to charge the battery when there is a surplus of energy and to discharge when the energy is scarce. This is depicted in Fig. 7. As it can be seen from the model in Fig. 8, power of flow battery should be kept within the interval $[P_{min}, P_{max}]$, and energy stored in the flow battery E_{FB} should stay within the limits of $[E_{min}, E_{max}]$. Controlling the flow battery's charge and discharge rate should compensate the high frequency part of the wind power variation from forecasted profile.

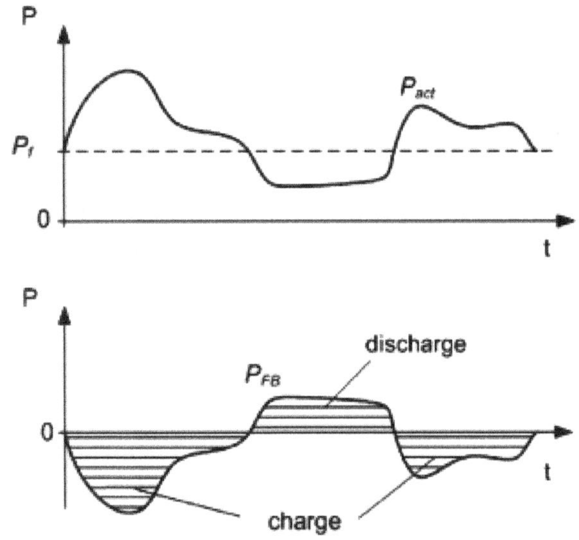

Figure 7. The control principle of the FB. P_{act} – actual wind power, P_f – forecasted wind power, P_{FB} – power of flow battery.

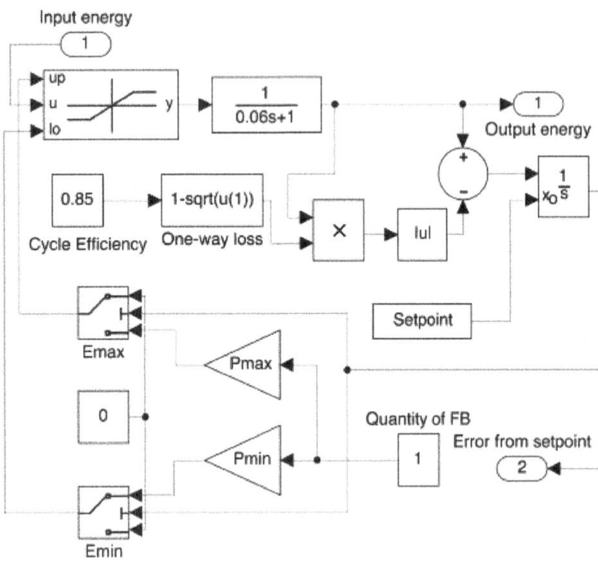

Figure 8. The structure of the flow battery model.

Also, the relative cycle losses are equally divided into charging and discharging losses using the following formula:

$$\eta_{one-way} = \sqrt{\eta_{cycle}}, \tag{5}$$

$$L_{one-way} = 1 - \eta_{one-way}, \tag{6}$$

$$L_{one-way} = 1 - \sqrt{\eta_{cycle}} \tag{7}$$

where $\eta_{one-way}$ is the efficiency of charging or discharging of the battery, η_{cycle} is the total cycle efficiency and $L_{one-way}$ are the losses associated with either charging or discharging of the battery.

Some assumption were made while designing dynamic model of the flow battery:

- The total efficiency of the flow battery cycle is η_{cycle}=85%[39],
- The inertia time constant is 0.06 s (considering inertia of power electronics) [40],
- The total discharge of the flow battery is allowed (E_{min}=0) [38].

Model of Hydro Power Plant

A hydro turbine was used to compensate the imbalance in the system that is left after TPP. The turbine output power follows the load variation trend and aims to reduce the error. It also helps to compensate losses associated with the FB (see Section 'Control strategy of hybrid power system'). The HPP was modelled in matlab using traditional governor controller/regulator popularly found in the literature and transfer functions [29]. In particular, a transfer function of hydro turbine is shown in Eq. (1) and the block diagram can be found in [30, p. 33]. The rest of the hydraulic turbine and speed regulator model parameters are given in Table 1. These parameters are chosen to match Kruonis hydro pumped storage power plant in Lithuania.

Table 1. Parameters of the hydro turbine

Parameter	Notation	Value
Permanent droop	R	0.06
Temporary droop	r	0.5
Temporary droop time constant	T_r	5 s
Auxilary servo motor time constant	T_f	0.2 s
Gate servo motor time constant	T_g	0.2 s
Water time constant	T_w	4 s

The power to be balanced by the HPP is calculated using the following equation:

$$P_{HPPin} = P_{TPPerror} \times \frac{1}{T_{hLPF}s + 1} \tag{8}$$

where P_{HPPin} is the power to be balanced by the HPP, $P_{TPPerror}$ is the power imbalance after TPP and the T_{hLPF} is the time constant of the LPF related to hydro power plant.

Model of Thermal Power Plant

Thermal transient process constrains and specific construction of thermal power plant could cause power to vary significantly slower than that of hydro power plant. It only follows the trend and does not reduce the error noticeably. Fig. 6 represents the whole system and contains TPP block. This block represents a general TPP model and has been created in Matlab according to model found in [29, p. 436].

The model of turbine consists of three cylinders: high pressure, intermediate and low pressure cylinders. Turbine is described with linear model and the transfer function of the turbine is:

$$W_T(s) = \frac{K_H(1+sT_{CO})(1+sT_R) + K_I(1+sT_{CO}+K_L)}{(1+sT_{SC})(1+sT_R)(1+sT_{CO})} \tag{9}$$

where K_H,K_I,K_L – the power of high, intermediate and low pressure cylinders per units; T_{SC},T_R,T_{CO} – time constants of steam chest, reheater and the crossover between intermediate and low pressure cylinders. Speed governor is modelled as a periodic link of servomotor and power change speed rate limiter which holds the speed within V_{max},V_{min} values.

The linear mathematical model of the regulating processes of steam turbine speed governor, could be described as second order transfer function W_{SG}:

$$W_{SG} = \frac{1}{(1+sT_{RM})(1+sT_{SM})} \quad (10)$$

where T_{RM} – time constant of speed relay, T_{SM} – time constant of servomotor.

A block diagram of steam turbine can be found in [29] and [31]. The parameters of the turbine that was used for the investigation are presented in Table 2.

Table 2. Parameters of the steam turbine

Parameter	Notation	Value
Steam chest time constant	*TSC*	0.25 s
Reheater time constant	T_R	5 s
Crossover piping time constant	*TCO*	0.5 s
Factor of high pressure section	K_H	0.3
Factor of intermediate pressure section	K_I	0.3
Factor of low pressure section	K_L	0.4
Speed relay time constant	*TSR*	0.1 s
Speed motor time constant	*TSM*	0.3 s

Overall boiler model's transfer function W_B, as boiler's pressure p_B and fuel flow m*FL*ratio, when considering the constant steam mass flow from boiler:

$$W_B(s) = \frac{e^{-sT_D}}{T_B s(1+sT_{FL})(1+sT_W)} \quad (11)$$

where T_D – fuel feed delay time constant, *TFL* – heat transfer inertia time constant to pipes during fuel burning and T_W – inertia time constant of pipes for heat transfer to water and steam.

A block diagram from [31, p. 2] was used to create Matlab model. Boiler parameters used for the investigation are presented in Table 3.

Table 3. Boiler parameters

Parameter	Notation	Value (s)
Fuel feed delay time constant	T_D	10
Boiler heat accumulation time constant	T_B	100
Heat transfer to pipes inertia time constant	*TFL*	7
Heat transfer to water and steam time constant	*TWP*	6

The power to be balanced by the TPP is calculated using the following equation:

$$P_{TPPin} = (P_{error} + PI_{out}) \times \frac{1}{T_{tLPF}s + 1} \quad (12)$$

where P_{TPPin} is the power to be balanced by the TPP, P_{error} is the initial power imbalance, P_{Iout} is the output from the PI controller and the T_{tLPF} is the time constant of the LPF related to TPP.

INVESTIGATION OF WIND POWER IMBALANCE

The actual data of wind farms installed in Lithuania was used in this research (Fig. 9). In this particular case, it is the forecasted data and the actual wind power data for the period of 1st to 10th of February 2014. The total installed capacity at that date was 222 MW. The mean absolute percentage error (MAPE) of the forecasted wind power during the investigated period was 32% with a standard deviation (SD) of 61 MW. A histogram of the initial errors can be seen in the results section. It should be noted that this period was chosen due to high variation in wind generation as well as high prediction mismatch in order to investigate wind balancing technique in extreme case.

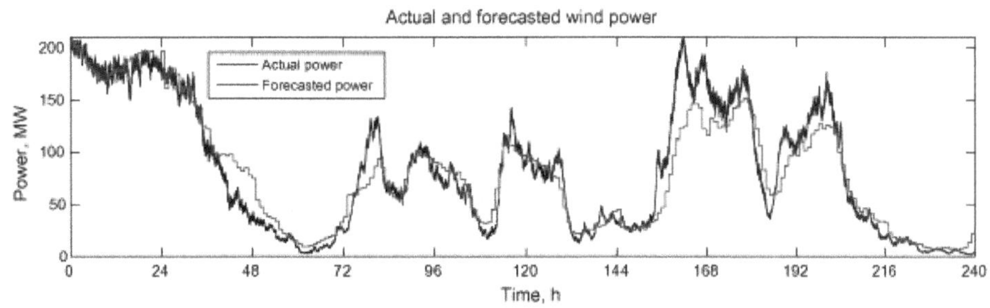

Figure 9. The forecasted and actual wind power for a installed capacity of 222 MW from 1st to 10th of February, 2014.

The Sensitivity Analysis of Low-Pass Filter cut-off Frequencies

In order to investigate the influence of the low-pass filters cut-off frequencies to the proposed balancing system, a sensitivity analysis of the low-pass filters time constants have been prepared. Thermal power plant low-pass filter time constant ranging from 0 to 10,000 s (step of 500 s) and hydro power plant low-pass filter time constant of 0–500 s (step of 25 s) have been tested. The main parameters of the proposed system, such as the required active power reserve and mean power rates of thermal and hydro power plants, also the required capacity and power ratings of flow battery have then been analysed.

The dependence of required TPP active power reserve, for different low-pass filter cut-off frequencies could be seen in Fig. 10, while mean power is shown in Fig. 11. The mean power of TPP could represent the total energy generated by the power plant as a regulating energy. It can be clearly seen from the figures that higher cut-off frequencies (lower low-pass filters time constants) increase active power reserve as well as regulating energy of the thermal power plant. It could also be noted that low-pass filter time constant of hydro power plant has minor influence on the thermal power plant specific parameters.

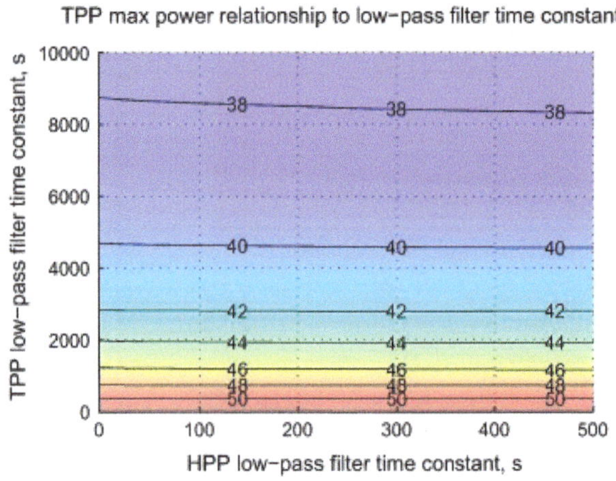

Figure 10. Relationship of required TPP active power reserve with respect to low-pass filter time constants.

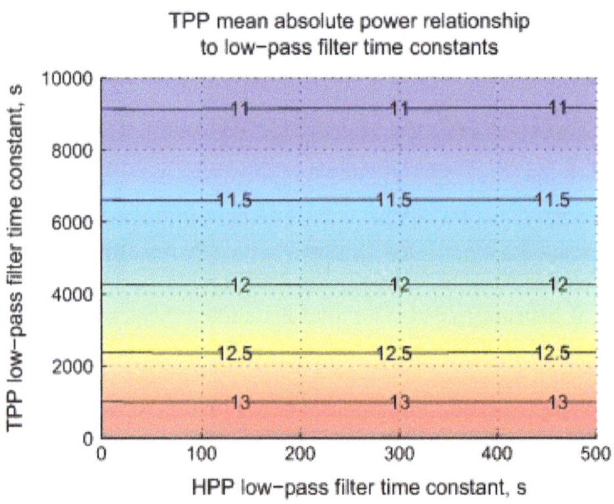

Figure 11. TPP mean absolute power relationship with respect to low-pass filter time constants.

Different situation could be identified in respect to HPP specific parameters. Both the HPP low-pass filter time constant and TPP low-pass filter time constants have appreciable influence. The required active power reserve of HPP is shown in Fig. 12, whereas Fig. 13 represents mean power of this type of plant.

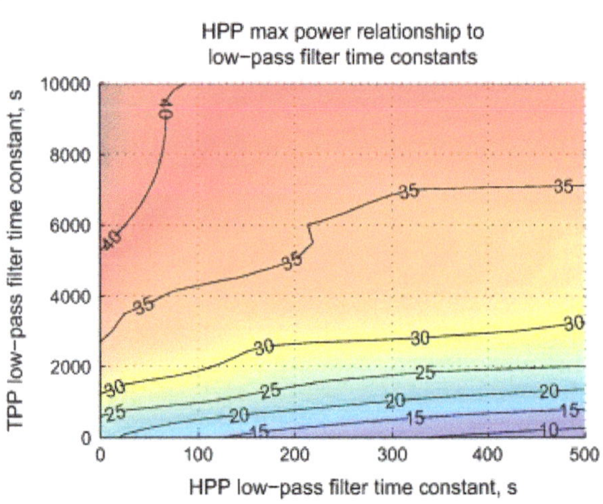

Figure 12. Relationship of required HPP active power reserve with respect to low-pass filter time constants.

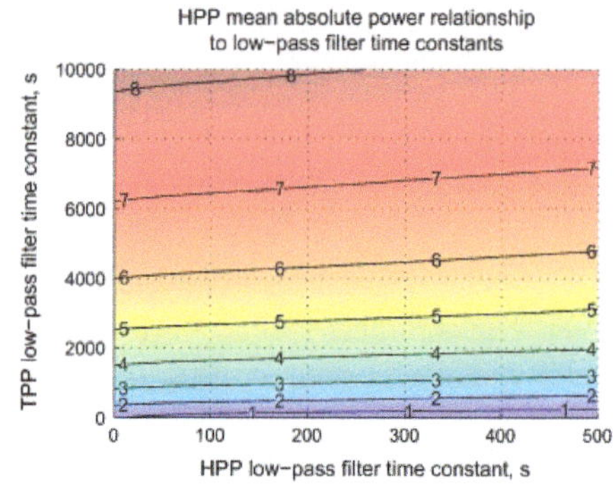

Figure 13. HPP mean absolute power relationship with respect to low-pass filter time constants.

Active power reserve of HPP increases when TPP low pas-filter parameter increases and HPP low pass-filter parameter decreases. It could be seen in Fig. 12.

A series of simulations of different low-pass filters time constants have also been run in order to get the flow battery specific parameters, in relationship with different cut-off frequencies. The required flow battery capacity and power range are represented in Fig. 14 and Fig. 15 respectively. Flow battery required parameters, as shown in the figures, depend on both low-pass filters' time constants. But the main influence is done by the one associated with HPP. As the parameter of HPP low-pass filter time constant increases, the flow battery specifications also increase. It can be noticed that the flow battery mean absolute power relationship to low-pass filters cut-off frequencies is also similar, and taken into account for the analysis.

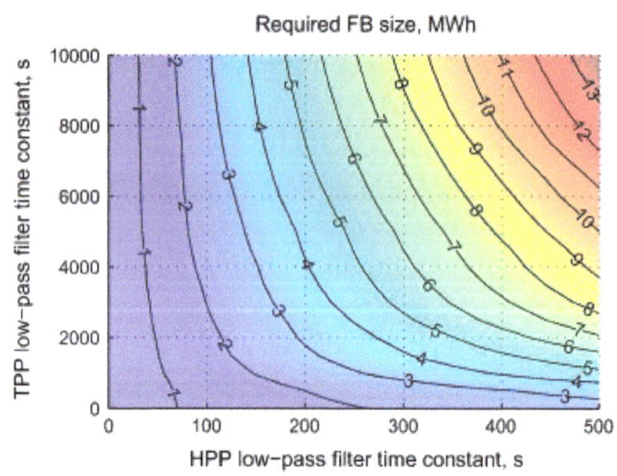

Figure 14. Relationship of required flow battery capacity with respect to low-pass filter time constants.

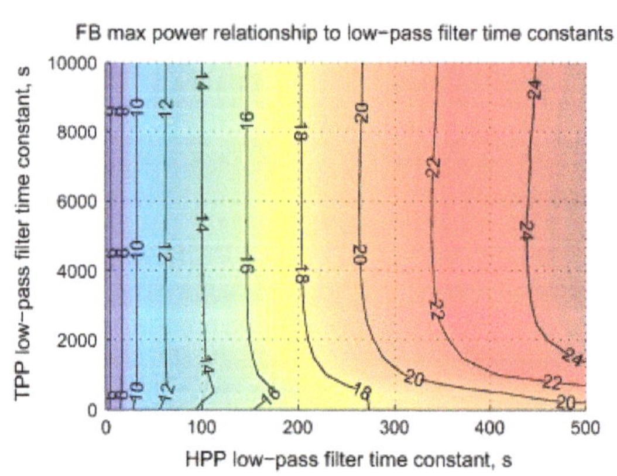

Figure 15. Relationship of required FB active power reserve with respect to low-pass filter time constants.

One of the most important things in this sensitivity analysis is to identify the relationships of thermal and hydro power plants' active power rates, which represent the intensity of power output variation of these power plants as well as the asset depreciation. Thermal power plant mean absolute power rate is represented in Fig. 16.

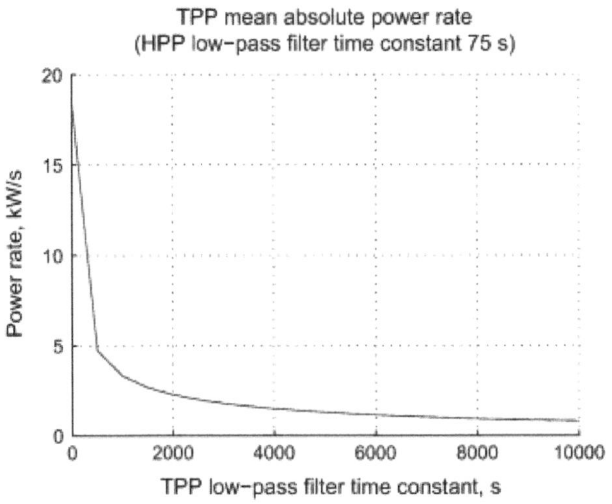

Figure 16. TPP mean absolute power rate with respect to low-pass filter time constant.

The significant mean absolute power rate downfall was identified when thermal low-pass filter time constant increase. In order to show the better visibility the low-pass filer time constant of HPP was kept constant at 75 s. However the full relationship on both filter parameters was assessed in the investigation.

Mean absolute power rate of HPP is shown in Fig. 17. Similar results to TPP mean power rate relationship can be observed. A significant drop was identified when hydro power plant's low-pass filter parameter increases. The TPP low-pass filter time constant was also kept constant at 5000 s due to better visualisation of the dependency, while the relationship on both filter parameters was taken into account in this analysis.

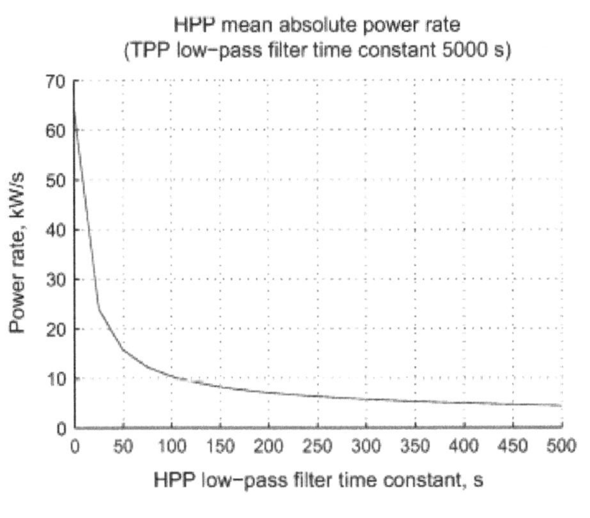

Figure 17. HPP mean absolute power rate with respect to low-pass filter time constant.

Optimal low-pass filter parameters identification

Sensitivity analysis presented in the previous section has shown various dependencies of model parameters and relationships. It was identified that required active power reserve in conventional power plant increases as low-pass filter time constant decreases. However specific parameters of the flow battery have opposite relationship as well as active power rates and intensity of traditional power plants regulation. Optimal low-pass filters time constants and cut-off frequencies are investigated in this section.

Objective function was prepared in order to investigate the optimal parameters and the best operation of proposed balancing system:

$$\begin{aligned} min \quad C = {} & C_1 \times P_{TPPmax} + C_2 \times P_{HPPmax} + \\ & C_3 \times P_{TPPmean} + C_4 \times P_{HPPmean} + \\ & C_5 \times R_{TPP} + C_6 \times R_{HPP} + \\ & C_7 \times P_{FBmean} + C_8 \times P_{FBmax} \\ & C_9 \times E_{FB} \end{aligned} \tag{13}$$

where P_{TPPmax} and P_{HPPmax} are required active power reserve of conventional power plants, $P_{TPPmean}$ and $P_{HPPmean}$ represents mean absolute power generated by thermal and hydro power plants, R*TPP* and R*HPP* are active power rates, P_{FBmean},P_{FBmax} and E_{FB} are specific parameters of flow battery – active mean power, active power range and required capacity respectively. Finally c_1,c_2,c_3,...,c_9 are relative price corresponding to each of previously mentioned parameters. Different price ratios were used in order to obtain optimal parameters and to avoid conflicts as the price is always controversial. Table 4 shows specific relative prices which were used in this investigation. The minimisation of objective function leads to identification of the optimal low-pass filter parameters and overall operation of proposed balancing system.

Table 4. Relative prices

Parameter	Price notation	Value
TPP reserve power	C_1	0.03
HPP reserve power	C_2	0.02
TPP energy	C_3	0.1
HPP energy	C_4	0.15
TPP mean absolute power rate	C_5	350
HPP mean absolute power rate	C_6	30
FB mean power	C_7	0.005
FB max power	C_8	0.05
FB capacity parameter	C_9	0.1

Results of low-pass filters time constants objective function are represented in Fig. 18. The region of the minimum objective function could be seen with the parameters – TPP low-pass filter time constant 5000 s and HPP low-pass filter time constant 75 s. These values will be used in time domain simulations of wind power balancing.

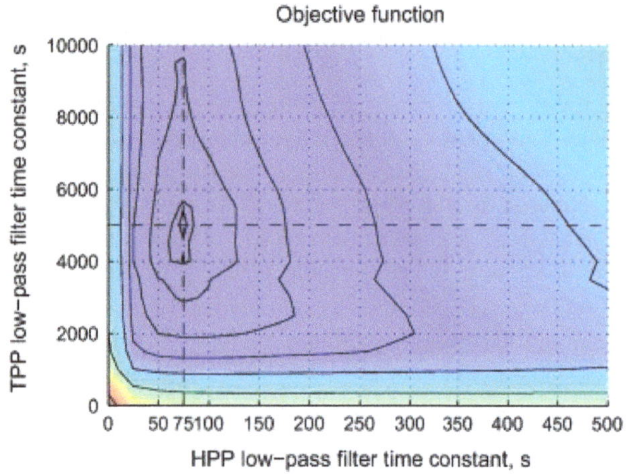

Figure 18. The output of the objective function. A minimum can be seen then time constants are 75 and 5000 for HPP and TPP respectively.

The relationship of the low-pass filter parameters could be expressed:

$$f_c = \frac{1}{2\pi\tau} \tag{14}$$

where f_c is the filter cut-off frequency and τ is the filter time constant.

Fig. 19 shows the initial imbalance of wind power and conventional power plants output generation as well as flow battery output in respect to signal frequency decomposition using identified optimal low-pass filter parameters. It can be clearly seen that low frequencies of initial wind power imbalance are covered by thermal power plant, while hydro power plant deals with mid-frequencies and the flow battery eliminates high frequencies.

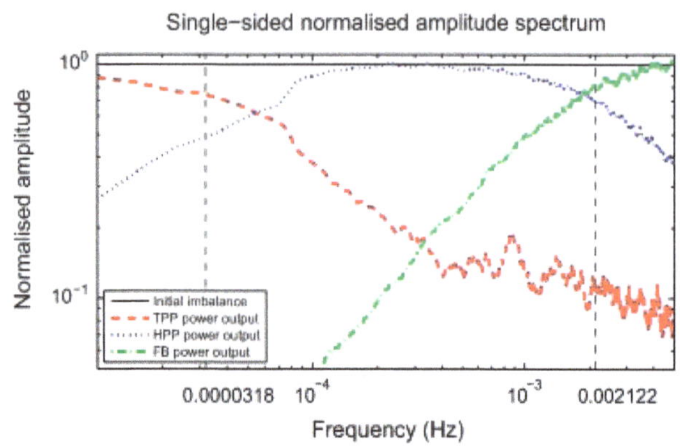

Figure 19. Frequency decomposition of initial wind power and TPP, HPP, FB output powers.

The results of proposed balancing technique

The simulations were run using the chosen optimal time constants of 5000 s and 75 s for TPP and HPP correspondingly and the data from Fig. 9. The results are shown in Fig. 20. It includes the initial power system imbalance, difference after both TPP and HPP and final imbalance as well as the operating power of TPP, HPP and FB. After the addition of TPP, the initial imbalance SD dropped from 19.5 MW to 9.2 MW. Similarly after the addition of HPP, system imbalance improved to SD of 1.65 MW. Finally, after FB the SD of imbalances were 0.1 MW. Fig. 21 shows the histogram of initial imbalances in green, imbalances after the TPP in red and imbalances after HPP in blue. From the same graph it can be observed how the spread of imbalances decrease while propagating through the system. It should be noted that addition of TPP and HPP gave positive results and, the mean over-generation of TPP and HPP is equal to mean losses in FB (about 80 kW). This justifies the validity of the model.

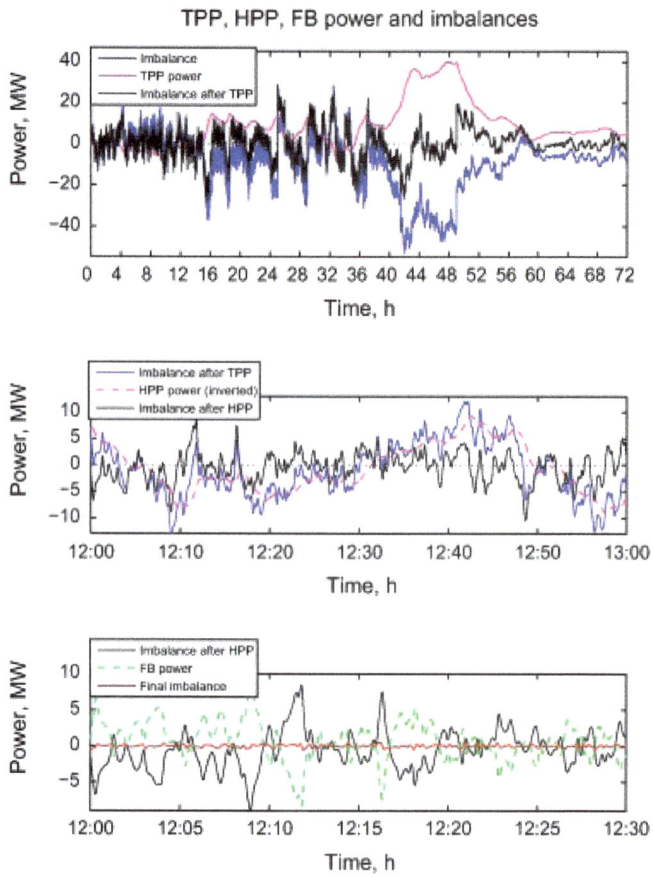

Figure 20. Balancing the difference between actual and forecasted wind power with thermal power plant, hydro power plant and flow battery.

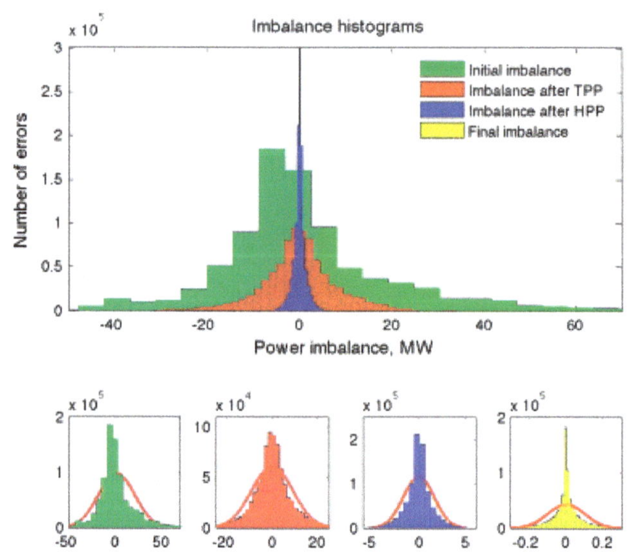

Figure 21. Histograms of imbalances at different stages in the system.

The top part of Fig. 20 shows a three day simulation period. This period is enough to demonstrate the performance of the system and the nature of TPP and HPP output. The middle part of the diagram also shows a magnified portion of errors (between the time of 12:00 and 13:00 h of the first day) and, it can be seen how the HPP power tracks the trend of the imbalance after TPP. On the other hand, the bottom part of the graph shows how FB follows the imbalance after HPP and the graph in red is the final imbalance (time from 12:00 to 12:30 of the first day). The MAPE has now dropped to 0.068%.

During the investigation, the required power and energy of the flow battery was determined to be 55 MW and 1.90 MW h respectively. This is about 66.4% of the mean wind power during the investigated period and about 24.9% of the total installed wind power capacity in the Lithuanian power system (the total wind power is about 222 MW). A more detailed discussion on why the required FB ratings might be chosen lower can be found in Section 'Limiting the maximum required FB power'. Overall, these results can be considered as feasible for implementation.

Fig. 22 shows the charge level and accumulative losses in the FB during the simulated period. The required battery size is recorded to be 1.90 MW h. This is the total required capacity in a perfect balance situation. If some power and energy spikes were ignored, the required battery power and size would be considerably lower. To cover the spikes, a super capacitor could potentially be used to serve the required ultra-short term power.

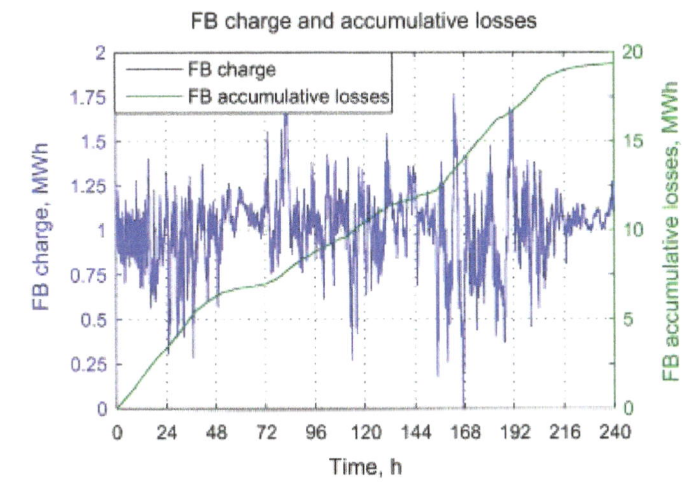

Figure 22. FB charge and accumulative losses.

Limiting the maximum required FB power

Considering the SD of power to be balanced by the FB and the three-sigma rule it is reasonable to limit the installed FB power because high power is required during a negligible amount of time. This section discusses the trade-off between FB power rating and the final balance of power.

The data from simulations was extracted to calculate the balanced portion of energy depending on battery power and low-pass filter time constants (Fig. 23). The graph shows the FB power and capacity ratings needed to balance different portions of initial energy.

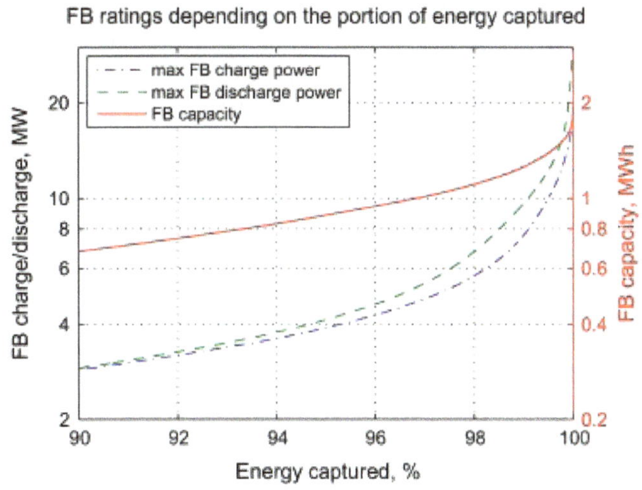

Figure 23. Relative balance dependancy on FB rated discharge and LPF time constant.

As it can be seen from Fig. 23, when 99.7% of energy is balanced, the required FB maximum power is about 14.1 MW. That is a significant drop in the required FB power rating. The required capacity of the FB would also become 1.47 MW h. On the other hand, this limits the power needed for balancing, which results in the increased final imbalance MAPE to 0.073% and a SD to 0.2 MW, but is still quite small compare to initial values (32% and 19.5 MW).

A further decrease in energy capture could be considered. The 95% would correspond to 4.1 MW FB power limit and 0.88 MW h required FB energy and would increase MAPE of final imbalance to 0.14% and SD to 0.58 MW, but is still 43% amount of imbalance improvement compared to the value without hybrid system operation.

CONCLUSIONS

The proposed hybrid wind power balancing technique, using TPP, HPP and FB control strategy, presented positive results in balancing the wind power. It also generalises a technique to find the optimal ratings for a flow battery in the Lithuanian power system. However this method could be applied to other electric power systems as well. It might be particularly important to power systems, which operate in island mode.

In this paper, a series of simulations were carried out to identify the cut-off frequencies for the low-pass filters, which optimally controls the power output of thermal and hydro power plants. The optimal cut-off frequency identification enabled the estimation of the required flow battery power and capacity ratings. Also, the paper mainly focuses on the technical side of the method instead of looking at the economical value in detail.

After implementing the proposed control strategy for the TPP, the initial imbalance decreased by about 24% (from 32% MAPE to 24.36%). A further imbalance decrease was reached after the HPP was added – from 24.36% MAPE to 1.44%. After introducing flow batteries, the system became fully balanced. However, it required FB size of 1.9 MW h and power of 55 MW. This constitute to about 25% of the total installed wind power in the Lithuanian power system. Most of the FB's power and capacity is required when the power system is in emergency state. It should be mentioned that such ratings were required when covering every moment in time.

On the other hand, these events are relatively rare, besides some types of flow batteries tolerate overloads for short periods without negative side effects. Making such assumptions the results of additional investigation showed that more reasonable flow battery ratings could be chosen. By covering 99.7–95% of the power imbalance, it is possible to reduce FB power rating by about 4–13 times and FB required capacity from about 23–54% (Table 5 and Table 6).

Table 5. Overview of results (Imbalances).

Imbalance stage	MAPE (%)	SD (MW)	ME (kW)
Initial imbalance	32.00	19.52	1291
After TPP	24.36	9.20	−78.9
After HPP	1.44	1.65	−81.5
100% final	0.068	0.10	−0.02
99.7% final	0.073	0.20	1.6
95% final	0.144	0.58	6.3

Table 6. Final balance level vs. required FB ratings.

Balance level (%)	FB rated power (MW)	FB capacity (MW h)
100	55.3	1.90
99.7	14.1	1.41
95	4.1	0.88

Many other storage technologies could be similarly modeled and investigated. In particular, high power and low capacity storage devices, such as super capacitors, could be added to compensate highest frequency imbalances thus highly improving results and reducing power requirements for the FB. Having many different power plants in the model it is then potentially useful to research control strategies in order to reach for the highest economical or environmental benefit.

ACKNOWLEDGEMENTS

The authors would like to acknowledge the funding support from EPSRC via Faculty of Science and Technology, Lancaster University, UK. The authors would also like to thank the Department of Power Systems, Kaunas University of Technology, Lithuania and National Transmission System Operator in Lithuania for providing the necessary data. The data generated in this work are available from the Lancaster University data archive[41].

REFERENCES

1. Gelazanskas L, Gamage KAA. Demand side management in smart grid: a review and proposals for future direction. Sustainable Cities and Society. http://dx.doi. org/10.1016/j.scs.2013.11.001.
2. Le HT, Santoso S, Nguyen TQ. Augmenting wind power penetration and grid voltage stability limits using ESS: application design, sizing, and a case study. IEEE Trans Power Syst 2012;27(1):161–71. http://dx.doi.org/10.1109/TPWRS. 2011.2165302.
3. Diaz-Gonzalez F, Sumper A, Gomis-Bellmunt O, Bianchi FD. Energy management of flywheel-based energy storage device for wind power smoothing. Appl Energy 2013;110(0):207–19. http://dx.doi.org/10.1016/j.apenergy.2013.04.029. .
4. Tewari S, Mohan N. Value of NAS energy storage toward integrating wind: results from the wind to battery project. IEEE Trans Power Syst 2013;28 (1):532–41. http://dx.doi.org/10.1109/TPWRS.2012.2205278.
5. Union for the coordination of the transmission of electricity. Operation Handbook; 2013.
6. Denholm P, Ela E, Kirby B, Milligan M. The role of energy storage with renewable electricity generation. Tech. Rep. NREL/TP-6A2-47187. A national laboratory of the U.S. Department of Energy; January 2010. .
7. Droste-Franke B, Paal BP, Rehtanz C, Sauer DU, Schneider J-P, Schreurs M. Balancing renewable electricity: balancing renewable electricity energy storage, demand side management and network extension from perspective an interdisciplinary; 2011.
8. Han X, Chen F, Cui X, Li Y, Li X. A power smoothing control strategy and optimized allocation of battery capacity based on hybrid storage energy technology. Energies 2012;5(5):1593–612. http://dx.doi.org/10.3390/en5051593. .
9. Soder L, Hamon C. Power balance regulation at large amounts of wind power. Tech. rep., Elforsk.
10. Lubosny Z, Bialek J. Supervisory control of a wind farm. IEEE Trans Power Syst 2007;22(3):985–94. http://dx.doi.org/10.1109/TPWRS.2007.901101.
11. Jiang Q, Wang H. Two-time-scale coordination control for a battery energy storage system to mitigate wind power fluctuations. IEEE Trans Energy Convers 2013;28(1):52–61. http://dx.doi.org/10.1109/TEC.2012.2226463.
12. Abbey C, Strunz K, Joos G. A knowledge-based approach for control of twolevel energy storage for wind energy systems. IEEE Trans Energy Convers 2009;24(2):539–47. http://dx.doi.org/10.1109/TEC.2008.2001453.
13. Wang P, Gao Z, Bertling L. Operational adequacy studies of power systems with wind farms and energy storages. IEEE Trans Power Syst 2012;27 (4):2377–84. http://dx.doi.org/10.1109/TPWRS.2012.2201181.
14. Khalid M, Savkin A. Minimization and control of battery energy storage for wind power smoothing: aggregated, distributed and semi-distributed storage. Renew Energy 2014;64(0):105–12. http://dx.doi.org/10.1016/j.renene.2013.09.043. .
15. Khalid M, Savkin A. A model predictive control approach to the problem of wind power smoothing with controlled battery storage. Renew Energy 2010;35(7):1520–6. http://dx.doi.org/10.1016/j.renene.2009.11.030
16. special Section: IST National Conference 2009. .
17. Yuan Y, Zhang X, Ju P, Qian K, Fu Z. Applications of battery energy storage system for wind power dispatchability purpose. Electric Power Syst Res 2012;93(0):54–60. http://dx.doi.org/10.1016/j.

epsr.2012.07.008. . Table 5 Overview of results (Imbalances). Imbalance stage MAPE (%) SD (MW) ME (kW) Initial imbalance 32.00 19.52 1291 After TPP 24.36 9.20 78.9 After HPP 1.44 1.65 81.5 100% final 0.068 0.10 0.02 99.7% final 0.073 0.20 1.6 95% final 0.144 0.58 6.3 Table 6 Final balance level vs. required FB ratings. Balance level (%) FB rated power (MW) FB capacity (MW h) 100 55.3 1.90 99.7 14.1 1.41 95 4.1 0.88 320 L. Gelazˇanskas et al. / Electrical Power and Energy Systems 74 (2016) 310–321

18. Ansari MMT, Velusami S. DMLHFLC (dual mode linguistic hedge fuzzy logic controller) for an isolated wind-diesel hybrid power system with BES (battery energy storage) unit. Energy 2010;35(9):3827–37. http:// dx.doi.org/10.1016/j. energy.2010.05.037. .
19. Kalantar M, Mousavi G SM. Dynamic behavior of a stand-alone hybrid power generation system of wind turbine, microturbine, solar array and battery storage. Appl Energy 2010;87(10):3051–64. http://dx.doi. org/10.1016/j. apenergy.2010.02.019. .
20. Mousavi G SM. An autonomous hybrid energy system of wind/tidal/ microturbine/battery storage. Int J Electrical Power Energy Syst 2012;43 (1):1144–54. http://dx.doi. org/10.1016/j.ijepes.2012.05.060. .
21. Ma T, Yang H, Lu L. A feasibility study of a stand-alone hybrid solar/ wind/battery system for a remote island. Appl Energy 2014;121(0):149–58. http://dx.doi. org/10.1016/j.apenergy.2014.01.090. .
22. Apostolopoulou D, Sauer PW, Domingeuz-Garcia AD. Automatic generation control and its implementation in real time; 2014.
23. Ram P, Jha AN. Automatic generation control of interconnected hydro-thermal system in deregulated environment considering generation rate constraints; 2010.
24. Gandhi NF, Mohan YK, Rao AV. Load frequency control of interconnected power system in deregulated environment considering generation rate constrains; 2012.
25. Cheng Y, Sahni M. Alti-ees automatic generation control study. Tech. rep., PwrSolutions Inc.; February 2012.
26. Bevrani H, Hiyama T. Intelligent automatic generation control. CRC Press; 2011.
27. Dagdougui H, Minciardi R, Ouammi A, Robba M, Sacile R. A dynamic decision model for the real-time control of hybrid renewable energy production systems. IEEE Syst J 2010;4(3):323–33. http://dx.doi.org/10.1109/ JSYST.2010.2059150.
28. Yoshimoto K, Nanahara T, Koshimizu G. New control method for regulating state-of-charge of a battery in hybrid wind power/battery energy storage system. In: Power systems conference and exposition, 2006. PSCE '06. 2006 IEEE PES, 2006. p. 1244–51. http://dx.doi. org/10.1109/PSCE.2006.296485.
29. Castronuovo E, Peas Lopes J. On the optimization of the daily operation of a wind-hydro power plant. IEEE Trans Power Syst 2004;19(3):1599–606. http:// dx.doi. org/10.1109/TPWRS.2004.831707.
30. Kundur P. Power system stability and control. Reading, MA: McGraw-Hill; 1994.
31. Tenorio LAL. Hydro turbine and governor modelling. Norwegian University of Science and Technology. .
32. Gao L, Dai Y. Modeling large modern fossil-fueled steam-electric power plant and its coordinated control system for power system dynamic analysis. In: International conference on power system technology, 2010. POWERCON 2010. http://dx.doi.org/10.1109/ POWERCON.2010.5666144. .
33. Mohamed MR, Sharkh SM, Ahmad H, Seman MNA, Walsh FC. Design and development of unit cell and system for vanadium redox flow batteries (VRFB). Int J Phys Sci 2012;7(7):1010–24. .
34. Li L, Kim S, Wang W, Vijayakumar M, Nie Z, Chen B, Zhang J, et al. A new vanadium redox flow batteru using mixed acid electrolytes.
35. Shigematsu T. Redox flow battery for energy storage. Tech. Rep. 73, Sumitomo Electric Industries; October 2011. .
36. Weber AZ, Mench MM, Meyers JP, Ross PN, Gostick JT, Liu Q. Redox flow batteries: a review. J Appl Electrochem 2011;41(10):1137–64. .
37. Barote L, Weissbach R, Teodorescu R, Marinescu C, Cirstea M. Technologies for energy storage – present and future: flow batteries; 2008. p. 407–12. .
38. Tokuda N, Kanno T, Hara T, Shigematsu T, Tsutsui Y, Ikeuchi A, Itou T, et al. Development of redox flow battery system. Tech. Rep. 50, Sumitomo Electric Industries; June 2000. .
39. Youa D, Zhanga H, Chen J. A simple model for the vanadium redox battery.
40. Teleke S, Baran ME, Huang AQ, Bhattacharya S, Anderson L. Control strategies for battery energy storage for wind farm dispatching, vol. 24; 2009. p. 725–32. http://dx.doi.org/10.1109/TEC.2009.2016000.
41. Connolly D. An investigation into the energy storage technologies available, for the integration of alternative generation techniques. Tech. rep., Department of Physics, University of Limerick; November 2007.
42. Gelazanskas L, Gamage K. Wind generation, 2015. http://dx.doi.org/10.17635/ lancaster/researchdata/20.

CHAPTER 2

PERFORMANCE COMPARISON ON REPOWERING OF A STEAM POWER PLANT WITH GAS TURBINES AND SOLID OXIDE FUEL CELLS

Masoud Rokni

Department of Mechanical Engineering, Thermal Energy Section, Technical University of Denmark, Nils Koppels Allé, Copenhagen 2800, Denmark

ABSTRACT

Repowering is a process for transforming an old power plant for greater capacity and/or higher efficiency. As a consequence, the repowered plant is characterized by higher power output and less specific CO_2 emissions. Usually, repowering is performed by adding one or more gas turbines into an existing steam cycle which was built decades ago. Thus, traditional repowering results in combined cycles (CC). High temperature fuel cells (such as solid oxide fuel cell (SOFC)) could also be used as a topping cycle, achieving even higher global plant efficiency and even lower specific CO_2 emissions. Decreasing the operating temperature in a SOFC allows the use of less complex materials and construction methods, consequently reducing plant and the electricity costs. A lower working temperature makes it also suitable for topping an existing steam cycle, instead of gas turbines. This is also the target of this study, repowering of an existing power plant with SOFC as well as gas turbines. Different repowering strategies are studied here, repowering with one gas turbine with and without supplementary firing, repowering with two gas turbines with and without supplementary firing and finally repowering using SOFC. Plant performances and CO_2 emissions are compared for the suggested repowered plants.

Keywords

solid oxide fuel cell (SOFC); repowering; hybrid cycles; combined cycles (CC); CO_2emission

INTRODUCTION

Due to the ever-increasing demand for more power production and distribution, one of the main topics of research and development in the field of electricity generation are improving efficiency and reducing pollutant emissions. Transforming an old power plant for a greater capacity and/or higher efficiency is known as repowering. One way of repowering is by adding one or more gas turbines into a plant as a topping cycle, thus converting an existing steam power plant into a combined cycle (CC) system. It would be ideal for an old steam plant in which steam turbine(s) after many years of operation still have considerable service life expectancy, but for example the boiler is ready to be replaced (due to a shorter lifecycle). The boilers are normally replaced or supplemented with gas turbines and heat recovery steam generators (HRSG), see e.g., [1].

Citation: Masoud Rokni, Performance Comparison on Repowering of a Steam Power Plant with Gas Turbines and Solid Oxide Fuel Cells, doi: 10.3390/en9060399.

Due to the rapid growth of electricity consumption, there is an increased interest in converting old coal fired steam plants into CC plants to increase their power output and efficiency, and at the same time, decrease their emissions, see e.g., [2,3]. However, it shall also be noted that since the share of renewable energy systems in electricity generation is increasing in some regions (such as in Europe) then the CC operating mode in those countries may start to shift (or has already shifted) from base- and medium-load to part-load. This is seen in some European countries such as in Denmark where the share of wind energy is increasing every year, see e.g., [4].

Currently repowering of steam plants can be achieved in two ways: feed water repowering and boiler repowering, see e.g., [5,6]. The first option uses heat from the turbine exhaust to raise the feed water temperature instead of bleeding steam. This means that increased steam flow has to be managed by the low pressure section of the original steam turbine, requiring either extensive modification of the steam turbine or impairing the repowered plant performance. The other option, boiler repowering, entails major steam generator redesign or replacement. Gas turbine exhaust gas is used as the heat source for the existing steam cycle. This increases plant efficiency close to that of new combined cycle plants. Such repowering has been performed on various old steam plants, see e.g., [3,7,8]. The second option is widely used across developed countries in which many steam plants are relatively old and are coal fired. This option is also investigated in this study. Economic analysis of repowering by means of gas turbines has also been investigated widely such as in [9,10,11].

Steam turbine units in older power stations generally have relatively low steam data and can easily be adapted for use in combined cycles as the bottoming cycle for a gas turbine (gas turbines). Depending on the steam plant data such as live steam temperature, pressure and mass flow, one needs to screen available gas turbines in the market and choose the one which can easily be adapted for the basic steam plant without changing its original configurations. If one gas turbine cannot supply the required heat and temperature then one may complement the repowering with a supplementary firing or by using more gas turbines. Both options will be used in this study.

In the present study, it is also suggested to use a third option which is repowering with solid oxide fuel cell stacks (SOFCs). Solid oxide fuel cell (SOFC) stacks may soon enter the commercialization phase and therefore it would be interesting to integrate such technology into repowering of old steam plants. SOFCs are one of the most promising types of fuel cells, particularly regarding energy production. They are expected to produce clean electrical energy at high conventional rates with low noise and pollutant emissions [12]. The exhaust temperatures of SOFCs are high due to the high operating temperature of the cells. Additionally, because the fuel utilization in the fuel cell is less than 100 percent, the unreacted fuel needs to be combusted in a burner. This combustion in turn produces even hotter off-gases that are perfectly suited for use in a steam generator to produce steam for the bottoming steam cycle.

Numerous studies in the literature have investigated SOFC-based power systems and suggested high thermal efficiencies. However, the majority of these studies use gas turbines as the bottoming cycle, see e.g., [13,14,15]. A steam turbine has also been used as a bottoming cycle [16,17], resulting in high plant efficiency, but no study was found on repowering an old steam plant with SOFCs, which is the basis of this study. At present, using the Brayton and Rankine cycles as bottoming cycles for SOFCs seems to be the most practical because of the maturity of these technologies. Given that the development trends suggest that the operating temperature of the SOFC will decrease, using gas turbines as bottoming cycles will become less beneficial over time.

The present work is an analytical study that conducts a thermodynamic investigation of repowering of old steam plants with SOFCs that also functions as a topping cycle for a steam plant using the heat from the off-gasses exhausted from the topping cycle. The results will be compared with the traditional repowering strategies using gas turbines as the driving heat source. One gas turbine with supplementary firing, two gas turbines as well as two gas turbines with supplementary firing will also be studied. The comparison will be carried out in terms of plant efficiency and CO_2 emission as the most important parameters. It should be noted that the SOFCs used in this study are based on a theoretical model with empirical coefficients calibrated from experimental data.

Despite numerous studies on SOFC-based power plants, no investigation on steam plant repowering with SOFCs has been found in the open literature, and therefore, the current investigation seems to be completely novel and might provide new ideas for designing more efficient and less pollutant energy systems for future applications. It should also be noted that the system presented here is studied thermodynamically and that the objective of this study is not to present the associated costs. However, a simple cost analysis is included to compare the additional cost of repowering with different techniques. The performances of the various plants are compared in terms of efficiency, CO_2 emission, fuel consumption and other related parameters.

ORIGINAL PLANT MODEL

The principal components of the plant are the burner, steam generator, high pressure turbine (HP), two intermediate pressure turbines (IP), low pressure turbine (LP), condenser and deaerator. The coal burner provides the heat needed for generating steam in a single pressure level steam generator through economizer (Eco), evaporator (Eva) and super heater (SH). Three steam extractions from the steam turbines are used to preheat (PH) the sub-cooled water after the condenser. Coal is supplied at point 61 in Figure 1 while ash is removed from the burner at point 62 in Figure 1. The coal composition (mass based) is assumed to be 0.7818 C (solid form), 0.0489 H_2, 0.0603 O_2, 0.0171 N_2, 0.0102 S (solid form) and 0.0817 water (liquid form). The net calorific and gross caloric values are 31,120 kJ/kg and 32,380 kJ/kg, respectively, with a mean mole mass of 10.34 kg.

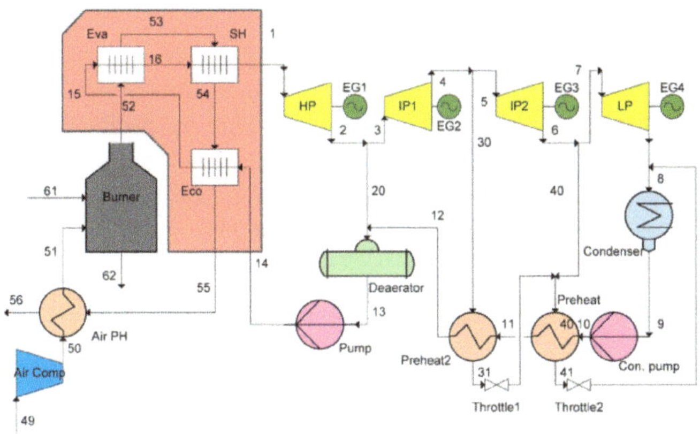

Figure 1. Kyndby original coal steam plant

The plant net output power is 264MW with an efficiency of 33% (LHV).Live steam temperature, pressure and mass flow (point 1 in Figure 1) is 500 °C, 80 bar and 280 kg/s, respectively. The power required to generate such steam is about 756.168 MW, which can be calculated from the enthalpy difference between the economizer inlet and super heater outlet (live steam) multiplied by the mass flow rate. This of course is lower than coal power input due to losses through air preheating and air compressor for the boiler. The main parameters for the turbines are summarized in Table 1.

Table 1. Turbine parameters

Turbine	Isoentropic Efficiency	Turbine Constant
High pressure turbine	0.885	97.75
Intermediate pressure turbine 1	0.878	766.3
Intermediate pressure turbine 2	0.811	2467
Low pressure turbine	0.7565	11,140

Ambient conditions are assumed to be 25 °C and 1.01 bar, both for air (point 49 in Figure 1) and coal (point 61 in Figure 1). The isentropic efficiency of the turbines is calculated from the real data of the existing plant by means of the turbine constant defined here. The turbine constant is a parameter that depends on the turbine mass flow, the inlet temperature as well as inlet and outlet pressure defined as (see e.g., [18]):

$$C_T = \frac{\dot{m}\sqrt{T}}{\sqrt{p_{in}^2 - p_{out}^2}} \quad (1)$$

As mentioned above, in the repowering studied here the current existing steam plant will not be changed and only the boiler will be replaced with a HRSG. Another important issue to be mentioned is that in the existing coal fired steam plant the steam temperature entering the boiler (point 14 in Figure 1) is rather high, 164 °C, and the reason is to decrease the heat demand from burning the coal. In the repowering technique used here (the existing plant balance of the steam cycle is not be changed at all), then such high a steam temperature in the boiler does not allow having high plant efficiency. The reason is that the off-gases temperature on the other side of the economizer must be higher and therefore off-gases will be exhausted at a temperature higher than 164 °C (economizer pinch temperature). However, in a new designed combined cycle the steam temperature entering the economizer will be kept as low as possible to recover as much off-gas energy as possible. In designing a new combined cycle the steam temperature to the HRSG shall be as low as possible and the reason is to recover the exhaust gas from the gas turbines as much as possible. Other important data are shown in Table 2.

Table 2. The main plant parameters

Parameter	Value	Node in Figure 1
Coal	-	61
Fuel consumption (mass flow), (kg/s)	25.75	-
Fuel consumption based on LHV, (MW)	801.146	-
Burner		
Air fuel ration, λ	1.2 times stoichiometric value	-
Exhaust		56
Temperature (°C)	140	-
Pressure (bar)	1.01	-
Pressure conditions		
Maximum pressure, (bar)	100	14
Minimum pressure, (bar)	0.081	9
Others		
Air preheater effectiveness, (%)	80	-
Generators efficiency, (%)	97	-
Power output, (MW)	263.932	-
Plant efficiency based on LHV, (%)	32.94	-

MODELLING

The modeling for SOFC and gas turbine will be briefly explained below while modeling of other components will be referred to previous publications.

SOFC Modeling

The SOFC model developed in [19] is used in this investigation, which is calibrated against experimental data for a planar SOFC type with a relative error of about ±3%. For the sake of clarity, it is briefly described here. In such modeling one must distinguish between electrochemical modeling, calculation of cell irreversibility (cell voltage efficiency) and the species compositions at outlet. For electrochemical modeling, the operational voltage (E_{cell}) was found to be:

$$E_{cell} = E_{Nernst} - \Delta E_{act} - \Delta E_{ohm} - \Delta E_{conc} \quad (2)$$

where E_{Nernst}, ΔE_{act}, ΔE_{ohm} and ΔE_{conc} are the Nernst ideal reversible voltage, activation polarization, ohmic polarization and concentration polarization, respectively. Assuming that only hydrogen is electrochemically converted, then the Nernst equation can be written as:

$$E_{Nernst} = \frac{-\Delta g_f^0}{n_e F} + \frac{RT}{n_e F}\ln\left(\frac{p_{H_2,tot}\sqrt{p_{O_2}}}{p_{H_2O}}\right) \quad (3)$$

$$p_{H2,tot} = p_{H2} + p_{CO} + 4p_{CH4} \quad (4)$$

where Δg_f^0 is the Gibbs free energy (for H_2 reaction) at standard pressure. The water-gas shift reaction is very fast and therefore the assumption of hydrogen as only species to be electrochemically converted is justified, see [20,21]. In the above equations p_{H2} and p_{H2O} are the partial pressures for H_2 and H_2O, respectively.

The activation polarization can be evaluated from the Butler-Volmer equation [22,23], which is isolated from other polarizations to determine the charge transfer coefficients and exchange current density from the experiment by the curve fitting technique. It follows that:

$$\Delta E_{act} = \Delta E_{act,c} + \Delta E_{act,a} = \frac{2RT}{F}\left[\sinh^{-1}\left(\frac{i+i_n}{2i_{0,a}}\right) + \sinh^{-1}\left(\frac{i+i_n}{2i_{0,c}}\right)\right] \tag{5}$$

where R, T, F and i are the universal gas constant, operating temperature, Faraday constant and current density, respectively. i_n is an internal current density added to the actual current density in order to account for the mixed potential caused by fuel crossover and electrons passing through the electrolyte. This value was adjusted when calibrating the electrochemical model against experimental data, found to be 2 mA/cm^2. The anodic [24] and cathodic [25] current densities are calculated from:

$$i_{0,a} = \gamma_a \left(\frac{p_{H_2,tot}}{p_a}\right)\left(\frac{p_{H_2O}}{p_a}\right)\exp\left(\frac{-E_{act,a}}{RT}\right) \tag{6}$$

$$i_{0,c} = \gamma_c \left(\frac{p_{O_2}}{p_c}\right)^{0.25}\exp\left(\frac{-E_{act,c}}{RT}\right) \tag{7}$$

where $E_{act,a} = E_{act,c} = 1.1 \times 10^5$ J/mol. The constants γ_a and γ_c are calibrated against experimental data and found to be 2.13×10^7 mA/cm^2 and 1.49×10^7 mA/cm^2, respectively [19].

The ohmic polarization [26] depends on the electrical conductivity of the electrodes as well as the ionic conductivity of the electrolyte. This was also calibrated against experimental data for a cell with anode thickness, electrolyte thickness, cathode thickness and interconnect thickness of 750, 40, 50 and 100 μm respectively and can be described as:

$$\Delta E_{ohm} = i(\delta_a\sigma_a + \delta_c\sigma_c + \delta_e\sigma_e + \delta_i\sigma_i) \tag{8}$$

$$\sigma_j = \sigma_{ohm,j}\exp\left(\frac{B_{ohm,j}}{T}\right), \quad j = \{a, c, e, i\} \tag{9}$$

The constants $\sigma_{ohn,a}$, $\sigma_{ohn,c}$, $\sigma_{ohn,e}$ and $\sigma_{ohn,i}$, are assumed to be 2.98×10^{-6} kΩ·cm, 8.11×10^{-6}kΩ·cm, 2.94×10^{-6} kΩ·cm and 0.1256×10^{-3} kΩ·cm respectively. The constants $B_{ohn,a}$, $b_{ohn,c}$, $B_{ohn,e}$ and $B_{ohn,i}$, are assumed to be −1392, 600, 10,350 and 4690 K, respectively [19].

The concentration polarization is dominant at high current densities for anode-supported SOFCs, wherein insufficient amounts of reactants are transported to the electrodes and the voltage is then reduced significantly. Again the concentration polarization was calibrated against experimental data by introducing the anode limiting current, [24], in which the anode porosity and tortuosity were also included among other parameters. Neglecting the cathode contribution (see e.g., [27,28]), it can be modeled as:

$$\Delta E_{conc} = \frac{RT}{n_e F}\left(-\ln\left(1 + \frac{p_{H2}(i+i_n)}{p_{H2O}i_{as}}\right) - \ln\left(1 - \frac{i+i_n}{i_{as}}\right)\right) \tag{10}$$

where $n_e = 2$, p_{H2} and p_{H2O} are the partial pressures for H_2 and H_2O respectively. The anode limiting current i_{as} was assumed to be 1000 mA/cm^2 in the calibration.

It should be mentioned that the SOFC model developed here aims to represent the performance of the second generation SOFC stacks developed by from Topsoe Fuel Cell A/S (TOFC) and the Fuel Cells and Solid State Chemistry Division at Risø—DTU (Technical University of Denmark, Roskilde, Denmark). This SOFC type is anode supported and the anode consists of Ni/YSZ, the electrolyte of YSZ and the cathode of LSM/YSZ [29]. As mentioned above, to fit the model with the desired stack performances, the electrochemical model is calibrated against experimental data, as shown in [19].

The fuel composition at anode outlet was calculated using the Gibbs minimization method as described in [30]. Equilibrium at the anode outlet temperature and pressure was assumed for the following species: H_2, CO, CO_2, H_2O, CH_4 and N_2. Thus the Gibbs minimization method calculates the compositions of these species at outlet by minimizing their Gibbs energy. The equilibrium assumption is fair because the methane content in this study is very low.

To calculate the voltage efficiency of the SOFC cells, the power production from the SOFC (P_{SOFC}) depends on the amount of chemical energy fed to the anode, the reversible efficiency (η_{rev}), the voltage efficiency (η_v) and the fuel utilization factor (U_F). It is defined in mathematical form as:

$$P_{SOFC} = \left(\text{LHV}_{H_2}\dot{n}_{H_2,in} + \text{LHV}_{CO}\,\dot{n}_{CO,in} + \text{LHV}_{CH_4}\,\dot{n}_{CH_4,in}\right)\eta_{rev}\,\eta_v\,U_F \tag{11}$$

where U_F was a set value and η_v was defined as:

$$\eta_v = \frac{\Delta E_{cell}}{E_{Nernst}} \tag{12}$$

The reversible efficiency is the maximum possible efficiency defined as the relationship between the maximum electrical energy available (change in Gibbs free energy) and the fuels LHV (lower heating value) as follows, see e.g., [31]:

$$\eta_{rev} = \frac{\left(\Delta\overline{g}_f\right)_{fuel}}{\mathrm{LHV}_{fuel}} \quad (13)$$

$$\begin{aligned}\left(\Delta\overline{g}_f\right)_{fuel} &= \left[\left(\overline{g}_f\right)_{H_2O} - \left(\overline{g}_f\right)_{H_2} - \tfrac{1}{2}\left(\overline{g}_f\right)_{O_2}\right] y_{H_2,in} \\ &+ \left[\left(\overline{g}_f\right)_{CO_2} - \left(\overline{g}_f\right)_{CO} - \tfrac{1}{2}\left(\overline{g}_f\right)_{O_2}\right] y_{CO,in} \\ &+ \left[\left(\overline{g}_f\right)_{CO_2} + 2\left(\overline{g}_f\right)_{H_2O} - \left(\overline{g}_f\right)_{CH_4} - 2\left(\overline{g}_f\right)_{O_2}\right] y_{CH_4,in}\end{aligned} \quad (14)$$

The partial pressures were assumed to be the average between the inlet and outlet as:

$$\begin{aligned}\overline{p}_j &= \left(\frac{y_{j,out}+y_{j,in}}{2}\right)\overline{p} \qquad j = \{H_2, CO, CH_4, CO_2, H_2O, N_2\} \\ \overline{p}_{O_2} &= \left(\frac{y_{O_2,out}+y_{O_2,in}}{2}\right)\overline{p}_c\end{aligned} \quad (15)$$

Additionally, equations for conservation of mass (with molar flows), conservation of energy and conservation of momentum were also included into the model. Table 3 displays the main parameters for the SOFC stacks used in this study.

Table 3. The main SOFC parameters used in this study

Parameter	Value	Unit
Fuel utilization factor	0.8	-
Current density	300	mA/cm^2
Cathode pressure drop ratio	0.1	bar
Anode pressure drop ratio	0.05	bar
Cathode inlet temperature	600	°C
Anode inlet temperature	650	°C
Outlet temperatures	780	°C
DC/AC converter efficiency	0.97	-

Modeling of Other Components

Modeling of other components such as desulfurization reactor, pre-refomer, *etc.* are adopted from the study of [32], by building a benchmark system consisting of SOFC, methanator, *etc.* and fed with different fuels such as natural gas, ethanol, methanol and di-methyl ether (DME). The obtained results agreed well with the corresponding data obtained by other researchers in the open literature for all fules and cases studied [32]. Standard modling for pumps, heat exchangers, tubines and compressor are used in this study, see e.g., [33,34].

Modeling of Selected Gas Turbine

In [35] the specification of all gas turbines currently available in the market is specified. All gas turbines in the data sheet are screened and based on the required temperature as well as heat for the steam cycle, the gas turbine chosen here is the Siemens SGT5 4000F (Orlando, Florida, FL, USA). The specifications in [35] are slightly different from the ones on the Siemens website [36], which could depend on improvements. Therefore, an average data is chosen which is summarized in Table 4. A gas turbine model based on these data is then developed here to capture all important specifications such as power output, efficiency, *etc.*

Table 4. Comparison between Siemens SGT5 4000F and the model developed here

Parameter	Datasheet Value	Model	Error (%)
ISO base rating, (MW)	288	290.95	1.0
Heat rate, (kJ/kWh)	9114	9114.4	0.0
Efficiency LHV (%)	39.5	39.5	0.0
Pressure ratio, (-)	18	18	-
Exhaust mass flow rate, (kg/s)	688	688	-
Turbine speed, (rpm)	3000	-	
Exhaust temperature, (°C)	580	580.0	0.0

As seen in the exhaust gas temperature if this gas turbine is well above live steam temperature of 500 °C, allowing for a large terminal temperature and consequently lower HRSG cost. In fact that the exhaust temperature of the gas turbine must be above 500 °C, eliminates the choice of many gas turbines listed in the screening process.

As mentioned above, the minimum power (heat) required for the steam plant is about 756 MW. From the gas turbine specifications one can calculate its exhaust power by:

$$Q_{exhaust,GT} = Q_{in,GT} - P_{GT} = P_{GT}\left(\frac{1}{\eta_{GT}} - 1\right) \quad (16)$$

which gives about 441 MW. This in turn means that one gas turbine alone will not be enough to generate the required steam and supplementary firing will be necessary. Another option would be using two gas turbines with or without supplementary firing.

In modeling ambient temperature and pressure are assumed to be 25 °C and 1.01 bar, respectively. Generator efficiency is assumed to be 97% which is a typical value. The calculated fuel mass flow and fuel consumption (based on LHV) are 16.08 kg/s respective 736.7 MW. The LHV of the fuel (natural gas) is assumed to be 45,810 kJ/kg.

The turbine inlet temperature set in the model does not correspond to the inlet temperature of the real gas turbine. In reality, during the expansion both gases and air cooling are mixed and results in a lower average temperature. However, in modeling cooling air is neglected and therefore the inlet temperature would be higher than the reality. Thermodynamically, the most important parameters would be gas turbine exhaust temperature, exhaust mass flow, fuel consumption, power production and efficiency which all are calculated correctly.

SUGGESTED REPOWERING CONFIGURATIONS

As mentioned above, the idea is to maintain the steam cycle as it was designed originally and replace the burner with a HRSG. Adding a gas turbine (or gas turbines) on the top of the steam and designing a CC is not new, but will be studied here for comparison with the new suggested plant. As seen in Figure 1, the existing steam plant is a simple plant without multi-pressure level steam generators and reheating and therefore the plant efficiency is also low (only about 33% for about 260 MW). Other newer steam plants could have much higher efficiencies (e.g., 45%) and higher power output (e.g., 500 MW), see e.g.,

[37]. In this study, the repowering is accomplished without changing the balance of the existing plant, meaning that when repowering the final plant efficiency will not be as high as that possible with a modern new designed combined cycle.

Repowering with Gas Turbine

The first option for the repowered plant is to use one GT with supplementary firing as shown inFigure 2. In the configuration, the HRSG is designed with one drum connected to the evaporator. The off-gases are sent out at point 55 in Figure 2. The component settings are not changed for the steam plant at all, allowing significantly less cost associated with repowering. Only the current burner is replaced with a heat recovery steam generator. Further, repowering of an existing plant without altering the existing steam cycle will have substantially lower plant efficiency compared to a new designed combined cycle, as discussed above.

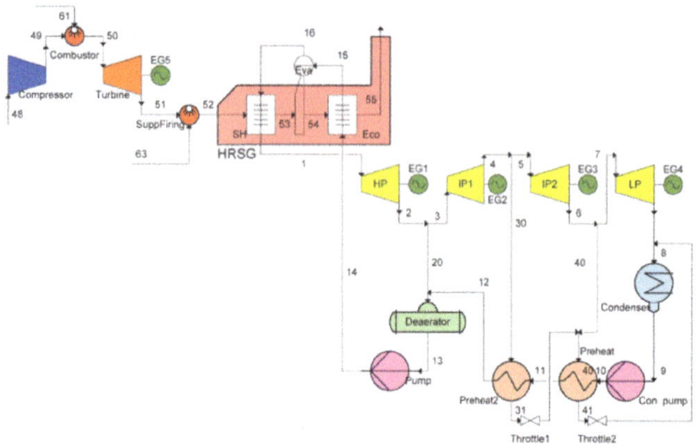

Figure 2. Natural gas fired CC plant with one GT and supplementary firing.

As shown in Figure 2, the fuel (NG) shall be supplied to the gas turbine chamber at point 61 as well as supplementary firing at point 63. With the estimations shown above the minimum power required from the supplementary firing will be about 315 MW, which in turn requires a large size burner as supplementary firing. Another option is to have two gas turbines with supplementary firing as displayed in Figure 3.

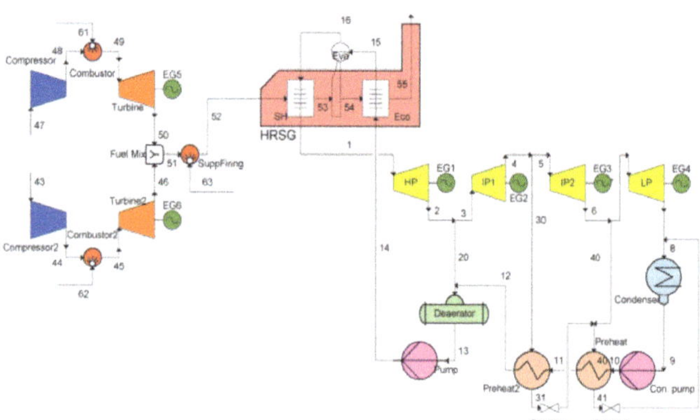

Figure 3. Natural gas fired CC plant with two gas turbines and with or without supplementary firing.

Generally, including a supplementary firing has the pro of allowing shutting down one gas turbine to undergo servicing without shutting down the entire plant. In this configuration, fuel is only supplied to the gas turbines' combustion chambers at points 61 and 62, if no supplementary firing is used. By including supplementary firing then fuel must be supplied to point 63 in addition to the gas turbines combustion chambers (points 61 and 62 in Figure 3). The HRSG design is similar to case with one gas turbine, allowing for fair thermodynamic comparison. It should be noted that in the case with two gas turbines the repowering cost is substantially higher than the corresponding case with one gas turbine.

New Hybrid Cycle

The new system suggested here for repowering is presented in Figure 4, which uses a natural gas fired SOFC system functioning as a topping cycle, while the steam cycle comprises the bottoming cycle. For the topping SOFC cycle, the ambient air at 25 °C is compressed to the working pressure of the SOFC (normal pressure) and then heated in the cathode air preheater (Cathode PH in Figure 4) to 600 °C before entering the cathode side of the SOFC stacks. For the anode side, the fuel was preheated in a heat exchanger (NG PH in Figure 4) before it was sent to a desulfurization unit to remove the sulfur content in the NG. This unit was assumed to be using a catalyst and operated at temperature of 200 °C. The heavier carbon contents in the desulfurized gas are cracked down in a Catalytic Partial Oxidation (CPO) type pre-reformer. Before that, the fuel must be preheated again to reach the operational temperature of the CPO catalyst, which is accomplished in the reformer preheater (Ref PH in Figure 4). The temperature of the pre-reformed gas is supposed to reach 650 °C which is high enough to be sent to the anode side of the SOFC.

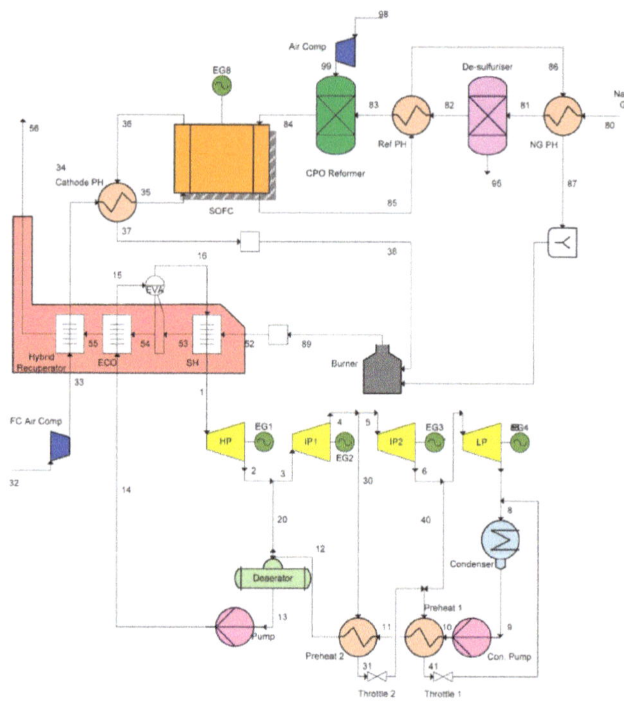

Figure 4. Natural gas fired SOFC—steam hybrid plant.

The off-fuel (the anode remaining fuel out of the fuel cell) is used to preheat the fuel in its path. The operating temperature of the fuel cell is assumed to be 780 °C which is enough to preheat the incoming gas. The entering temperatures mentioned above are the minimum entering temperatures and are essential requirements for the proper functioning of SOFC stacks, not only to initiate the chemical reactions but also to avoid cell thermal fractures. The burner is implemented because all of the fuel will not be reacted in the fuel cell stacks due to SOFC fuel utilization factor.

The off-gases from the burner have a high heat quality, which can be used to generate steam in a HRSG through economizer, evaporator and super heater. As discussed in [16,38], the off-gases out of HRSG maintain a high quality heat, which can be used to preheat the air after the compressor in the SOFC cycle. In other words, heat is recycled back to the topping cycle, and therefore, this technique is called hybrid recuperation. Such hybrid recuperator (HR) is shown to be very efficient and can increase the plant efficiency significantly. It increases the energy supplied to the SOFC cycle which in turn decreases the duty of the cathode pre-heater. Therefore, the energy from the SOFC off-fuel will be higher, allowing for more heat to be available after the burner. Other parameters assumed for the SOFC plant are summarized in Table 5.

Table 5. System operating input parameters.

Parameter	Value	Unit
Air compressor isentropic efficiency	0.8	-
Air compressor mechanical efficiency	0.95	-
Heat exchangers air side pressure drops	0.1	bar
Hybrid recuperator gas side pressure drop	0.08	bar
Hybrid recuperator effectiveness	0.9	-
Heat exchangers fuel side pressure drops	0.05	bar
Reformer compressor isentropic efficiency	0.85	-
Reformer compressor mechanical efficiency	0.95	-
Reformer inlet temperature	330	°C
Desulfurizer pressure drop	0.05	bar

RESULTS AND CONCLUSIONS

Natural gas is assumed to be pressurized and have the following compositions (molar fraction based); CH_4 = 0.87, C_2H_6 = 0.081, C_3H_8 = 0.01, C_4H_{10}-N = 0.006, CO_2 = 0.02925, H_2S = 0.00375 with LHV of 45,810 kJ/kg.

The calculations show that using two gas turbines without supplementary firing, the heat produced from the two gas turbines is not enough to satisfy the heat required by the steam cycle. The exhaust gases out of the gas turbines are mixed but the total heat provided in the exhausts is still not enough to satisfy the request from the steam cycle and small amount of supplementary firing is thus needed. Due to the high excess air in the gas turbines, the oxygen necessary for the supplementary firing is already bounded in the off gases. It is also noted that the two gas turbines are able to provide more than 90% of the heat needed by the steam cycle, and therefore the contribution from the supplementary firing will be small.

The performance comparison among repowering plants presented above is shown in Figure 5, in terms of plant net power output. Note that SF stands for supplementary firing in the figure. As can be seen, already with one gas turbine and supplementary firing (CC – 1 GT + SF in Figure 5) the net power can be increased by more than 210% compared to the base case (original steam plant). Adding two gas turbines and supplementary firing (CC – 2 GT – SF in Figure 5) increases the net power by about 325%. So far, all such repowering lies withing the CC plants category. The suggested repowering with SOFC without supplementary firing (SOFC in Figure 5) increases the net power more than 375% (almost four times larger). Thus the suggested hybrid SOFC-ST plant perfoms far better than the CC category plants.

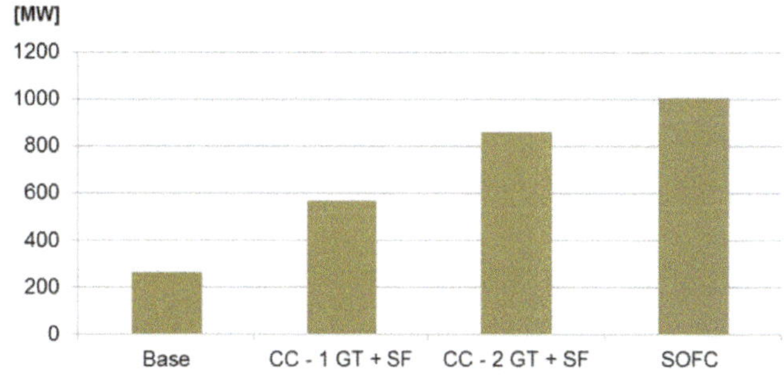

Figure 5. Comparison performance of the repowering plants in terms of power. SF = supplementary firing, CC = repowered combined cycles, GT = gas turbine, SOFC = solid oxide fuel cell.

Another interesting performance comparison could be plant thermal efficiency and CO_2emissions. These are displayed in Figure 6. As demonstrated in the figure, among CC categories, repowering with two gas turbines and supplementary firing has the best efficiency which is about 53%. This is of course lower than in a new designed CC plant which can reach about 59%. The reason is that the steam plant is not designed for the gas turbines but instead the gas turbines are fitted with an existing steam plant. The CC – 1 GT + SF (repowered plant with onegas turbine and supplementary firing) cannot achieve 50% plant efficiency. However, repowering with SOFC results in plant efficiency more than 60% and can compete with a new designed CC plant. Again, the efficiency of repowering with SOFC is well below a new designed SOFC-ST (*cf.* [16]).

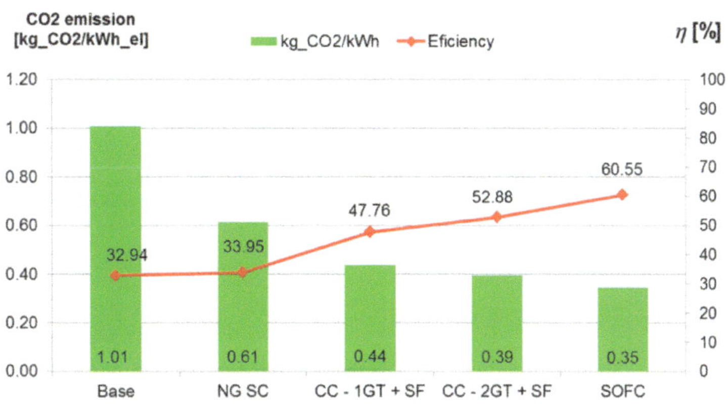

Figure 6. Comparison performance of the repowering plants in terms of plant thermal efficiency and CO_2 emission.

Here, the specific CO_2 emission is defined as:

$$e_{CO2} = \frac{\dot{m}_{CO2}}{P_{el}} (kg_{CO2}/kWh_{el}) \tag{17}$$

which is the mass flow of CO_2 per net power output (electricity). Note that the composition and mass flow of the off-fuel and off-air entering the catalytic burner are known. Thus the composition and mass flow of the off-gases (after the catalytic burner) are also known through chemical reactions and mass balance. Therefore the amount of CO_2 emission can easily be calculated.

As seen in Figure 6, the specific emission of CO_2 for the base case (coal fired steam plant) is about 1.01 kg/kWh. The results clearly demonstrate that the higher the efficiency is the lower CO_2emission will be. For the CC category, the combined cycle with two gas turbines and supplementary firing (CC – 2 GT + SF) has the lowest CO_2 emissions of about 0.39 kg/kWh which is only about 40% of the original plant. This of course is much lower than the base case, however, all the decrease is not only due to plant efficiency but also because the fuel is changed from coal to natural gas. Therefore, another case is included in the figure which is the same steam plant but fired with natural gas instead of coal (NG SC in the figure). A natural gas fired steam plant has emissions that are about 60% of those of the coal fired plant. The suggested repowering with SOFC plant performs best with an emission of about 35% compared to the original plant, but also about 60% of the original plant fired when natural gas instead.

Other plant calculated data such as fuel consumption, total fuel rate, auxiliary power and supplementary fuel mass flow are shown in Table 6. As can be seen the auxiliary power consumption for the repowered plants (CC category plants) are much higher than the others, which is due to the compressor of the gas turbines. The auxiliary power consumption for the presented SOFC plant is relatively very low compared to the other repowered plants, which is due to the non-pressurized SOFC stacks. Supplementary fuel consumption for the repowered plant with two GT and supplementary firing is only 3.34 kg/s which is very low compared to the main fuel rate, explaining that the two gas turbines are able to generate about 90% of the heat required from steam plant.

Table 6. Plant performance for different configurations

Parameter/Configuration	Base	CC – 1 GT + SF	CC – 2 GT + SF	SOFC
Fuel consumption, (kW)	801.15	1191.35	1626.38	1655.81
Total fuel flow rate, (kW)	25.75	26.00	35.50	36.14
Auxiliary power, (kW)	17.639	297.254	590.993	118.014
Supplementary fuel rate, (kg/s)	-	9.92	3.34	-

In order to study the composition of the off-gases Figure 7 shows the comparison between the base case (coal fired) and suggested repowering with SOFC. The CO_2 emission is significantly lowered (by about 2.9 times) but also oxygen flow rate is considerably increased (by about 5.8 times). Note that small amount of sulfur (SO_2) exists in the chimney of the coal fired base plant while no sulfur exits in the repowered SOFC plant because the fuel was desulfurized.

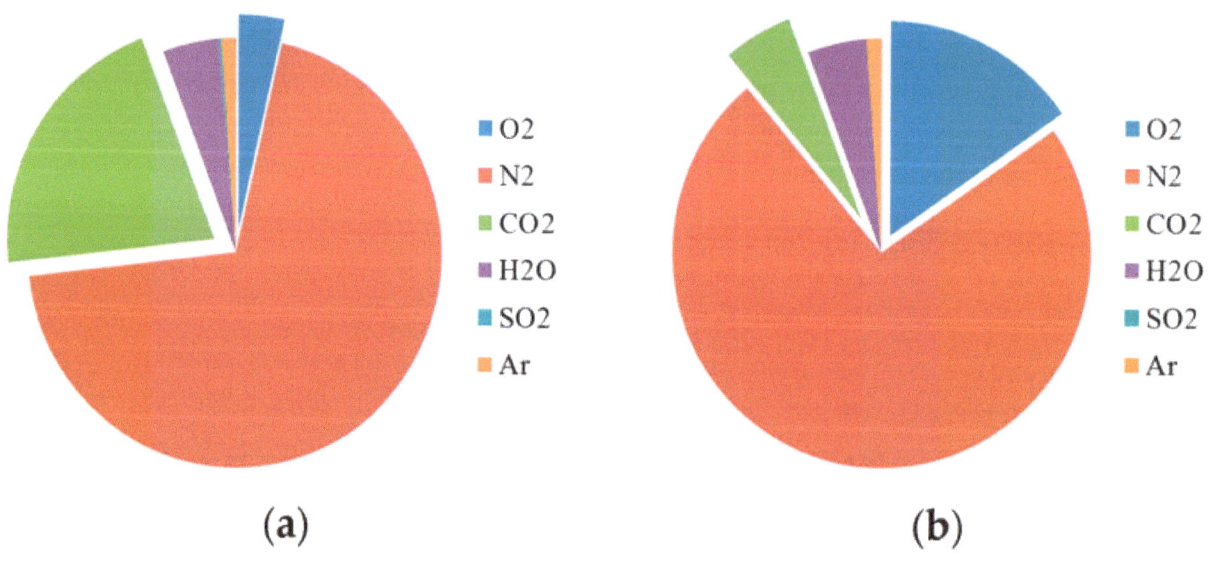

Figure 7. Specific off-gases emissions for (**a**) base case; and for (**b**) repowering with SOFC.

T–Q Diagram

To study the reason why the presented repowering with SOFC (SOFC-ST hybrid) is superior when compared to the other repowered plants in terms of plant efficiency, one need to analyze the temperature–heat diagram for HRSG among others. Such diagram for repowering with SOFC is shown in Figure 8, in which the temperature and heat of each component in the HRSG is shown with the corresponding node number appearing in Figure 4.

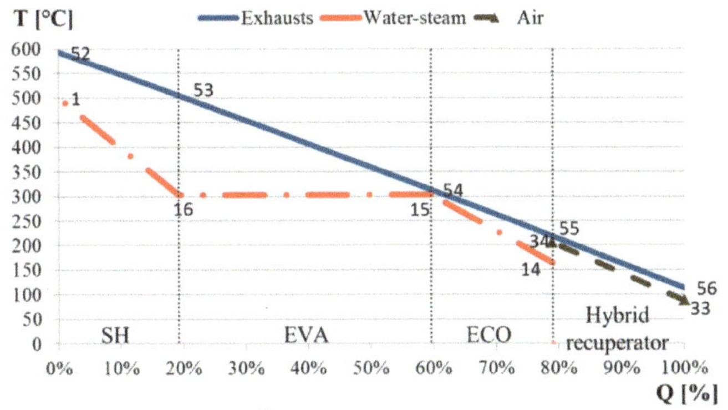

Figure 8. The heat-temperature diagram for HRSG of the SOFC-ST hybrid system. SH = Super Heater, EVA = Evaporator, ECO = Economizer.

For all heat exchangers, the pinch temperature (difference between nodes 54 and 15) is set to 10 °C. The terminal temperature difference (difference between node numbers 52 and 1) is about 93 °C and the gases leave the HRSG at temperature about 113 °C. Significant energy has been recovered by hybrid recuperation, from 218 °C to 113 °C, which corresponds to 20% of total energy recovered in the HRSG. Without hybrid recuperator, the majority of this energy would be lost to the ambient. Moreover, both the super heater (SH) and economizer (ECO) each uptake 20% of the total energy recovered by HRSG. About 40% of the total energy in the HRSG is allocated in the evaporator (EVA).

Similarly, the temperature–heat diagram is also shown for the combined cycle with two gas turbines and with supplementary firing, Figure 9. The reason that CC – 2 GT + SF is chosen is that this combined cycle performs best among all combined cycles studied here. The gases leave the HRSG at a temperature of about 189 °C which is significantly higher than the case with SOFC repowering. Thus lower energy has been recovered in HRSG for the CC case compared with SOFC case.

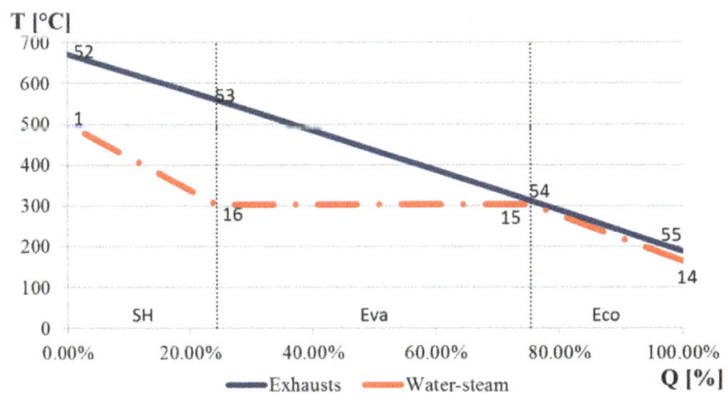

Figure 9. The heat-temperature diagram for HRSG of the CC – 2G + SF combined cycle.

The terminal temperature difference is about 170 °C which is also higher in this case when compared with the previous case (ca. 93 °C). Here both the super heater and economizer each uptake about 25% of the energy in the HRSG, which is slightly higher than the SOFC case. The evaporator absorbs about 50% of the energy from HRSG, which is also higher when compared with SOFC repowering. All these together make the area between the off-gases temperature line (red line) and the water-steam temperature line (blue line) to be larger than the case with SOFC repowering. This in turn means that less exergy has been wasted in HRSG when SOFC repowering is used. In other word, for the case with SOFC repowering, the off-gases temperature line approaches the corresponding line for water-steam and therefore resulting in less exergy loss and higher effectiveness for HRSG. The effectiveness (also called as recovery efficiency) of HRSG can be defined as (see e.g., [38]):

$$\varepsilon_{HRSG} = \frac{T_{gas,in} - T_{gas,out}}{T_{gas,in} - T_{ambient}} \tag{18}$$

then the HRSG effectiveness (recovery efficiency) can be calculated as 78.4% and 84.5% for the CC – GT + SF combined cycle respective SOFC-ST hybrid system.

Effect of SOFC Current Density and Utilization

Increasing SOFC current density decreases plant efficiency, while increasing SOFC operating temperature is in favor for plant efficiency, as shown in Figure 10. Generally, increasing SOFC current density decreases the power generated by SOFCs and thereby decreases plant efficiency as well. This is of course also true for SOFC powering, see Figure 10a. Decreasing the current density below 100 mA/cm^2 will not be realistic since the cell voltage reaches to a very high value (0.9374 V) which is close to open circuit voltage. At this current density, plant efficiency reaches to 67.4% and 64.09% for $U_F = 0.8$ respective $U_F = 0.7$, which are significantly higher than the corresponding efficiency at 300 mA/cm^2. The figure also demonstrates that the higher the current density is the lower the difference between two utilization factors will be.

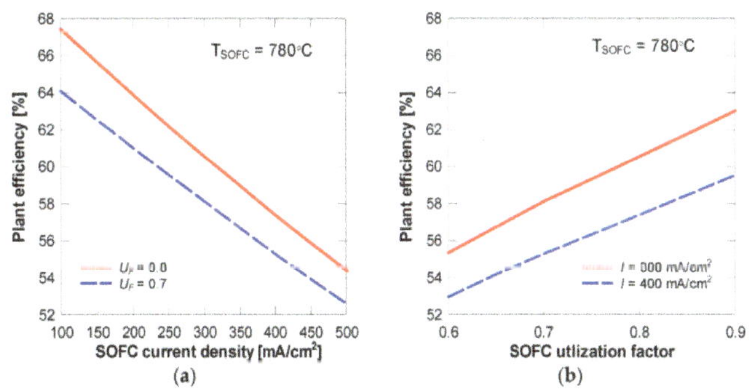

Figure 10. Effect of SOFC (**a**) current density and (**b**) utilization factor on plant efficiency of repowering with SOFC.

Similarly, increasing the fuel cell utilization factor increases plant efficiency when current density is assumed to be unchanged, see Figure 10b. For example, increasing the utilization factor from 0.6 to 0.9, increases plant efficiency from 53.0% to 59.5%, respectively, when the current density is assumed to be 400 mA/cm^2. Increasing the utilization factor beyond 90% is not beneficial for the cell, since the ionic concentration increases significantly inside the porous electrolyte and thereby blocks their pathway and consequently decreases cell voltage. However, for current density of 300 mA/cm^2, plant efficiency will be 63.0% when utilization factor is anticipated to be 0.9, as displayed in Figure 10b.

In fact, there exists an optimum operating temperature at which the plant efficiency is maximum. This is in contrast with the study of [38]. Note that in the study of [38] the steam plant was designed and adapted to the SOFC topping cycle while here the topping SOFC cycle is adapted and designed to provide required heat for an existing steam plant. These are completely different situations which affect the plant efficiency dramatically in terms of plant efficiency and net power output. For a current density of 300 mA/cm², this maximum efficiency is calculated to be about 62% at a temperature of 740 °C, as revealed in Figure 11a. Decreasing current density to 200 mA/cm², then the maximum plant efficiency happens at an operating temperature of 730 °C, and the maximum point is also shifted towards lower temperature, see Figure 11b. Here the plant efficiency will be as high as 66.2% for $U_F = 0.8$ while corresponding value for $U_F = 0.7$ is 63.4%, although slightly lower when utilization factor is lowered. Both these temperatures are lower than the current temperature technology of 780 °C, which in turn means companies' endeavor to decrease the SOFC operating temperature, is in agreement for such repowering system.

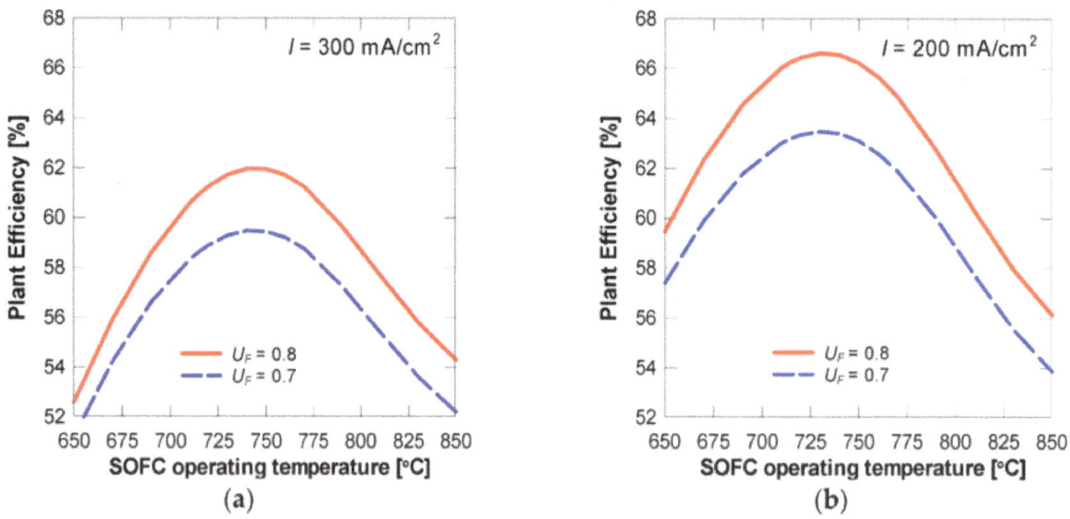

Figure 11. Effect of SOFC operating temperature on plant efficiency of repowering with SOFC, (**a**) For current density of 300 mA/cm²; (**b**) For current density of 200 mA/cm².

Effect of SOFC Operating Temperature

On the other hand, increased SOFC operating temperature is not always in favor for plant efficiency as established in Figure 11. In fact, there exists an optimum operating temperature at which the plant efficiency is maximum. This is in contrast with the study of [38]. Note that in the study of [38] the steam plant was designed and adapted to the SOFC topping cycle while here the topping SOFC cycle is adapted and designed to provide required heat for the existing steam plant. These are completely different situations which affect the plant efficiency dramatically in terms of plant efficiency and net power output. For current density of 300 mA/cm², this maximum efficiency is calculated to be about 62% at a temperature of 740 °C, as revealed in Figure 11a. Decreasing current density to 200 mA/cm², then maximum plant efficiency happened at operating temperature of 730 °C, also the maximum point is shifted towards lower temperature, see Figure 11b. Here the plant efficiency will be as high as 66.2% for $U_F = 0.8$ while corresponding value for $U_F = 0.7$ is 63.4%, although slightly lower when utilization factor is lowered. Both these temperatures are lower than the current temperature technology of 780 °C, which in turn means companies' endeavors to decrease the SOFC operating temperature, are in agreement for such repowering systems.

Effect of Hybrid Recuperator on Plant Performances

As discussed in [38], introducing a hybrid recuperator into the plant (inserting a heat exchanger between the off-gases after the stack and air prior to the cathode preheater and utilizing off-gases energy) increases the energy supplied to the SOFC cycle which in turn decreases the duty of the cathode pre-heater and consequently the energy of the off-fuel out of SOFC will be higher and thus more heat will be available after the burner and bottoming cycle. Thus, plant efficiency increases significantly. The effect of hybrid recuperator is shown in Table 7.

Table 7. Effect of HR on plant efficiency and power

Parameter/Configuration	With HR	Without HR	Without HR with SF
SOFC Utilization factor, (-)	0.76	0.76	0.8
Current density, (mA/cm²)	300	300	300
SOFC operating temperature, (°C)	780	780	780
Fuel consumption, (MW)	1727.37	2253.68	2332.90
Auxiliary power, (MW)	102.568	119.690	194.150
Net power production, (MW)	924.146	1231.12	1308.20
Plant efficiency, (%)	59.57	54.63	56.08
Electric power production, (MW)	1026.71	1350.81	1502.35
SOFC cell voltage, (V)	0.828	0.825	0.835

Note that the SOFC utilization factor is decreased to 0.76 because the burner outlet temperature otherwise would not be high enough to generate steam at 500 °C in the super heater, see Figure 12. The terminal temperature difference (difference between HRSG inlet gas temperature and generated steam temperature) is allowed to be 30 °C, which is within the recommended value so that the size of HRSG is not too large (see e.g., [1]). As revealed, fuel consumption decreases by about 23% and similarly the auxiliary power is decreased by about 14%. The net power will decrease by about 307 MW which is mainly due to the lower electricity from SOFC cells (*ca.* 324 MW). Despite less net power, plant efficiency is increased by about 5 point percent.

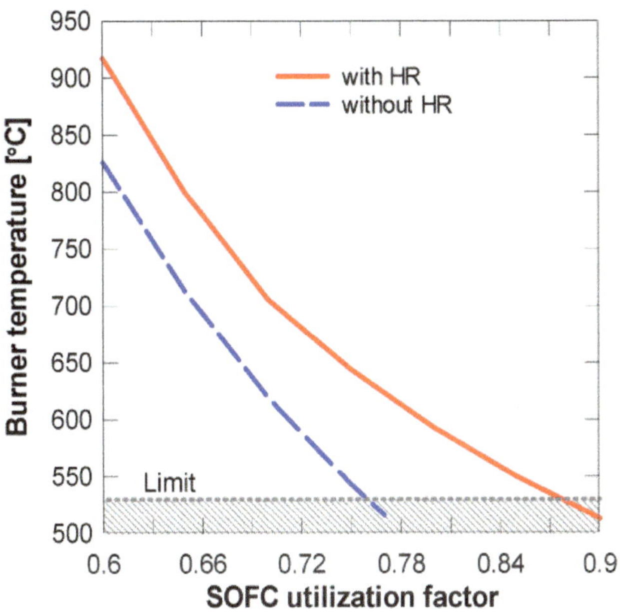

Figure 12. Burner outlet temperature as function of SOFC utilization factor.

Table 7 also shows that by bypassing the hybrid recuperator it would be possible to generate additional power of 384 MW which is about 40% larger than the case with hybrid recuperator (in expense of lower plant efficiency). Thus plant flexibility is increased dramatically and plant net power production can be altered without undergoing part-load conditions, e.g., under pick hours.

Figure 12 shows that by decreasing SOFC utilization factor burner temperature will be increased which is obvious because less fuel is consumed in the SOFCs and therefore more fuel will be burned in the combustor. Figure 12 also demonstrates that by removing hybrid recuperator the burner temperature drops significantly, for example for U_F values above about 0.78 the burner temperature will not be sufficient to generate steam at 500 °C. Taking into account the terminal temperature of 30 °C then U_F value shall be below 0.76 (see intersection between the blue line and limit line).

Effect of Fuel Mass Flow

Another parameter to be studied is the fuel mass flow (natural gas). As noted above the reformer inlet and outlet temperatures are fixed indicating that the fuel mass flow will be calculated to achieve such specified temperatures. Allowing fuel flow rate to be changed means that either the reformer inlet temperature or its outlet temperature should be variable. Such study is performed and the results are displayed in Figure 13. Increasing fuel mass flow increases cell voltage of the SOFC and therefore more power will be produced by the SOFC stacks. As result plant efficiency will also be increased, see Figure 13. As fuel mass flow is changed then CPO reformer air flow will also be varied to keep either its inlet or its outlet temperature at the specified value.

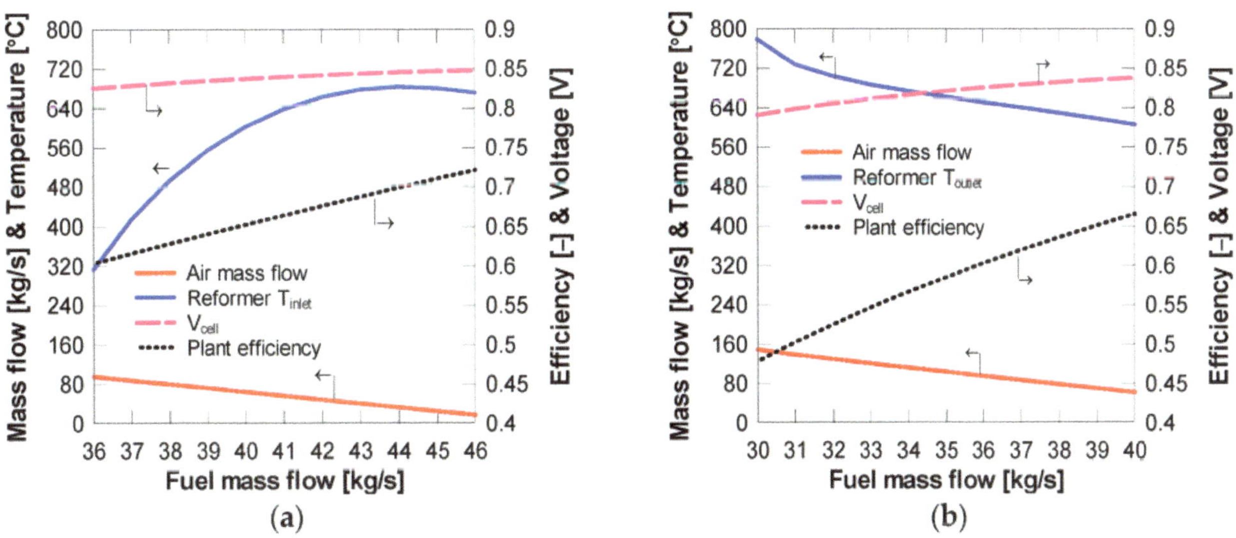

Figure 13. Effect of fuel mass flow on plant performance, (**a**) reformer outlet temperature is fixed and (**b**) reformer inlet temperature is fixed.

For the case with constant reformer outlet temperature, as more fuel is entering the CPO reformer then the reformer inlet temperature will be increased significantly, see Figure 13a. Reformer outlet temperature corresponds to SOFC anode inlet temperature which is specified at a value of 650 °C meaning that a temperature difference of 130 °C is applied. This is the maximum temperature difference over the anode, assumed in this study.

Too high temperature difference over the anode causes cell thermal stresses and thereby cell fatigue. Note that CPO inlet temperature and its air mass flow shall be read from the left side axis while plat efficiency and SOFC cell voltage shall be read from the right side axis. Reformer inlet temperature varies from 320 °C to about 640 °C as fuel mass flow is increased from 36 kg/s to 46 kg/s. Plant efficiency may reach over 70% at high fuel mass flows without compromising reformer inlet temperature. On the other hand if CPO reformer inlet temperature is kept constant (e.g., specified at 330 °C) then its outlet temperature will be decreased if the fuel mass flow is increased, see Figure 13b. Reformer outlet temperature varies from approximately 780 °C to 605 °C when fuel mass flow is changed from 30 kg/s to 40 kg/s. For fuel flows above 36 kg/s, reformer outlet temperature (same as anode inlet temperature) will be below the assumed value of 650 °C.

SIMPLE COST ANALYSIS

A simple cost analysis is performed to estimate the additional cost required for the proposed repowering plants. The costs of each component such as compressor, gas turbine, generator, heat exchangers, pre-reformer, desulfurizer and inventor are taken from the previous studies [39,40,41]. The idea to calculate the additional cost associated with repowering of each plant with each other and therefore the common components are not included in the cost analyses. These common components are assumed to be the existing Rankine cycle (remains the same without any changes), HRSG (must be replaced anyway), and burner and supplementary firing which are usually integrated with the HRSG.

In this study techno-economy is used to investigate the price of different plants in terms of kW/$. First, the purchased cost of different components is estimated from the previous studies and thus the so called Purchased Equipment Cost (PEC) is obtained. Then, the additional costing such as direct and indirect cost is added to each equipment cost. Direct cost includes onsite and offsite costs. In onsite cost, installation, piping, instrumentation and electrical equipment costs are included, while in the offsite cost architectural work and service facilities are accounted for. Indirect costs include engineering, construction and contractors' profit plus contingency. Finally, the cost of a plant is divided by the plant size to obtain plant cost in terms of kW/$. Detailed techno-economic cost analysis for this study can be found in [42]. Note that the cost of SOFC stacks is assumed to be valid for future serial production. The cost of each uninstalled SOFC stack is calculated to be about 188 $/kW for serial production in the future scenario which corresponds to installed valued of 778 $/kW (including installation costs). This agrees very well with the study of [43] which reported the installed cost of SOFC stack is about 770–850 $/kW.

The final cost of each plant is presented in Table 8. Again in these costs, existing steam plant (unchanged), HRSG (must be replaced and common in each plant) and supplementary firing (usually integrated with HRSG) is not included. In the table total cost corresponds to the installed cost (PEC plus installation costs). The interesting result is that the cost of repowering with SOFC is less than repowering with two gas turbines. Note that again the installed cost of each SOFC stack is assumed for a future serial production scenario (about 770 $/kW) while at the moment this cost is about four to five time larger, which means that the repowering plant with SOFCs would be about 470–600 $/kW which is much larger than that reported with one GT or two GTs.

Table 8. Total cost of additional required components for repowering

Plant Configuration	Plant Net Power (MW)	Total Cost ($)	Cost
Repowered plant with 1 GT	569	54,131,342	95 $/kW
Repowered plant with 2 GT	860	108,621,716	126 $/kW
Repowered plant with SOFC and HR	1003	119,863,704	120 $/kW

CONCLUSIONS

Repowering of an existing coal fired power plant is studied and different repowering designs are analyzed. In addition, a new repowering with SOFCs plant is also suggested. CO_2 emissions from different plant designs are calculated and compared with each other. The following conclusions can be drawn:

- Repowering with SOFCs produces the highest power.
- Repowering with SOFCs has the highest plant efficiency which is about 60% and competes with a new CC plant in terms of efficiency.
- CO_2 emissions from the SOFC-repowered plant are far less than those of the original coal fired plant, but also much lower than if the original plant was powered with natural gas instead of coal.
- Among repowered plants, the repowering with two Siemens SGT5 4000F gas turbines with small supplementary firing has the best performance, with an efficiency of 53%, which is considerably lower than that of a newly designed CC plant.
- There exists an optimum SOFC operating temperature at which plant efficiency is maximized. This temperature is calculated to be 740 °C for 300 mA/cm^2.
- It is possible to reach plant efficiencies above 65% if the current density is low enough.

REFERENCES

1. Kehlhofer, R.; Hannemann, F.; Stirnimann, F.; Rukes, B. *Combined-Cycle Gas and Steam Turbine Power Plants*, 3rd ed.; PennWell: Tulsa, OK, USA, 2009.
2. Danish Energy Agency. Energy in Denmark, 2014. Available online: http://www.ens.dk/en/info/facts-figures/energy-statistics-indicators-energy-efficiency/annual-energy-statistics(accessed on 27 January 2016).
3. Walters, A.B. Power plant topping cycle repowering. *Energy Eng.* **1995**, *92*, 49–71.
4. Kovacik, J.M.; Stoll, H.G. Economics of repowering steam turbines. *Natl. Eng.* **1990**, *95*, 27–37.

5. Carapellucci, R.; Milazzo, A. Repowering combined cycle power plants by a modified STIG configuration. *Energy Convers. Manag.* **2006**, *48*, 1590–1600.
6. Chellini, R. Repowering a steam power plant with a gas turbine. *Diesel Gas Turbine Worldw.* **1986**, *18*, 28–30.
7. Donatelli, R.U.; Ondryas, I.S.; Fransson, T.H. *Combustion Turbine Repowering at the Lauderdale Plant*; ASME, International Gas Turbine Institute (Publication): New York, NY, USA, 1990; pp. IGTI 5:201–IGTI 5:206.
8. Waller, H.; Scherer, V.; Scherer, D. The GT26 gas turbine in a combined gas-steam turbine power plant: Repowering the Rheinhafen Steam Turbine Power Plant of the Badenwerk AG. *VGB PowerTech* **1996**, *76*, 571–576.
9. Brander, J.A.; Chase, D.L. Repowering application considerations. *Eng. Gas Turbines Power* **1992**, *114*, 643–652.
10. Liszka, M.; Szargut, J. Simulation analysis of a repowered double fuel CHP plant including a non-evaporative heat recovery boiler. *Int. J. Energy Res.* **2004**, *28*, 661–682.
11. Modesto, M.; Nebra, S.A. Analysis of a repowering proposal to the power generation system of a steel mill plant through the exergetic cost method. *Energy* **2006**, *31*, 3261–3277.
12. Calise, F.; d'Accadia, D. M.; Palombo, A.; Vanoli, L. Simulation and exergy analysis of a hybrid solid oxide fuel cell (SOFC)—Gas turbine system. *Energy* **2006**, *31*, 3278–3299.
13. EG & G and G Technical Services Inc. *Fuel Cell Handbook*, 7th ed.; U.S. Department of Energy, Office of Fossil Energy, National Energy Technology Laboratory: Washington, DC, USA, 2004.
14. Riensche, E.; Achenbach, E.; Froning, D.; Haines, M.R.; Heidug, W.K.; Lokurlu, A.; Adrian, S. Clean combined-cycle SOFC power plant-cell modeling and process analysis. *Power Sources* **2000**, *86*, 404–410.
15. Haseli, Y.; Dincer, I.; Natere, G.F. Thermodynamic modeling of a gas turbine cycle combined with a solid oxide fuel cell. *Hydrog. Energy* **2008**, *33*, 5811–5822.
16. Rokni, M. Thermodynamic analysis of an integrated solid oxide fuel cell cycle with a Rankine cycle. *Energy Convers. Manag.* **2010**, *51*, 2724–2732.
17. Rokni, M. Plant characteristics of an integrated solid oxide fuel cell and a steam cycle. *Energy* **2010**, *35*, 4691–4699.
18. Traupel, W. *Thermische Turbomaschinen*, 3rd ed.; Spinger: Berlin, Germany, 1982.
19. Bang-Møller, C.; Rokni, M. Thermodynamic performance study on biomass gasification, solid oxide fuel cell and micro gas turbine hybrid systems. *Energy Convers. Manag.* **2010**, *51*, 2330–2339.
20. Holtappels, P.; DeHaart, L.G.J.; Stimming, U.; Vinke, I.C.; Mogensen, M. Reaction of CO/CO_2 gas mixtures on Ni-YSZ cermet electrode. *Appl. Electrochem.* **1999**, *29*, 561–568.
21. Matsuzaki, Y.; Yasuda, I. Electrochemical oxidation of H_2 and CO in a H_2-H_2O-CO-CO_2 system at the interface of a Ni-YSZ cermet electrode and YSZ electrolyte. *Electrochem. Soc.* **2000**, *147*, 1630–1635.
22. Chan, S.; Khor, K.; Xia, Z. A complete polarization model of a solid oxide fuel cell and its sensitivity to the change of cell component thickness. *Power Sources* **2001**, *93*, 130–140.
23. Keegan, K.M.; Khaleel, M.; Chick, L.A.; Recknagle, K.; Simner, S.P.; Diebler, J. *Analysis of a Planar Solid Oxide Fuel Cell Based Automotive Auxiliary Power Unit*; SAE Technical Paper Series 2002, No. 2002-01-0413; SAE International: Warrendale, PA, USA, 2002.
24. Costamagna, P.; Selimovic, A.; Del Borghi, M.; Agnew, G. Electrochemical model of the integrated planar solid oxide fuel cell (IP-SOFC). *Chem. Eng.* **2004**, *102*, 61–69.
25. Achenbach, E. Three-dimensional and time-dependent simulation of a planar solid oxide fuel cell stack. *Power Sources* **1994**, *49*, 333–348.
26. Zhu, H.; Kee, R.J. A general mathematical model for analyzing the performance of fuel-cell membrane-electrode assemblies. *Power Sources* **2003**, *117*, 61–74.
27. Kim, J.W.; Virkar, A.V. The effect of anode thickness on the performance of anode-supported solid oxide fuel cell. In Proceedings of the Sixth International Symposium on SOFCs, (SOFC-VI), PV99-19, Hawaii, HI, USA, 17–22 October 1999; Electrochemical Society: Pennington, NJ, USA, 1999; pp. 830–839.
28. Prentice, G. *Electrochemical Engineering Principles*; Prentice Hall International: Houston, TX, USA, 1991.
29. Christiansen, N.; Hansen, J.B.; Holm-Larsen, H.; Linderoth, S.; Larsen, P.H.; Hendriksen, P.V.; Hagen, A. Solid oxide fuel cell development at Topsoe fuel cell and Risø. *Electrochem. Soc.* **2007**, *7*, 31–38.
30. Smith, J.M.; Van Ness, H.C.; Abbott, M.M. *Introduction to Chemical Engineering Thermodynamics*, 7th ed.; McGraw-Hill: Boston, MA, USA, 2005.
31. Winnick, J. *Chemical Engineering Thermodynamics*; John Wiley & Sons: New York, NY, USA, 1997.
32. Rokni, M. Thermodynamic analysis of SOFC (solid oxide fuel cell)—Stirling hybrid plants using alternative fuels. *Energy* **2013**, *61*, 87–97.
33. Incropera, F.P.; DeWitt, D.P.; Bergman, T.L.; Lavine, A.S. *Introduction to Heat Transfer*, 5th ed.; Wiley: Danvers, MA, USA, 2006.
34. Cengel, Y.A.; Boles, M. *Thermodynamics: An Engineering Approach*, 4th ed.; McGraw-Hill: New York, NY, USA, 2002.
35. *Gas Turbine World: 2007 Performance Specs*, 24th ed.; Victor de Biasi: London, UK, 2007.
36. Siemens. Siemens Energy. Available online: http://www.energy.siemens.com (accessed on 17 May 2016).
37. Kishimoto, M.; Yoshihiro, M.; Takayanagi, K.; Umaya, M. *Operating Experience of Large Ultra Super Critical Steam Turbine with Latest Technology*; PWR-1994; American Society of Mechanical Engineers: New York, NY, USA; Volume 26, pp. 43–47.
38. Rokni, M. Thermodynamic investigation of an integrated gasification plant with solid oxide fuel cell and steam cycles. *Green* **2012**, *2*, 71–86.

39. Bejan, A.; Tsatsaronis, G.; Moran, M. *Thermal Design and Optimization*; John wiley and Sons Inc.: New York, NY, USA, 1996.

40. Braun, R.J. Techno-economic optimal design of solid oxide fuel cell systems for micro-combined heat and power applications in the U.S. *J. ASME Fuel Cell Technol.* **2010**, *7*.

41. Vialetto, G.; Rokni, M. Innovative household systems based on solid oxide fuel cells for a northern European climate. *Renew. Energy* **2015**, *78*, 146–156.

42. Rokni, M. Techno-economy of different solid oxide fuel cell based hybrid cycles. In Proceedings of the ASME 2014 International Mechanical Engineering Congress and Exposition, Montreal, QC, Canada, 14–20 November 2014. IMECE14:36858.

43. Staffell, I.; Green, R. The cost of domestic fuel cell micro-CHP systems. *Hydrog. Energy* **2013**,*38*, 1088–1102.

CHAPTER 3

THERMAL SYSTEM ANALYSIS AND OPTIMIZATION OF LARGE-SCALE COMPRESSED AIR ENERGY STORAGE (CAES)

Zhongguang Fu, Ke Lu and Yiming Zhu

Key Laboratory of Condition Monitoring and Control for Power Plant Equipment, North China Electric Power University, Ministry of Education, Changping District, Beijing 102206, China

ABSTRACT

As an important solution to issues regarding peak load and renewable energy resources on grids, large-scale compressed air energy storage (CAES) power generation technology has recently become a popular research topic in the area of large-scale industrial energy storage. At present, the combination of high-expansion ratio turbines with advanced gas turbine technology is an important breakthrough in energy storage technology. In this study, a new gas turbine power generation system is coupled with current CAES technology. Moreover, a thermodynamic cycle system is optimized by calculating for the parameters of a thermodynamic system. Results show that the thermal efficiency of the new system increases by at least 5% over that of the existing system.

Keywords

Compressed air energy storage; gas turbine; Brayton cycle

INTRODUCTION

The problem of electric power supply shortages has recently been alleviated somewhat with the accelerated pace of global power construction. This shortage has restricted economic development and long-term residential electricity use levels. Nonetheless, electricity cannot be stored on a large scale; power must be generated instantly as needed. A peak-valley difference in electricity is observed on the user side of a power grid, which requires the power supply to change with electricity load. Figure 1 shows the typical daily load curve of a region and highlights the peak and valley on the user side.

Citation: Zhongguang Fu, Ke Lu and Yiming Zhu, Thermal System Analysis and Optimization of Large-Scale Compressed Air Energy Storage (CAES), doi: 10.3390/en8088873.

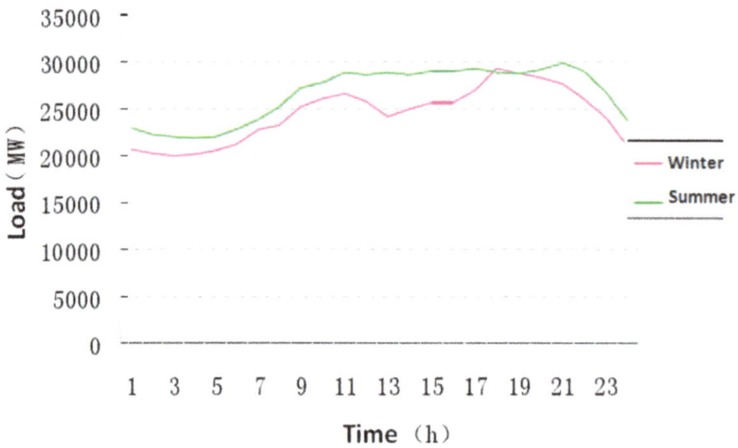

Figure 1. Typical daily load curve.

The peak load capacity on the power generation side cannot meet cities' increasing electricity load demands. The pressure of regulating the peak of the main power grid has increased with the growth of the proportion of residential electricity consumption, as well as the increased use of air conditioners and popular household appliances [1].

Moreover, the utilization of renewable sources, such as solar energy and wind energy, has been associated with a large-scale power grid problems given its intermittence, randomness, and instability; these sources limit grid stability and security [2,3]. Due to the presence of various obstacles, much of the electricity generated by renewable energy power plants cannot connect to the power grid. This problem induces to the phenomena of "abandoned wind" and "abandoned solar energy", which waste much renewable energy and investment [4].

Compressed air energy storage (CAES) is a large-scale industrial energy storage system that stores the energy generated at one time via compressed air. This energy can then be used in another situation. At a utility scale, the energy generated during periods of low-energy demand (off-peak) can be released to meet high-demand (peak load) periods. The study of energy storage can therefore promote the development of the current power grid.

Currently, the main problem of the CAES power plants is that high-pressure compressed air energy is not fully utilized, which is due to the limit of the inlet pressure of gas turbines. The purpose of this paper is to improve the energy efficiency of CAES system, by improving the thermodynamic system, optimizing parameters of the thermodynamic system, for the fuller use of compressed air energy [5].

TECHNICAL PRINCIPLES, ANALYSIS, AND OPTIMIZATION OF COMPRESSED AIR ENERGY STORAGE

As depicted in Figure 2, a CAES power plant consists of an energy storage system and an energy release system. The energy storage system uses air compressors to store energy by converting electric energy into the pressure energy of compressed air when the amount of generated energy is greater than the needs of the power grid. The energy release system later releases energy from compressed air in the form of electricity to the power grid when the grid side has a high demand [6,7,8]. In an actual runtime, one to two expanders and combustions are set up in a CAES power plant, such as the McIntosh Power Plant. A regenerator can be used to maximize the heat of the turbine exhaust [9].

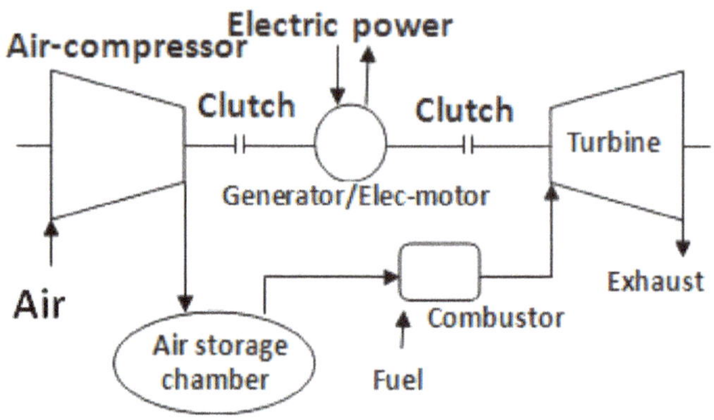

Figure 2. Compressed air energy storage (CAES) power generation system.

Two CAES power plants operate commercially: the Huntorf Power Plant, which was operationalized in Germany in 1978, and the Mcintosh Power Station, which began operating in the United States in 1991 [10,11]. In the 1990s, Japan committed to the building a 35 MW experimental CAES power plant that was eventually operationalized in 1997. Other countries, such as Israel, Finland, Great Britain, and Italy, also plan to conduct feasibility studies and demonstration projects related to such an initiative. Since the first CAES patent was published in 1949, two generations of CAES plants have been developed. An outstanding feature of the advanced, second-generation CAES system is that the storage system applies high pressure and the release system employs advanced gas turbines to generate electricity [12,13,14,15]. The energy conversion factor can exceed 50%.

Characteristics of a Compressed Air Energy Storage Thermal System

A CAES generation system has the following features [16,17,18,19]:

- Compression pressure is high. The pressure of air that is compressed by the compressor and stored in the gas storage device can exceed 100 bar.

- Storage temperature in the gas-storing devices is low. At present, most gas storage devices are located in underground caves at room and even lower temperatures.
- Storage air pressure decreases linearly with the consumption of compressed air during power generation.

Based on these characteristics, the Brayton Cycle can therefore be optimized and efficiency can be improved further.

Analysis and Improvement of a Compressed Air Energy Storage Thermal System

Thermal System Calculation and Analysis of a Second-Generation Compressed Air Energy Storage Power System

If the use of an air expander in the energy release system of a CAES power plant defines the first-generation CAES systems, then the utilization of an entire gas turbine as the energy release system characterizes the second-generation CAES systems. This enhancement significantly increases the energy transformation efficiency of the second-generation systems over that of the first-generation systems.

A schematic of the second-generation CAES power system is illustrated in Figure 3.

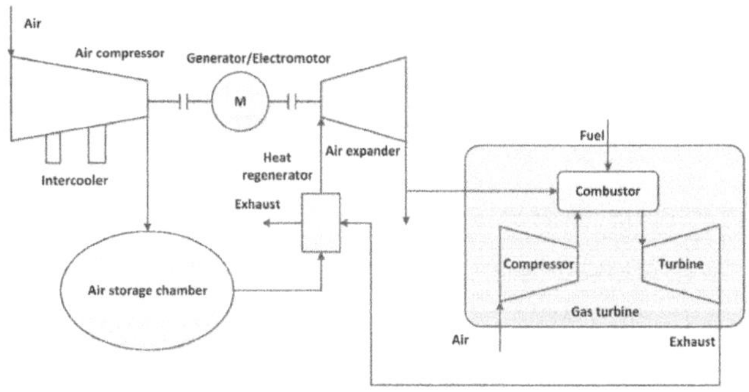

Figure 3. Schematic of the second-generation CAES power system.

A thermodynamic model is built for the system using the software Aspen Plus with the PENG-ROB property method. The initial isentropic efficiency of the compressor and the turbine is 0.88, and the initial mechanical efficiency of these components is 0.99. Both isentropic and mechanical efficiencies decrease, and the rate of decrease rises with pressure. The gas turbine combustor is assumed to be adiabatic and exerts a constant pressure. The difference in the terminal temperature of the heat exchanger ranges between 10–15 °C. The fuel is a natural gas composed of 97.6% methane, 0.62% ethane, 0.41% propane, 0.21% butane, 0.01% pentane, 0.05% hexane, 0.65% carbon dioxide, and 0.45% nitrogen. The ratio of C/H is 3.95. For convenient modeling analysis, the system does not consider blade cooling. The systems are compared and analyzed on the basis of theoretical aspects. The results are presented as Table 1.

Table 1. Simulation results of the second-generation compressed air energy storage (CAES) system

Parameters	Gas pressure	Flow Rate of Air	Pressure Ratio of a Gas Turbine	Inlet Temperature of a Turbine	Exhaust Temperature
Values	100 bar	600 kg/s	15.4	1300 °C	383.4 °C

The second-generation system can be used as the reference system, and the improved system is compared with this system to reduce the effect of selecting the calculation methods and the physical parameters that comprise the calculation results.

The results are as follows:

- When the pressure of the compressed air is high (above 100 bar), the exhaust temperature of the expander decreases if high-pressure energy is utilized further although air temperature increases after heat regeneration. Thus, expander efficiency is reduced.
- In the release system, the gas turbine circle belongs to the simple Brayton cycle. An efficient gas turbine and system can be generated by optimizing this cycle.

Improvement and Optimization of the Thermodynamic Cycle of a System

(1) Use of a new type of high-pressure reheat gas turbine

The pressure of the air originating from the air storage chamber is too high. Thus, air must be expanded to attain an ordinary pressure and flow into the combustion chamber for the turbine to operate effectively. This new system adopts a new type of high-pressure reheat gas turbine that uses high-pressure energy, and the values of the gas parameters increase.

In addition, the new type gas turbine includes a "reheat" process. This process is characterized by splitting the combustion process into two stages that are separated by an expansion to an intermediate pressure level. In this process, energy is added part way through the expansion process, thus resulting in high gas turbine efficiency and high power density.

The reheating process can raise the average heat absorption temperature, thus enhancing system efficiency. Figure 4 shows the T-S (Temperature-Entropy) chart of the common Brayton cycle, whereas Figure 5 indicates the temperature-entropy chart of the reheat Brayton cycle. The average heat absorption temperature of the reheat Brayton cycle is higher than that of the previous cycle, as is efficiency. Hence, overall efficiency improves.

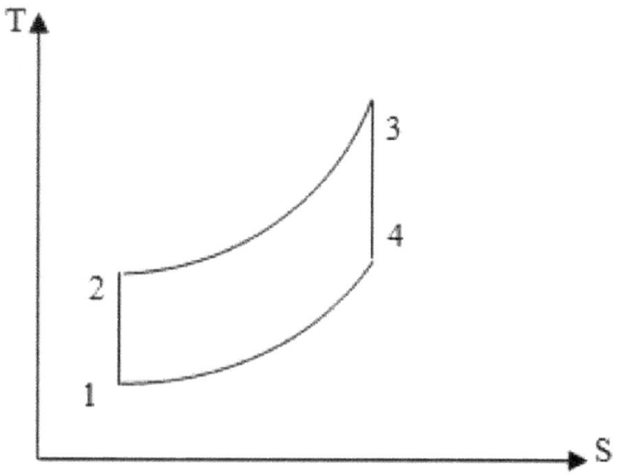

Figure 4. T-S (Temperature-Entropy) chart of the Brayton cycle.

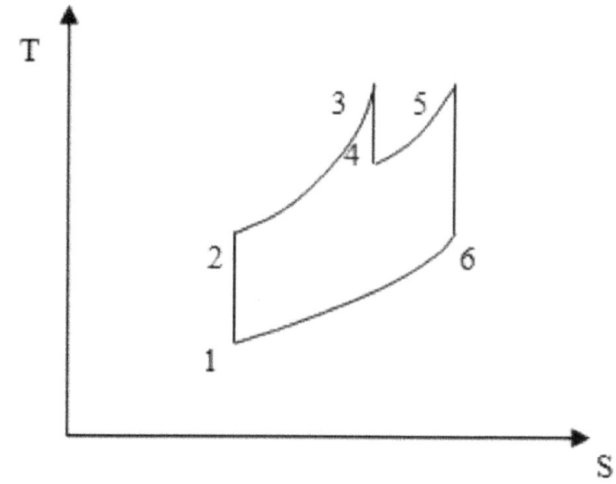

Figure 5. T-S chart of the reheat Brayton cycle.

(2) Using the regenerative cycle

Based on the calculations, the power consumption of the compressor that does not utilize inter-cooling devices is almost 600 MW. This consumption is the result of a sharp increase in air temperature during the compression process. To maximize the gas turbine exhaust, a regenerative cycle is generated to preheat low-temperature air from the gas storage chamber. Hence, the temperature of the gas turbine exhaust drops to improve energy efficiency.

Improved System

The model of the improved CAES system, which is shown in Figure 6, is built for calculation and analysis purposes.

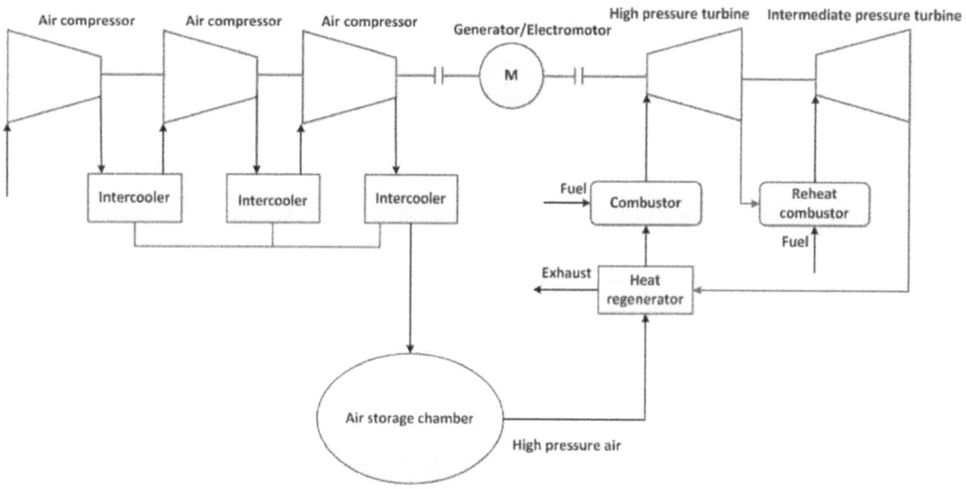

Figure 6. Schematic of the improved CAES system.

When air expands in operation, the air enters the regenerator for preheating. In this stage, multiple grades of heat can be used, thus maximizing the temperature of high-pressure air before entering the combustor. Consequently, fuel consumption can be reduced.

The improved system adopts a new gas turbine power generation technology that allows high-parameter air to enter the gas turbine. This technology can utilize the pressure energy of high-pressure air further.

The improved system uses the "reheat" combustion form. In the intermediate-pressure portion, the ordinary F-class gas turbine is used. The high-pressure combustion stage is added on this basis. The new inlet pressure of the high-pressure reheat gas turbine exceeds 60 bar, thus enabling the use of high-pressure energy. High-pressure air that originates from the air storage chamber is heated by the heat stored during the inter-cooling process and re-heated by the regenerator. The air enters the high-pressure gas turbine, is heated in a combustor to 1300 °C, and enters a high-pressure turbine (HPT) to expand partially. The HPT exhaust enters the reheat combustor for reheating and then transitions to the intermediate pressure turbine (IPT) to expand completely. The complete working process is thus described. The new gas turbine system adopts a split-shaft arrangement, and the intermediate-pressure portion employs the ordinary, widely-used F-class gas turbine. As a result, the system arrangement is simplified. The new reheat gas turbine technology can maximize the pressure energy-compressed air directly and utilize the reheat technology to further improve system efficiency.

RESULTS AND DISCUSSION

Effect of Pressure on Efficiency

Effect of High-Pressure Turbine Inlet Pressure on Efficiency

The inlet pressure of IPT is maintained at 15.4 bar, and the inlet pressure of the HPT is modified. Subsequently, the corresponding efficiency is calculated. The result is shown in Figure 7. The power source of the CAES system is composed of both the chemical energy from fuel and the mechanical energy from compressed air; hence, energy conversion factor (Equation (1)) is calculated as follows:

$$\xi = \frac{P_e}{P_c + G_f \cdot LHV} \quad (1)$$

where P_e is total system output power; P_c is compressor power consumption; G_f is the flow rate of fuel; and LHV is the low heat value of the fuel.

To economically evaluate CAES system generation, heat rate (HR) (Equation (2)) and electric power consumption rate (ER) (Equation (3)) are computed as follows:

$$HR = \frac{G_f \cdot LHV}{P_e} \quad (2)$$

$$ER = \frac{P_c}{P_e} \quad (3)$$

The results are depicted in Figure 7 and Figure 8. As indicated in Figure 7, system efficiency increases gradually at first and then decreases steadily with an increase in HPT (P_H) pressure to a maximum of approximately 140–160 bar. Figure 8 a indicates that when P_H increases, system HR decreases. In the low-pressure range (below 200 bar), HR decreases rapidly; however, the rate of this decrease slows down in the high-pressure range. Meanwhile, Figure 8b shows that system ER increases linearly with P_H.

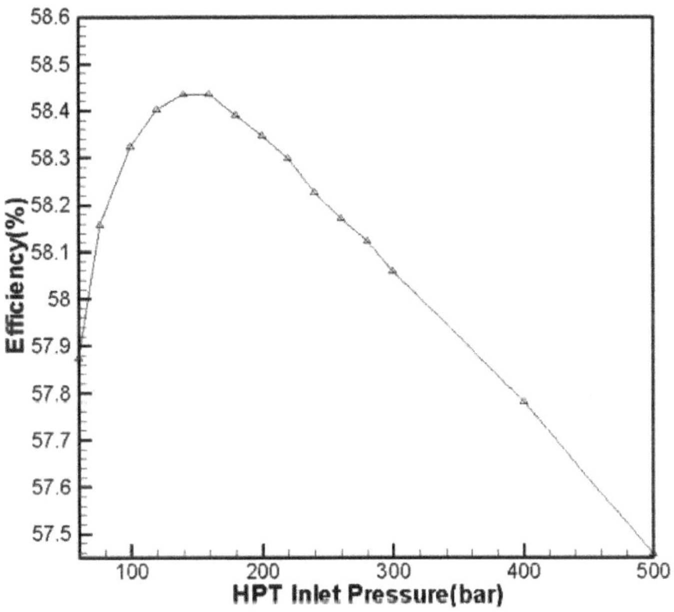

Figure 7. Effects of high-pressure turbine (HPT) inlet pressure on efficiency.

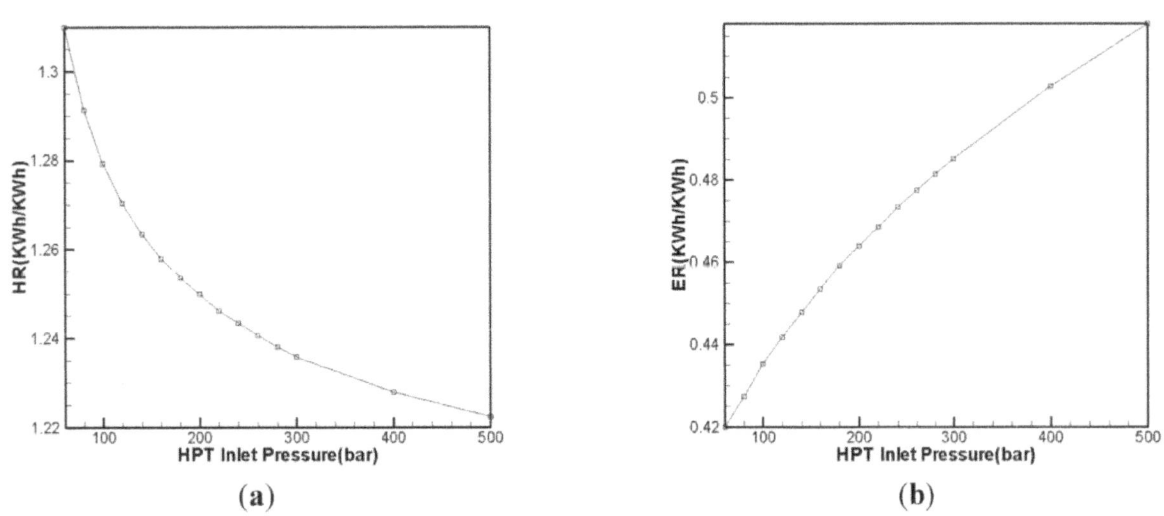

Figure 8. Effects of HPT inlet pressure on (**a**) heat rate (HR); (**b**) electric power consumption rate (ER).

The trends of HR and ER in Figure 8a,b highlight why system efficiency initially increases and eventually decreases. When compression pressure increases, compressor energy consumption increases exponentially, thus enhancing system ER. The work capacity per unit of fuel gradually increases as well when pressure increases. This increase in capacity is significant in the low-pressure range and plateaus in the high-pressure range. The reduction of HR in the low-pressure range is greater than the increase in ER; hence, system efficiency gradually increases. In the high-pressure range, however, the increase in ER is significantly greater than the reduction of HR reduction. Thus, system efficiency decreases considerably.

To explore the effect of high pressure on efficiency under different IPT (P_I) inlet pressures further, experiments are conducted in these conditions.

As shown in Figure 9, system efficiency increases initially and then eventually decreases with an increase in P_H given different P_I values. When P_I increases, the P_H value that corresponds to the peak point of system efficiency gradually increases.

As indicated in Figure 10a, HR declines with an increase in P_H given different P_I values. In addition, the higher P_H, the smaller HR.

Overall ER increases with P_H given different P_I values (Figure 10b). When the P_I value is high, overall ER initially decreases and then increases in a low-pressure range. When P_H is low and P_I is excessively high, gas cannot expand completely in HPT. Thus, the output power is low although compressor power consumption and ER are high. When P_H increases, the work capacity of the high-pressure gas turbine increases and ER declines. At a pressure of 10 bar, work capacity is limited relative to an increase in compressor power consumption, and ER is greater than other high pressures because a low P_I does not match combustion temperature.

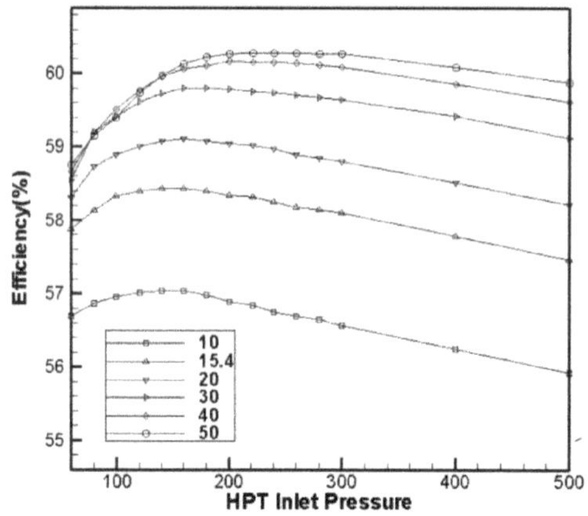

Figure 9. Effects of HPT inlet pressure on efficiency under different intermediate pressure turbine (IPT) inlet pressures.

Effect of Intermediate-Pressure Turbine Inlet Pressure on Efficiency

P_H is set to certain values, and P_I value is modified the corresponding efficiency, HR, and ER are then calculated. The results are displayed in Figure 11 and Figure 12.

As per Figure 11, under different P_H, P_I values also reach an optimum pressure value. System efficiency increases and the corresponding optimum pressures of middle pressure gradually rise as well with P_H value. Figure 12a shows the relationship between P_I and HR and indicates that HR declines with an increase in P_I. A high P_H value corresponds to a low HR given different P_I values. As depicted in Figure 12b, ER gradually increases with P_I with different P_H values. When P_I is low and P_H is high, ER initially increases and then decreases. When the HPT pressure measures 60 bar, ER increases significantly with P_I. This phenomenon can be attributed to the low P_H and high P_I, that is, the inlet pressure is close to the outlet pressure in the high-pressure stage. In this case, work apply is low, work capacity increases slightly, and ER increases considerably.

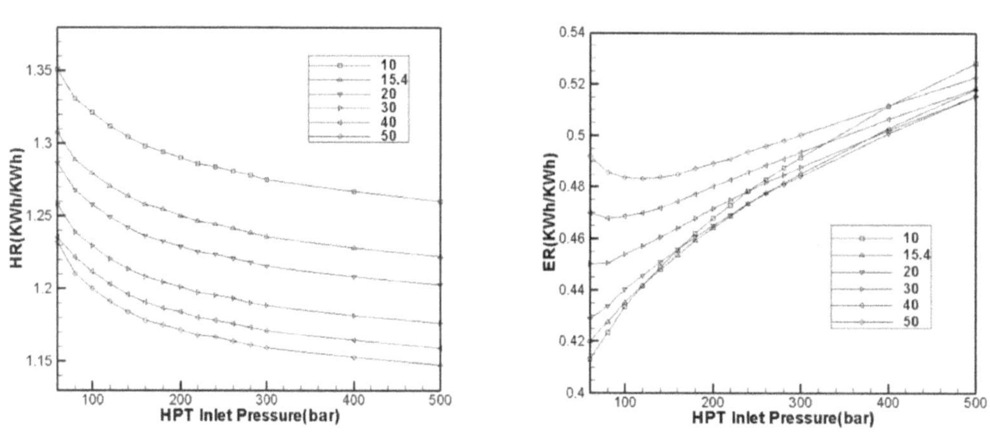

(a) (b)

Figure 10. Effects of HPT inlet pressure on (a) HR and (b) ER under different IPT inlet pressures.

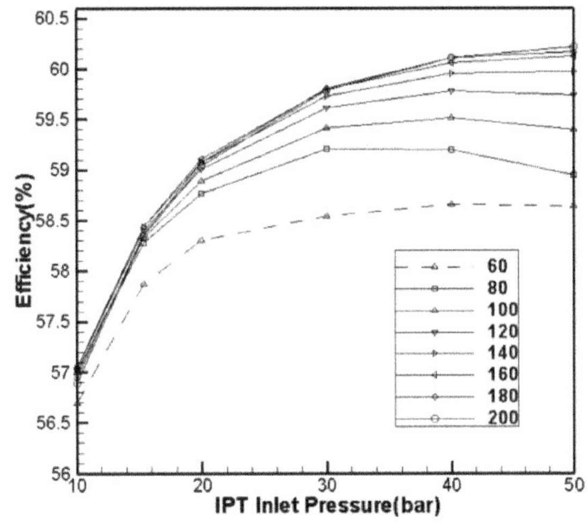

Figure 11. Effect of IPT inlet pressure on efficiency given different HPT inlet pressures.

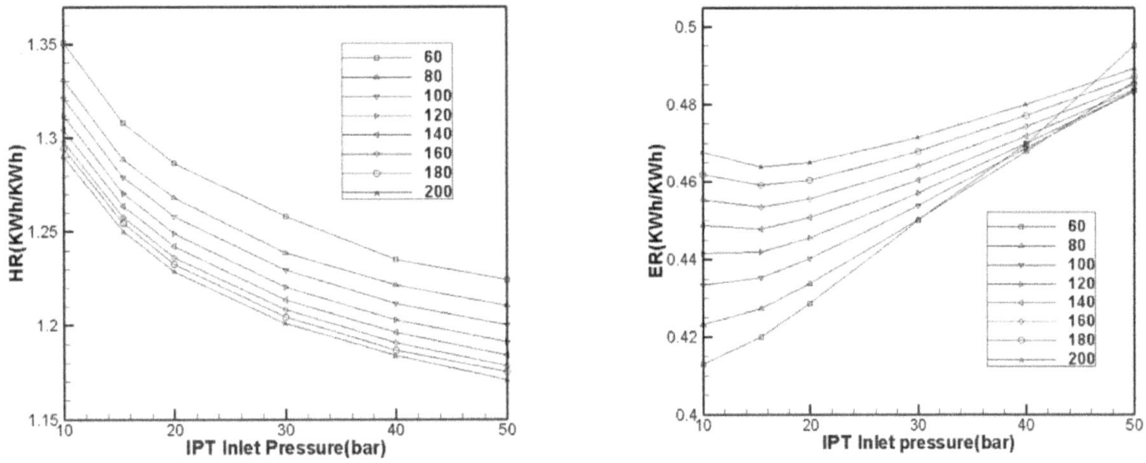

(a) (b)

Figure 12. Effect of IPT inlet pressure on (**a**) HR and (**b**) ER given different HPT inlet pressures.

Effect of High-Pressure Turbine Inlet Temperature on Efficiency

As exhibited in Figure 13, the IPT inlet temperature of the turbine is fixed at 900, 1100, and 1300 °C. The relationship between HPT inlet temperature and efficiency can be established by changing the HPT inlet temperature. The purpose of this study is to determine whether or not system efficiency improvement relies heavily on the IPT pressure-level turbine.

Figure 13 suggests that system efficiency increases with HPT inlet temperature of. The efficiency of the proposed system is higher than that of the existing second-generation CAES system even in the low-temperature region. Efficiency also increases in the high-temperature region. Hence, a highly efficient operation process does not rely significantly on the temperature level of the middle pressure-level turbine. Moreover, efficiency is enhanced considerably even when the E gas turbine (900 °C) is used in the reheating process.

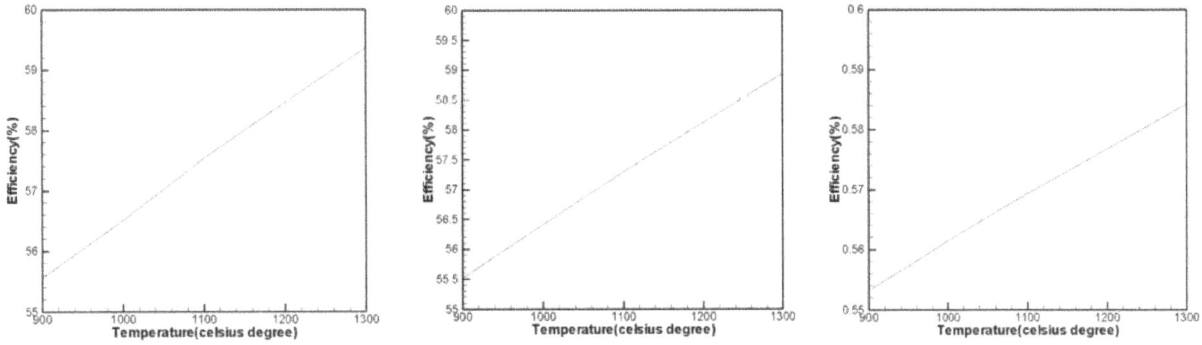

(a) (b) (c)

Figure 13. Effect of HPT inlet temperature on efficiency While IPT inlet temperature is: (**a**) 900 °C; (**b**) 1100 °C; (**c**) 1300 °C.

Performance Under Various Conditions

In CAES power plants, the outlet air pressure of an air storage chamber decreases gradually. Therefore, CAES plants actually function in sliding-pressure operation. As depicted in Figure 14, the efficiencies of the second-generation CAES system and of the improved system are calculated under various conditions.

As indicated in Figure 14, system efficiency fluctuates under various conditions. Nonetheless, the efficiency of the improved CAES system is always significantly higher than that of the second-generation CAES system. Furthermore, this efficiency does not decline significantly at low pressure. Therefore, the improved CAES system is highly practical.

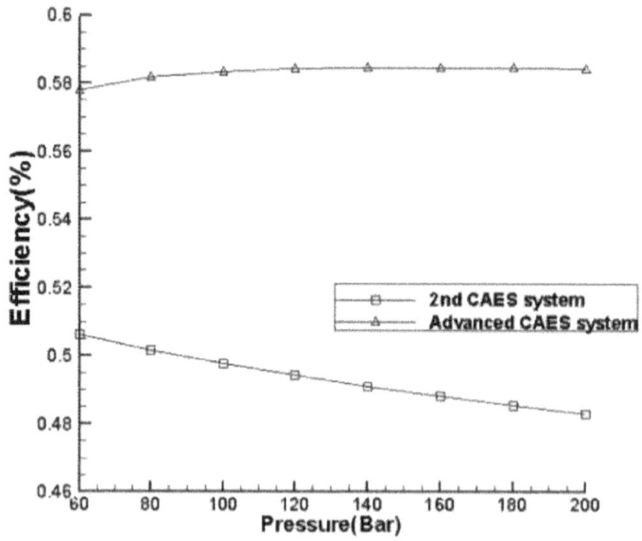

Figure 14. Effect of changes inair storage chamber pressure on efficiency.

Calculation Results

The analysis results regarding the influence of P_H and P_I on the efficiency of the improved system suggest that the maximum system efficiency ranges between 140–150 bar. Thus, setting 140 bar as the P_H value is appropriate. In addition, a system is efficient when P_I value is lower than 20 bar; therefore, the conventional F-class pressure of 15.4 bar is applied. The use of the conventional pressure in IPT implies that the existing gas turbine can be adopted and that system construction is simplified. Table 2 presents a comparison of the improved system with the second-generation system with respect to selected parameters. The table shows that the efficiency of the improved CAES is 9% higher than that of the second-generation CAES system. Furthermore, ER and HR are lower in the former than in the latter. Hence, the pressure energy of the stored air is applied efficiently.

Table 2. Comparison of the parameters of second-generation CAES systems with those of the improved CAES system.

System/Parameters		Second-Generation CAES	Improved System
Air storage chamber parameters	Pressure (bar)	140	
	Temperature (°C)	50	
Compressor outlet parameters	Pressure (bar)	140	
	Temperature (°C)	50	
Air flow (kg/s)		600	600
Turbine expansion work (MW)	HPT (air turbine)	157.3	396.6
	IPT (gas turbine)	474.7	492.0
Compressor power consumption (MW)		398	398
Turbine inlet pressure (bar)/temperature (°C)	HPT	-	133/1300
	IPT (gas turbine)	15.4/1300	15.4/1300
Flow rate of the fuel (kg/s)		17.74	22.36
Power output (MW)		632.0	888.6
Efficiency		49.06%	58.44%
HR (kWh/kWh)		1.410	1.265
ER (kWh/kWh)		0.629	0.447

Investment Estimation

The construction costs of an F-class CAES power plant are referenced to estimate the investment into an improved CAES system. The results are presented in Table 3. The construction costs of the improved CAES system increase with the requirement for new research facilities, but these costs decrease once the facilities mature. Cost per KWh did not increase significantly because power capacity increases when reheat technology is used.

Table 3. Investment estimation (million dollars)

Project	Reference F-Class CAES Power Plant (2 MW × 600 MW)	Mature Reheating Combustion Equipment (2 MW × 800 MW)	New Reheat CAES Equipment (2 MW × 800 MW)
Electromechanical equipment	153.4	271.08	355.55
High-pressure gas turbine	-	67.84	135.67
Low-pressure gas turbine	41.90	45.22	45.22
Compressor	17.08	17.08	32.55
Gas turbine generator	8.87	14.07	14.07
Intercoolers	2.66	2.66	2.66
Instrument and control system	3.91	5.86	7.03
Spare parts fee + freight	78.90	118.35	118.35
Regenerator	25.48	25.48	25.48
Main transformer supply	5.31	8.09	8.09
252 KV GIS supply	2.09	3.18	3.18
Gas pressure regulating station and gas replacement	2.05	2.96	2.96
Total equipment cost	188.34	310.78	395.25
Civil engineering cost and gas storage chamber	398.21	495.60	566.51
252 KV GIS supply	2.09	3.18	3.18
Gas pressure regulating station and gas replacement	2.05	2.96	2.96
Total equipment cost	188.34	310.78	395.25
Civil engineering cost and gas storage chamber	398.21	495.60	566.51
Crew. including heating pipe network	17.94	24.94	24.94
Gas transmission pipeline	17.73	25.59	25.59
Static total investment	622.22	865.90	1012.29
Static investment per kW ($/KW)	518	536	633
Dynamic total investment	691.36	952.12	1124.77
Dynamic investment per kW ($/KW)	576	595	703

CONCLUSIONS

The conclusions drawn from the analysis and calculations are as follows:

- The efficiency of the improved CAES system increased by 9% with the adoption of a high-pressure gas turbine that includes a reheat process and the utilization of heat energy in the inter-cooling process based on the second-generation CAES system.
- In the improved system, efficiency is related to HPT inlet pressure. When pressure increases, efficiency increases at first and then decreases. An optimum value was thus obtained.
- System efficiency increases initially and decreases eventually with an increase in IPT inlet pressure. Optimal pressure also increases with HPT inlet pressure.
- Efficiency is related to HPT inlet temperature, and an increase in temperature can improve the thermal efficiency of a cycle to some extent.
- System efficiency changes with air storage chamber pressure. Despite changes in pressure, the efficiency of the improved system is always significantly higher than that of the existing second-generation CAES system.
- In the improved system, not only did the parameter values increase, but capacity increased as well. System capacity reached 800 MW, thus indicating that the system can be used on a large scale.
- The costs of the F-class gas turbine equipment and the equipment for the improved system are estimated, and the cost per kW of the latter do not increase significantly.

ACKNOWLEDGMENTS

This work was supported by The Fundamental Research Funds for the Central Universities (13XS09 and 2014ZZD04) by Chinese Ministry of Education. China Guodian New Energy Technology Research Institute gave technical support. The authors express their sincere gratitude for supporting this important work.

AUTHOR CONTRIBUTIONS

Zhongguang Fu conceived and designed the improved system and the research; Ke Lu performed the Calculating process, data analysis and wrote the original manuscript; Yiming Zhu did the data processing, illustrations modification and helped during the editing, rewriting and review process.

REFERENCES

1. Ouyang, C.Y. To attach great importance to the development of power grid peaking problems. *Electr. Power Technol. Econ.* **2002**, *1*, 10–15. (In Chinese).
2. Grazzini, G.; Milazzo, A. Thermodynamic analysis of CAES/TES systems for renewable energy plants. *Renew. Energy* **2008**, *33*, 1998–2006.
3. Abbaspour, M.; Satkin, M.; Mohammadi-Ivatloo, B.; Lotfi, F.H.; Noorollahi, Y. Optimal operation scheduling of wind power integrated with compressed air energy storage (CAES).*Renew. Energy* **2013**, *51*, 53–59.
4. Li, C. Progress in renewable energy problem in China. *Forum Sci. Technol. China* **2008**, *2*, 111–120. (In Chinese).

5. Jubeh, N.M.; Najjar, Y.S. Green solution for power generation by adoption of adiabatic CAES system. *Appl. Therm. Eng.* **2012**, *44*, 85–89.
6. Raju, M.; Khaitan, S.K. Modeling and simulation of compressed air storage in caverns: A case study of the Hundorf plant. *Appl. Energy* **2012**, *89*, 474–481.
7. Liu, W.; Yang, Y.; Zhang, X.; Xin, Y. Present Situation and Development Trend of Compressed air energy storage (CAES) power plant. *Shandong Electr. Power* **2007**, *2*, 10–14. (In Chinese).
8. Kushnir, R.; Dayan, A.; Ullmann, A. Temperature and pressure variations within compressed air energy storage caverns. *Int. J. Heat Mass Transf.* **2012**, *55*, 5616–5630.
9. Nakhamkin, M.; Chiruvolu, M. Available Compressed Air Energy Storage (CAES) Plant Concepts. Available online: http://espcinc.com/mobile/library/PowerGen_2007_paper.pdf(accessed on 18 August 2015).
10. Proczka, J.J.; Muralidharan, K.; Villela, D. Guidelines for the pressure and efficient sizing of pressure vessels for compressed air energy storage. *Energy Convers. Manag.* **2013**, *65*, 597–605.
11. Crotogino, F.; Mohmeyer, K.U.; Scharf, R. Huntorf CAES: More than 20 Years of Successful Operation. In Proceedings of the Spring 2001 Meeting, Orlando, FL, USA, 15 April 2001.
12. Van der Linder, S. The case for CAES. Modern Power Systems. Available online: http://www.modernpowersystems.com/features/featurethe-case-for-caes/ (accessed on 18 August 2015).
13. *Concept Screening of Coal Gasification CAES Systems*; EPRI Research Report; Electric Power Research Institute: Palo Alto, CA, USA; May; 1979.
14. Schoenung, S.M. Utility energy storage applications studies. *IEEE Trans. Energy Convers.* **1996**, *11*, 658–664.
15. Thumann, A.; Mehta, D.P. *Handbook of Energy Engineering*, 5th ed.; The Fairmont Press Inc: Anaheim, CA, USA, 2001; pp. 391–418.
16. Liu, W.Y.; Yang, Y.P.; Song, Z.P. Compressed air energy storage and performance simulation. In Proceedings of the China Engineering Thermophysics engineering thermodynamics and energy utilization Conference, Xi'an, China, 8 November 2004; pp. 288–296. (In Chinese).
17. Liu, W.Y.; Yang, Y.P.; Song, Z.P. Optimization and Performance Simulation of Different CAES Systems. *J. Eng. Thermophys.* **2005**, *6*, 25–28. (In Chinese).
18. Huang, Q.H. *Steam Turbine and Gas Turbine Principles and Applications*; Southeast University Press: Nanjing, China, 2005; pp. 190–205. (In Chinese)
19. Kushnir, R.; Ullmann, A.; Dayan, A. Compressed air flow within aquifer reservoirs of CAES plants. *Transp. Porous Media* **2010**, *81*, 219–240.

CHAPTER 4

A NOVEL MODELING OF MOLTEN-SALT HEAT STORAGE SYSTEMS IN THERMAL SOLAR POWER PLANTS

Rogelio Peón Menéndez[1], Juan Á. Martínez[2], Miguel J. Prieto[2], Lourdes Á. Barcia[3] and Juan M. Martín Sánchez[4]

[1]Group TSK, 33203 Gijón, Asturias, Spain

[2]Department of Electrical Engineering, Universidad de Oviedo, 33203 Gijón, Asturias, Spain

[3]González Soriano S.A., 33420 Llanera, Asturias, Spain

[4]ADEX, S.L., 28031 Madrid, Spain

ABSTRACT

Many thermal solar power plants use thermal oil as heat transfer fluid, and molten salts as thermal energy storage. Oil absorbs energy from sun light, and transfers it to a water-steam cycle across heat exchangers, to be converted into electric energy by means of a turbogenerator, or to be stored in a thermal energy storage system so that it can be later transferred to the water-steam cycle. The complexity of these thermal solar plants is rather high, as they combine traditional engineering used in power stations (water-steam cycle) or petrochemical (oil piping), with the new solar (parabolic trough collector) and heat storage (molten salts) technologies. With the engineering of these plants being relatively new, regulation of the thermal energy storage system is currently achieved in manual or semiautomatic ways, controlling its variables with proportional-integral-derivative (PID) regulators. This makes the overall performance of these plants non optimal. This work focuses on energy storage systems based on molten salt, and defines a complete model of the process. By defining such a model, the ground for future research into optimal control methods will be established. The accuracy of the model will be determined by comparing the results it provides and those measured in the molten-salt heat storage system of an actual power plant.

Keywords

solar thermal power plant; thermal energy storage; process modeling

INTRODUCTION

Thermal solar plants use mirrors to focus the energy coming from the sun on a point where a heat transfer fluid (HTF) is heated; this is usually referred to as concentrated solar power (CSP) [1,2]. The fluid heated in this way is then used in a thermodynamic cycle (usually a water-steam cycle) to produce electricity. Most thermal power stations nowadays use parabolic trough collector (PTC) technologies [3,4,5] to heat some kind of oil, which plays the role of HTF. This is a very mature technology that is being used, for instance, in the 160-MW power plant that Sener, Acciona and TSK are installing in Ouarzazate, Morocco, and which has an overall cycle efficiency of 40%. This paper deals with this kind of power plants.

Citation: Rogelio Peón Menéndez, Juan Á. Martínez, Miguel J. Prieto, Lourdes Á. Barcia and Juan M. Martín Sánchez, A Novel Modeling of Molten-Salt Heat Storage Systems in Thermal Solar Power Plants, doi: 10.3390/en7106721.

However convenient as it may seem, solar energy is only available during certain times of day and the electric power this energy can originate might not match that demanded. Therefore, as well as defining an efficient and economical way to produce electric energy, it is also important to store thermal energy in an efficient and economical way so that it can be used when demanded [6]; this is usually referred to as thermal energy storage (TES).

Although several possibilities have been defined to carry out this TES (kinetic energy storage, potential energy storage, pressure energy storage, chemical energy storage), most plants store thermal energy without previous conversion into other type of energy, thus increasing the overall efficiency of the process. There are also different ways to store thermal energy [7]: directly storing HTF before the turbine supply, storing pressurized vapor [8] or storing thermal energy in an external system. The latter is the solution most widely implemented, especially when high energy is to be stored. Most TES use an external system based on molten salt [9,10], which has been identified as the preferred possibility, especially when nitrate salt is used for the storage medium [11]. In this type of systems, the molten salt typically moves between two tanks: a cold-salt tank and a hot-salt tank. When necessary, the HTF will meet the molten salt in a heat exchanger so as to give rise to the heat transfer required in each case: from the HTF to the molten salt or vice versa [12,13]. It must be noticed that, although the technology used in these heat exchangers is mature, there is not a standard disposition for such exchangers.

Given the complexity of thermal solar power plants, a distributed control system (DCS) must be implemented with control variables and algorithms distributed between several controllers [14,15]. This is a very common solution for all types of complex industrial processes. The molten-salt system is usually managed by a specific controller within the DCS. The DCS receives the field signals requested by the TES control system and sends them to this specific controller, which generates the adequate actions. Although many different algorithms have been used in the literature to control heat transfer in exchangers (mainly predictive control algorithms [16,17,18], but also neural networks [19], controllers designed using the coefficient diagram method (CDM) [20] and control based on fuzzy models [21,22] or on nonlinear models [23,24]), the control of TES systems is somewhat different, and the special features related to heat transfer in this kind of systems result in proportional-integral-derivative (PID) control systems being most commonly used.

Some authors pursue optimization of the power plant as a whole [25,26,27] and state that, because energy storage systems represent only one part of a greater energy system, it is critical to consider the entire system, and not the storage in isolation [28]. In this paper, however, the thermal energy storage process is considered by itself. This alternative is also used in other works [1,25,29] that consider that each process in the power plant should be individually optimized in order to have an overall operation as optimum as possible.

Optimization of the TES process in order to make it become as efficient and economical as possible involves developing more sophisticated control systems and, in order to do so, it is necessary to have an accurate model of the process associated to the performance of TES systems. In most cases, the TES system is simply defined in terms of how many hours the plant can rely on it [25]; some others [30] define the parameters to be included in an online simulation software [31]; others [1,29] describe the equations related to these systems but either they do it too thoroughly or they are again only used to determine how much the solar share of the power plant can be increased as a function of the storage capability of the system.

The goal of this paper is to develop a model that can be used by designers to optimize the energy storage control without the need to interfere with normal plant operation. Experience has shown that, although a correct estimation of temperature values is important, the main feature of the model to develop must be its capability to accurately represent the delays introduced by the large elements in the system, since these delays must be considered when designing the control strategy. The precision of the model developed will be tested by comparing the results it provides to those measured in the molten-salt heat storage system of an actual power plant.

It must be pointed out that, even though the heat exchanger is a very important part of the TES, this paper does not aim to provide a very accurate model of this single device. As far as the performance of the TES system is concerned, it is not so important to have great accuracy in the temperatures estimated by the model as it is to have the model provide reliable information of the times involved in the TES operation.

MATERIALS AND METHODS

In order to accurately model the TES process, its operation modes must be first clearly defined. Conventional systems consist of: one cold salt tank, one hot salt tank, an HTF-salt exchanger, pumps (to move the molten salt from one tank into the other), pipes and control valves. Figure 1shows a typical schematic that includes a single train formed by three exchangers and Table 1shows the control elements (and their respective control variables) identified in this schematic; flow directions in this schematic depend on whether the heat storage system is being charged (HTF flows from the hot side to the cold side, salt flows from the cold tank to the hot tank) or discharged. The model to develop must include the set points of the control variables as inputs, since the state of the elements they control will determine the performance of the TES.

Similarly, the model must have several outputs that allow users checking whether the response of the system is the expected one or not. Process outputs are monitored in actual systems by means of several sensors placed at the inputs and outputs of the heat exchanger as shown in Figure 2. These elements, which are summarized in Table 2, are the model outputs to consider during the different operation modes of the energy storage system.

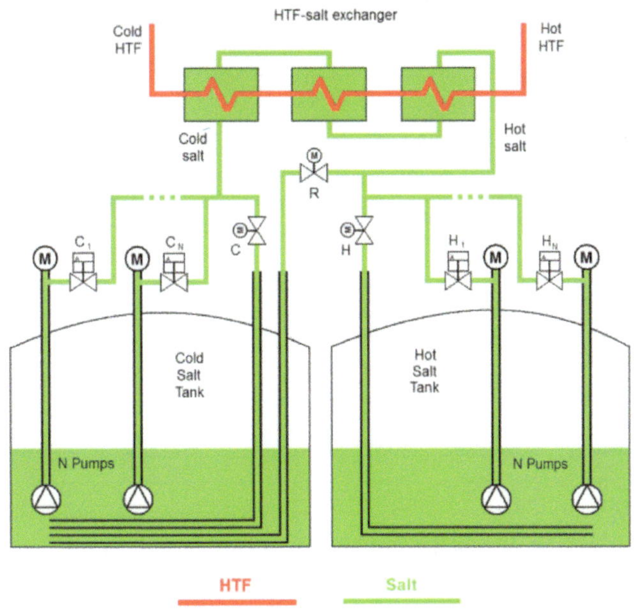

Figure 1. Schematic of a thermal energy storage (TES) system.

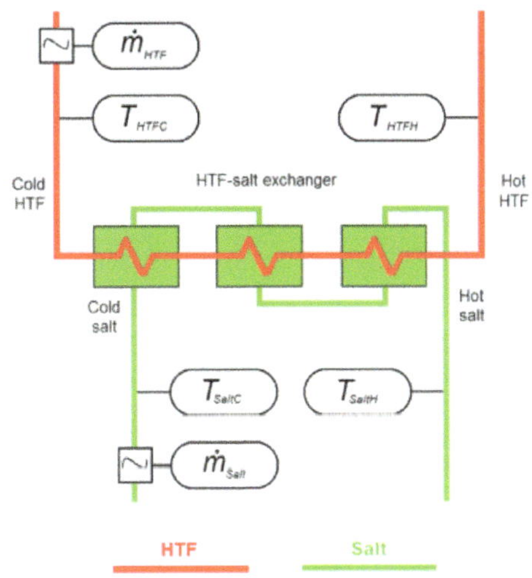

Figure 2. Schematic of the heat exchanger.

The disposition of the TES in a thermal solar plant is schematized in Figure 3. This or similar schematics are widely accepted in literature [1,25,32] to represent the three main circuits in plants of this kind: the solar field (SF), through which the heat transfer fluid (HTF) circulates, the power block, which circulates water and steam, and the TES system.

Table 1. Control elements

Control Element			Control Variable		
Symbol	Component	Description	Symbol	Magnitude	Unit
C_1 to C_N	Pump	Cold-salt pump	SPV_{C1} to SPV_{CN}	Velocity	rpm
C	Valve	Main valve in the cold-salt tank	SPO_C	Opening	%
H_1 to H_N	Pump	Hot-salt pump	SPV_{H1} to SPV_{HN}	Velocity	rpm
H	Valve	Main valve in the hot-salt tank	SPO_H	Opening	%
R	Valve	Recirculation valve	SPO_R	Opening	%

Table 2. Monitoring elements

Monitoring Element		Output Variable		
Element	Description	Symbol	Magnitude	Unit
Thermocouples	Temperature sensors for cold HTF	T_{HTFC}	Temperature	°C
Thermocouples	Temperature sensors for hot HTF	T_{HTFH}	Temperature	°C
Ultrasonic flow meter	Measurement of the flow in cold HTF	$\dot{m}_{HTF}$	Mass flow	kg/s
Thermocouples	Temperature sensors for cold salt	T_{SaltC}	Temperature	°C
Thermocouples	Temperature sensors for hot salt	T_{SaltH}	Temperature	°C
Ultrasonic flow meter	Measurement of the flow in cold salt	$\dot{m}_{Salt}$	Mass flow	kg/s

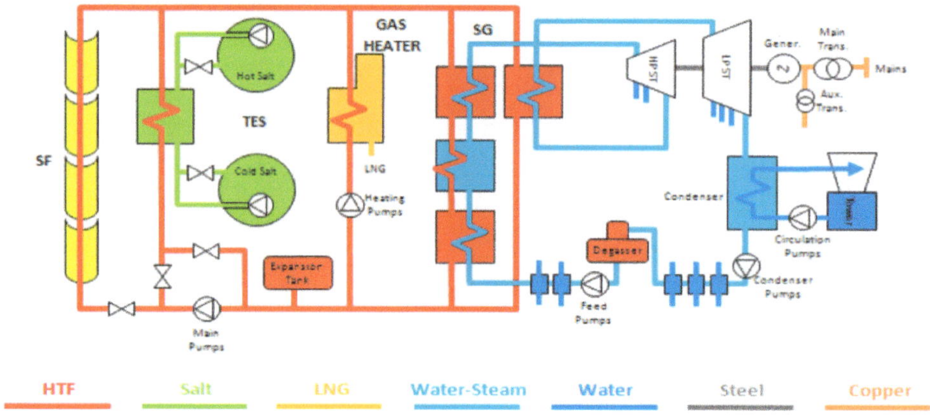

Figure 3. Schematic of a thermal solar power plant.

Three main operation modes can be identified in the performance of a TES system: charge, discharge and idle mode. Each of these modes might occur under different conditions as described below.

Charging of the TES System

Charging begins with recirculation of cold salt: it is pumped into the heat exchanger and returned to the tank through valve *R* (see Figure 1). Then hot HTF passes through the exchanger until the salt is hot enough to be sent to the hot-salt tank; at this moment, the recirculation valve *R*is closed and the main valve in the hot-salt tank (*H*) is opened instead.

The mass flow of salt is first controlled by the opening of valve *H*. When this valve has reached its preset opening value, the remaining pumps will be started one by one, thus controlling the flow. Once all the pumps are started, the control is carried out through the frequency converters of the pumps.

Table 3, Table 4 and Table 5 include all the variables involved in the process described above are: input variables (used to control the elements included in Table 1), perturbation variables affecting the process and output variables (quantifiable by means of the monitoring elements indicated inTable 2).

Table 3. TES charging input variables

Charging of the TES System	
Input Variable	**Description**
SPV_{Ci}	Set Point for the speed defined by the frequency converter of the *i*-th cold-salt pump
SPO_H	Set Point for the opening of the main valve in the hot-salt tank
SPO_R	Set Point for the opening of the recirculation valve

Table 4. TES charging perturbations

Charging of the TES System	
Perturbation	**Perturbation**
T_{HTFH}	Hot HTF temperature at the input of the heat exchanger
T_{SaltC}	Cold salt temperature at the input of the heat exchanger
$\dot{m}_{HTF}$	HTF mass flow through the heat exchanger

Table 5. TES charging output variables

Charging of the TES System	
Output Variable	**Description**
T_{HTFC}	Cold HTF temperature at the output of the heat exchanger
T_{SaltH}	Hot salt temperature at the output of the heat exchanger (control variable)
$\dot{m}_{Salt}$	Molten salt mass flow through the heat exchanger

The goal during this operation mode is regulating temperature T_{SaltH} by adequately setting the values of SPV_{Ci}, SPO_H and SPO_R while taking perturbations into account and having the other output variables provide insight into what is actually happening. The model to develop must make all these variables available. Figure 4 shows a block representation of the charging process.

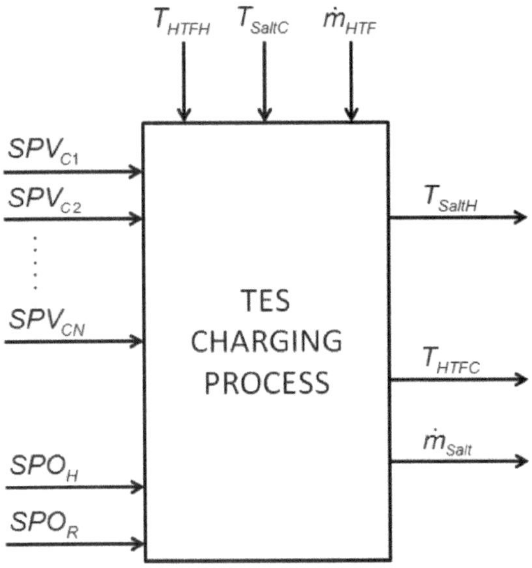

Figure 4. Process block associated to TES charging.

It must be noticed that, depending on whether it is the solar field (SF) or the gas heater that supplies the thermal energy storage (TES) system (and perhaps the steam generator system, SGS, as well), charging of the thermal energy storage system can take place in either of the following ways.

From SF to SGS + TES

In some summer days, around noon, the thermal energy produced by the solar field (SF) is greater than that required to operate the turbine at nominal power. This excess of energy must be stored in the molten salt while the steam generation system (SGS) keeps operating at nominal power.

From SF to TES

In winter days, when the thermal power provided by the solar field is not enough to operate the turbine with an acceptable efficiency, the choice is storing all of this energy in the molten salt.

From Gas Heater to TES

When the molten salt needs to be heated but there is not enough energy coming from the solar field, the gas heater can be used for this purpose.

The difficulty associated to the control of each of these charging processes varies according to the variations of HTF mass flow and temperature during the charge.

Discharging of the TES System

Discharge begins by pumping hot salt through the heat exchanger and into the cold-salt tank through its main valve (*C*). Then cold HTF passes through the exchanger to be heated. As HTF flow increases, salt flow is increased too by opening valve *C*. When this valve has reached its preset opening value, the remaining pumps will be started one by one, thus controlling the flow. Once all the pumps are started, the control is carried out through the frequency converters of the pumps.

The variables involved in the process described above are summarized in Table 6, Table 7 andTable 8:

Table 6. TES discharging input variables

Discharging of the TES System	
Input Var.	**Description**
SPV_{Hi}	Set Point for the speed defined by the frequency converter of the *i*-th hot-salt pump
SPO_C	Set Point for the opening of the main valve in the cold-salt tank

Table 7. TES discharging perturbations

Discharging of the TES System	
Perturbation	**Description**
T_{HTFC}	Cold HTF temperature at the input of the heat exchanger
T_{SaltH}	Hot salt temperature at the input of the heat exchanger
$\dot{m}_{HTF}$	HTF mass flow through the heat exchanger

Table 8. TES discharging output variables

Discharging of the TES System	
Output Var.	**Description**
T_{HTFH}	Hot HTF temperature at the output of the heat exchanger (control variable)
T_{SaltC}	Cold salt temperature at the output of the heat exchanger
$\dot{m}_{Salt}$	Molten salt mass flow through the heat exchanger

The goal in this case is regulating temperature T_{HTFH} by adequately setting the values of $SPV_{Hi}$and SPO_C while taking perturbations into account and having the other output variables provide insight into what is actually happening. A possible block representation of the discharging process that presents all the variables described above is shown in Figure 5.

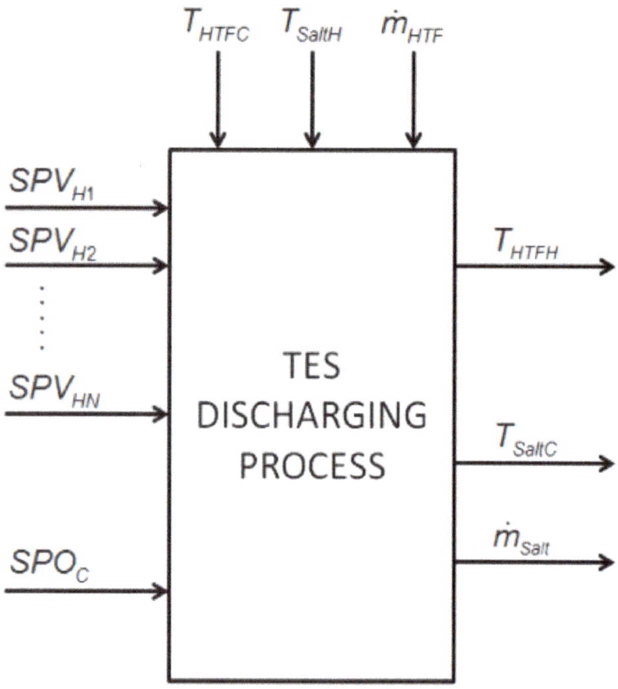

Figure 5. Process block associated to TES discharging.

Similarly to the charging process, discharging the thermal energy storage system can also take place in either of the following ways.

From SF + TES to SGS

The energy provided by the solar field is not enough to have the steam generation system provide the power required. The thermal energy storage system supplies additional energy to make the SGS operate as close as possible to nominal power conditions.

From TES to SGS

All the energy provided by the generator comes from the TES (typical situation at nights). The power obtained is slightly lower than nominal power.

From TES + Gas Heater to SGS

Sometimes the TES system cannot provide enough energy to operate the steam generation system at the required power. In these cases, the gas heater can be used to supply additional energy.

Regulation of both the charge and discharge processes experiences several problems, namely: process perturbations, measurement errors and non-linearity. As already stated, optimization of all these processes cannot be carried out unless an accurate model of the process is available.

TES System in Idle Mode

This mode corresponds to the idle times between charge and discharge. No regulation is required during this mode.

After the TES system has been fully charged, it enters idle mode, with the hot-salt tank full, until the SF cannot keep the turbine operating at nominal power.

After the TES has been completely discharged, it enters idle mode, with the hot-salt tank empty, until the SF provides an excess of thermal power above that required by the turbines (typically the following day).

MODEL DEFINITION

Up to this point, two models have been described: one for the charge and another for the discharge of the TES system. It might be argued that having two process models is not functional since the system is the same in both cases. Despite this being true, the fact that internal flows are opposite during charge and discharge, and that relevant input/output variables change for each case, has led authors to consider it more convenient to have two separate models. In addition, charge and discharge are completely independent process (either the TES system is charging or it is discharging) and can therefore be optimized separately.

Whatever the model considered (charge or discharge), only three basic equations are needed to define the performance of a thermal energy storage system: mass conservation, energy conservation and heat transmission. Other than that, the models must accurately define the mass flows within the system.

The rest of the paper focuses on the description of a model associated to the charge process of a TES to be included in a simulation oriented visual environment. Although Matlab-Simulink blocks are shown, any other simulation environment of these characteristics could be used. The model will be developed by carrying out a top-down partitioning of the system in order to have it implemented bottom-up later. The model associated to the discharge can be similarly obtained.

Figure 6 shows the simplified model associated to the charge process in the TES system. As can be seen, the set points of the valves are not directly provided in actual systems. Instead, there is a simple internal PID regulator that implements mass-flow control through the frequency converters of the pumps. This is represented in the figure and considered to be part of the model to be derived.

From this point on, the aforementioned top-down analysis is carried out so that the subsystems included in each block are defined. To start with, the TES charge process is modeled by the pumps controlling the mass flow entering the trains of exchangers as indicated in Figure 7. The TES discharge model will only differ from this in the input and output variables (refer to Figure 5 to see these variables).

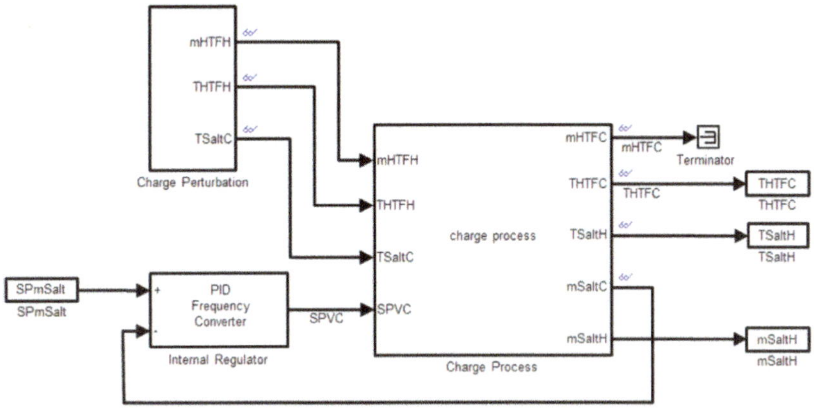

Figure 6. Simulink model for TES charge simulations.

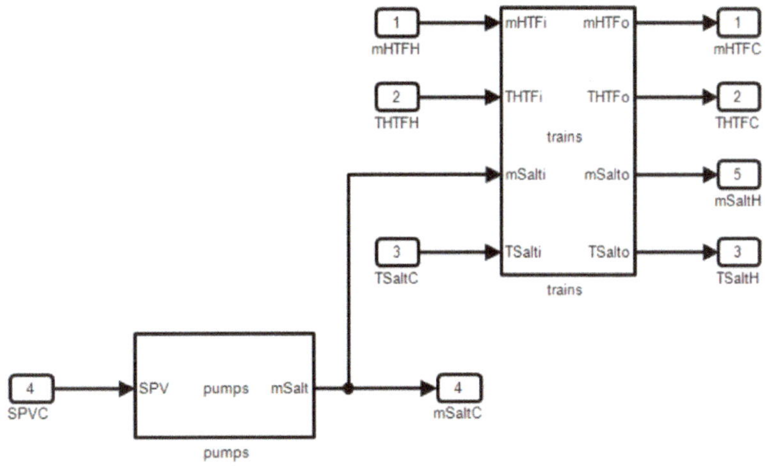

Figure 7. Diagram for TES charge process.

The main block in this TES charge model is the trains block. This block is meant to define the performance of the HTF-Salt heat exchanger, which may consist of several series-connected modules. In the case considered for this work, a system with several trains connected in parallel has been defined. In turn, each of these trains includes several heat exchangers connected in series. The trains block must accurately model all the different parts in a heat exchanger and their relations. Accurate description of the flows inside the system is needed in order to obtain a valid model. Figure 8 shows these parts.

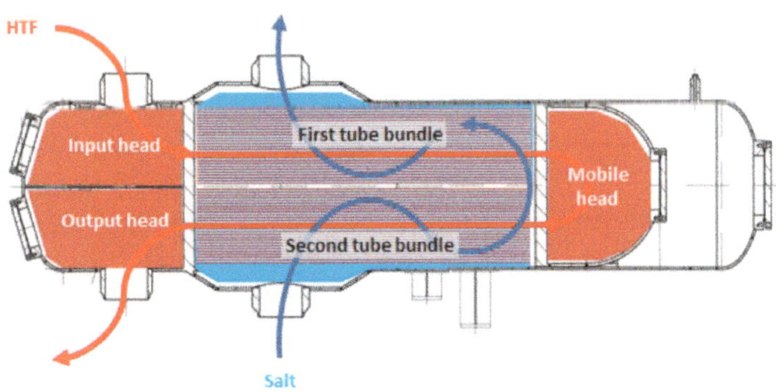

Figure 8. Components of an HTF-Salt heat exchanger.

Shell-and-tube heat exchangers are very common in process industry and their transient modeling has been subject of numerous publications [33,34,35]. Accurate modeling of this kind of devices involves considering the shell-side convective heat transfer over the tube bundle in detail. This is typically done via the Bell-Delaware method [36,37,38], which is an empirical method based on numerous experiments with shell-and-tube heat exchangers. Recently, there have also been published some numerical studies that reconfirmed its validity [39,40,41].

However, the model of the thermal energy storage system does not require a very accurate heat-exchanger model to adequately define the heat transfer. The heat exchanger model should accurately represent the dynamic effects that provide the TES model with a good estimation of the time-evolution of the system operations. Different heat transfer models for the heat-exchangers have been tried by the authors with negligible impact on the simulation results, and with complex heat transfer models increasing the computational time due to the complexity of the overall model. In addition, 2-D or 3-D models have been tried for the heat exchangers sections or volumes, with negligible impact on the simulation results. In fact, experience has shown that, rather than a high accuracy in temperature values, it is necessary to have a system that adequately models the delays produced in very large heat exchangers like the ones used in these power plants. It must be taken into account that these heat exchangers might consist of up to six trains connected in series (as is the case in Andasol I, a 150-MW thermal power plant in Andalusia, southern Spain), which results in relatively long delays before the response is obtained at the output. Should these delays not be properly modeled, the parameters of the PID regulator that controls the overall system cannot be adequately calculated, thus giving rise to abnormal operation.

Thus, a simple 1-D model has been used to define the heat exchanger. The flux exchange is considered to be unidirectional and the bundles are divided into smaller discrete parts where the basic heat transfer equations will be solved. No 3-D simulation of the shell is deemed necessary.

The interface variables for the heat exchanger are shown in Table 9 and Table 10.

Table 9. Heat exchanger input variables

Heat Exchanger	
Input Var.	**Description**
$\dot{m}_{HTFi}$	Input HTF mass flow
T_{HTFi}	Input HTF temperature
T_{HTFc}	Counterflow input HTF temperature
$\dot{m}_{Salti}$	Input salt mass flow
T_{Salti}	Input salt temperature
T_{Saltc}	Counterflow input salt temperature

Table 10. Heat exchanger output variables

Discharging of the TES System	
Output Var.	**Description**
$\dot{m}_{HTFo}$	Output HTF mass flow
T_{HTFo}	Output HTF temperature
$\dot{m}_{Salto}$	Output salt mass flow
T_{Salto}	Output salt temperature

Figure 9 shows the block diagram for the exchanger depicted in Figure 8. The connection between the different blocks (input head, first bundle, mobile head or turn, second bundle and output head) represents actual flow inside the exchanger (assuming a one-dimensional simplification), thus contributing to obtaining an accurate model of the overall process.

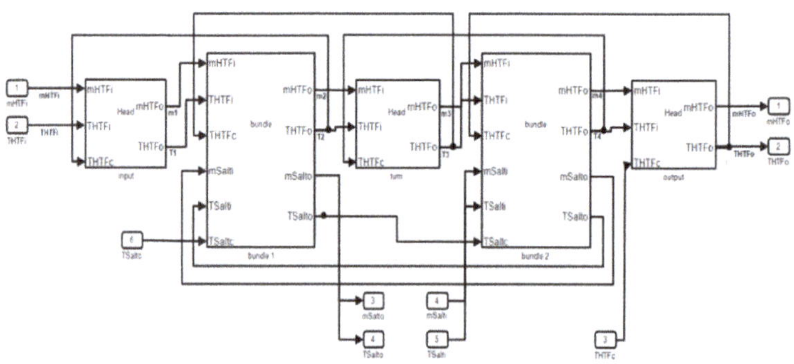

Figure 9. Diagram for the heat exchanger.

Similarly, the bundle blocks must include information of how many tubes form part of the bundle and how the HTF flows through them. This can be achieved by defining block "tube" and using several of them adequately connected to define the block "bundle". The number of tubes and the connections between them depends on the actual exchanger being considered: there will be a different and unique model for each type of exchanger.

It is inside the exchanger that energy transfer between the fluids is carried out. In order to model this phenomenon equation blocks must be included both in the bundle model and in the head model. Both these models implement equations related to mass conservation, Equation (1), and energy conservation, Equation (2), in order to derive output mass flows and output temperatures:

$$\frac{dM}{dt} = -\oiint \rho \cdot \vec{v} \cdot \overrightarrow{dS} \quad (1)$$

$$\frac{dE}{dt} = -\oiint \rho \cdot \left(h + \frac{v^2}{2} + g \cdot z\right) \cdot \vec{v} \cdot \overrightarrow{dS} - \dot{Q} \quad (2)$$

These equations can be simplified considering actual constraints of the process within the control volume considered, such as constant density, incompressible flow, similar input and output mass flows or negligible potential and kinetic energy. Under these considerations, the mass conservation equation to be implemented is:

$$\dot{m}_o = \dot{m}_i - V \cdot \frac{d\rho_{FLUID}(T)}{dt} \quad (3)$$

where $\dot{m}_i$ and $\dot{m}_o$ are the input and output mass flows, V is the control volume considered and ρ_{FLUID} is the fluid density.

As far as the energy conservation equation is concerned, the assumptions above result in an enthalpy equation, which is the one to be implemented in the model:

$$\frac{dH}{dt} = -\oiint \rho \cdot h \cdot \vec{v} \cdot \overrightarrow{dS} - \dot{Q} \quad (4)$$

where ρ is the fluid density, h is the fluid enthalpy, v is the fluid velocity, S is the external surface of the control volume and $\dot{Q}$ is the heat transmission out of the control volume.

Additionally, the model of the bundle, implements Equation (5) in order to derive the heat transmission between the control volume of the HTF and that of the molten salt through the steel tubes:

$$\dot{Q} = U \cdot A \cdot (\overline{T_{HTF}} - \overline{T_{Salt}}) \quad (5)$$

where U is the global heat transmission coefficient, A is the exchange area and the temperatures used are average values corresponding to HTF and salt; average values are used instead of logarithmic values to avoid problems derived from the fact that it is possible to have the temperature difference inverted during transients. The global heat transmission coefficient, U, included in this equation considers thermal conductivity of the steel and of the fluids involved in the heat exchange as well as superficial heat transmission both inside and outside of the bundles:

$$U = \frac{1}{\frac{D_e}{D_i \cdot h_i} + \frac{D_e \cdot ln(D_e/D_i)}{2 \cdot k_{STEEL}} + \frac{1}{h_e}} \quad (6)$$

where D_i is the tube internal diameter, D_e is the tube external diameter, k_{STEEL} is the steel thermal conductivity, h_i is the film coefficient inside the tube and h_e is the film coefficient outside the tube.

The parameters associated to superficial heat transmission inside the tubes (h_i) are all related to the HTF (Nusselt number, Reynolds number, Prandtl number, fluid speed, flow, viscosity). The Nusselt number for the HTF inside a round tube can be calculated as:

$$Nu_{HTF} = 0.023 \cdot Re_{HTF}^{\ 0.8} \cdot Pr_{HTF}^{\ n} \quad (7)$$

where n equals 0.33 when salt is being charged, and 0.4 when it is being discharged; the other parameters (Reynolds and Prandtl numbers) depend on the type of HTF used.

The same parameters related to the molten salt are used to determine heat transmission on the external surface of the tubes (h_e). In this case, and taking into account the one-dimensional assumption indicated above [42], the Nusselt number for the salt going through the bundle of tubes is calculated to be:

$$Nu_{Salt} = 0.41 \cdot Re_{Salt}^{\ 0.6} \cdot Pr_{Salt}^{\ 1/3} \cdot \left(\frac{Pr_{Salt}}{Pr_{Salt_0}}\right)^{1/4} \cdot \varphi_1 \quad (8)$$

where φ_1 equals 1.12 as corresponds to a situation in which the tube to tube pitch normal to flow is longer than twice the tube to tube pitch parallel to flow; the other parameters in Equation (8) are the Reynolds number of the salt, Re_{Salt}, the Prandtl number of the salt, Pr_{Salt}, and the Prandtl number of the salt on the outer surface of the tubes, Pr_{Salt0}.

By adequately combining the above-defined blocks, it is possible to model any complete TES charge process. The blocks defined have been kept simple so that it is easy to determine how to link one to another. The fact that they represent actual physical phenomena or system disposition also contributes to simplifying the task of adapting them to model a given process. A dynamic model like the one presented here is necessary in order to optimize the TES process as a whole, especially its regulation. Nowadays, the control of this process is performed by means of simple, non-optimized control methods. The model presented in this paper might contribute to improving the state of the art by allowing advanced control techniques to be tested so that more convenient algorithms can be implemented in this part of the power plant and, hence, improve the overall performance of the plant.

MODEL VALIDATION

Validation of the model presented was carried out in two stages. Firstly, simple simulations were run in order to find out whether the results obtained were consistent with the behavior expected. Secondly, the information provided by the model was compared to that measured in an actual plant.

Two simulations were run considering the two most usual control strategies to regulate this kind of processes: a so-called semiautomatic control and a PID regulator. These control methods are very popular and widely used for TES regulation. Since they are well-known strategies, they will be used together with the model developed in order to validate its performance. The main features of these control methods are sketched below.

In the case of semiautomatic control, an operator is in charge of observing the values of the output variables and defining new set points for the input variables. Thus, the regulation will be carried out by means of steps, the accuracy of which depends only on the operator's experience. For a charge process, this person will define the set point for the salt mass flow (SP_{Salt}) while observing the evolution of the output salt temperature (T_{SaltH}). Internal PID regulators will be in charge of operating the pumps so that the set point fixed by the operator is achieved. Figure 10shows a block diagram of the whole system.

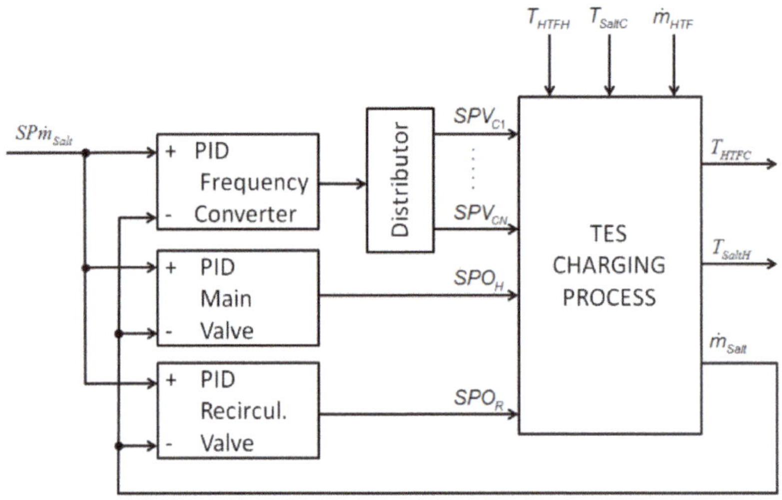

Figure 10. Semiautomatic control of a TES charge process.

In a PID control, the person in charge is substituted by an external regulation loop. A simple PID regulator is usually enough in these cases, and only the set point for the desired output salt temperature must be provided. This system is depicted in Figure 11.

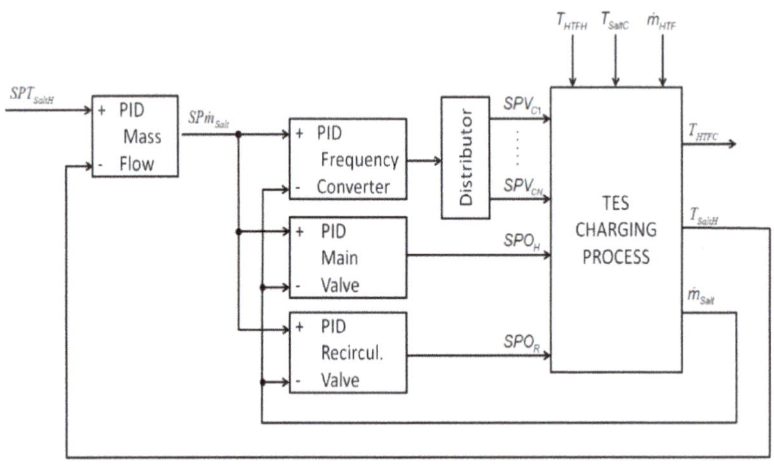

Figure 11. PID regulator in a TES charge process.

A complete TES model implemented with MatLab-Simulink was used to determine whether the results obtained matched the expected ones. The performance of the system during a period of five hours was simulated considering a semiautomatic control and a PID control.

Figure 12 shows a representative 3000 s sample of the evolution of the error made in the hot salt temperature (eT_{SaltH}) when simulating the performance of the model together with the control strategies described above; this error is defined as the difference between the temperature obtained and the set point established (SPT_{SaltH} = 384.5 °C): $eT_{SaltH} = T_{SaltH} - SPT_{SaltH}$. Variations of the perturbations in the system, T_{HTFH} and ${}_{HTF}$ (the cold salt temperature was considered to have a constant value T_{SaltC} = 298 °C) during the time interval considered are represented in Figure 13. These perturbations are actual values measured in a real power plant and applied to the model developed. The variations shown are related to the regulator not being ideal (in the case of T_{HTFH}, although a ±2 °C variation can be considered to be a very good result) or to solar insolation conditions (which affect the value of $\dot{m}_{HTF}$).

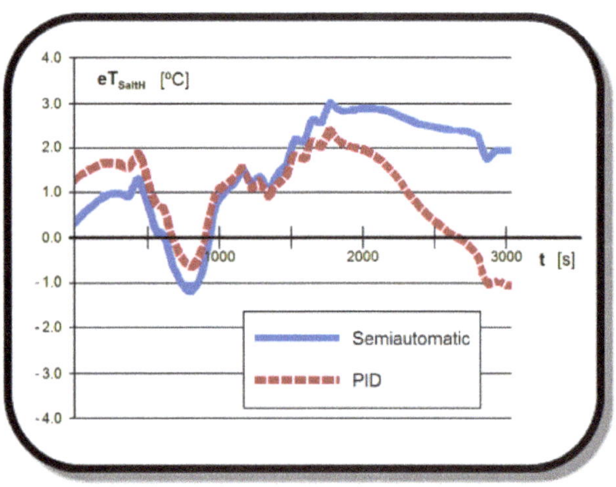

Figure 12. Simulation of the hot salt temperature error obtained during the charge process.

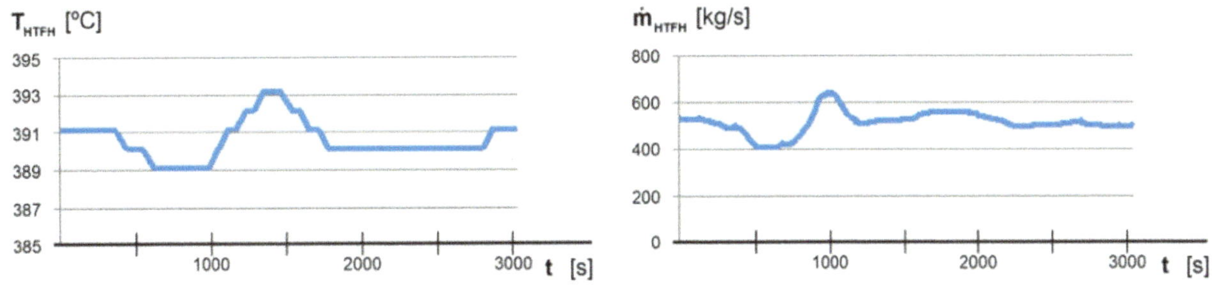

Figure 13. Variation of the perturbations considered when calculating the error represented in Figure 12.

The results obtained in the simulations seem to indicate that the hot salt temperatures calculated by the model follow the set points established. As was to be expected, the accuracy obtained with PID regulation is higher than that obtained with the semiautomatic control, which validates the performance of the model.

For the sake of convenience, only a small fraction of the simulation has been represented, but the accuracy mentioned above can also be seen in Table 11, where μ represents the error average value and σ is the error standard deviation calculated for the whole set of data obtained during the five hours; as already indicated, the variations of the perturbations considered throughout the simulation are actual values measured in a real plant. As was to be expected, both values are lower when using a PID regulator.

Table 11. Numerical results of the simulation test

Control Strategy	$eT_{SaltH} = T_{SaltH} - SPT_{SaltH}$		
	μ	σ	Confidence Interval
Semiautomatic control	0.72	1.74	[−2.76, 4.20]
PID regulator	0.38	1.17	[−1.96, 2.72]

Once simulation results have validated the performance of the model, further tests were carried out using actual measurements from a real thermal power plant in order to compare them with the results provided by the model. The plant chosen was "SAMCA2-La Dehesa", a 50-MW PTC thermal plant built by Group TSK in 2010, with a TES capacity of 7 h. The comparison was established by considering the net energy generated by the thermal plant during a discharge of the TES. Using the set of actual data collected in the plant (and further measurements taken when necessary), the model developed was supplied with the information of the actual values used for the input variables. The outputs provided by the model under these conditions were compared to those stored in the thermal plant. Table 12 shows how both results (actual and simulated) compare when used to determine the net energy produced after seventeen complete discharges. Results are provided for every hour of the discharge process, which always takes place during the night, when the solar field is not providing energy to the plant. Production for the 17 discharges is added for every hour in Table 12.

Table 12. Comparison of the net energy produced after 17 discharges

Time Interval Considered	Net Energy (MW·h)					
	Semiautomatic Control			PID Regulator		
	Real ($E_{N,Real}$)	Simulated ($E_{N,Simulated}$)	Difference (%)	Real ($E_{N,Real}$)	Simulated ($E_{N,Simulated}$)	Difference (%)
23 h–00 h	1435	1432	+0.2%	1453	1450	+0.2%
00 h–01 h	1363	1368	−0.3%	1381	1377	+0.3%
01 h–02 h	1148	1148	0.0%	1163	1166	−0.3%
02 h–03 h	789	786	+0.5%	799	801	−0.2%
03 h–04 h	431	434	−0.8%	436	437	−0.2%
04 h–05 h	215	217	−1.0%	218	217	+0.7%
05 h–06 h	72	74	−3.0%	73	71	+2.0%
Total	5453	5459	−0.1%	5522	5518	+0.1%

Since each of these tests lasted one day and only one of the control strategies could be analyzed at a time, the experiments were conducted according to the following criteria:

- The tests should not interfere with current production.
- All the tests were performed during the summer in order to have similar radiation conditions.
- Pairs or consecutive days, one for each of the control strategies, were chosen for the analysis so that both of them would take place under as similar conditions as possible.
- The real net energy produced in the plant was measured before and after the discharge, and then compared with that calculated by using the model.

The results obtained further confirm that the model described in this paper accurately represents the behavior of thermal energy storage systems in solar power plants. Nevertheless, it must be pointed out that the model presented in this paper is not actually meant to estimate the energy produced in the solar plant. The main use of this model is designing the regulation of the thermal storage system; Table

12 has only been used as a way to validate the accuracy of the results provided by this model.

CONCLUSIONS

A simulation dynamic model of the molten-salt thermal energy storage system included in thermal power plants has been developed. The model has been described to be used in simulation oriented visual environments, so that physical aspects or component disposition in the system to be modeled can be easily described.

This model accurately determines delays in the system that must be taken into account when designing the control strategy associated to the heat storage system. This is especially important when large heat exchangers consisting of several trains connected in series are included in the thermal solar power plant.

The precision of the model has been validated by simulation and by comparison with actual data obtained from a real PTC thermal plant. In both cases, the results obtained proved the accuracy of the response offered by the model developed.

The availability of the TES model described in this paper opens new possibilities in optimization of the charge/discharge process of heat storage systems and, hence, improvement of the overall efficiency of the solar thermal plant.

AUTHOR CONTRIBUTIONS

This paper is part of the PhD Thesis developed by Rogelio Peón Menéndez, who has therefore carried out most of the work presented here. Professors Juan Á. Martínez and Miguel J. Prieto were the supervisors of this work, whereas Lourdes Á. Barcia and Juan M. Martín Sánchez respectively assisted Rogelio Peón with Simulink simulations and regulation topics.

REFERENCES

1. Yang, Z.; Garimella, S.V. Cyclic operation of molten-salt thermal energy storage in thermoclines for solar power plants. *Appl. Energy* 2013, *103*, 256–265.
2. Usaola, J. Operation of concentrating solar power plants with storage in spot electricity markets. *IET Renew. Power Gener.* 2012, *6*, 59–66.
3. Weinrabe, G.; Ortmanns, W. Solar thermal power plants. In *Renewable Energy*; Kaltschmitt, M., Streicher, W., Wiese, A., Eds.; Springer: Berlin, Germany, 2007; pp. 171–228.
4. Schnatbaum, L. Solar thermal power plants. *Eur. Phys. J. Spec. Top.* 2009, *176*, 127–140.
5. Wu, S.Y.; Xiao, L.; Cao, Y.; Li, Y.R. A parabolic dish/AMTEC solar thermal power system and its performance evaluation. *Appl. Energy* 2010, *87*, 452–462.
6. Ter-Gazarian, A.G. Thermal Energy Storage. In *Energy Storage for Power Systems*; IET Press: Stevenage, UK, 2011; pp. 55–76.
7. Sangster, A.J. Intermittency buffers. In *Green Energy and Technology. Energy for a Warming World*; Sangster, A.J., Ed.; Springer: Berlin, Germany, 2010; pp. 81–123.
8. Steinmann, W.D.; Laing, D.; Tamme, R. Storage systems for solar steam. In Proceedings of ISES World Congress 2007 (Vol. I – Vol. V), Springer, Berlin, Germany; 2007; pp. 2736–2740.
9. Laing, D.; Steinmann, W.D.; Tamme, R. Sensible heat storage for medium and high temperatures. In Proceedings of ISES World Congress 2007 (Vol. I – Vol. V), Springer, Berlin, Germany; 2007; pp. 2731–2735.
10. Goods, S.H.; Bradshaw, R.W. Corrosion of stainless steels and carbon steel by molten mixtures of commercial nitrate salts. *J. Mater. Eng. Perform.* 2004, *13*, 78–87.
11. *USA Trough Initiative. Thermal Storage Oil-to-Salt Heat Exchanger Design and Safety Analysis*; Task Order Au. No. KAF-9-29765-09; Nexant Inc.: San Francisco, CA, USA, 2000.
12. Annaratone, D. *Engineering Heat Transfer*; Springer: Berlin, Germany, 2010.
13. Massoud, M. *Engineering Thermofluids*; Springer: Berlin, Germany, 2005.
14. Damsker, D.J.; Sandberg, C. Towards advanced concurrency, distribution, integration, and openness of a power plant distributed control system (DCS). *IEEE Trans. Energy Convers.* 1991, *6*, 297–302.
15. Bong-Kuk, L.; Yong-Hak, S. The integrated monitoring and control system for the combined cycle power plant. In Proceedings of the International Conference on Control, Automation and Systems, Seoul, Korea, 14–17 October 2008; pp. 1479–1483.
16. Rajasekaran, S. A Simplified predictive control for a shell and tube heat exchanger. *Int. J. Eng. Sci.* 2010, *2*, 7245–7251.
17. Garcia, C.E.; Prett, D.M. Model predictive control: Theory and practice—A survey.*Automatica* 1989, *25*, 335–348.
18. Lim, K.W.; Ling, K.V. Generalized predictive control of a heat exchanger. *IEEE Control Syst. Mag.* 1989, *1*, 9–12.
19. Nelles, O. *Nonlinear System Identification: From Classical Approaches to Neural Networks and Fuzzy Models*; Springer: Berlin, Germany, 2001.
20. Imal, E. CDM based controller design for nonlinear heat exchanger process. *Turk. J. Electr. Eng. Comput. Sci.* 2009, *17*, 143–161.
21. Skrjanc, I.; Matko, D. Predictive functional control based on fuzzy model for heat-exchanger pilot plant. *IEEE Trans. Fuzzy Syst.* 2000, *8*, 705–712.
22. Hooshang Mazinan, A.; Sadati, N. Fuzzy predictive control based multiple models strategy for a tubular heat exchanger system. *Appl. Intell.* 2010, *33*, 247–263.
23. Badgwell, A.B.; Qin, S.J.; Kouvaritakis, B. Review of nonlinear model predictive control applications. *IEE Control Ser.* 2001, *3*, 3–32.
24. Parker, R.S.; Gatzke, E.P.; Mahadevan, R.; Meadows, E.S.; Kouvaritakis, F.J.; Doyle, B. Nonlinear model predictive control: Issues and applications. *IEE Control Ser.* 2001, *5*, 34–57.

25. Llorente García, I.; Álvarez, J.L.; Blanco, D. Performance model for parabolic trough solar thermal power plants with thermal storage: Comparison to operating plant data. *Sol. Energy*2011, *85*, 2443–2460.

26. Niknia, I.; Yaghoubi, M. Transient simulation for developing a combined solar thermal power plant. *Appl. Therm. Eng.* 2012, *37*, 196–207.

27. Rolim, M.M.; Fraidenraich, N.; Tiba, C. Analytic modeling of a solar power plant with parabolic linear collectors. *Sol. Energy* 2009, *83*, 126–133.

28. Powell, K.M.; Hedengren, J.D.; Edgar, T.F. Dynamic optimization of a solar thermal energy storage system over a 24 h period using weather forecasts. In Proceedings of the American Control Conference, Washington, DC, USA, 17–19 June 2013; pp. 2946–2951.

29. Powell, K.M.; Edgar, T.F. Modeling and control of a solar thermal power plant with thermal energy storage. *Chem. Eng. Sci.* 2012, *71*, 138–145.

30. Kibaara, S.; Chowdhury, S.; Chowdhury, S.P. Analysis of Cooling Methods of Parabolic Trough Concentrating Solar Thermal Power Plant. In Proceedings of the 2012 IEEE International Conference on Power System Technology, Auckland, New Zealand, 30 October–2 November 2012; pp. 1–6.

31. Solar Advisor Model. Available online: https://sam.nrel.gov (accessed on 24 July 2014).

32. Martín, L.; Zarzalejo, L.F.; Polo, J.; Navarro, A.; Marchante, R.; Cony, M. Prediction of global solar irradiance based on time series analysis: Application to solar thermal power plants energy production planning. *Sol. Energy* 2012, *84*, 1772–1781.

33. Fraser, K.F.; Hollands, K.G.T.; Brunger, P. An empirical model for natural convection heat exchangers in SDHW systems. *Sol. Energy* 1995, *55*, 75–84.

34. Vera-García, F.; García-Cascales, J.R.; Gonzálvez-Maciá, J.; Cabello, R.; Llopis, R.; Sanchez, D.; Torrella, E. A simplified model for shell-and-tubes heat exchangers: Practical application.*Appl. Therm. Eng.* 2010, *30*, 1231–1241.

35. Ravagnani, M.A.S.S.; Caballero, J.A. A MINLP model for the rigorous design of shell and tube heat exchangers using the tema standards original. *Chem. Eng. Res. Des.* 2007, *85*, 1423–1435.

36. Green, D.W.; Perry, R.H. *Perry's Chemical Engineer's Handbook*, 6th ed.; McGraw-Hill: New York, NY, USA, 1984; Sections 10–11.

37. Kakaç, S.; Liu, H. *Heat Exchangers, Selection, Rating and Thermal Design*; CRC Press: Boca Raton, FL, USA, 1995; p. 53.

38. Peters, M.S.; Timmerhaus, K.; West, R.E. *Plant Design and Economics for Chemical Engineers*, 5th ed.; McGraw-Hill Professional: New York, NY, USA, 2003; pp. 716–724.

39. Tan, X.H.; Zhu, D.S.; Zhou, G.Y.; Yang, L. 3D numerical simulation on the shell side heat transfer and pressure drop performances of twisted oval tube heat exchanger. *Int. J. Heat Mass Transf.* 2013, *65*, 244–253.

40. You, Y.; Fan, A.; Huang, S.; Liu, W. Numerical modeling and experimental validation of heat transfer and flow resistance on the shell side of a shell-and-tube heat exchanger with flower baffles. *Int. J. Heat Mass Transf.* 2012, *55*, 7561–7569.

41. You, Y.; Fan, A.; Lai, X.; Huang, S.; Liu, W. Experimental and numerical investigations of shell-side thermo-hydraulic performances for shell-and-tube heat exchanger with trefoil-hole baffles. *Appl. Therm. Eng.* 2013, *50*, 950–956.

42. Zukauskas, A. Heat transfer from tubes in crossflow. In *Advances in Heat Transfer*; Irvine, T.F., Jr., Hartnett, J.P., Eds.; Academic Press, Inc.: New York, NY, USA, 1987; Volume 18, pp. 87–159.

CHAPTER 5

ANALYZING THE TECHNOLOGY OF USING ASH AND SLAG WASTE FROM THERMAL POWER PLANTS IN THE PRODUCTION OF BUILDING CERAMICS

A G Malchik[1], S V Litovkin[1], P V Rodionov[1], V V Kozik[1,2] and M A Gaydamak[1]

[1]Yurga Institute of Technology, National Research Tomsk Polytechnic University Affiliate, 652050, Yurga, ul. Leningradskaya 26

[2]National Research Tomsk State University, pr. Lenina 36, Tomsk, 634050 Russia

ABSTRACT

The work describes the problem of impounding and storing ash and slag waste at coal thermal power plants in Russia. Recovery and recycling of ash and slag waste are analyzed. Activity of radionuclides, the chemical composition and particle sizes of ash and slag waste were determined; the acidity index, the basicity and the class of material were defined. The technology for making ceramic products with the addition of ash and slag waste was proposed. The dependencies relative to the percentage of ash and slag waste and the optimal parameters for baking were established. The obtained materials were tested for physical and mechanical properties, namely for water absorption, thermal conductivity and compression strength. Based on the findings, future prospects for use of ash and slag waste were identified.

INTRODUCTION

The development of fuel and energy complex and construction industry, building of new towns and population centers, environmental and related social issues directly or indirectly depend on the disposal of coal-burning residuals - fly ash and slag. In the time of the Soviet Union, fly ash and slag were not separated and disposed of, using a hydro transport system, in impoundments and landfills, causing large amounts of slag waste within or close to cities. This problem is particularly pressing in Russia because of harsh climatic conditions, as well as disagreement of production time and potential consumption of coal-burning byproduct solids.

The share of coal burnt for producing electrical energy is 94% in Poland, 93% in South Africa. 78% in India, 77% in Australia, 70% in China, 67% in the Czech Republic, 62% in Greece, 52% in Germany, 50% in the USA, 17% in Denmark, 33% in the UK, over 27% in 15 European countries.

The share of coal in energy production was 37% worldwide ten years ago. Such a low share of coal in the energy production was associated with the signing of the Kyoto Protocol, which required reducing emissions of carbon dioxide into atmosphere. However, due to unstable prices for oil and gas and their limiting availability, this share is now about 40%, with an apparent trend to grow [1,2,3].

According to the Energy Strategy of Russia for the period up to 2020 with the fundamental principles thereof approved

Citation: A G Malchik, S V Litovkin, P V Rodionov, V V Kozik and M A Gaydamak, Analyzing the Technology of Using Ash and Slag Waste from Thermal Power Plants in the Production of Building Ceramics, doi: 10.1088/1757-899X/127/1/012024

by the government in 2000, a steady growth of electricity consumption is scheduled as much as 2-3% annually. It is planned to develop environmentally friendly power plants using renewable energy sources, including wind, tidal, geothermal and biofuel power stations, micro hydro power stations, and etc. But totally these sources will give 0.01% of generated energy. Thermal power plants will remain the foundation of the power generation industry, with their relative share in the total power capacity being at 67 - 70%.

The energy production using fossil fuels such as coal, shale oil, peat and other is inevitably accompanied by the formation of ash and slag waste; and fly ash is usually 70 to 90% of total balance in terms of waste solids formed during the coal burning. The share of costs estimated for handling waste is about one third of the cost of energy produced [1].

The problem of processing and recycling ash and slag waste, as well as making researches with the purpose of innovative processing technologies is being tackled in many countries. In Russia, the following experts are involved in dealing with ash and slag waste: I.S. Kozhukhovsky (CJSC "Energy Forecasting Agency"), V.Ya. Putilov (MPEI), V.V. Zyryanov (Nanopowder Technology Ltd.), B.L. Vishnya (Ural Energy Engineering Center), Yu. K. Tselykovsky (All-Russia Thermal Engineering Institute), R.V. Shevtsov (Territorial Generation Company 11). Currently, no more than 5- 10 % of ash and slag waste is used in various construction and production sector countrywide. The residue is being stored in the ash and slag disposal areas without use. With this, the disposed ash and slag waste is not reduced, and it is likely to increase taking into account the growing demand for electricity and insufficient development of alternative sources for its generation.

Ash and slag materials in terms of their chemical and mineralogical composition are largely identical to natural mineral raw materials. Their use in the industrial sector, construction industry and agriculture is one of the strategic solutions to environmental problems in the area of operation of thermal power plants.

The aim of this work is to study physical and chemical properties of ash and slag waste, to determine its possible use as a secondary energy source to reduce anthropogenic impacts on the environment, to explore the possibility of obtaining ceramic materials with adding ash and slag waste.

Equipment and Devices used in Studies

Samples for the study were heat treated in a muffle furnace LOIP LF-15/11 – 61; a laboratory mixer BL-10 was used for mixing samples. Particle size analysis was performed using a device Analizette 22 MicroTec Fritsch GmbH (Germany). The samples were tested for ultimate compression strength in accordance with the requirements of GOST 10180-90 using a laboratory press PM-20MG4. The activity of radioactive nuclides was determined using a gamma-radiometer RKG-AT1320. Weighting was performed on a laboratory balance BK-600. The samples were dried in an electric oven SNOL-F-67/350-I2P before baking, and a cubic mould FBS-70 was used for forming the samples.

RESULTS AND DISCUSSION

A composition of ash and slag waste is determined by the quantitative ratio of contained in the material minerals that are dependent on initial mineralogical composition of fuel used. The chemical composition of ash gives an idea of the composition of minerals in coal. Coal ash is typically presented by the overwhelming majority (96%) of oxides of silicon, aluminum, iron, calcium and magnesium. It contains sodium and potassium compounds only in small amounts. Some coals and ashes have a little precious metal (gold, silver, and platinum) as well as rare and trace elements.

The main compounds of ash are oxides SiO_2, Al_2O_3, FeO, Fe_2O_3, CaO, MgO. Sulfates $CaSO_4$, $MgSO_4$, $FeSO_4$ are presented in small amounts. Phosphates and alkali metal oxides K_2O, Na_2O are contained in ash to an even less degree. Ash and slag waste may have biogenic (fluorine, manganese, cobalt, lead, copper, etc.). and toxic (boron, vanadium, arsenic, strontium, beryllium, etc.) elements. Chemical compositions of ash produced in burning different solid fuels vary within a wide range of limits,%: SiO_2 10 – 68; Al_2O_3 10 – 40; Fe_2O_3 2 – 30; CuO_2 2 – 70; MgO 0 – 10; Na_2O and K_2O 0 – 10. Besides, ash contains a small amount of compounds of germanium, vanadium, arsenic, mercury, beryllium fluoride partially passing into water.

Knowing a chemical composition of ash and slag waste is a prerequisite for discussing its properties and making a conclusion on its possible use in various economy sectors. To determine the chemical composition of ash and slag waste, the ash samples were taken from electric filters and dumped ash according to the procedure RD 34.09.603-88 "Guidelines for the organization of monitoring compositions and properties of ash and slag supplied by thermal power stations to consumers." The samples were tested to determine ash chemical composition and activity of radioactive nuclides.

The radionuclide activity was determined in accordance with GOST 30108-94 "Building materials and products. Determination of specific activity of natural radioactive nuclei". A gamma-radiometer PKG-AT1320 controlling the radio nuclides of potassium (40K), radium (226Ra) and thorium (232Th) was used to determine the specific activity. The radioactivity measured at the Yurga thermal power plant showed the following results: ^{40}K – 526 Bq/kg, ^{232}Th – 72 Bq/kg, ^{226}Ra – 37 Bq/kg. The specific activity is calculated in accordance with the following formula:

$$A_{eff} = A_{Ra} + 1,31A_{Th} + 0,085A_K \quad (1)$$

where A_{Ra} , A_{Th} , A_K are the specific activities of radium, thorium and potassium respectively, Bq/kg.

The calculated value of ash and slag waste activity at the Yurga thermal power plant is 175 Bq/kg that does not exceed the requirements of construction standards, qualifying ash and slag waste as the first class building material; this means it can be used in all types of construction.

Chemical analysis was performed using an X-ray fluorescence spectrometer Quan`X of the Common Use Center at Tomsk Polytechnic University. The results are shown in Table 1.

Table 1: Chemical composition of ash from Yurga power generating plant

Compound	Ash waste		Electric filter	
SiO2	55,7	56,25	50,4	50,74
CaO	6,8	6,84	13,96	13,57
Al_2O_3	21,83	21,84	20,52	20,6
MgO	1,95	1,65	1,55	1,67
MnO	0,09	0,1	0,1	0,09
Fe_2O_3	7,44	7,4	8,55	8,53
FeO	6,69	6,66	7,69	7,68
K_2O	3,53	3,44	1,35	1,31
TiO_2	1,11	1,28	0,97	0,81
SO_3	0,72	0,68	0,87	0,82
BaO	0,44	-	0,5	0,59
P_2O_5	0,38	0,4	1,08	1,1
SrO	-	0,1	0,16	0,19

The main chemical compounds contained in the ash are oxides of silicon and aluminum; the content of iron oxide present is high as well, the content of calcium oxide in the ash waste samples is lower than that in the electric filter samples. Free calcium oxide is likely to react with the carbon dioxide, which is dissolved in the water used to transport the ash through a pipeline, to form calcium carbonate

With the purpose to specify the possible use of ash and slag waste, it is required to define such criteria of classification as: acidity index, basicity, silica module and quality factor the acidity index $M_к$ is calculated on the basis of the following formula:

$$M_к = \frac{SiO_2 + Al_2O_3}{Fe_2O_3 + CaO + MgO} \quad (2)$$

The acidity index is 2.9 for the electric filter samples and 4.7 for the ash and slag waste samples. The basicity module M_o is defined according to the base-acid ratio:

$$M_O = \frac{CaO + MgO + K_2O + Na_2O}{SiO_2 + Al_2O_3} \quad (3)$$

The basicity module is 0.2 for the electric filter samples and 0,16 for the ash and slag waste samples. The silica module M_c is calculated as the ratio of silicon oxide to the sum of aluminum and iron oxides:

$$M_C = \frac{SiO_2}{Al_2O_3 + Fe_2O_3} \quad (4)$$

The silica module is 1.7 for the electric filter samples and 1,9 for the ash and slag waste samples. The quality factor K shows the ratio of oxides increasing the hydraulic activity to oxides reducing it:

$$K = \frac{CaO + Al_2O_3 + MgO}{SiO_2 + TiO_2} \quad (5)$$

The quality factor is 0.7 for the electric filter samples and 0.54 for the ash and slag waste samples. The results obtained suggest that the ash and slag waste is of an acidic type of ash. Acidic ashes are characterized by unstable chemical composition, a small amount of free calcium oxide and a high content of silicon oxide. Such types of ash do not have distinct bonding properties which are possible with calcium oxide that forming the hydroxide in water. Drying in air, hydroxide absorbs carbon dioxide from the air and converts into calcium carbonate that makes the product hard. Particle-size distribution is one of the main indicators of raw materials. The greater the content of micro-dispersed particles, the higher ductility of a material is. Consequently, raw material has a higher bonding property that positively influences strength performance of products made; and a particle size composition is important for determining the adsorption capacity of a material [2,4,5].

The results of the particle-size distribution are represented in Figure 1. The analysis showed that 60% of particle size is 10 to 70 microns.

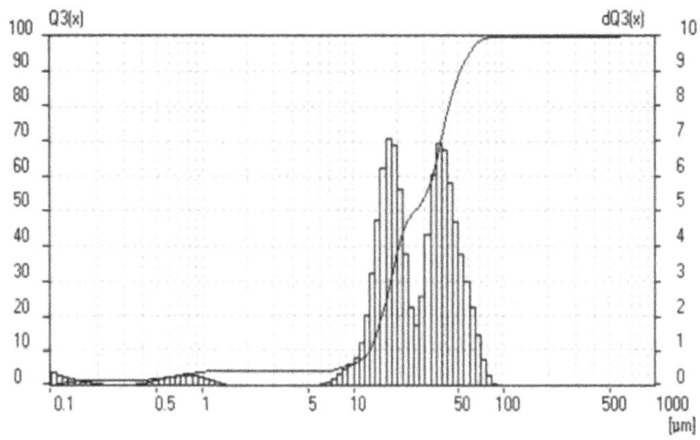

Figure 1. Particle-size distribution of ash and slag waste

The next stage of the work was to make samples of ceramic bricks with the addition of ash and slag waste. Under laboratory conditions, ceramic bricks with different percentage of ash and slag waste were made by the soft–mud moulding process followed by baking at various temperatures. For illustrative purposes, all the data are summarized in Table 2 and Figures 2, 3 and 4. The figures are the graphs to show the dependency of baking temperatures on proportions.

Table 2. Physical and mechanical properties of ceramic bricks at various baking temperatures and different percentages of ash and slag waste

Content of ash in clay, %	Heat conductivity, W/m °C				Water absorption, %				Ultimate compression strength, kg/cm²			
	Temperature, °C				Temperature, °C				Temperature, °C			
	700	800	900	1000	700	800	900	1000	700	800	900	1000
0	0,84	0,85	0,85	0,85	11	10	9	8	60	80	90	110
5	0,83	0,84	0,82	0,85	17	17	15	10	70	96	100	115
10	0,83	0,81	0,82	0,81	18	18	16	12	72	104	110	120
15	0,77	0,79	0,74	0,76	20	19	17	14	75	106	115	125
20	0,77	0,69	0,73	0,71	24	23	18	16	70	90	105	115
25	0,55	0,6	0,57	0,61	27	25	20	18	65	85	95	110

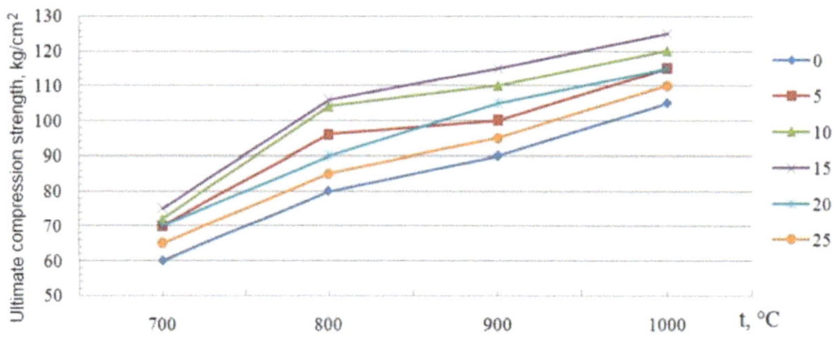

Figure 2. Backing temperature dependence of the ultimate compression strength.

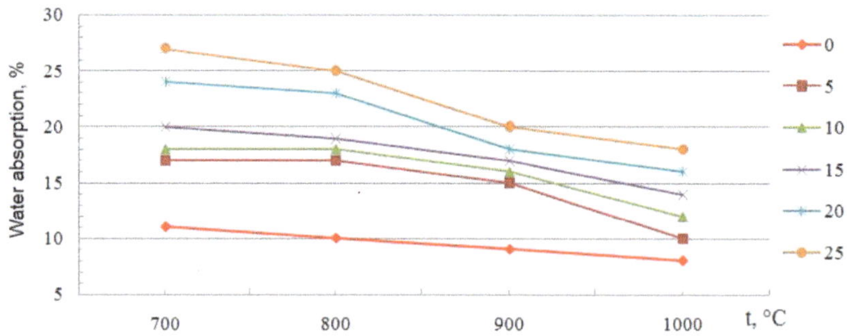

Figure 3. Backing temperature dependence of the water absorption capacity.

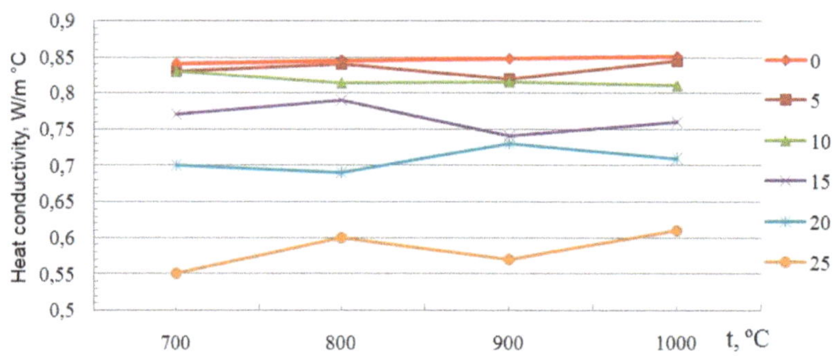

Figure 4. Backing temperature dependence of the heat conductivity capacity.

CONCLUSIONS

The studies found that the heat conductivity, the ultimate compression strength and the water absorption capacity depend on both the amount of ash added and baking temperatures. The greater the ash and slag content in bricks, the lower their heat conductivity is. The water absorption capacity increases with an increase in ash content. The ultimate compressive strength also decreases with increasing the ash content in ceramics. The optimum percentage of adding ash and slag waste is fifteen percent at a baking temperature of 1000°C. It has been established, if ash and slag waste is of an acidic type, it does not have self-sustained bonding properties and require adding curing agents. The analysis of the particle-size distribution showed that 60% of particle size is 10 to 70 microns; the material is finely dispersed and can be used as a filler in the production of construction mixes and ceramics.

REFERENCES

1. Putilin E I Obzornaja informacija otechestvennogo i zarubezhnogo opyta primenenija othodov ot szhiganija tverdogo topliva na TJeS
2. Survey of domestic and foreign experience in using waste from burning solid fuels by thermal power stations. / E I Putilin, V S Tsvetkov. - Moscow: Soyuzdornii, 2003 - 60 p Zyryanov V V Zola-unosa – tehnogennoe syr'e
3. Fly ash – industry-related raw materials. / V V Zyryanov, D V Zyryanov Moscow: OOO IPTz "Maska", 2009 - 320 p.
4. Dik E P, Mashkovich K I, Vasilchenko Z A, Komponentnyj sostav zoloshlakovyh othodov TJeC // Novoe v rossijskoj jelektrojenergetike Chemical composition of ash waste from thermal power plant // New in Russian power sector. 2003 No.5. – 40 p
5. Spirin E K, Lugovtsova N Yu, Khomich O I Special features of VP-1Ap anion exchanger behavior under deep silicifi cation // Russian Journal of Applied Chemistry. – 2012 – Vol. 85 – p.921-925.
6. Torosyan V F , Torosyan E S , Sorokin P D, Telitsyn A A Updating of sewage – purification facilities of electroplating enterprises with counterflow ion-exchange filters // IOP Conference Series: Materials Science and

Engineering. – 2015 – Vol. 91, Article number 012077. – pp 1-8

7. Bespalov V I, Prirodoohrannye tehnologii na TJeS: uchebnoe posobie Environmentally freindly technologies at thermal power plants: Workbook. / V I Bespalov, S U Bespalova, M A Wagner. - Tomsk: Publishing House of Tomsk Polytechnic University, 2010 – 240 p.
8. Putilov V Ya O roli i meste nauki v reshenie problemy obrashhenija s zoloshlakami TJeS The role and place of science in addressing the issue of treating ash and slag from thermal power plants. / V Ya Putilov, I V Putilova. Warsaw: Energoizdat, 2010 – 320 p.
9. Gavrilov V I Toplivno-transportnoe hozjajstvo i zoloshlakoudalenie na TJeS Fuel and transport facilities and ash and slag removal at TPP. /V I Gavrilov Moscow: Energoizdat, 1987. 168 p.
10. Borisenko L F Perspektivy ispol'zovanija zoly ugol'nyh teplovyh jelektrostancii Prospects for the use of ash from coal-fired power plants. / L F Borisenko, L M Deltsin, A S Vlasov. - Moscow: ZAO "Geoinformmark", - 2001 -68 p.
11. Tymoshchuk N S Primenenie zoly i shlaka Burshtynskoj GRJeS v zhelezobetonnyh izdelijah dlja dorozhnogo stroitel'stva Use of ash and slag from Burshtyn TPP in concrete products for road construction / N S Tymoshchuk, I.S. Bobyk. - Moscow: Mir, 1991 – 70 p.
12. Bazhenov Yu M, Schebenkin P F, Dvorkin L I Primenenie promyshlennyh othodov v proizvodstve stroitel'nyh materialov Use of industrial wastes in the production of construction materials.. - Moscow: Stroyizdat, 1986 -120 p.
13. Bozhenov P I Kompleksnoe ispol'zovanie mineral'nogo syr'ja dlja proizvodstva stroitel'nyh materialov Multiple use of mineral raw materials for the production of construction materials. / P.I. Bozhenov. - Moscow: Stroyizdat, 1963 - 170 p
14. Tabin Rushad S 2011 Experimental Studies on Lime-Soil-Fly Ash Bricks (International journal of civil and structural engineering, volume 1, № 4)
15. Obada Kayali 2005 High Performance Bricks from Fly Ash. (USA, Kentucky, Lexington World of Coal Ash, April 11-15)
16. Henry Liu 2009 Environment Properties of Fly Ash Bricks, (USA in Lexington World of Coal Ash Conference – May 4-7)
17. Kevan D Sharp. 1993 Fly ash for soil improvement: proceedings of sessions sponsored by Committees on Soil Improvement and Geosynthetics of the Geotechnical Engineering Division of the American Society of Civil Engineers in conjunction with the ASCE Convention in Dallas (Texas, October 24-28)

CHAPTER 6

THE USE OF FLY ASH THE THERMAL POWER PLANTS IN THE CONSTRUCTION

R S Fediuk and A M Yushin

Far Eastern Federal University, Vladivostok, Russia

ABSTRACT

The problems of ecological and radiation safety of the construction of man-made waste like fly ash thermal power plants were researched. The chemical composition of TPPs ashes of Primorsky Territory was studied, defined their specific effective activity of natural radionuclides. The most modern research methods were used - differential thermal analysis, thermogravimetry, X-ray analysis. It was revealed that the ash of the Primorskaya TPP and Partizanskaya TPP has exceed the permissible parameters of radioactivity, so not suitable for use in construction. Ashes of Vladivostok TPP-2 and Artem TPP of Primorsky Region on parameters radioactivity and chemical composition have suitable for use as a filler in the concrete.

INTRODUCTION

The use of technogenic waste in addition to saving material and natural resources, protects the environment. Therefore, during the design phase of the construction of the object should be possible to use a man-made waste. At the same time, once made a focus on the environmental aspect, you need carefully to select building materials.

In this paper we will focus on the study of the properties of fly ash thermal power plants as an additive in cement composite binder, and special attention will be paid to the assessment of its radioactivity.

Citation: R S Fediuk and A M Yushin, The use of fly ash the thermal power plants in the construction, doi:10.1088/1757-899X/93/1/012070.

THE USE OF FLY ASH OF THERMAL POWER PLANTS

Waste heat power plants mainly divided into two categories: ash and fly ash, different way to remove [1-3]. According to [4], fly ash, it is a more effective additive in cement compositions than ash. In this case, the amount alite phase is increases and aluminate is decreases compared with a control sample, respectively, increases the strength of the cement stone [5-10].

The objects of study are fly ash of the largest thermal power plant Primorye Territory: Vladivostok TPP, TPP Artem, Primorye SRPS and Partizansk power station. An important factor is the ability to separate the selection of dry ash that is currently implemented in these thermal power plants [11-15].

The composition and structure and ash depends on a set of concurrent factors: the type and morphological features of combusted fuel fineness in the course of its preparation, the ash content of the fuel, the chemical composition of the mineral part of the fuel, the temperature in the combustion zone, the residence time of the particles in this zone, and others (figure 1).

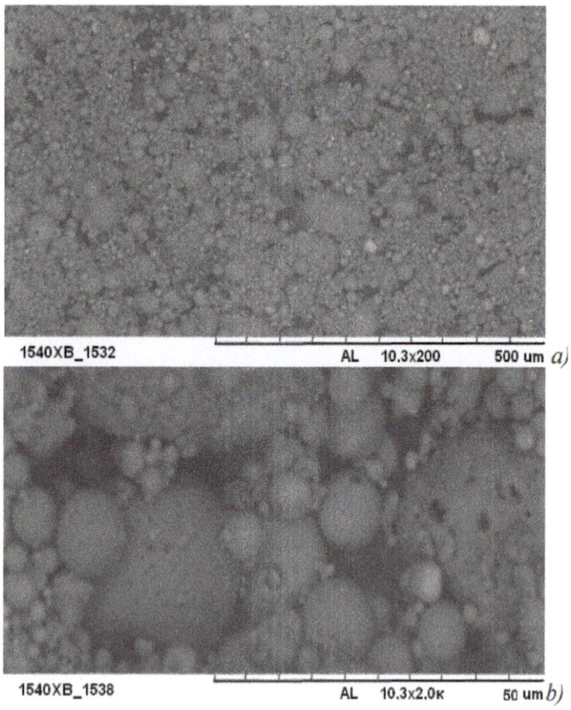

Figure 1. The micrographs of fly ash Vladivostok TPP-2. Magnification 200 and 2000 times

The laboratory research have set the basic characteristics of the fly ash, determine the possibility of its use for the production of polymer-mineral construction materials. The main criteria for the selection of fly ash for study the possibility of its use as a component of the composite binder:

- Shape of the particles;
- Particle size distribution;
- Specific surface area;
- Packing density;
- Chemical composition.

Samples of fly ash produced from a large silo capacity, where it could be shipped for delivery to consumers or, depending on the filling of the silo through the mixing device in the pressure system to move to the ash dump. After the selection of the private trials were combined, mixed thoroughly and quartering reduced to laboratory sample. Laboratory samples of fly ash were sieved through 008; as a result of the screening on the sieve remained a significant part of the unburned coal particles, magnetite beads, and microspheres; ash, pass through a sieve, it was selected for the study and preparation of concrete. Chemical analysis of the sol (Table. 1) is made in accordance with GOST 10538-87.

This standard applies to brown and black coal, anthracite, oil shale, peat and coke and establishes methods for determining the fuel in the ash of silicon dioxide (SiO_2), iron oxide (Fe_2O_3), aluminum oxide (Al_2O_3), magnesium oxide (MgO), calcium oxide (CaO), potassium oxide (K_2O), sodium oxide (Na_2O), phosphorous oxide (P_2O_5), titanium dioxide (TiO_2), sulfur trioxide (SO_3), mixed manganese oxide (Mn_3O_4).

Table 1. The chemical composition of ashes TPPs the Primorsky Territory

		TPPs			
		Primorskaya	Vladivostok	Artem	Partizansk
The predominant type of coal		Luchegorsk and Bikin	Primorsky lignite (Pavlovsky section)	Black coal	Neryungri black coal
Content of elements on an oxide basis, %	SiO_2	55,3	63,0	48.1	47,4
	TiO_2	0,5	0,5	0,0	0,9
	Al_2O_3	12,6	21,4	29,3	22,3
	Fe_2O_3	10,7	7,5	6,5	19,6
	CaO	12,5	3,4	9,7	4,8
	MgO	3,5	2,1	1,8	2,8
	K_2O	1,0	1,3	1,2	0,1
	Na_2O	0,4	0,3	0,2	0,4
	SO_3	3,4	0,6	2,3	1,62
	CaO_{free}	1,0	0,4	<0,1	no
	LOI	2,3	1,4	0,6	<5

The data on the chemical composition of the ash indicate that the contents of the individual oxides in the fly ash derived from the combustion of various kinds of pulverized coal has significant deviation.

It determines the difference in the properties of ash and possible areas of use in the manufacture of building materials. Ash depending on the total effective specific activity of natural radionuclides Aeff less 370 Bq / kg is applied to the production of materials, products and structures, residential and public buildings.

The results of determination of specific effective activity of fly ash on spectrometric complex "USC Gamma Plus" are shown in Table. 2.

Table 2. Determination of specific effective activity of fly ash TPPs the Primorsky Territory

Parameter	The result of the measurement (A), Bq / kg			
	Primorskaya TPP	Vladivostok TPP	Artem TPP	Partizansk TPP
Activity ^{40}K	496.9±101	392±89	342±68	516.9±101
Activity ^{232}Th	153.6±20.3	31.5±19.7	29.5±15.7	193.2±22.3
Activity ^{226}Ra	163.1±9.36	37.63±6.32	27.23±5.93	113.1±6.37
$A_{eff} = A_{Ra} + 1,31A_{Th} + 0,085A_K$	>398	80±30	93±20	>410

Ashes of Vladivostok TPP and TPP Artem belong to the first class of materials (less than 370 Bq / kg) in accordance with GOST 30108-94 “Building materials and products. Determination of the specific effective activity of natural radionuclides”. This material can be used for all kinds of construction work. Ashes of Primorskaya TPP and Partizansk TPP areexceed the permissible parameters of radioactivity, so not suitable for use in construction.

Thus, by GOST 25592-91 “Mixes slag thermal power plants for concrete” the ashes of Vladivostok TPP and TPP Artem are selected for further research. The high ash content in Al_2O_3 (to 29.3%) and SiO_2 (63%) can cause crystallization like mullite compounds. The loss on ignition of less than 1.5% say a small amount of residual fuel that minimizes manufacturing processes shrinkage during firing.

Thermal studies were performed on raw by thermogravimetric analyzer Shimadzu DTG- 60H at a rate of temperature rise of 20°C / min, in the range of 20-1100°C. The results of thermal analysis are presented graphically in figure 2.

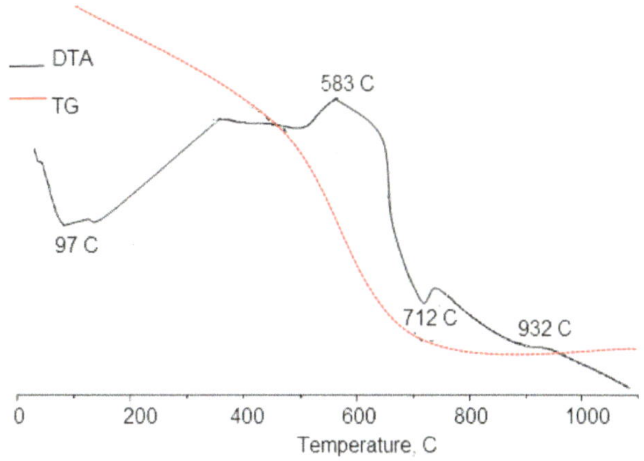

Figure 2. The results of DTA and TG of Vladivostok TPP fly ash

During heat treatment the ash in the range of 40-200 ° C, is the loss of water adsorbed surface area of the particles. Carbonate decomposition is observed at a temperature of 712°C. Burning residual fuel - 500-700°C. The nature and intensity of loss of mass during this indicates an amount of unburnt residues are particles of coal and coke and semi-coke residue. Relatively small exotherm with a maximum of 932°C reflects the crystallization of compounds in mullitopodobnyh aluminosilicate phase XRD analysis showed that the ash except quartz is determined clearly expressed mullite phase (figure 3).

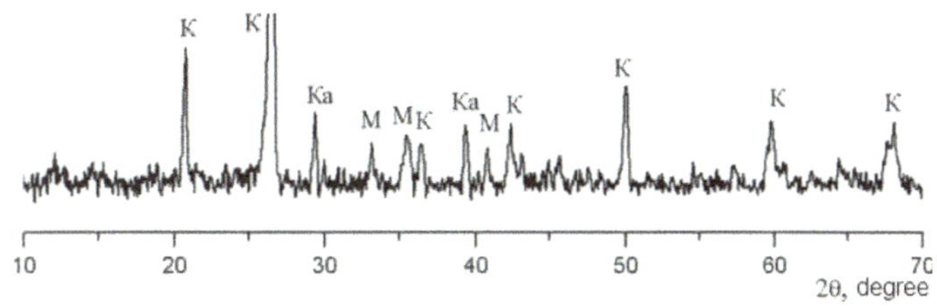

Figure 3. XRD results of Vladivostok TPP fly ash. K - quartz, Ca - calcite, M - mullite

CONCLUSION

The test material is relatively homogeneous, multicomponent. According to the composition closest to aluminosilicates, due to the high content of silicon and aluminum oxides to 80-90%, of which about 2/3 of silicon oxide. At the same time, fly ash has almost no unburned particles, in which tend to concentrate harmful components. Ash consists of crystalline and amorphous phases.

The crystalline phase contains quartz, feldspar, mullite, et al., The amorphous phase is represented primarily in the form of glass. Thus, the fly ash consists of minerals that are widely used separately as filler dispersed in composite binders. Therefore, we can assume that the ash CHP Primorsky Region on chemical composition suitable for use as a filler in the cement composition.

According to the classification of GOST 24640-91 selected are low-calcium fly ash (acidic) and serve as additives, the material composition of the components (active mineral additives having pozzolanic properties).

According to the classification committee RILEM, you can add that both selected fly ash (Vladivostok TPP and TPP Artem) pozzolanic have normal activity. Hydration products binders have high adhesion beans silica component, which has a significant amount of structural defects, which also contributes to the growth of a preliminary mechanical activation in the preparation of the composite binder, which allows us to strengthen and improve the processes of hydration of clinker minerals, and leads to the formation of more tumors by increasing the active nucleation sites on the particle surface.

REFERENCES

1. Ichiko Misumi, Jun-ichiro Kitta, Hiroyuki Fujimoto, Satoshi Gonda, Yasushi Azuma, Keisuke Maeda, Tomizo Kurosawa, Yoshiyasu Ito, Kazuhiko Omote, Yoshinori

Nakayama and Hiroki Kawada 2012 Meas. Sci. Technol. 23 015002

2. M Tabuchi, H Tameoka, T Kawase and Y Takeda 2011 1757-899X 24 012007
3. H Nakao, Y Yamasaki, J Okamoto, T Sudayama, Y Takahashi, K Kobayashi, R Kumai and Y Murakami 2014 1742-6596 502 012015
4. R V Ovchinnikov diss. ... cand. tech. sciences 2014
5. J del Cerro, J Manchado, F J Romero and M C Gallardo 2012 Meas. Sci. Technol. 23 035003
6. Camila Barreneche, Aran Solé, Laia Miró, Ingrid Martorell, A Inés Fernández and Luisa F Cabeza 2012 Meas. Sci. Technol. 23 085606
7. Md Jashim Uddin, N H Md Yusoff, O Anwar Bég and Ahamd Izani Ismail 2013 Phys. Scr. 87 025401
8. Hideyuki Nakano, Akira Kouchi, Shogo Tachibana and Akira Tsuchiyama 2003 0004-637X 592 p 1252
9. J M Cuevas, R Rubio, J M Laza, J L Vilas, M Rodriguez and L M León 2012 Smart Mater. Struct. 21 035004
10. Zhi Chen, Dai Mochizuki, Masato M Maitani and Yuji Wada 2013 Nanotechnology 24 265602
11. Chadwick A.V., Düvel A., Heitjans P., Pickup D.M., Ramos S., Sayle D.C. and Sayle T.X.T. 2015 IOP Conference Series: Materials Science and Engineering 80
12. Polat R., Demirbofa R. and Khushefati W.H. 2015 Construction and Building Materials 81 pp 268-275.
13. Raheek I Ibrahim Wong Z.H. and Mohammad A.W. 2015 IOP Conference Series: Materials Science and Engineering 80
14. Baeza F.J., Galao O., Zornoza E. and Garcés P. 2013 Materials 6 (3) pp 841-855.
15. Al-Rahman L.A. and Raja R.I. 2012 Attenuation of Noise By Using Absorption Materials and Barriers: A Review 2 (7) pp 1207-1217.

CHAPTER 7

HIGH-FLUX/HIGH-TEMPERATURE SOLAR THERMAL CONVERSION: TECHNOLOGY DEVELOPMENT AND ADVANCED APPLICATIONS

Manuel Romero and José González-Aguilar

IMDEA Energy Institute, Ramón de la Sagra 3, 28935 Móstoles, Spain

ABSTRACT

Solar Thermal Power Plants have generated in the last 10 years a dynamic market for renewable energy industry and a pro-active networking within R&D community worldwide. By end 2015, there are about 5 GW installed in the world, most of them still concentrated in only two countries, Spain and the US, though a rapid process of globalization is taking place in the last few years and now ambitious market deployment is starting in countries like South Africa, Chile, Saudi Arabia, India, United Arab Emirates or Morocco. Prices for electricity produced by today's plants fill the range from 12 to 16 c€/kWh and they are capital intensive with investments above 4000 €/kW, depending on the number of hours of thermal storage. The urgent need to speed up the learning curve, by moving forward to LCOE below 10 c€/kWh and the promotion of sun-to-fuel applications, is driving the R&D programmes. Both, Industry and R&D community are accelerating the transformation by approaching high-flux/high-temperature technologies and promoting the integration with high-efficiency conversion systems.

INTRODUCTION

Solar Thermal Power Plants (STPP) have generated, in the last 10 years, a dynamic market for renewable energy industry and a pro-active networking within R&D community worldwide. By end 2015, there are about 5 GW of STPP installed in the world, most of them still concentrated in only two countries, Spain and the US, though a rapid process of globalization is taking place in the last few years and now ambitious deployments are starting in countries like South Africa, Chile, Saudi Arabia, India, United Arab Emirates or Morocco [1]. However, Concentrating Solar Power (CSP) technologies are extremely urged to speed up their learning curves because of the high competition and drastic reduction of costs of other renewable energy technologies and the need of optimizing dispatch and grid integration for markets with high penetration of solar and wind.

Prices for electricity produced by today's STTP fill the range from 12 to 16 c€/kWh depending on the irradiation level

Citation: Manuel Romero, José González-Aguilar, High-flux/high-temperature solar thermal conversion: technology development and advanced applications, Renew. Energy Environ. Sustain. 1, 26 (2016)

and most importantly on the financing conditions and they are capital intensive with investments above 4000 €/kW, depending on the number of hours thermal storage [2]. However, most commercial initiatives are still scaled-up projects based upon technological developments and concepts that matured in the 1980s and 1990s. They are projects still based on conservative schemes and technologies, which do not exploit the enormous potential of concentrated solar energy. Commercial projects use technologies of parabolic troughs with low concentration in two dimensions and linear focus, or systems of central tower and heliostat fields, operating with thermal fluids at relatively modest temperatures [3]. The most immediate consequence of these conservative designs is the use of systems with efficiencies about 20% nominal in the conversion of direct solar radiation to electricity; the tight limitation in the use of efficient energy storage systems; the high water consumption and land extension due to the inefficiency of the integration with the power block; the lack of rational schemes for their integration in distributed generation architectures and the limitation to reach the temperatures needed for the thermochemical routes used to produce solar fuels like hydrogen [4].

The urgent need to speed up the learning curve, by moving forward to LCOE below 10 c€/kWh and the promotion of sun-to-fuel applications, is driving the R&D programmes. Both, industry and R&D community are accelerating the transformation by approaching high-flux/high-temperature technologies and promoting the integration with high-efficiency conversion systems.

CURRENT STATUS OF CONCENTRATING SOLAR POWER

From the 1970s to the 1990s, the development of solar thermal electricity (STE) technologies remained restricted to few countries and only few, though important, research institutions and industries were involved. The situation dramatically changed in 2007 with the approval of specific feed-in-tariffs (FIT) or power purchase agreements (PPA) in Spain and the US. Both countries with more than 4 GW of projects already in operation at the end of 2014 have been the key drivers for market implementation (Fig. 1). Other countries have recently started specific policies to support CSP favouring the globalization of the technology. Subsequently a number and variety of engineering and construction companies, consultants, technologists and developers committed to STE were rapidly growing in the period 2008–2014.

As depicted in Figure 2, the market development process has three clear periods. First period between 2007–2013 has been totally led by Spain because of the adequate FIT policy leading to the construction of 50 STTP with total power of 2.3 GW. Typical unitary sizes of projects were 50 MWe with parabolic trough being the technical solutions for most of them. This early market period served to demonstrate routine operation by choosing conservative designs and the validation of first commercial projects implementing molten salt thermal storage of about seven hours.

The second period led by the US has the peak between 2012 and 2015 and its distinct feature is the significant scaling up of the systems beyond 100 MWe (Ivanpah, Solana, Crescent Dunes, Mojave or Genesis) and the strong need of optimizing dispatchability value because of the highly demanding PPA by the utilities. Typical PPA with utilities (time of delivery value) is imposing a selling price $0.14/kWh. Commissioning and routine operation become more complex (3–4 years from groundbreaking to commissioning). Environmental impact becomes essential in the design of the projects: visual impact, glint and glare, birds, water.

The third market wave represents globalization of the technology, being additional challenges the indigenisation of components, the competitive bidding and the hybridization with other energy sources. Main countries leading the projects nowadays are Morocco and MENA, Chile, China, South Africa and India. Figure 3 illustrates current CSP projects around the world according to Solar Power And Chemical Energy Systems (SolarPACES), an implementing agreement of the International Energy Agency (IEA) [5]. SolarPACES estimates that worldwide overall capacity of CSP is 10.1 GW, including operational plants as well as plants under construction and in development stages.

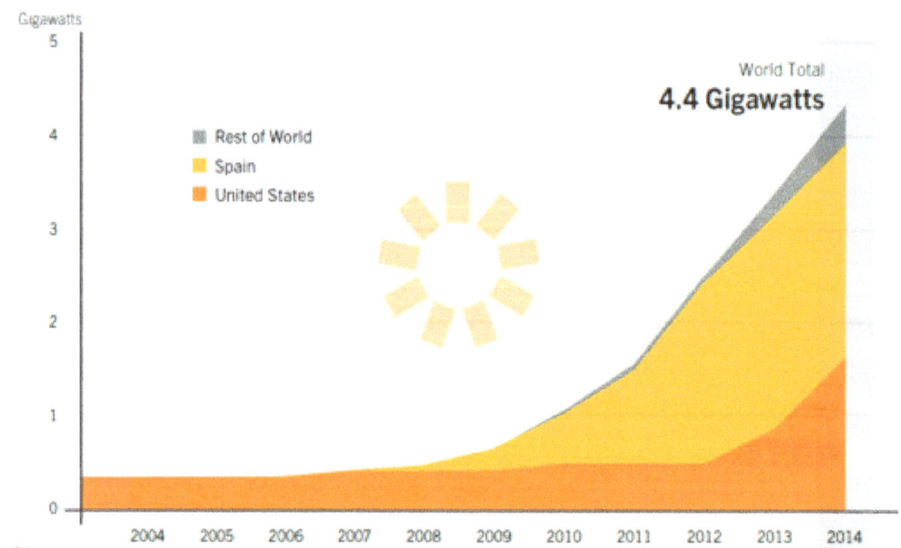

Figure 1. Accumulated capacity installed [1]

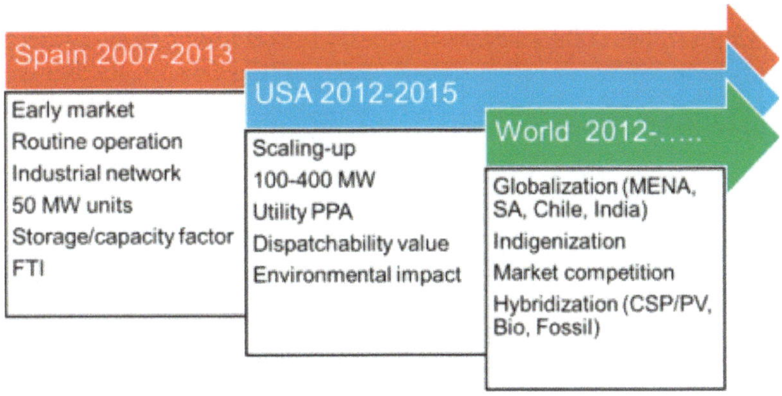

Figure 2. Concentrating Solar Power (CSP) market development waves since 2007.

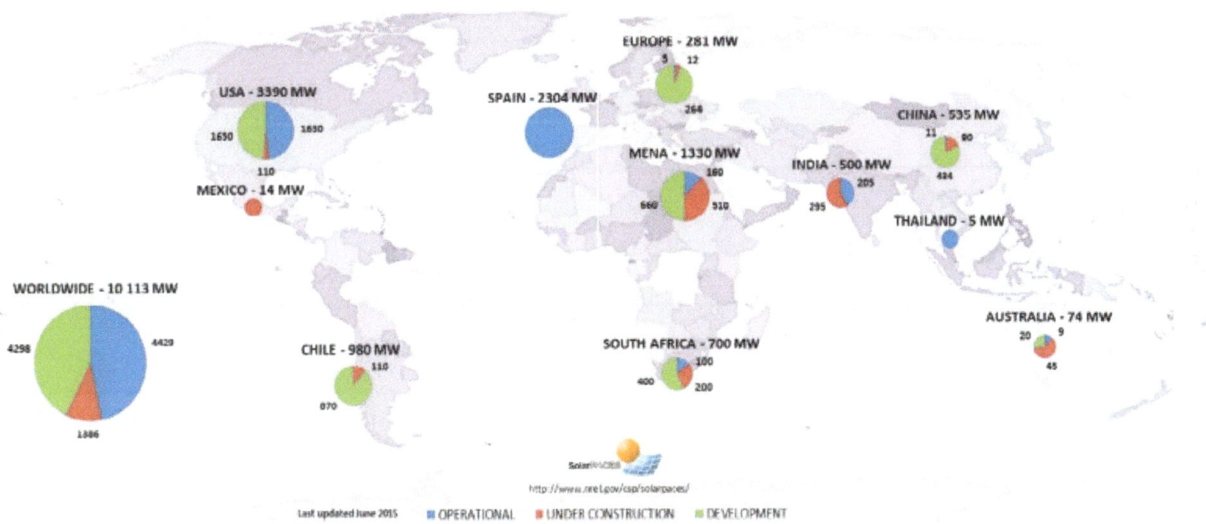

Figure 3. Accumulated capacity of Concentrating Solar Power (CSP) worldwide.

COST REDUCTION

In 2010, the IEA elaborated a first report including the globalization of STE commercial deployment for the future energy scenario. At that time, IEA started to consider that STE would play a significant role among the necessary mix of energy technologies for halving global energy-related CO_2 emissions by 2050 [6]. This 2010 scenario would require capacity addition of about 14 GW/year (55 new STPP of 250 MW each). In 2014, the IEA presented an updated technology roadmap on STE [7]. Although it confirms that STE has grown less than expected in the first IEA CSP roadmap, STE roadmap maintained a STE›s share of global electricity of 11% in 2050, which means 1 TW of installed CSP capacity by 2050.

However, the competition with photovoltaics, which has become cost competitive with fossil fuels without subsidies in an increasing number of countries, has introduced an important stress to the developers of STE. They have been forced to move from strategies oriented to early commercialization markets based upon special tariffs, to strategies oriented to a massive production of components and the development of large amounts of projects with less profitable tariffs. This situation has been speeding up the implementation of larger capacity STE plants and/or second-generation technologies, even though in some cases still some innovations are under assessment in early commercialization plants or demonstration projects.

LCOE value may be reduced up to an additional 30% when moving to future sites with very high direct normal irradiance.Figure 4 shows the normalized analysis carried out by the European Industry Association of Solar Thermal Electricity (ESTELA). The analysis reveals the impact on cost reduction after the implementation of 30 GW where it is predicted 25-year PPA prices of 10 c€/kWh by 2020 for a 150 MW reference plant with four-hour thermal storage [8]. The projected evolution of levelized electricity costs of different STE technologies is illustrated in Figure 5. The reduction in electricity production costs should be a consequence, not only of mass production but also of scaling-up and R&D. The technology roadmap promoted by the European Industry Association in 2010 states that when most of the improvements currently under development are expected to be implemented in new plants, energy production boosts greater than 10% and cost decreases up to 20% are expected to be achieved. Furthermore, economies of scale resulting from plant size increase will also contribute to reduce plants' CAPEX per MW installed up to 30%. STE deployment in locations with very high solar radiation further contributes to the achievement of cost competitiveness of this technology by reducing costs of electricity up to 25%.

All these factors can lead to electricity generation cost savings up to 50% by 2025, reaching competitive levels with conventional sources (e.g. coal/gas with stabilized electricity costs <10 c€/kWh). This objective is nowadays promoted by the European Commission in Horizon2020 for the year 2020 in consultation with industry sector and R&D centers for the revision of the SET-Plan.

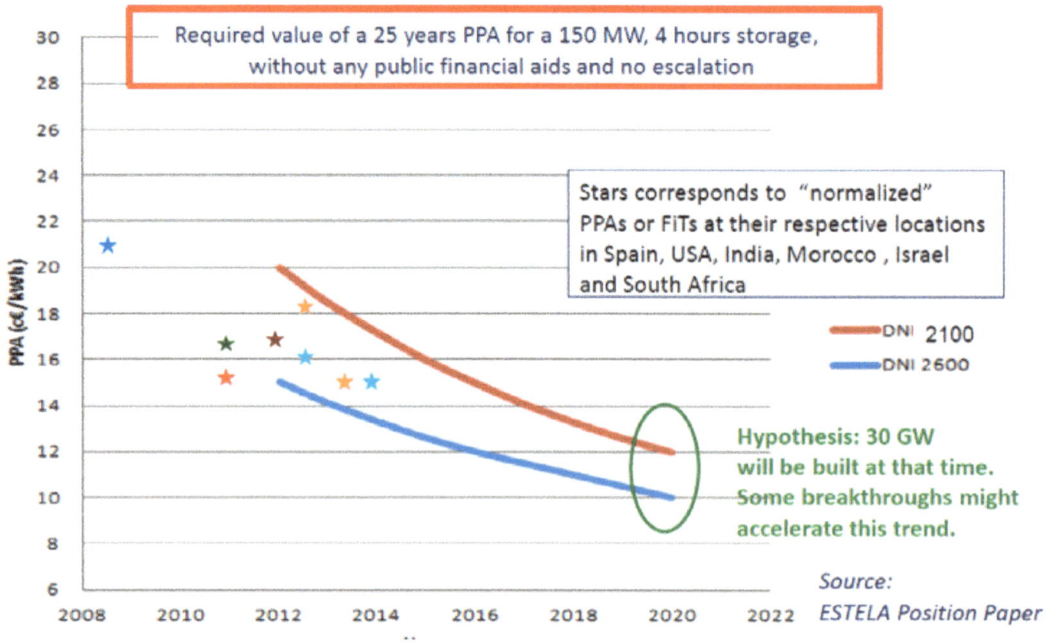

Figure 4. Cost reduction: the view from the European Industry (source: ESTELA).

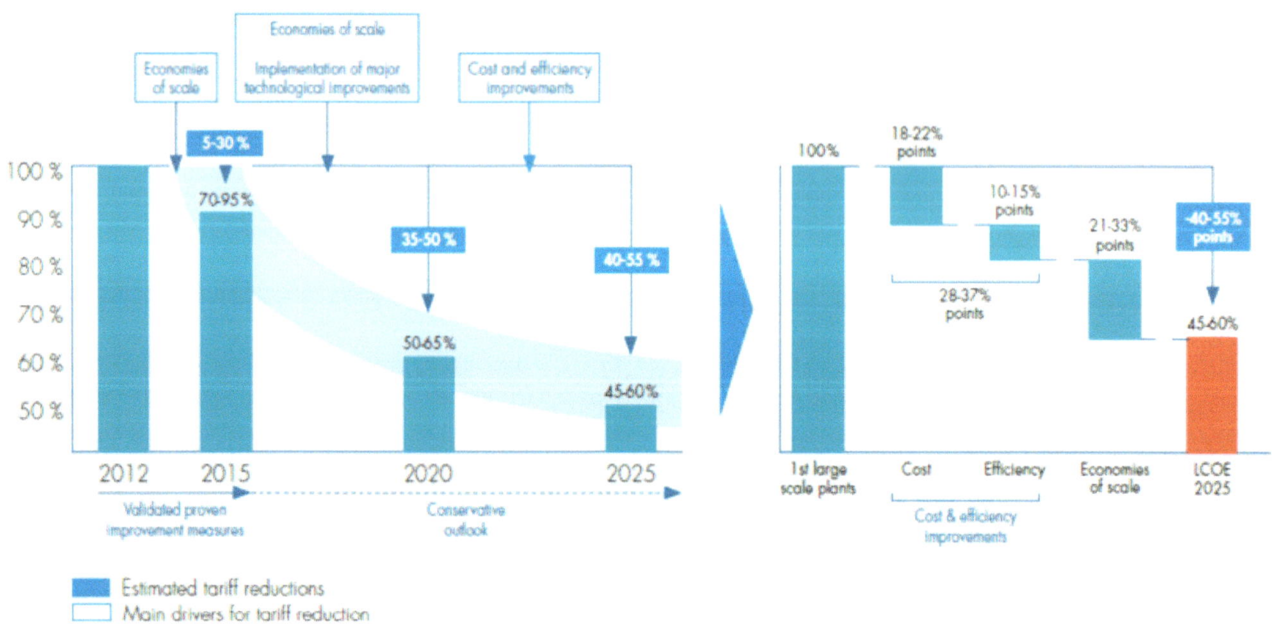

Figure 5. Expected LCOE reduction from 2012 to 2025 [8,9].

R&D CHALLENGES IN CONCENTRATING SOLAR POWER

Some reports have recently been published addressing the R&D challenges in CSP. The ESTELA report on STE: strategic research agenda 2020–2025 is probably the most relevant indicating guidelines in order to reduce STE costs. R&D topics are oriented towards three main objectives: increase efficiency and reduce generation, operation and maintenance costs; improve dispatchability and improve environmental profile [8]. IEA have also provided some hints in its technology roadmap on STE [7]. There is a consensus among stakeholders that next generation technologies should be based upon the use of higher flux and temperature to increment the thermal efficiency conversion (Fig. 6), this implies the use of ceramic materials and the development of solar volumetric receivers, particle receivers and reactors, as well as the use of supercritical fluids [10].

In optical engineering, the technology evolution is heading to ultra-modular concentrators of plug-and-play type with high degree of automatism, reliability and flexible unmanned operation that might be subject to integration in different environments, including semi-urban. At component level, new mirrors will require R&D activities on light reflective surfaces, antisoiling coatings and high reflective.

In heat transfer, the R&D is accelerating the development of new thermal fluids for operation at higher temperatures; new thermal (sensible heat and phase change) and thermochemical energy storage systems; solar receivers with advanced volumetric absorbers or clouds of particles for higher apparent absorptivity and development of dry-cooling at power block for avoiding water consumption.

Energy conversion is looking for the use of new turbomachinery and cycles making use of supercritical fluids as well as the integration of direct conversion systems like TPV (thermal PV), high-temperature thermoelectric converters or thermionic energy conversion.

In particular, STE based on central receiver technologies will need improvements in the heliostat field as a result of better optical properties, lower cost structures, and better control. In general terms, optical performance and durability of existing heliostats is acceptable (95% availability and beam quality below 2.5 mrad), therefore R&D resources should focus basically on cost reduction.

Development of water/superheated steam and advanced air-cooled volumetric receivers using both wire-mesh absorbers and ceramic monoliths is the subject of various projects. Dual-aperture receivers for water/steam and volumetric receivers for air need further development for scale-up, materials durability and thermal efficiency.

Heat storage is another key issue for CRS development. The new developments in air-cooled receivers have led to the development of advanced thermocline storage systems making use of packed-bed ceramic materials. This system has shown excellent performance for small units of a few MWh but pressure losses and design restrictions appear when size is increased.

Finally, more distributed control architectures, system integration and hybridization in high-efficiency electricity production schemes should be developed as already mentioned for trough systems.

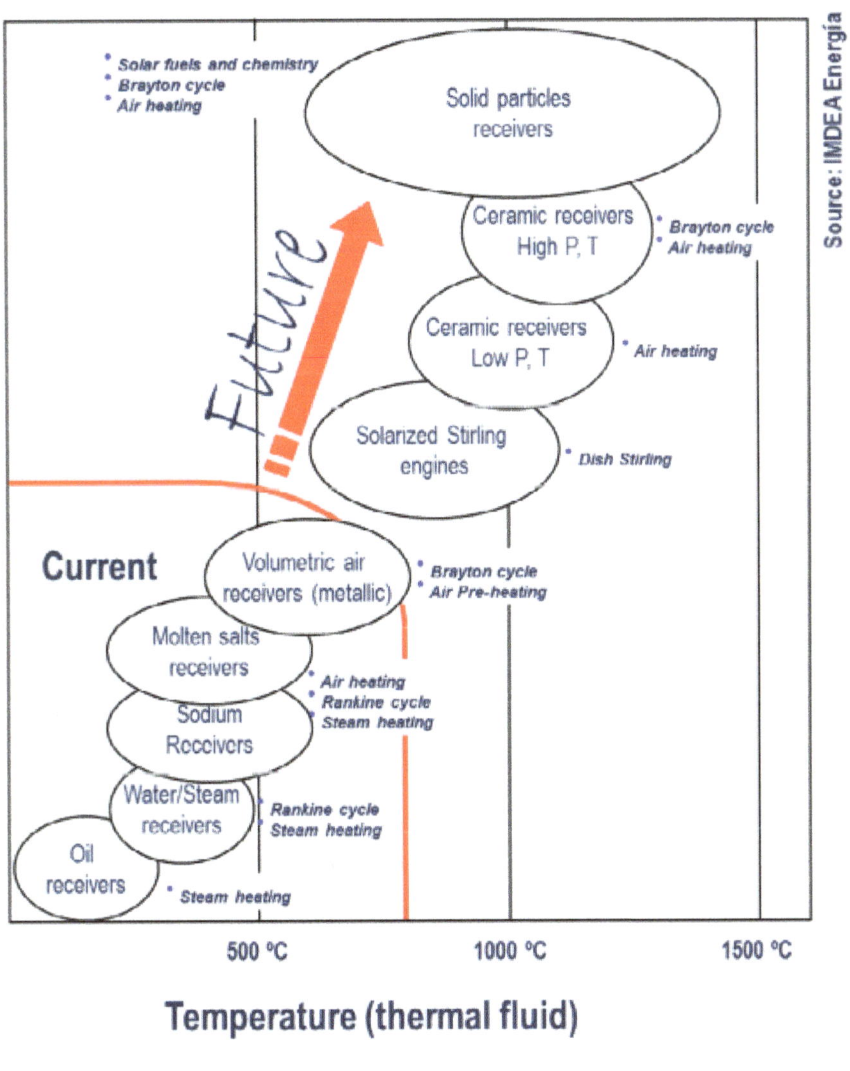

Figure 6. Concentrating Solar Power (CSP) towards high-flux/high-temperature technologies with efficient conversion.

R&D PROGRAMS

Europe – EERA – Integrated Research Program STAGE-STE

The Strategic Energy Technology Plan (SET-Plan) establishes the policy at European level in order to accelerate the development and deployment of cost-effective low carbon technologies for the achievement of Europe's 2020 targets and vision on greenhouse gas emissions, renewable energy and energy efficiency. The European Energy Research Alliance (EERA), which was launched in 2008, is one of the initiatives of the SET-Plan. Within EERA, the Joint Programme on Concentrated Solar Power (EERA JP-CSP), which is currently coordinated by CIEMAT, was formally approved in April 2011. From February 2012, the Integrated Research Program STAGE-STE (Scientific and Technological Alliance for Guaranteeing the European Excellence in Concentrating Solar Thermal Energy) supports the EERA JP-CSP. STAGE-STE gathers 42 members, all EU research institutions partners of EERA JP-CSP plus a significant number of additional organizations, including those from non-EU countries. The overall project budget is 21.198.352 €, where the European Commission Contribution is 9.997.207 €.

Other European actions exist through the EU Framework Programme for Research and Innovation: the pan-European STE research infrastructure "EU-Solaris" within the European Strategy Forum on Research Infrastructure (ESFRI) launched in April 2002 on new research infrastructure includes; 'Solar ERANET' within the European Research Area Net (ERANET) that focuses on establishing joint actions in the form of public-private partnerships; or the 'Solar Facilities for the European Research Area' (SFERA).

Sunshot – USA

The SunShot Imitative was launched in 2011 and supervised by the US Department of Energy (DOE). Its goal is to reduce the price of solar energy systems by about 75% between 2010 and 2020 and then to make solar energy cost competitive with other forms of energy, without subsidies, by the end of the decade. Figure 7 illustrates the strategy of cost reduction through technical improvements due to R&D programs. It is particular relevant the targets defined by DOE on CSP and that establish the R&D calls.

Any development on solar fields searches dramatically reducing the cost of the collector field (<$75/m^2, compare to 2010 ~$250–$300/m^2) while improving or maintaining optical efficiency (optical error <3 mrad, sustain wind speed >85 mph, and lifetime >30 years). Thus main research topics in this area concern the development of high optical accuracy reflectors, the reduction of collector structure weight and material, the development of clean and rapid manufacture, assembly and installation methods and highly efficient tracking and control and accurate metrology tools and strategies to reduce operations and maintenance (O&M) costs.

A second requirement concerns solar receivers that must significantly increase operating temperatures (>720 °C), efficiency (>90%) and lifetime (>10,000 cycles) at costs smaller than $150/kWth. Therefore, R&D activities focus on the development of novel solar selective coatings and new receiver designs.

Electricity generation through high-temperature power cycles requires a net cycle efficiency higher than 50%, they should involve dry cooling and investment cost lower than $900/kWe. Then R&D is oriented to develop high-temperature power cycles, solid-state power conversion techniques and hybrid cycles.

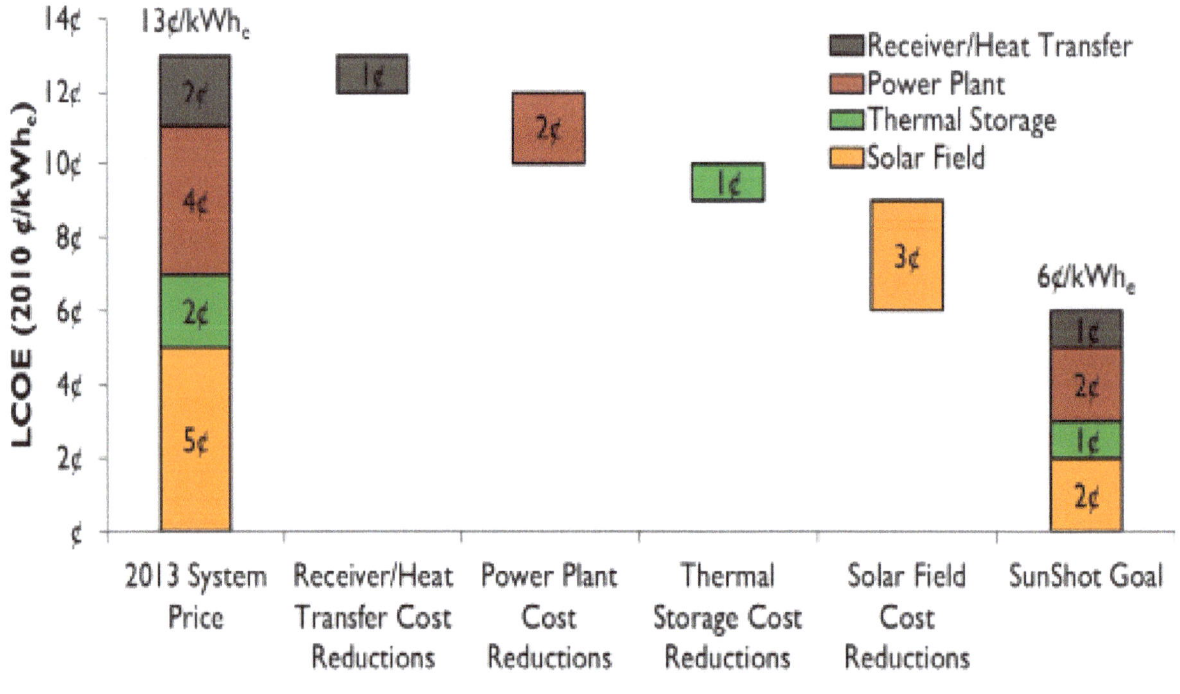

Figure 7. Expected LCOE reduction in SunShot Concentrating Solar Power (CSP) program.

Asia – Solar Mission (India), Australia

In India, the Jawaharlal Nehru National Solar Mission was launched on the 11th January, 2010 with the ambitious target of deploying 20,000 MW of grid connected solar power by 2022. In order to achieve this goal, the Solar Mission is supported by a R&D program led to developed excellence R&D centres on solar energy, pilot scale STE plants, amongst others.

China has a significant number of public centers like the Institute of Electrical Engineering of the Chinese Academy of Science and industry related with CSP like Himin Solar or Supcon. Recently, September 2015, the China National Energy Administration's (NEA) has published a notice on the organization of 1 GW of CSP Demonstration Projects during 2016–2017.

Australian Solar Thermal Research Initiative (ASTRI) is a consortium composed by Australian research institutions collaborating to deliver cost reductions and dispatchability improvements. It is the tool funded by the Australian Renewable Energy Agency (ARENA, a commercially oriented agency, established on 1st July 2012 by ARENA Act 2011) to position Australia in concentrating solar thermal (CST) power. ASTRI target is to lower the cost of solar thermal power from 26.5 to 12 cents a kilowatt-hour by 2020. The research program is established on four main objectives or nodes: reduce capital, increase capacity factor, improve efficiency, and add product value.

AFRICA – MENA, SOUTH AFRICA

In 2011, South Africa promulgated the Integrated Resource Plan (IRP) 2010–2030. The main objective was to add 56,500 MW of electricity by 2030, compared with current capacity of about 38,000 MW, mostly produced by Eskom (the national utility) coal-fueled power stations. From the new capacity, 21,534 MW (38%) is allocated to renewable energy technologies, and 1200 MW (2.1%) would assigned to CSP. The implementation of the IRP is mainly carried out by the Renewable Energy Independent Power Producers' Procurement Programme (REIPPPP). Here private developers are able to bid for allocations and the renewable electricity will be purchased by Eskom and supplied to the national grid. The program is driving R&D with relevant stakeholders like the University of Stellenbosch.

Moroccan Solar Plan (MSP) targets at least 2 GW of solar capacity by 2020 and a simultaneous development of a local solar industry. Morocco in setting up its own CSP R&D centre, under the auspices of Masen, the Moroccan Agency for Solar Energy. MENA region is especially active in R&D with the creation of relevant institutions like MASDAR in Abu Dhabi, QEERI in Qatar or KA CARE in Saudi Arabia.

Latin America Chile and Mexico

Demonstration projects and associated R&D are emerging in Mexico and Chile and mobilizing other countries like Brazil and Argentina. In Mexico, there is a strong R&D community with special relevance of the Institute of Renewable Energy from UNAM in Cuernavaca and the Central Receiver Test Center in Sonora. According to the National Energy Strategy (ENE) 2013–2027 of the Mexico's Government, the country should diversify and optimize the electric generation park. Based on this, the ENE includes a goal to increase the participation of RE sources in electricity generation to 35%. The policy in the country is advancing to promote CSP plants through liberalization of the market and R&D networking through several initiatives funded by CONACYT.

Chile has a National Energy Strategy 2012–2030 that establishes the targets in energy policy for the next decades. Among the objectives, it is relevant to mention up to 25 GW of renewable energy power by 2025. Several projects accounting for about 1 GW of STPP are already in construction in North Chile where very high solar irradiance is guaranteed. The public R&D is nucleated in the Solar Energy Research Center SERC Chile that integrates most of the universities and foundations related with solar energy research in the country. A specific facility promoted for CSP research is the Solar Platform of Atacama Desert in Antofagasta.

ACKNOWLEDGMENTS

The authors gratefully acknowledge "Comunidad de Madrid" and "European Social Fund" for its financial support to the ALCCONES Project through the Programme of Activities between Research Groups (S2013/MAE-2985). Thanks are also due to the support of the European Union's Seventh Framework Research Programme STAGE-STE (Scientific and Technological Alliance for Guaranteeing the European Excellence in Concentrating Solar Thermal Electricity, ENERGY.2013.10.1.10).

REFERENCES

1. REN21, *Renewables 2015 Global Status Report* (REN21 Secretariat, Paris, 2015). ISBN: 978-3-9815934-6-4.
2. International Renewable Energy Agency (IRENA), Concentrating solar power, Renew. Energy Technol. Cost Anal. Ser.**1** (2012).
3. M. Romero, J. González-Aguilar, Solar thermal CSP technology, WIRE›s Energy Environ. **3**, 42 (2014)
4. M. Romero, A. Steinfeld, Concentrating solar thermal power and thermochemical fuels, Energy Environ. Sci. **5**, 9234 (2012)
5. SolarPACES:
6. IEA, *Energy technology perspectives 2010 – Scenarios and strategies to 2050* (IEA Publications, Paris, 2010). ISBN: 978-92-64-08597-8
7. IEA, *Technology roadmap: solar thermal electricity* (IEA Publications, Paris, 2014).
8. European Solar Thermal Electricity Association (ESTELA), Concentrating solar power on the road to 2030, ESTELA a.s.b.l. report (2015).
9. A.T. Kearney, Inc., *Solar thermal electricity 2025, Clean electricity on demand: attractive STE cost stabilize energy production* (A.T. Kearney GmbH, Dusseldorf, Germany, 2010).
10. M. Romero, J. Gonzalez-Aguilar, E. Zarza, Chapter 42: Concentrating Solar Power, *Energy Efficiency and Renewable Energy Handbook*, 2nd ed., edited by D.Y. Goswami and F. Kreith (CRC Press, Boca Raton, Florida, 2015), 1237– 1346, Online ISBN: 978-1-4665-8509-6, Print ISBN: 978-1-4665-8508-9

CHAPTER 8

INTELLIGENT PREDICTION OF FAN ROTATION STALL IN POWER PLANTS BASED ON PRESSURE SENSOR DATA MEASURED IN-SITU

Xiaogang Xu[1], Songling Wang[1], Jinlian Liu[1] and Xinyu Liu[2]

[1] School of Energy Power and Mechanical Engineering, North China Electric Power University, 619 Yonghua North Street, Baoding, 071003, China

[2] Department of Mechanical Engineering, McGill University, 817 Sherbrooke Street West, Montreal, Quebec H3A 0C3, Canada

ABSTRACT

Blower and exhaust fans consume over 30% of electricity in a thermal power plant, and faults of these fans due to rotation stalls are one of the most frequent reasons for power plant outage failures. To accurately predict the occurrence of fan rotation stalls, we propose a support vector regression machine (SVRM) model that predicts the fan internal pressures during operation, leaving ample time for rotation stall detection. We train the SVRM model using experimental data samples, and perform pressure data prediction using the trained SVRM model. To prove the feasibility of using the SVRM model for rotation stall prediction, we further process the predicted pressure data via wavelet-transform-based stall detection. By comparison of the detection results from the predicted and measured pressure data, we demonstrate that the SVRM model can accurately predict the fan pressure and guarantee reliable stall detection with a time advance of up to 0.0625 s. This superior pressure data prediction capability leaves significant time for effective control and prevention of fan rotation stall faults. This model has great potential for use in intelligent fan systems with stall prevention capability, which will ensure safe operation and improve the energy efficiency of power plants.

INTRODUCTION

The fan is a common type of air flow control component, widely used in many impeller machines, such as ventilators, compressors, and pumps. These machines are employed in almost every field of industry and consume a significant amount of electricity. In thermal power plants, fans are the power sources of smoke exhaust systems, and blowing fans and induced draft fans of boilers account for approximately 30% of the plant electric consumption [1]. Thus, the fans in smoke exhaust systems should be targeted as an important candidate machinery for energy saving and emission reduction. The operation state of fans in the smoke exhaust systems also has significant effects on the safe and economic operation of a power plant.

According to the statistics of the Electric Power Research Institute (EPRI), 30%~50% of all power plant outage failures are caused by rotating machinery, such as steam turbine generators, fans and pumps. Based on the Chinese statistics of power plant operations in 2007, blower fan and induced draft fan failures are the second frequent cause of unscheduled shutdowns of thermal power plants, which causes significant

Citation: Xu, Xiaogang; Wang, Songling; Liu, Jinlian; Liu, Xinyu. 2014. "Intelligent Prediction of Fan Rotation Stall in Power Plants Based on Pressure Sensor Data Measured In-Situ." Sensors 14, no. 5: 8794-8809. DOI: 10.3390/s140508794

economic losses. Therefore, effective approaches for online diagnosis and active control of common faults in power plant fans are urgently needed to ensure the reliable operation of the plant. Avoiding operations of the fans in their close-to-fault modes will also improve the energy efficiency of the impeller machinery.

Rotation stalls are the most common cause of fan faults. One or several stall cells will form in the fan impeller passage when the flow of impeller machinery reduces to a certain limit. The occurrence of rotation stalls changes the stable flow field in the impeller, generates additional loads, and can even cause fatigue and fracture of the blades. If the rotation stall cannot be effectively contained, a flow surge may occur and lead to decreased unit efficiency and even vibrations of the fan body and connection pipes, which could possibly cause damages. Therefore, particular attention should be paid to the weak-stall mode of fans during operation, and active prediction and subsequent control of early-stage rotation stall in fans should be pursued.

The mechanism and detection of fan rotation stall have been actively studied in the industrial fan and compressor community. The mechanism of rotation stall has been investigated through signal processing methods or experimental approaches. For instance, Longley et al. [2] used spatial Fourier analysis of pressure signals in order to detect changes in the amplitude and phase of modal wave. McDougall et al. detected the transient process of rotation stalls through detailed measurements [3]. The continuous wavelet transform method was also introduced by Lin et al.for signal analysis of compressor stalls [4,5], allowing the analysis of the disturbance propagation process in the rotation stall development process of axial flow compressors.

For rotation stall detection, Cameron and Morris [6] proposed a cross-correlation method of circumferential space sensor signals, which effectively detected the abrupt onset of rotation stalls. Bianchi et al. [7] developed a stall detection system for induced draft fans of coal-fired power plants, in which a new methodology was proposed for the early detection of stalls in low-speed axial-flow fans used for tunnel ventilation. Teolis et al. [8] presented algorithms to detect stalls for gas turbine engines using data from eddy current sensors. Li et al. [5] enhanced the characteristics of unsteady rotor tip flows with the help of time- and scale-averaged wavelet spectra of a high-speed, single-stage compressor and a low-speed, three-stage compressor. Sheard et al. [9] presented a methodology involving a symmetrized dot pattern (SDP) technique that is capable of differentiating between stall conditions that constitute a mechanical risk and those that do not. Shiomi et al. [10] carried out an experiment in a semi-open propeller fan using a hot-wire anemometer, and two types of different periodical fluctuations whose frequencies were not the same as the rotor rotation frequency were found in the relatively high flow-rate region and low flow-rate region.

Most of the aforementioned research work targeted rotation stalls of the compressors of aero-engines, studied the unsteady characteristics of rotation stalls, and explored approaches for expanding the steady state operation and detecting stalls online. However, compared with the axial flow compressors, research on the rotation stalls of centrifugal fans is limited. There are significant differences in the rotation stall mechanisms between centrifugal fans and axial flow compressors [11]. More importantly, most techniques existing in the literature were focused on detecting the starting point of the rotation stall and cannot predict, ahead of a certain time period, the actual stall occurrence. This limits the application of these techniques in active control of fans to avoid rotation stalls. Detecting instability signs in the process of fan operations as early as possible and taking appropriate measures to prevent fan stalls are important for improving the efficient operation and avoiding unnecessary breakdowns of power plants. In addition, rotation stall prediction will also improve the energy efficiency of the fans, improve the efficiency of the thermal power plants, and thus reduce their CO_2 emissions.

This article presents a novel intelligent model, based on support vector regression machine (SVRM), for predicting the rotation stalls of centrifugal fans from pressure data measured online. Many experiments are performed under different running conditions on a centrifugal fan setup to obtain pressure signals of the fan operation process from normal state to rotation stall. The frequency characteristics of the pressure signals before and during rotation stalls are first analyzed using Fourier transform, and the pressure signals are then used to train the SVRM prediction model. The trained prediction model can accurately predict the fan pressure data 0.0625 s in advance, based on which the rotation stalls can be reliably detected. This model allows the active control of rotation stalls before their occurrence, which represents a major advantage over existing stall prediction techniques.

EXPERIMENTAL SETUP OF A CENTRIFUGAL FAN SYSTEM

The experiments were conducted on the 4-73No8D centrifugal fan, which is a common model from the 4–73 series widely used in Chinese thermal power plants. The experimental setup, as schematically shown in Figure 1A, consists of the centrifugal fan (nominal discharge coefficient: $\bar{q}_v = 0.205$; accuracy of speed adjustment: 0.3 rotations) attached to an alternating-current (AC) motor, inlet and outlet pipelines connected to the fan, an inlet flow divider between the inlet pipeline and the fan, and a signal conditioning unit connected with a computer. Five piezoresistive pressure sensors (measurement range: 20 kPa; accuracy: 10 Pa) were arranged on the inner surface of the fan casing with an angular spacing of 60° (Figure 1B), to measure the aerodynamic pressures of the air flow inside the fan. The pressure signals were recorded during the fan operation from a normal state to a rotation stall state, with a rotational speed of 1300 rpm. The experiments were repeated four times when the openings of the guide vanes were set to be 0°, 15°, 30° and 45°. The transition of the fan from the normal state to the rotation stall state was controlled by adjusting the resistance of the pipe network using a throttling cone installed in the outlet pipe. A photograph of the experimental setup is shown in Figure 1C. The signal conditioning unit (9118DG; ADLINK Technology, Taipei, Taiwan) was used to simultaneously acquire the pressure signals from the five pressure sensors at 320 Hz.

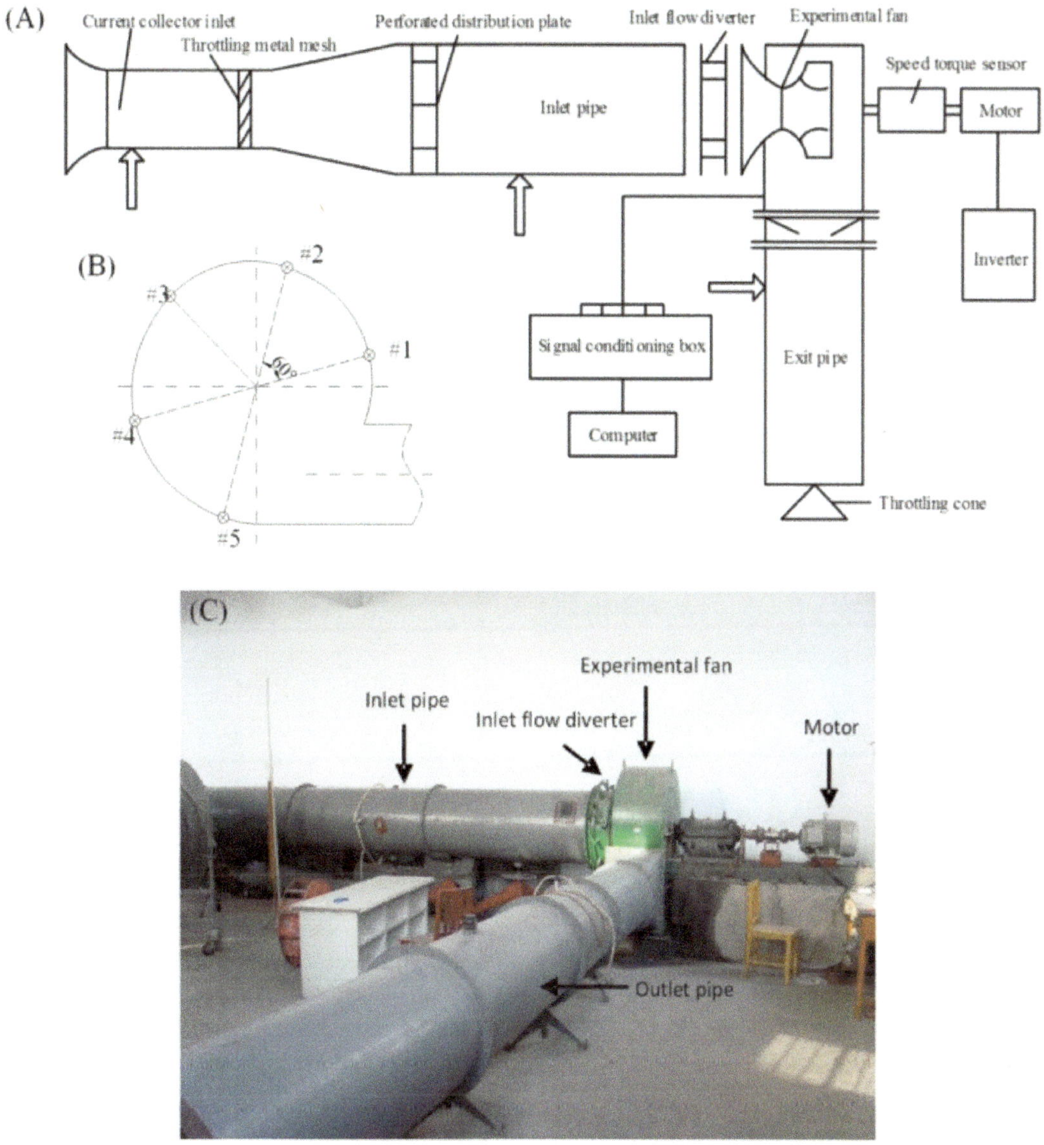

Figure 1: Experimental setup for internal pressure measurements of a centrifugal fan operated at normal conditions and under rotating stalls.

CHARACTERISTIC ANALYSIS OF PRESSURE SIGNALS FROM FAN NORMAL OPERATION AND ROTATION STALLS

Fourier Transform Frequency-Domain Analysis of Pressure Data

For pressure data analysis, the Fourier transform method was first adopted to extract the frequency distribution, the fundamental frequency, and the harmonic frequencies of the pressure data obtained in normal operations and fan rotation stall conditions. These results helped us understand the frequency characteristics of the fan rotation stalls. However, the Fourier transform results cannot reflect the gradual development of fan rotation stall, and thus cannot be used for rotation stall prediction. In recent years, with the advances in unsteady flow research, wavelet analysis has been applied to the time-frequency characteristic analysis of the mechanically unsteady flows of impellers. In this research, we applied the wavelet analysis to decipher the time-frequency characteristics of the pressure signals obtained during the transition process from normal operation to rotation stall, to quantitatively reveal the gradual development of rotation stalls, and to formulate a novel method for predicting the onset of rotation stalls.

The signals from all the five pressure sensors were analyzed using fast Fourier transform (FFT), and the FFT spectra were used to: (i) quantify the fundamental frequencies of the fan rotor and the rotating stall group, and (ii) reveal the graduate development of the stall group. Using the experiment with a full opening (0°) of the guide vane as an example, Figure 2 shows the FFT spectra of the pressure data from sensor #2 (Figure 1B) at different stages of the stall development. Pressure data obtained in 3.125 s (1000 data points) were used for FFT.

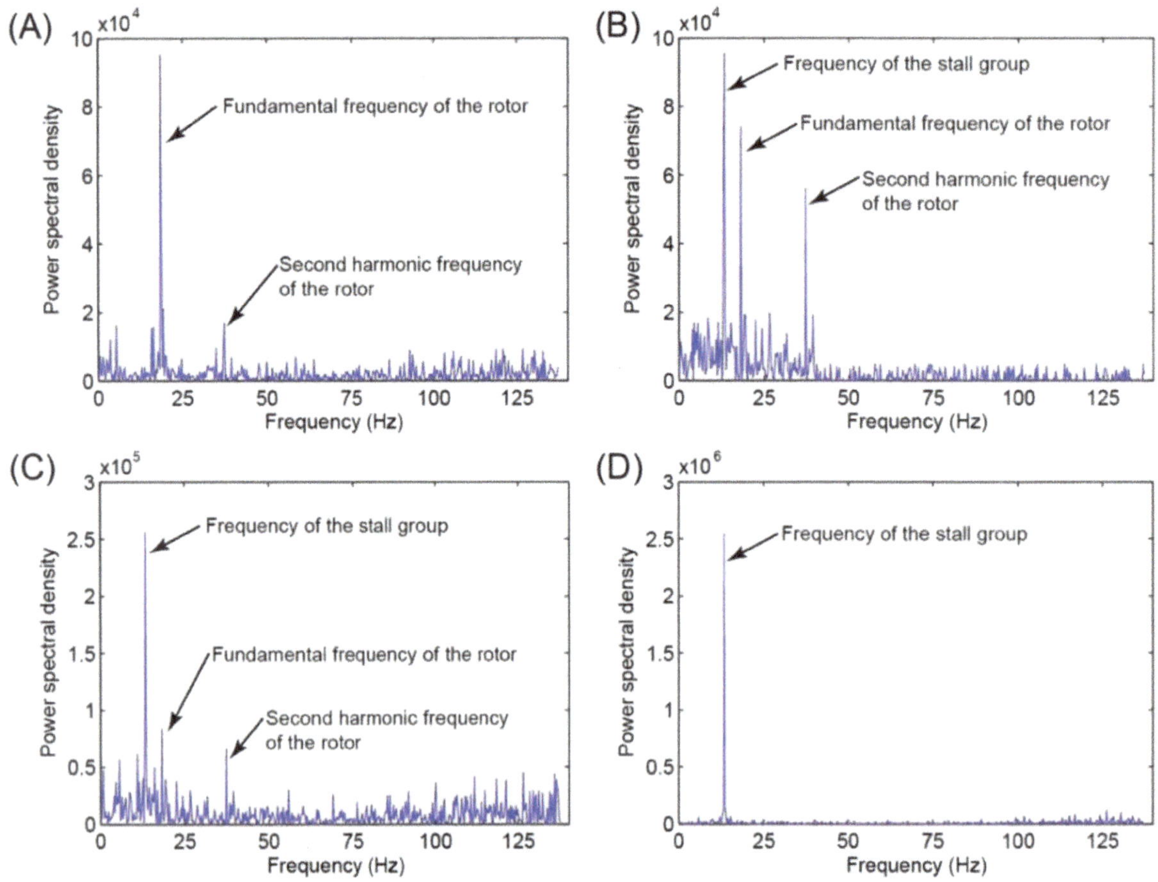

Figure 2: The spectrum diagram of pressure signal (from sensor #2) in the development of a rotation stall. From (**A**) to (**D**), the flow rate was gradually decreased by adjusting the throttling cone, corresponding a discharge coefficient ($\bar{q}_v$) of 0.172, 0.154, 0.151, 0.147, respectively. Starting from (**B**), the rotation stall occurred and developed from a weak stall in (**B**) into a strong stall cell in (**C**) and (**D**).

From Figure 2, one can make the follow observations; (1) Under the normal operation conditions, the spectra of the pressure signal mainly included the rotational (fundamental) and harmonic frequencies of the rotor; (2) With the decrease of the flow rate (indicated by $\bar{q}_v$), the fan deviated from the normal operating conditions into the stall region. As shown in Figure 2B, the weak stall group appeared in the spectra at 14.4 Hz (approximately 2/3 of the fan fundamental frequency of 21.7 Hz); (3) The weak stall regime corresponds to the range of low flow rates with small perturbations. With the further development of the rotation stall from the weak regime (Figure 2B) to the strong regime (Figure 2C,D), the magnitude (i.e., energy) of the stall group component increased substantially. During this transition, the flow perturbations, reflected in the pressure signals, were also magnified significantly.

Time-Frequency Analysis of Pressure Data Using Daubechies Orthogonal Wavelet

Through Fourier transform of the pressure signals, we have clearly observed the gradual development of the rotation stall and its frequency characteristics. However, the Fourier transform cannot reveal the time-domain characteristics of rotation stalls, and thus cannot be used for stall prediction. Differently, wavelet transform can identify signal components with different frequencies and the time associated with these frequencies, and has been widely applied to analysis of non-stationary signals [12]. The wavelet transform can decompose a time-domain signal into basis functions which are known as wavelets, and the decomposition process is computed separately for a series of frequency sub-domains, representing a multi-resolution analysis capability. To apply wavelet transform to signal analysis, we chose as the wavelet basis function the Daubechies orthogonal wavelet, which has the advantages of compact support, regularity, multi-resolution analysis ability, and high sensitivity to non-stationary signals. Daubechies wavelets are a family of orthogonal wavelets for discrete wavelet transform with a maximum number of vanishing moments [13].

In this work, the Daubechies orthogonal wavelet was utilized for multi-resolution decomposition of fan pressure signals and obtaining of the wavelet coefficients of the relevant frequency bands, which were used for detecting the rotation stall from the pressure data. The pressure signal (denoted by S) was decomposed, using Daubechies orthogonal wavelet, into the Nth-level approximation component (denoted by A_N, where N is the number of wavelet layers) and N number of detail components (denoted by D_i, i = 1, 2, ..., N) covering different frequency bands. The components A_N and D_N decomposed at different scales have different resolutions of time and frequency. The decomposition equation can be expressed by $S = A_N + D_N + \ldots + D_2 + D_1$. Changes in the frequency and amplitude of the pressure signal during fan operation were examined by analyzing a specific detail component D_j with a frequency band containing the rotation stall frequency of the pressure signal. Daubechies orthogonal wavelet is generally denoted by dbN where N = 1, 2, ..., 10, and the effective support length of the wavelet function and scale function of dbN is 2N−1. The vanishing moment order of the wavelet function is N. The bigger the value of N used, the longer and smoother corresponding wavelet will be. We selected the common db4 wavelet to analyze the dynamic pressure signals of the fan, which was proved, through decomposition trials, to be sufficient for the current application.

The pressure signals during the gradual development of fan rotation stall were obtained at 1300 RPM, and the opening of the guide vane was set to be 0°, 15°, 30° and 45°. The wavelet transform results of the pressure data from sensor #2 (guide vane opening: 15°), obtained during the rotating stall development, are shown in Figure 3. Daubechies wavelet transform of pressure signals from other sensors showed similar trends during rotation stall development. Because the sampling frequency was 320 Hz, the wavelet analysis frequency was set to be 160 Hz. The frequency band of the detail coefficient of the third layer (or third-layer detail coefficient) is 20–40 Hz, and the rotational frequency (21.7 Hz) of the fan lies in this band. The frequency band of the detail coefficient of the fourth layer (or fourth-layer detail coefficient) is 10–20 Hz, and the rotating stall frequency (14.4 Hz; obtained from the FFT analysis in Section 3.1) of the fan lies in this band.

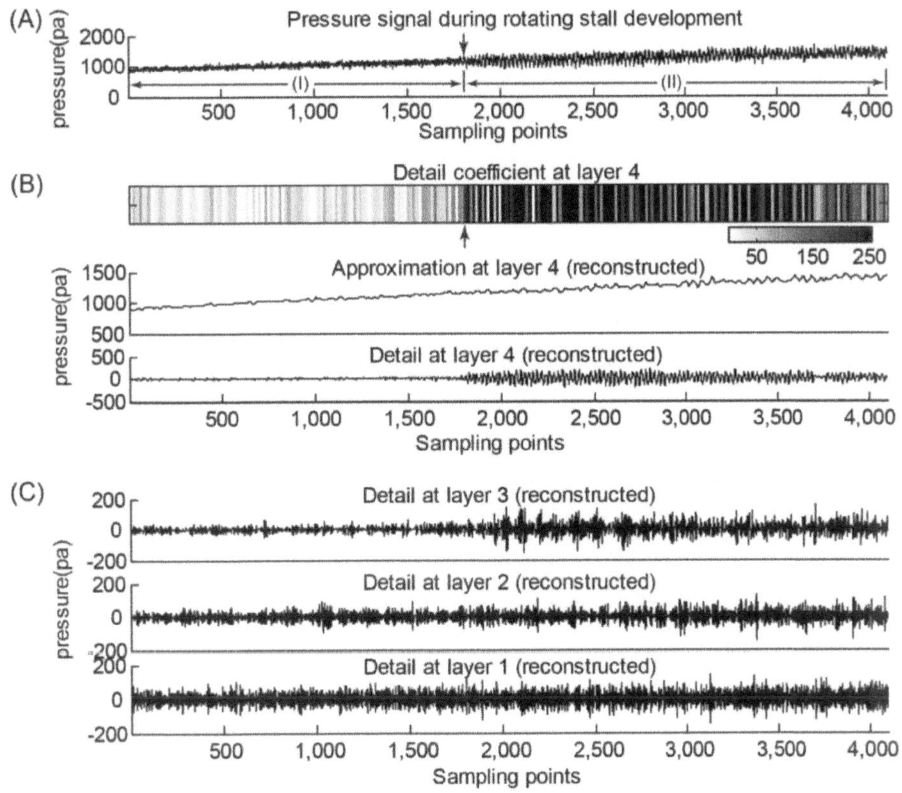

Figure 3: Fan casing pressure signal (from sensor #2) and its time-frequency map of wavelet analysis when βbeta; βequals; 15βdeg;. (**A**) Original pressure signal. (**B**) Decomposed detail component of the pressure signal at layer 4 (10–20 Hz). The detail coefficient at layer 4 indicates the energy of the pressure component caused by the fan rotation stall. (**C**) The decomposed detail components of the pressure signal at layers 1–3.

Thus, we used the four-layer detail coefficient as a quantitative indicator for detecting the onset of a rotation stall. For the pressure data from sensor #2 (guide vane opening: 15°), we empirically set a threshold of 200 Pa for the fourth-layer detail coefficient, and any pressure data point yielding a four-layer detail coefficient above 200 Pa indicates the presence of the rotation stall. This empirically-chose threshold was proven to be robust for other pressure data sets collected from sensor #2 at the guide vane opening of 15°. In Figure 3A, the arrow-labelled pressure data point (t = 5.6 s) was identified as the starting point of the rotation stall. One can observe that, during the development of rotation stalls, the internal pressure of the fan casing constantly increased with the gradual decrease of the flow rate and some decentralized energy components gradually emerged in the fourth-layer frequency band of the pressure signal, reflecting the onset of rotation stall. Two ranges (I) and (II) labelled in Figure 3A refers to the normal operation state and rotation stall stage, respectively.

THE SVRM MODEL FOR PREDICTING PRESSURE DATA IN THE DEVELOPMENT OF FAN ROTATION STALLS

At the initial stage of fan rotation stall, the formed stall groups are not stable and tend to circularly occur, rupture, disappear, and re-develop with changes of the flow field [11]. If the onsite of a rotation stall can be accurately predicted in advance, active control methods can be used to adjust the flow field conditions, thus avoiding the occurrence of rotation stalls, and guaranteeing the stable working operation of the fan [11,14].

To effectively predict the fan rotation stall, we proposed a support vector regression machine (SVRM) model for multi-step prediction of the pressure data. These predicted data were then used to predict the fan rotation stall via wavelet transform. For initial training of the SVRM model, sample data of the pressure signals were obtained during the transition process from normal operation to rotation stall. Since the SVM model can predict the pressure data of the fan by a certain time advance, the rotation stall can then be predicted rather than detected. This work was focused on the theoretical development and experimental validation of the SVRM prediction model, and the follow-up fan active control will be pursued in our future research.

The proper selection of a suitable prediction model is the key to ensure the precision and robustness of the rotation stall prediction. Existing regression models for data prediction mainly include support vector machines [15,16], neural networks [17,18], chaos theory [19,20], and so on. The SVRM has advantages over other prediction models in the following aspects: SVRM is a supervised learning method particularly powerful in addressing pattern recognition problems with small sample sizes, nonlinearity, and high dimensions. It has also been applied to function curve fitting and other machine learning problems. In comparison to neural networks, SVRMs have better generalizability in regression. Additionally, SVRMs dissociate the number of input vectors and the computational complexity. Moreover, the solution of SVRM models can be converted into the solution of convex optimization problems, which guarantees the global optimality of the algorithm and avoids local-minimum traps commonly existing in the neural network models.

For the training of SVRMs, the feature extraction and selection are crucial tasks. Feature extraction is difficult in many prediction applications, and feature selection is largely subjective and usually induces lack or redundancy of information. If we simply take the time series as the input vector for SVRM training, the input vector contains less information and will result in low accuracy of prediction. Therefore, we employed a phase space transformation approach based on chaos theory for preprocessing the time series and reconstructing their phase space which was then used as the input vector for SVRM training.

The chaos characteristics of the pressure data were first analyzed using chaos theory [19,21]. The SVRM prediction model was then set up by following steps: (1) To determine the input dimension and the training data set for the SVRM model. The embedding dimension and delay time are critical parameters to the phase space transformation, which were carefully selected to reconstruct the phase space of the time series; (2) To train the SVRM model using the training sample data selected in the step 1. In order to verify the accuracy and generalizability of the prediction model, the pressure signals of the rotating stall gradual process were taken as the training data when the opening of guide vane was 15°, and the pressure signals under the other openings of the guide vane were taken as the testing data.

The Phase Space Reconstruction of Pressure Signal

The evolution of any signal in a system is determined by its interactions with other signal components. Thus, the evolution of one signal includes the information of other related signals. The characteristic of the system can be extracted from a set of time series data of any signal in the system, and this characteristic can be represented by a trajectory in a high-dimensional space. Takens [22] proved that a space can be found to restore the original characteristic of the system using chaos theory.

Using x(1), x(2), ..., x(n) to denote a set of pressure data measured during the rotating stall development, the reconstructed phase space of the pressure signal obtained can be expressed by:

$$\begin{cases} X_1 = [x(1), x(1+\tau), \cdots, x(1+(m-1)\tau)] \\ X_2 = [x(2), x(2+\tau), \cdots, x(2+(m-1)\tau)] \\ \vdots \\ X_M = [x(M), x(M+\tau), \cdots, x(M+(m-1)\tau)] \end{cases} \tag{1}$$

where M is the dimension of the pressure signal phase space: M = n − (m − 1)τ, m is the embedding dimension, and τis the time delay. The training input for the prediction model is then X = [X_1, X_2, ..., X_M] and the training output is Y = [x(2 + (m − 1) τ), ..., x(n)].

The selection of the delay time and the embedding dimension is vital to successful phase space reconstruction of the pressure data. We employed the popular autocorrelation function method [23] to determine the delay time to be τ = 2 s, which was defined by the time corresponding to the minimal data points. To select an effective embedding dimension, we used the Cao method [24] which can effectively distinguish the difference between the random signals and deterministic signals. The embedding dimension was determined to be m = 3.

Training of the SVRM Prediction Model

Using the reconstructed phase space of the pressure time series (X = [X_1, X_2, ..., X_M]) as the input space, the training of the SVRM model becomes an optimization problem to find a function of the input vector, Y(X), which can accurately calculate (or predict) the next pressure sampling point y based on the reconstructed phase space of the previously measured pressure time series X_i. Based on the SVR theory, the function Y(X) has a general linear format:

$$Y(X) = (w^T \varphi(X) + b) \tag{2}$$

where φ is a kernel function, w is a weighing vector with the same dimension as φ, and b is a bias.

The function Y(X) is determined by solving the following optimization problem:

$$\min_{w,b,\xi} \frac{1}{2} w^T w + C \sum_{i=1} \xi_i \tag{3}$$

subject to $y_i(w^T \phi(x_i) + b) \geq 1 - \xi_i$, where $\xi_i \geq 0$.

Here, training vectors X_i are mapped into a higher dimensional space by the kernel function φ. SVRM finds a linear separating hyperplane with the maximal margin in this higher dimensional space. For the kernel function, we chose the widely used radial basis function (RBF) kernel:

$$\phi(x_i, x_j) = (\gamma x_i^T x_j + r)^d, \quad \gamma > 0 \tag{4}$$

Accurate determination of the parameters C and γ is critical to guarantee the regression effectiveness. We performed a grid search to find out the values of C and γ. Different combinations of (C, γ) values were iterated and the one yielding the best regression accuracy was selected. Exponentially growing sequences of C and γ (i.e., 2^{-10}, 2^{-9}, ... , 2^{10}) were used for iteration. To reduce the computation complexity, we first performed a coarse grid search to identify a region including the optimal combination of C and γ, and then conducted a fine grid search in that region. After C and γ were found, the parameters w, b, and ξ were calculated from Equation (3) and the function Y(X) (Equation (2)) was determined.

EXPERIMENTAL VALIDATION OF THE SVRM MODEL FOR ROTATION STALL PREDICTION

After being trained using the sample pressure data, the SVRM model is ready for rotation stall prediction. The major advantage of this SVRM prediction model is that it can provide accurately predicted pressure data ahead of a certain time, which enables prediction rather than detection of rotation stalls. This model can be combined with some popular time-domain stall detection techniques [25,26] for online stall prediction (that is, the SVRM model predicts pressure data and the stall detection technique determines the starting time point of the emerging stall). The scope of this paper is to establish the SVRM prediction model and validate the feasibility of using this SVRM model for rotation stall prediction. Thus, we opted to use offline wavelet transform as the stall detection method to detect stalls from both the measured and the predicted pressure data. As we will present in this section, the stall detection results (i.e., starting point of the stall), based on both measured and predicted data sets, agree well with each other. This proves the feasibility of using this SVRM model for rotation stall prediction. To achieve online stall prediction, one could adopt these time-domain stall detection techniques [25,26] to further process the predicted pressure data, in real time (computation time shown in Table 1), using the SVRM model.

With a fixed sampling frequency, the time advance by which the rotation stall can be predicted mainly depends on the number of prediction steps performed by the SVRM model. We performed pressure data prediction in single and multiple steps. In single-step prediction, we fed five measured pressure data points into the SVRM model to predict the next pressure data point. In multi-step prediction, we iterated the predicted pressure data points as part of the input data of the SVRM model for next prediction. The number of predicted data points used for next prediction was defined as the number of prediction steps. Multi-step predictions are desired to increase the time advance of the stall prediction. However, the prediction errors are accumulated in these multiple steps, and we must verify the maximum number of steps by which the SVRM model can predict the pressure data accurately enough to guarantee reliable stall prediction. In the experiments, we predicted the pressure data at 1–6 steps and demonstrate that the pressure data predicted by up to five steps are reliable for wavelet-transform-based stall prediction.

To further increase the time advance of stall prediction, we also increased the time interval of pressure data sampling by decreasing the sampling frequency from 320 Hz to 80 Hz. Since the rotational frequency of the centrifugal fan is 21.7 Hz, the new sampling frequency of 80 Hz still satisfies the sampling frequency theorem. The time-series pressure data were obtained from pressure sensor #2 at 80 Hz when the centrifugal fan was operated at 1300 rpm with guide vane openings of 0°, 15°, 30° and 45°. Two hundred and fisty pressure data points were obtained for each guide vane opening.

Single-Step Prediction

The time-series pressure data were first converted into their reconstructed phase space $X = [X_1, X_2, ..., X_M]$ usingEquation (1), which were then fed into the SVRM model to predict the pressure value of the next sampling point $Y' = x(i + 1 + (m - 1)\tau$. We generated the whole time series of the predicted pressure data using the measured pressure data, in which five measured pressure data points were used to predict the next data point. Using a personal computer (Pentinum Dual-Core CPU T4200@2.0 GHz; 2 GB RAM, Dell Computer, Round Rock, TEX, USA), the average computation time of the SVRM model was 0.00096 s.

Figure 4 shows the following results: (i) the raw measured pressure data from sensor #2 (at guide vane openings of 0°, 15°, 30° and 45°); (ii) the predicted pressure data by the SVRM model; and (iii) the detail coefficient of the fourth layer whose frequency band (10–20 Hz) includes the rotating stall frequency (14.4 Hz) obtained through wavelet transform. We investigated the fourth-layer detail coefficient and used a threshold of 200 as the indicator of rotation stall, and the data points labelled with an arrow (in the four-layer detail coefficient plot) are the predicted starting point of the rotation stall. One can observe that the predicted pressure data sets (at all four guide vane openings) lead to the same prediction results (i.e., the same starting points of rotation stall) as the measured pressure data sets. This demonstrates the feasibility of using our SVRM model for single-step rotation stall prediction.

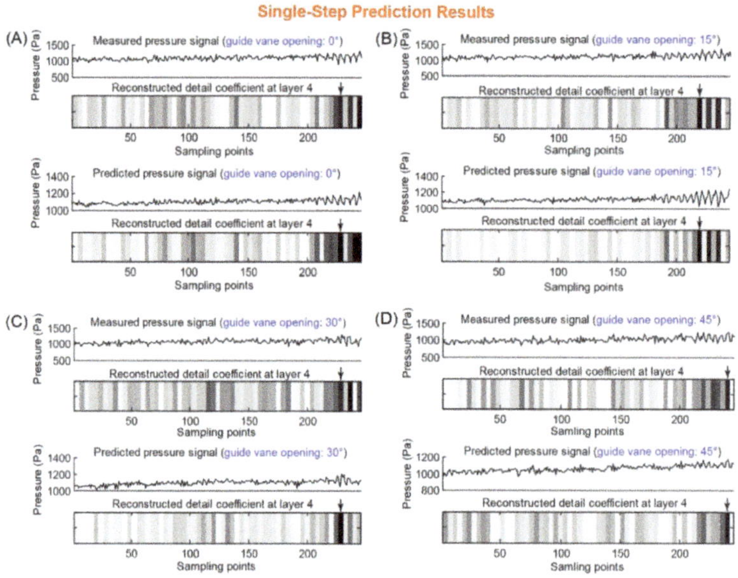

Figure 4: Single-step prediction results of the fan rotating stall at guide vane openings of 0°, 15°, 30° and 45° (panels **(A)**–**(D)**). In each panel (**(A)**–**(D)**), the experimental measured pressure signal and its wavelet analysis result (i.e., fourth-layer detail coefficient) are shown on the top, and the SVRM-predicted pressure signal and its wavelet analysis result (i.e., fourth-layer detail coefficient) are shown on the bottom. The arrows on the fourth-layer detail coefficient plot indicate the starting points of the rotation stall.

Multi-Step Prediction

The time advance of rotation stall prediction is the most important parameter of any stall prediction technique. We determined the longest time advance of our SVRM prediction model by examining the maximum number of prediction steps the model can perform. Multi-step prediction was performed by iterating the previously predicted pressure data points as part of the input vector of the SVRM model. The experimental results show that, by up to five prediction steps, the SVRM model can predict pressure data accurate enough to yield reliable wavelet-transform-based stall prediction.

Figures 5 and 6 show the results of five- and six-step predictions. One can observe that the five-step prediction (Figure 5) accurately predicted the starting point of the rotation stall for all four guide vane openings (compared with the stall detection results of the measured pressure data). However, for the six-step prediction (Figure 6), the predicted pressure data did not yield an accurate prediction of the starting point of the rotation stall (the predicted and actual starting points of the rotation stall are not the same in Figure 6A,B,D), which was mainly due to the intolerable accumulation of multi-step prediction errors. The five-step advance means a prediction time advance of 0.0625 s, corresponding to approximately two rotor revolutions for axial compressors and 10 for industrial fans. This means they are, according to typical rotating frequencies, ranging from $O(10^{-3}\ s)$ to $O(10^{-1}\ s)$ [11]. Thus, the time advance of our prediction model is fairly significant and enough to allow a rotation stall control mechanism to react and prevent the stall development.

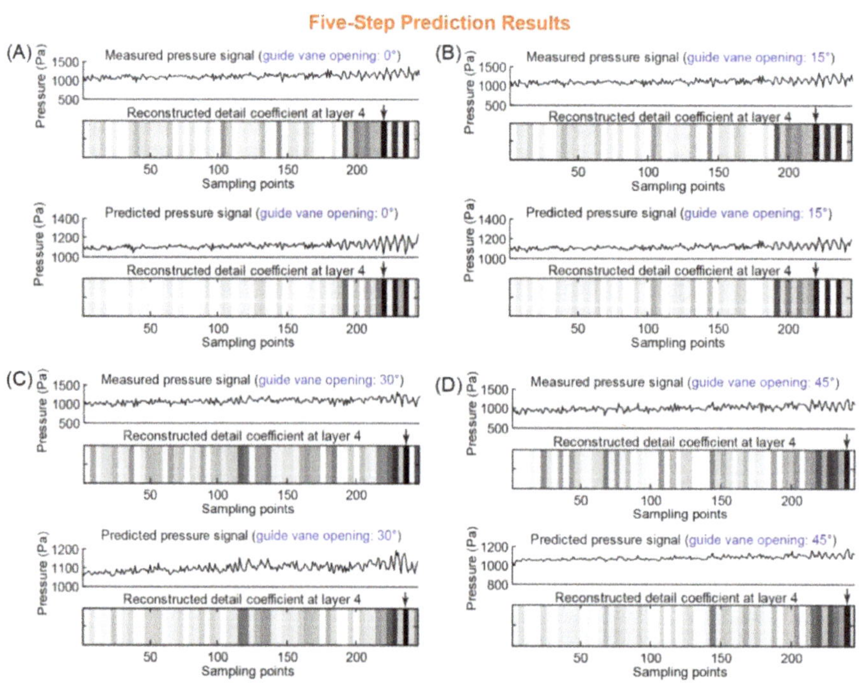

Figure 5: Five-step prediction results of the fan rotation stall at guide vane openings of 0°, 15°, 30° and 45° (panels **(A)**–**(D)**). In each panel (**(A)**–**(D)**), the experimental measured pressure signal and its wavelet analysis result (i.e., fourth-layer detail coefficient) are shown on the top,

and the SVRM-predicted pressure signal and its wavelet analysis result (i.e., fourth-layer detail coefficient) are shown on the bottom. The arrows on the fourth-layer detail coefficient plot indicate the starting points of the rotation stall.

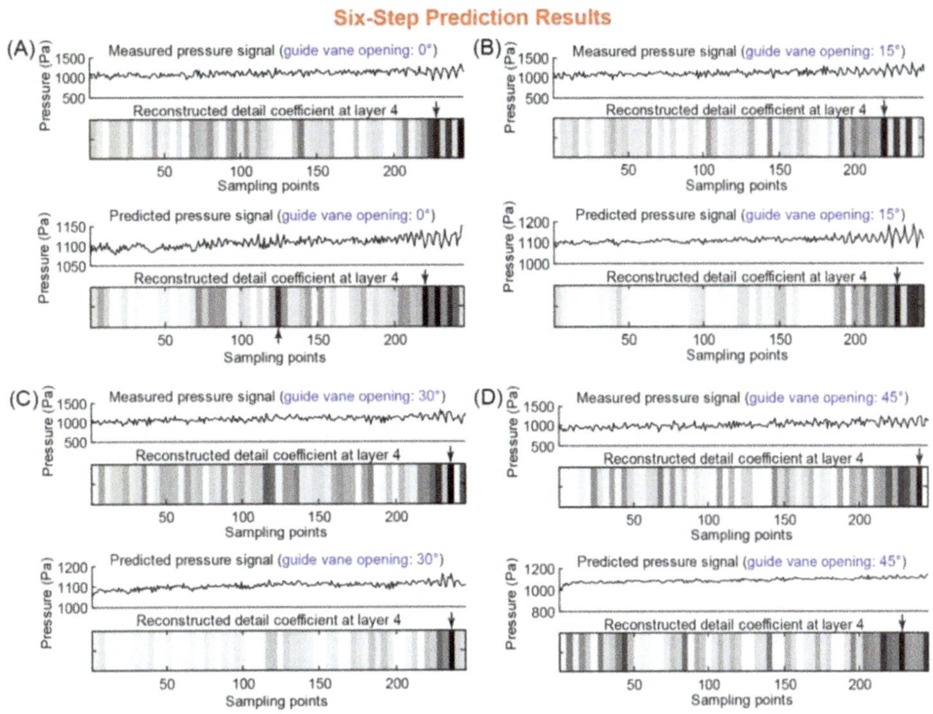

Figure 6: Six-step prediction results of the fan rotation stall at guide vane openings of 0°, 15°, 30° and 45° (panels (**A**)–(**D**)). In each panel ((**A**)–(**D**)), the experimental measured pressure signal and its wavelet analysis result (i.e., fourth-layer detail coefficient) are shown on the top, and the SVRM-predicted pressure signal and its wavelet analysis result (i.e., fourth-layer detail coefficient) are shown on the bottom. The arrows on the fourth-layer detail coefficient plot indicate the starting points of the rotation stall.

CONCLUSIONS

This paper presents the development and experimental validation of a SVRM model for intelligent prediction of rotation stalls of centrifugal fans in power plants. The internal pressure of the fan casing was used as the experimental data for rotation stall prediction. The SVRM prediction model was established using chaotic phase space transformation and a SVRM model, and was demonstrated to perform single and multi-step (up to five steps) prediction of the pressure data which can be further processed by real-time stall detection techniques. To validate the feasibility of using the SVRM model for rotation stall prediction, offline wavelet transform was performed for rotation stall prediction using both the measured and predicted pressure data sets. Through comparison of the stall prediction results obtained from the measured and predicted pressure data, we confirmed that the SVRM model is capable of predicting the pressure data by up to five steps and yielding reliable stall prediction. The effective time advance of stall prediction was 0.0625 s, which is significant enough for a stall control mechanism to react and suppress the stall development. The experimental studies were made on a 4-73No8D centrifugal fan, which is one of the most widely used centrifugal fans in Chinese power plants. We believe that this SVRM prediction model will find important applications in experimental rotation stall prediction and control of centrifugal and other types of fans, and therefore improve the power efficiency of combustion-based power plants.

ACKNOWLEDGMENTS

This work is supported by the Fundamental Research Funds for the Central Universities (12MS116).

Conflicts of Interest

The authors declare no conflict of interest.

REFERENCES AND NOTES

1. Wang, S.; Lei, Y.; Li, C.; Huang, X. Numerical analysis of cubage loss of centrifugal fan in power plant. J. North China Electric Power University. 2006, 1, 1–13. .
2. Longley, J.P. Inlet Distortion and Compressor Instabilities; Cambridge University: Cambridge, UK, 1988. .
3. McDougall, N.M.; Cumpsty, N.A.; Hynes, T.P. Stall inception in axial compressors. J. Turbomach. 1990, 112, 116–123. .
4. Lin, F.; Chen, J.; Li, M. Wavelet analysis of rotor-tip disturbances in an axial-flow compressor. J. Propuls. Power2004, 20, 319–334. .
5. Feng, L.; Jingyi, C.; Meilin, L. Practical issues of wavelet analysis of unsteady rotor tip flows in compressors. Proceedings of the 38th aiaa/asme/sae/asee Joint Propulsion Conference & Exhibit, American Institute of Aeronautics and Astronautics, Indianapolis, IN, USA, 7–10 July 2002.

6. Cameron, J.D.; Morris, S.C. Spatial Correlation Based Stall Inception Analysis. In ASME Turbo Expo 2007: Power for Land, Sea, and Air; ASME: Montreal, Quebec, Canada, 2007. .
7. Bianchi, S.; Corsini, A.; Sheard, A.G. Demonstration of a stall detection system for induced draft fans. Proc. Inst. Mech. Eng. Part A: J. Power Energy 2013, 227, 272–284. .
8. Teolis, C.; Gent, D.; Kim, C.; Teolis, A.; Paduano, J.; Bright, M. Eddy current sensor signal processing for stall detection. Proceedings of the Aerospace Conference, 2005 IEEE, Big Sky, MT, USA, 5–12 March 2005; pp. 3479–3495.
9. Sheard, A.G.; Corsini, A.; Bianchi, S. Stall warning in a low-speed axial fan by visualization of sound signals. J. Eng. Gas Turbines Power 2010, 133, 041601–041601. .
10. Shiomi, N.; Kinoue, Y.; Jin, Y.-Z.; Liu, P.; Setoguchi, T. Unsteady flow at low flow-rate region in a semi-open propeller fan (velocity fluctuation outside of blade tip). J. Therm. Sci. 2011, 20, 218–223. .
11. Bianchi, S.; Corsini, A.; Sheard, A.G.; Tortora, C. A critical review of stall control techniques in industrial fans.ISRN Mech. Eng. 2013, 2013, 1–18. .
12. Daubechies, I. The wavelet transform, time-frequency localization and signal analysis. Inf. Theory, IEEE Trans.1990, 36, 961–1005. .
13. Daubechies, I. Ten Lectures on Wavelets; Society for Industrial and Applied Mathematics: Philadelphia, PA, USA, 1992. .
14. Paduano, J.D.; Epstein, A.H.; Valavani, L.; Longley, J.P.; Greitzer, E.M.; Guenette, G.R. Active control of rotating stall in a low-speed axial compressor. J. Turbomach. 1993, 115, 48–56. .
15. Drucker, H.; Burges, C.J.; Kaufman, L.; Smola, A.; Vapnik, V. Support vector regression machines. Adv. Neural Inf. Process. Syst. 1997, 9, 155–161. .
16. Smola, A.J.; Schölkopf, B. A tutorial on support vector regression. Stat. Comput. 2004, 14, 199–222. .
17. Specht, D.F. A general regression neural network. Neural Netw. IEEE Trans. 1991, 2, 568–576. .
18. Cheng, B.; Titterington, D.M. Neural networks: A review from a statistical perspective. Stat. Sci. 1994, 9, 2–30. .
19. Sprott, J.C. Chaos and Time-Series Analysis; Oxford University Press: Oxford, UK; New York, NY, USA, 2003; p. 507. .
20. Farmer, J.D.; Sidorowich, J.J. Predicting chaotic time series. Phys. Rev. Lett. 1987, 59, 845–848. .
21. Haykin, S.; Xiao, B.L. Detection of signals in chaos. Proc. IEEE 1995, 83, 95–122. .
22. Takens, F. Detecting Strange Attractors in Turbulence. In Dynamical Systems and Turbulence, Warwick 1980; Rand, D., Young, L.-S., Eds.; Springer: Berlin/ Heidelberg, Germany, 1981; Volume 898, pp. 366–381. .
23. Xu, P. Differential phase space reconstructed for chaotic time series. Appl. Math. Modell. 2009, 33, 999–1013. .
24. Cao, L. Practical method for determining the minimum embedding dimension of a scalar time series. Phys. D Nonlinear Phenom. 1997, 110, 43–50. .
25. Dhingra, M.; Breeze-Stringfellow, A.; Shin, H.-W.; Neumeier, Y.; Prasad, J.V.R.; Szucs, P.N. A stochastic model for a compressor stability measure. J. Eng. Gas Turbines Power 2006, 129, 730–737. .
26. Tryfonidis, M.; Hendricks, G.J.; Etchevers, O.; Paduano, J.D.; Epstein, A.H. Prestall behavior of several high-speed compressors. J. Turbomach. 1995, 117, 62–80. .

CHAPTER 9

MODELING AND OPTIMIZATION OF THE MEDIUM-TERM UNITS COMMITMENT OF THERMAL POWER

Shengli Liao, Zhifu Li, Gang Li , Jiayang Wang and Xinyu Wu

Institute of Hydropower System & Hydroinformatics, Dalian University of Technology, Dalian 116024, China

ABSTRACT

Coal-fired thermal power plants, which represent the largest proportion of China's electric power system, are very sluggish in responding to power system load demands. Thus, a reasonable and feasible scheme for the medium-term optimal commitment of thermal units (MOCTU) can ensure that the generation process runs smoothly and minimizes the start-up and shut-down times of thermal units. In this paper, based on the real-world and practical demands of power dispatch centers in China, a flexible mathematical model for MOCTU that uses equal utilization hours for the installed capacity of all thermal power plants as the optimization goal and that considers the award hours for MOCTU is developed. MOCTU is a unit commitment (UC) problem with characteristics of large-scale, high dimensions and nonlinearity. For optimization, an improved progressive optimality algorithm (IPOA) offering the advantages of POA is adopted to overcome the drawback of POA of easily falling into the local optima. In the optimization process, strategies of system operating capacity equalization and single station operating peak combination are introduced to move the target solution from the boundary constraints along the target isopleths into the feasible solution's interior to guarantee the global optima. The results of a case study consisting of nine thermal power plants with 27 units show that the presented algorithm can obtain an optimal solution and is competent in solving the MOCTU with high efficiency and accuracy as well as that the developed simulation model can be applied to practical engineering needs.

INTRODUCTION

In the past several decades, China's power system has experienced rapid development along with rapid economic growth. Among all electric power sources, thermal power, which mainly consists of coal-fired plants, accounts for the largest proportion of China's power system. By the end of 2012, China's total installed generation capacity reached 1146.76 GW, of which 819.68 GW (approximately 71.5%) was from thermal power, and the total electric energy production reached 4986.5 TWh, with 3925.5 TWh (approximately 78.7%) coming from thermal power. Simultaneously, the utilization hours of the installed capacity for thermal power plants are larger than those of hydro power or wind power. The statistical results of 2012 showed that the annual utilization hours of

Citation: Liao, S.; Li, Z.; Li, G.; Wang, J.; Wu, X. Modeling and Optimization of the Medium-Term Units Commitment of Thermal Power. Energies 2015, 8, 12848-12864. DOI: 10.3390/en81112345

power plants with an installed capacity over 6 MW are 4982 h for thermal power, 3591 h for hydropower and 1929 h for wind power [1,2,3,4,5]. However, because the preheating and cooling of the stream-turbine requires more time than other resource units, the coal-fired thermal unit is inflexible between start-up and shut-down resulting in a slow response to system load demands. Moreover, the start-up and shut-down of coal-fired thermal units cause high fuel costs [6,7,8,9]. Generally, thermal plants are usually scheduled to meet base load demands to ensure that the generation process runs smoothly and reduce the start-up and shut-down times of thermal units. Furthermore, the different values of power system demand become larger among adjacent days due to various factors, including heavy weather, minor vacation, and transition periods between dry and wet season for hydro power [10,11]. This situation causes remarkable difficulties in preparing the generation schedule for power systems, especially for ones with numerous thermal plants. Consequently, the power dispatch centers of China are in charge of the medium-term optimal commitment of thermal units (MOCTU), which usually span from one month to half a year and take one day as the time interval in calculations to obtain the numbers and times of the start-up and shut-down units for every thermal plant each day before actual short-term and daily-ahead scheduling.

The objective of MOCTU is to find the optimal set of thermal generating units or the boot capacity (sum capacity of units that putting into operation in the first operation of a plant) scheme of a power system to satisfy the system load demand, operational restrictions, reliability constraints, and security requirements in each time period. Thus the MOCTU is a form of unit commitment (UC) problem [8,9,12]. In contrast with short-term UC in power systems, which involves determining a start-up and shut-down schedule of units to meet the required demand during a short-term period such as one day or one week [13,14], MOCTU is responsible for dispatching load to each plant for maintenance scheduling and hydrothermal coordination. The main difference between medium-term and short-term thermal UC optimization is that medium-term optimization takes each plant's operating capacity or unit numbers as decision variables instead of scheduling a single unit by short-term operation [15]. It is well-known that the utilization rate of power generation equipment reflects the generation ability of a thermal plant, and the standby condition of generation equipment reflects the ability of a thermal plant to deal with sudden accidents, and utilization hours of the installed capacity for thermal power plants can reasonably reflect the two abovementioned factors, so that operators can operate and manage plants better. Hence, many power grids take equal accumulated operating hours of installed capacity for all thermal power plants as the primary objective of MOCTU during the selected period. However, due to thermal power's main characteristics, including slow start-up and shut-down, complex constraints, large-scale as well as large complicated demands by power grid, developing and establishing computer simulation and optimization models for thermal UC optimization have been most challenging and complex issues [16,17,18]. Furthermore, to achieve the best possible optimal solution and the largest benefit for thermal power plants and power grid, the model should be simulated as close to reality as possible to suit the practical needs of power dispatch centers, especially in China [19,20,21]. However, according to our retrieval results, there are few studies on MOCTU, especially on the development of simulation models for MOCTU that meet the demands of practical engineering [22,23]. In this paper, a medium-term thermal UC optimization model with equal capacity utilization hours for all thermal power plants is established. The model not only respects the basic principles of electric power dispatch (equality, impartiality and transparency), but also adopts a reward principle that gives extra operation hours to the thermal plants with low energy consumption and dust discharge and high efficiency, namely reward hours.

The work described in this paper is originated from the real, practical demands for power dispatch centers of China to develop a medium-term optimal commitment of thermal units. The aim of this paper is to develop a simulation model for MOCTU to satisfy the needs of practical engineering and present a feasible and effective algorithm to optimize the model. MOCTU is a multi-stage decision problem that involves a highly nonlinear and computationally expensive objective function with a large number of constraints. The progressive optimality algorithm (POA) has been shown to be an effective method for solving multi-stage optimization problems by decomposing a multi-stage decision problem into a series of non-linear programming two-stage problems [24,25,26,27,28], and it is suitable for solving MOCTU. However, it is a difficult task to find feasible solutions for a large-scale MOCTU problem using POA due to its drawback, namely the easily encountered local optimum for complex problems. Therefore, in this paper, an improved progressive optimality algorithm (IPOA) is proposed for a large-scale MOCTU problem with nine thermal plants and 27 units in Yunnan Province of China to validate the effectiveness and practicality of the developed simulation model as well as to improve the quality of optimal solutions of POA. The original optimization problem is first optimized by using a heuristic method and an initial feasible solution is obtained, then POA is adopted to search the optimal solution, and finally two strategies are utilized to adjust the solution from the constraint boundaries into the feasible zone's interior to continue to search for the global solution. Actually, the mid-term boot scheduling plan of thermal plants has been generated by operators in Yunnan power grid (YNPG) using this proposed simulation model and method since 20 May 2013. Over a two-year implementation period, it has given operators more confidence to use it.

This paper is organized as follows: Section 2 gives the formulation of the mathematical model of MOCTU; Section 3analyzes the characteristics of this problem and presents the IPOA used to solve it; Section 4 applies the model to a real-world scenario with nine thermal plants and 27 units in Yunnan province of China; Finally, Section 5 presents the conclusions.

MATHEMATICAL MODEL

Objective Function

MOCTU is an important and complex task for medium-term scheduling of power grids. On the one hand, reasonable MOCTU results should not only satisfy the complicated operation requirements of thermal plants, but respond to system load changes rapidly. Furthermore, they should meet the requirements of the national energy conservation policy as well as balance the interests between different thermal plants. In actual management and operation of thermal power in China, utilization hours of installed capacity for thermal plants can reflect the utilization rate of power equipment and standby application of thermal plants. Thus, controlling the utilization hours of installed capacity for thermal plants has always been the goal that needs to be considered in MOCTU for many power grids. Based on the abovementioned information, taking the equal utilization hours of installed capacity for thermal plants as the optimization goal, and at the same time, the concept of reward-hours is introduced in the model, though which power grids can reward stations that with small coal consumption flexibly according to energy saving principles. Assuming the total number of plants participating in the calculation is m, the objective function can be expressed as follows:

$$h_i = h_j \ (i \neq j) \tag{1}$$

where h_i and h_j are the accumulated operating hours of installed capacity for plants i, j, respectively, in the selected period.

Although the above objective function (Equation (1)) is conceptually clear, it is very difficult to solve directly because the objective formulation is related to all plants and the target result cannot be obtained by computing a single mathematical expression. Considering the fact that variance can be used to describe the dispersion degree of data series, the original objective function is replaced by minimizing the variance of vector $h=[h_1,h_2,\cdots,h_i,\cdots,h_m]$, which is composed of plants' capacity utilization hours. The replacement function can be represented as:

$$\min f = \frac{1}{m}\sum_{i=1}^{m}(h_i - \overline{h})^2 \tag{2}$$

$$\overline{h} = \frac{1}{m}\sum_{i=1}^{m} h_i \tag{3}$$

$$h_i = h_i^P + h_i^T - h_i^G \tag{4}$$

$$h_i^T = \frac{1}{N_i} \times \overline{C_i} \times T \times 24 \times r_i \tag{5}$$

$$\overline{C_i} = \frac{1}{T}\sum_{t=1}^{T} C_i^t \tag{6}$$

where the following notations are used:

i, j	plant i and plant j, $i, j = 1, 2, \cdots m$;
T	Number of time steps. $t = 1, 2, \cdots, T$;
h_i	Accumulated operating hours of installed capacity of plant i in the selected period;
h_i^P	Operating hours of installed capacity of plant i accumulated in pre-calculation period;
h_i^T	Operating hours of installed capacity of plant i accumulated in the calculation period;
h_i^G	Extra award hours assigned to plant i;
N_i	Installed capacity of plant i;
$\overline{C_i}$	Average operating capacity of plant i in the calculation period;
r_i	Given load factor of plant i;
C_i^t	Operating capacity of plant i in period t. It is also the decision variable to be solved.

In this model, the values of h_i^P and h_i^G are given and do not require special treatment while solving the problem. It should be noted that, under the condition of h_i^P being unequal to h_i^G, to equalize h_i, h_i^T has to be made unequal. Thus, it realizes the differentiation of capacity utilization hours of all thermal power plants in the calculation and ensures the flexibility of the model.

Constraints

For MOCTU, each individual thermal plant is subjected to its own set of constraints, while the power system is subjected to system power balance constraints. Specifically, we consider the following constraints:

(1) Unit number constraints:

$$\underline{n_i^t} \leqslant n_i^t \leqslant \overline{n_i^t} \tag{7}$$

where nti— and nti‾ are the minimum and maximum unit number of plant i in period t, respectively; nti is the active unit number of plant i in period t.

(2) System power balance constraints:

$$C_S^t \times \underline{r} \times \Delta h_t \leqslant P_D^t \times \Delta h_t \leqslant C_S^t \times \overline{r} \times \Delta h_t \tag{8}$$

Where C_S^t is the operating capacity of the system in period t in MW; Δh_t is the number of hours in period t in h, and here $\Delta h_t = 24$; P_D^t is the load demand of the system in period t in MW; $\underline{r}$ and $\overline{r}$ are, respectively, the minimum and maximum load factors of thermal power, which are calculated from the long-term real-world operation of power grid, $\underline{r}, \overline{r} \in [0,1]$, and $r_m = (\underline{r} + \overline{r})/2$ is recorded as the mean-value of the load factor of thermal power.

(3) Peak and valley duration constraints:

The medium-term thermal UC optimization only dispatches system load to plants, and there is no need to distribute load among units. Therefore, we transform minimum up- and down-time constraints into minimum peak and valley time constraints:

$$\begin{cases} T_i^{\text{up}} \geqslant \underline{T}_i^{\text{up}} \\ T_i^{\text{down}} \geqslant \underline{T}_i^{\text{down}} \end{cases} \tag{9}$$

Where T_i^{up} is the minimum duration time during the peak in the operating capacity process of plant i in calculation period; $\underline{T}_i^{\text{up}}$ is the limit value of the minimum duration time in the peak in the operating capacity process of plant i in the calculation period; T_i^{down} is the minimum duration time in the valley period in the operating capacity process of plant i; $\underline{T}_i^{\text{down}}$ is the limit value of the minimum duration time in the valley period in the operating capacity process of plant i.

(4) Boundary condition constraints:

The plants' generating scheme is not only related to the calculation periods, but also relevant to the boot capacity scheme in previous periods, namely pre-calculation periods, where the start-up mode has been determined, because the boot capacity scheme in the pre-calculation period has an impact on the generating scheme in current periods. Therefore, the target start-up mode in calculation periods should link up with existing p days start-up mode in calculation periods. The relationship among all periods is shown in Figure 1.

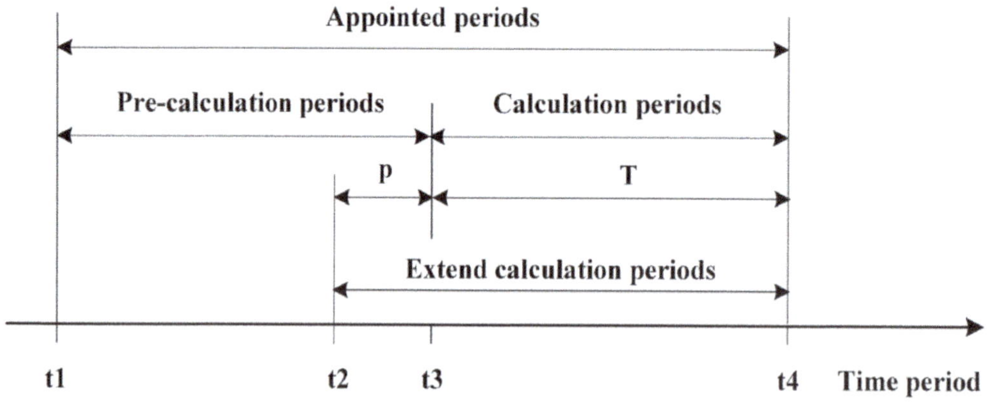

Figure 1: Sketch map of the time relationship.

Note that the actual unit number and operating capacity of a plant are variables to be calculated, and the others are given input data.

MODEL SOLUTION

Solution Approach

MOCTU is a form of the UC problem with a large number of constraints, including unit number constraints (Constraint Group 1), system power balance constraints (Constraint Group 2), peak and valley duration constraints (Constraint Group 3), and boundary condition constraints (Constraint Group 4). It is obvious that Constraint Groups 1 and 2 are single-period constraints that can be treated to reduce the search space. Although Constraint Groups 3 and 4 are multi-period constraints, Constraint Group 4 can be satisfied when Constraint Group 3 is satisfied in extended calculation periods. On the other hand, MOCTU is a high dimensional problem to be optimized. For a power grid with two thermal plants, each having three units. For one plant with three units, it has (3 + 1) kinds of combination in one period, and for two plants, it reaches $(3 + 1)^2$, so the total number of unit combinations will reach $[(3 + 1)^2]^{30} \approx 1.329\times10^{36}$ for a horizon of one month with 30 periods, and the overhead of optimization will be computationally expensive. Considering the complex constraints, especially Constraint Group 3, the solving process can be divided into two procedures:

Procedure I: Obtaining the feasible solution space $S=[S_1,S_2,\cdots,S_t,\cdots,S_T]$, where St represents all of the possible unit-commit combinations that satisfy Constraint Groups 1 and 2 in period t. First, combine all units of all plants to obtain all possible unit-commit combinations in each period. Second, obtain the solution space S by filtering out the combinations that cannot satisfy Constraint Groups 1 and 2 period-by-period. The solution for the UC scheme in the calculation period can be acquired from the unit-commit combinations in S. The above process is relatively simple.

Procedure II: Search for the global optimal solution that can meet Constraint Group 3. Because the solution in S from Procedure I already satisfies Constraint Groups 1 and 2, and as well as Constraint Group 4 being automatically satisfied after satisfying Constraint Group 3, the aim of this search process is to find the solution that satisfies Constraint Group 3. The following discussion will introduce how to find the optimal scheduling that meets Constraint Group 3 in the solution space S. First, generate an initial feasible solution in S that satisfies Constraint Group 3 by using heuristic method. Second, produce the global optimal solution by the POA. However, POA is sensitive to initial trajectories and sometimes cannot guarantee a global optima, which is the reason that restricts its application. Thus, POA easily converges to a local optimum while being directly used for MOCTU, which is a strong constrained optimization problem. For the medium-term thermal UC optimization, a local optimum means that the system power reaches the upper or lower limit boundary constraints or that the duration periods of peak and valley in the operating capacity process are equal to the limit value of the minimum duration periods. In this case, it is very difficult to further optimize the result. Therefore, an improved POA combining POA with two adjustment strategies is presented to overcome the mentioned demerits.

Initial Feasible Solution Generation

As Procedure I has achieved the solution space S that meets Constraint Groups 1 and 2, the initial feasible solution can be acquired by searching the solution space S by heuristic method only when the operating load rate approximates to the given target load rate that obtained from long-term schedule of power dispatch centers in each period. The feasible region in the latter period will be sharply contracted when the generating scheme in the previous period is determined and the duration period constraints (Constraint Group 3) are considered. To avoid the infeasible region, this paper suggests r_m as the target load rate. Before searching, sort the generating schemes in St in every period according to the ascending order of the index value $|r_k-r_m|$. Where r_k represents the operating load rate of the kth generating scheme in St, and its formula is as follows:

$$r_k = \sum_{i=1}^{m} C_i^k / \sum_{i=1}^{m} N_i \quad (10)$$

The generating scheme S't, which is sorted according to the ascending order of the index value $|r_k-r_m|$ in every period, can be obtained by using Equation (10). Then, the feasible solution can be obtained from the sorted solution space $S' = [S'_1, S'_2, \cdots, S'_t, \cdots, S'_T]$ by the heuristic method. The detailed procedure is as follows:

- Step 1: Construct an integer array ks of length T. Set ks[1] = 1 and t = 1. Make sure that the combination of boot capacity in period 1 is the first element of the solution space S'_t.
- Step 2: Set t = t + 1. If t > T, go to Step 6. Otherwise, set ks[1] = 1 and set the combination of boot capacity in period t as the first element of the solution space S'_t.
- Step 3: Verify whether the start-up mode lying in the interval of [0,1] satisfies the constraints of Equation (3). If it does satisfy them, go to Step 2; otherwise, go to Step 4.
- Step 4: Set ks[t] = ks[t] + 1. If ks[t] is greater than the number of elements in solution space S'_t, go to Step 5. If not, replace the combination of boot capacity in period t with the ks[t]th element of the solution space S'_t and go to Step 3.
- Step 5: Set t = t + 1, and go to Step 1.
- Step 6: Output the result.

Optimization Process of Progressive Optimality Algorithm (POA)

Considering the complex requirements by power grids and the constraints mentioned above, the present optimization problem exhibits multi-stage, large-scale and high dimension characteristics. A suitable solution algorithm is required for solving the MOCTU. Based on Bellman's Principle, the POA, which is proposed by Howson and Sancho for reducing dimensionality difficulties by decomposing a multi-stage decision problem into a series of non-linear programming two-stage problems [29], has been shown to have great advantages over classical optimization methods as one of the most widely used techniques for hydroelectric generator scheduling and water resources problems [30,31]. The advantages of POA over other optimization techniques are that it can decompose a multi-state decision problem into several nonlinear programming sub-problems to reduce the dimensionality. More specifically, the merits of POA have clearly been elaborated in [32], including no need to discretize the state variables, no resolution or linearization of nonlinear objective functions and constraints and minimal storage requirements.

To understand the main principle of POA, a general multi-stage optimization example was given. In this example, a set of state variables $x_0^*, x_1^*, x_2^*, \cdots, x_{T-1}^*, x_T^*$ is given to determine the optimal solution by minimizing objective function $f(x_0, x_1, x_2, \cdots x_T)$, where x_0 and x_T are given as the initial state values. The algorithm starts with an initial trajectory $x_0, x_1^0, x_2^0, \cdots, x_{T-1}^0, x_T$ which is gained in a certain way. Then, this multi-stage decision problem would be decomposed into two problems, each of which minimizes $f(x_{i-1}^j, x_i^j, x_{i+1}^j)$ to obtain optimal $x_i^{j,*}$ at stage i during the jth iteration by fixing values of x_{i-1}^j and x_{i+1}^j. In other words, x_0^j and x_2^j are fixed to optimize x_1^j to yield $x_1^{j,*}$, as shown in Figure 2a. Then $x_1^j = x_1^{j,*}$ and x_3^j are fixed to optimize x_2^j for $x_2^{j,*}$, as shown in Figure 2b, in turn, the jth iteration isn't finished until $x_{T-1}^{j,*}$ is obtained, as shown in Figure 2c,d. Based on the optimized results from the last iteration, another iteration is restarted again. The process isn't over until the difference between the last two iterations meets the predefined precision limit. Figure 2 illustrates the optimization process during the jth iteration.

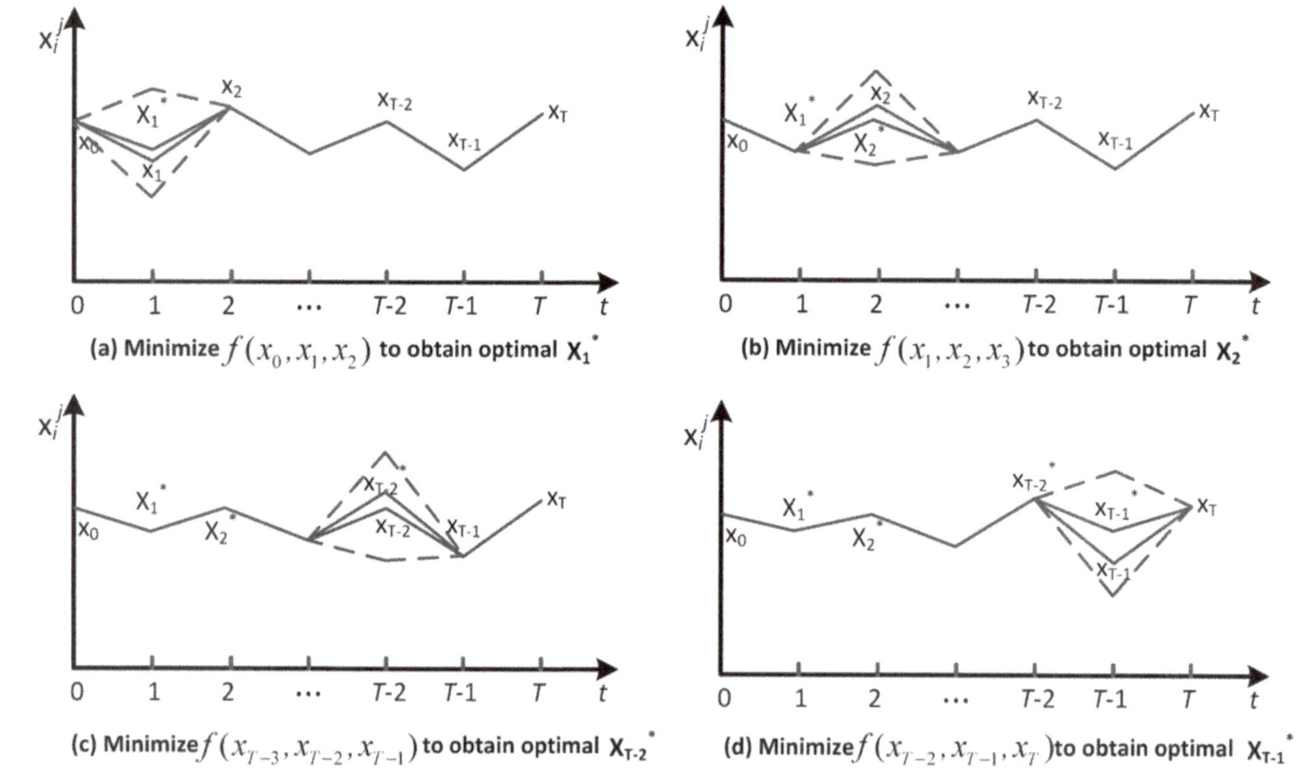

Figure 2: Optimization process of progressive optimality algorithm (POA).

The initial feasible solution obtained in Section 3.3 is used for POA to search for optimal solution. The general process can be mainly described in the following steps:

- Step 1: Set the generating scheme to an initial feasible solution generated by the heuristic search and an initial objective function value value. Set t = 1 and k = 1.
- Step 2: Obtain a generating scheme by replacing the generating scheme in period t with the kth element of the solution space St. If the new generating scheme meets Constraint Group 3, go to Step 3. If not, go to Step 4.
- Step 3: Calculate the objective function value value′ according to Equation (5). If value′<value, adjust the results by using the equalization operation and combination operation and set value′=value.
- Step 4: Set k = k + 1. If k is greater than the number of elements in solution space St, go to Step 5. If not, go to Step 2.
- Step 5: Set t = t + 1 and k = 1. If t < T, go to Step 2. If not, it means one iteration has been performed, then check whether the objective value has been improved over the previous iteration. If so, set the initial solution as the acquired result and go to Step 2. If not, go to Step 6.
- Step 6: Output the result.

Equalization Operation and Combination Operation

Although possessing advantages of quick and strong convergence, POA easily falls into sub-optimal results because each iterative calculation is merely associated with a single period. Unable to make full use of the multi-stage information of the original problem, the sub-optimal results are easily attained at the bounds of constraints. To enhance the searching space and obtain the global optimal result, two strategies are presented including: equalization operation of the system operating capacity and single station boot peak combination operation.

The equalization operation of the system operating capacity is for making the start-up units' capacity equal to the shut-down units' capacity as much as possible by means of adjusting the ratio of the system load demand for each time and operating capacity to approach rm. The equalization operation has two purposes: (1) to make the system generating scheme move from the power constraint boundary into the feasible solution's interior; and (2) to make it more convenient to move and merge the peak of the operating capacity. The equalization process is equivalent to solving the following programming problem:

$$\min f = \sum_{t=1}^{T} \left| C_s^t r_m - P_D^t \right| \tag{11}$$

The constraints are the same as those we have previously mentioned. The approach needs to traverse the generating scheme of each plant in turn and to find whether it satisfies all constraints when performing start-up (shut-down) of one unit in period t and shut-down (start-up) of one unit in period t+j (j>0). If it satisfies the constraints, the problem can be optimized with Equation (11) as the target. In the search process, the system's operating capacity remains unchanged, and it can guarantee that the capacity utilization hours of every plant are unchanged. The search process is shown in Figure 3.

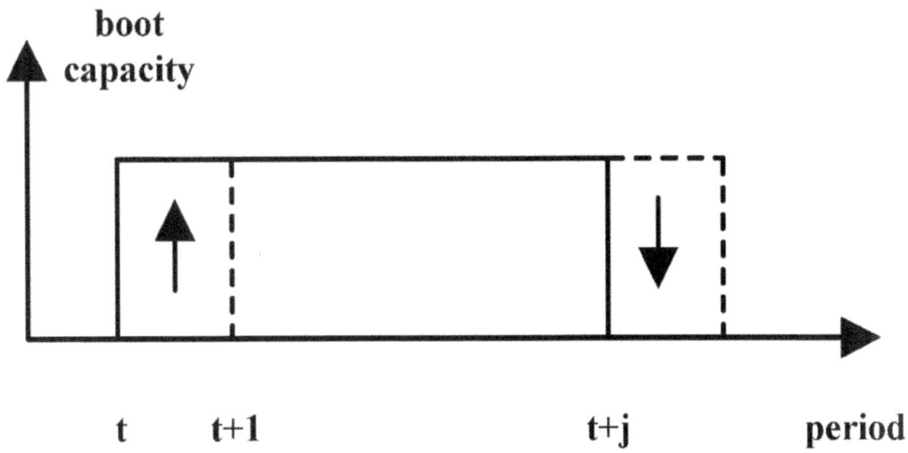

Figure 3: Sketch map of the system's boot capacity equalization.

As the equalization operation makes the start-up unit's capacity equal to the shut-down unit's capacity as much as possible, it provides the peak moving of a single plant's operating capacity some space for adjustment. The single station boot peak combination process searches the single station boot capacity in turn for the peak with the minimum number of duration periods and finds whether it should move and merge it with other peaks. If so, it executes the combination operation. Without changing the objective function value, this method can make an adjustment to the operating capacity process, which requires that the operating durations of peak and valley periods equalize to the demanded minimum duration, and forces it to leave the local optimum. The peak combination operation is shown in Figure 4.

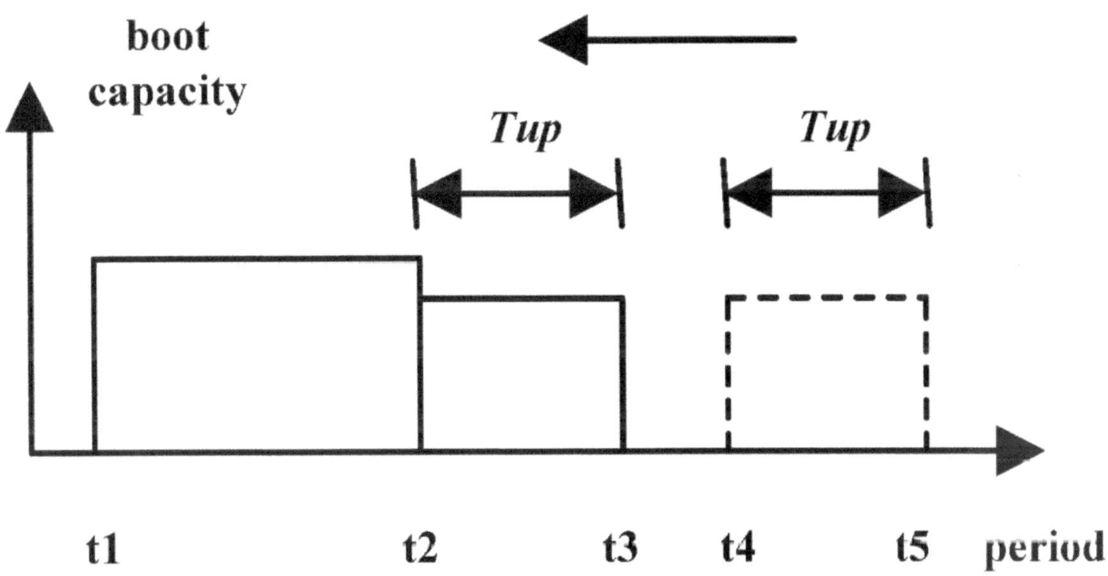

Figure 4: Sketch map of the single station boot peak combination.

From Equations (2)–(6), it can be seen that the installed capacity utilization hours of a thermal power plant is relative to the accumulative total but not the process of the scheme. Utilizing the two operations mentioned above, the results from POA can be converted by making the generating scheme move from the constraint boundary to the feasible region internal area under the condition of the unchanged objective function. As a result, POA acquires its new search space and will not stop iterating until convergence is obtained. The search principle of POA with the two strategies is shown in Figure 5. The lines of $f=F_0$, $f=F_1$ and $f=F_2$ represent the isopleths of the objective function value. According to Equation (5), the objective function value is zero for the best solution x^*, namely $f(x^*) = 0$. Two main steps are defined in the optimization process. Step α is defined as the optimization process of POA and β presents the adjustment process of equalization operation and merge operation. First, the initial feasible solution x_0 converges to the local optimal solution x_1 through Step α which cannot be further optimized because x_1 lies on the constraint boundary. Secondly, through Step β the local optimal solution x_1 can be moved along the target isopleth $f=F_1$ to x_2 which is still in the feasible solution internal area, then step α is reutilized and solution x_3 is obtained. If x_3 still lies on the constraint boundary, then step β is utilized again and the local optimal solution x_3 is moved along the target isopleth $f=F_2$ to x_4. Based on the optimized results from the adjustment process, another iteration is restarted again until the global optimal solution x^* is obtained.

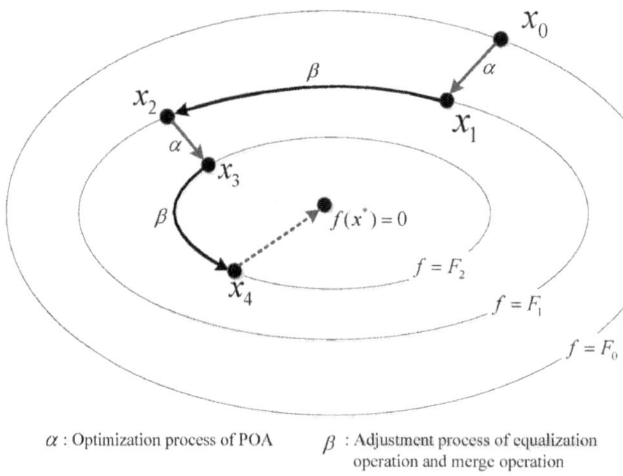

Figure 5: Sketch map of the search principle.

Solution Architecture

With a large amount of constraints and complex requirements, MOCTU is a challenging optimization problem that involves a highly nonlinear and computationally expensive objective functions. The optimization process is divided into two procedures: generation of an initial feasible solution and optimization of POA. However, POA is sensitive to initial trajectories and easily converges to local optima. Thus any changes from the decision or state variables may have a slight improvement on the solutions. Therefore, IPOA is adopted with the equalization operation and the combination operation to adjust the results acquired from POA. By doing this, it moves the solutions from constraint boundaries into the feasible zone's interior and then continues to search for the optimal solution with POA. The solution flowchart of MOCTU is shown in Figure 6 and the general process for the IPOA procedure can be mainly described in the following steps:

Step 1: Obtain the initial feasible solution by heuristic search and calculate the objective function value;

Step 2: Obtain the optimal results by using POA and calculate the objective function value′ (the detailed process is described in Section 3.3);

Step 3: Judge whether the current isopleths of the objective function value are equal to zero, if not, use the two strategies to move the optimal results from the constraint boundary to the feasible region internal area to search again. Otherwise, go to Step 4.

Step 4: Output optimal scheduling.

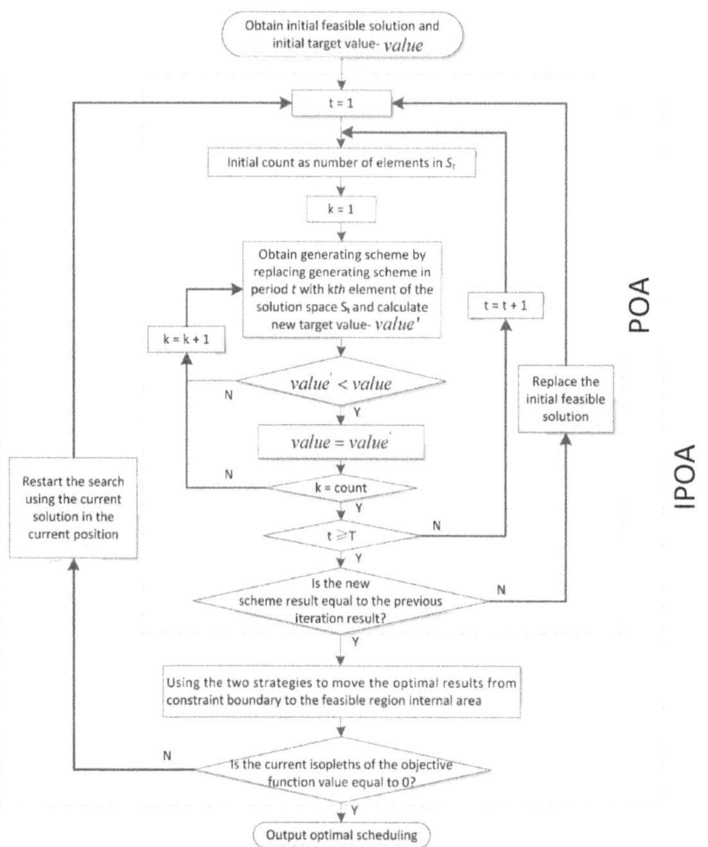

Figure 6: Solution flowchart of medium-term optimal commitment of thermal units (MOCTU) by improved progressive optimality algorithm (IPOA).

SIMULATION RESULTS

The proposed simulation model and method have been applied to the YNPG in China. Currently, it is used as the primary tool to determine the mid-term boot scheduling of thermal plants by the operators of YNPG. By the end of 2013, the total installed capacity of YNPG had reached 26.1 GW among which hydro power was responsible for 16.7 GW and thermal power was responsible for 8.935 GW. The reasonable scheduling of thermal power is beneficial for optimizing hydropower systems and system security.

The thermal power system in the YNPG consists of nine thermal power plants and 27 units. As mentioned above in Section 3.3, it is a highly dimensional and complex problem to optimize. Therefore, it is a substantial challenge to determine the medium-term operating policies for these thermal plants. The dispatch center in the YNPG is in charge of these thermal power systems. Table 1 lists the basic data of these plants. IPOA has been implemented in Java on a PC with an Intel® Core™2 Duo CPU, operating at 2.93 GHz, with 2 GB of memory. A real scheduling for 2013 in the YNPG is used to test the validity and computational efficiency of the proposed method.

October, a typical month and the beginning of the dry season, was selected to demonstrate the actual availability of the simulation model as well as the practicality and efficiency of the method. In October, the hydro-power system comes to its lowest output and the demand on the thermal system increases. As a result, the operating scheduling of thermal power plants will obviously change. The simulation results and rationality of the algorithm will be presented in the paper. According to the real-world operating experience and users› actual demands, the parameters are set as follows:

1) Given load rate of plant i, $r_i = 0.8$;
2) Minimum load rate of thermal power, $\underline{r} = 0.7$;
3) Maximum load rate of thermal power, $\overline{r} = 0.9$;
4) Minimum duration periods of peak in the operating capacity process of plant i during the calculation period, $T_i^{\text{up}} = 7d$;
5) Minimum duration periods of valley in the operating capacity process of plant i during the calculation period, $T_i^{\text{down}} = 3d$;
6) Maximum unit number of plant k in period t, set $\overline{n_i^t}$ as the number of installed units
7) Minimum unit number of plant i in period t, $\underline{n_i^t} = 1 (i = A \sim H)$, except $\underline{n_i^t} = 0$ while $i = I$;
8) Actual capacity utilization hours of plant i in the pre-calculation period, $h_i^{\text{P}} = 0$;
9) Extra award hours of plant i, $h_i^{\text{C}} = 0$;
10) System load demands in each period and the first ten days' boot process ($p = 10d$) are all given.

Table 1: Thermal power system in Yunnan power grid (YNPG)

Thermal plant	Units (number × capacity, MW)	Capacity (MW)
A	4 × 600	2400
B	6 × 300	1800
C	4 × 300	1200
D	2 × 200 + 2 × 300	1000
E	2 × 300	600
F	2 × 300	600
G	2 × 300	600
H	2 × 300	600
I	1 × 135	135

Comparison of each power plant's capacity utilization hours calculated by IPOA in October is shown in Table 2.

Table 2: Comparison of each power plant's capacity utilization hours in October (h)

Items	Heuristic search	POA	First adjustment	IPOA
Plant A	547.2	422.4	422.4	422.4
Plant B	297.6	416.0	416.0	422.4
Plant C	369.6	422.4	422.4	422.4
Plant D	240.0	422.4	422.4	422.4
Plant E	489.6	422.4	422.4	422.4
Plant F	489.6	432.0	432.0	422.4
Plant G	499.2	422.4	422.4	422.4
Plant H	528.0	432.0	432.0	422.4
Plant I	422.4	422.4	422.4	422.4
Average value	431.5	423.8	423.8	422.4
Max-min difference	307.2	16.0	16.0	0
Objective value (h^2)	10283	23.0	23.0	0

First adjustment means the results from the first equalization operation and combination operation.

Figure 7 and Figure 8, respectively, show system energy balance maps and boot modes of plant F by different methods (the horizontal axis represents the time period). As can be seen from Figure 7, the results of the simulation model match the actual characteristics of thermal power system in October that the load demand on the thermal system is increasing. It demonstrates that the proposed simulation model for MOCTU is very practical and can satisfy the actual project requirements.

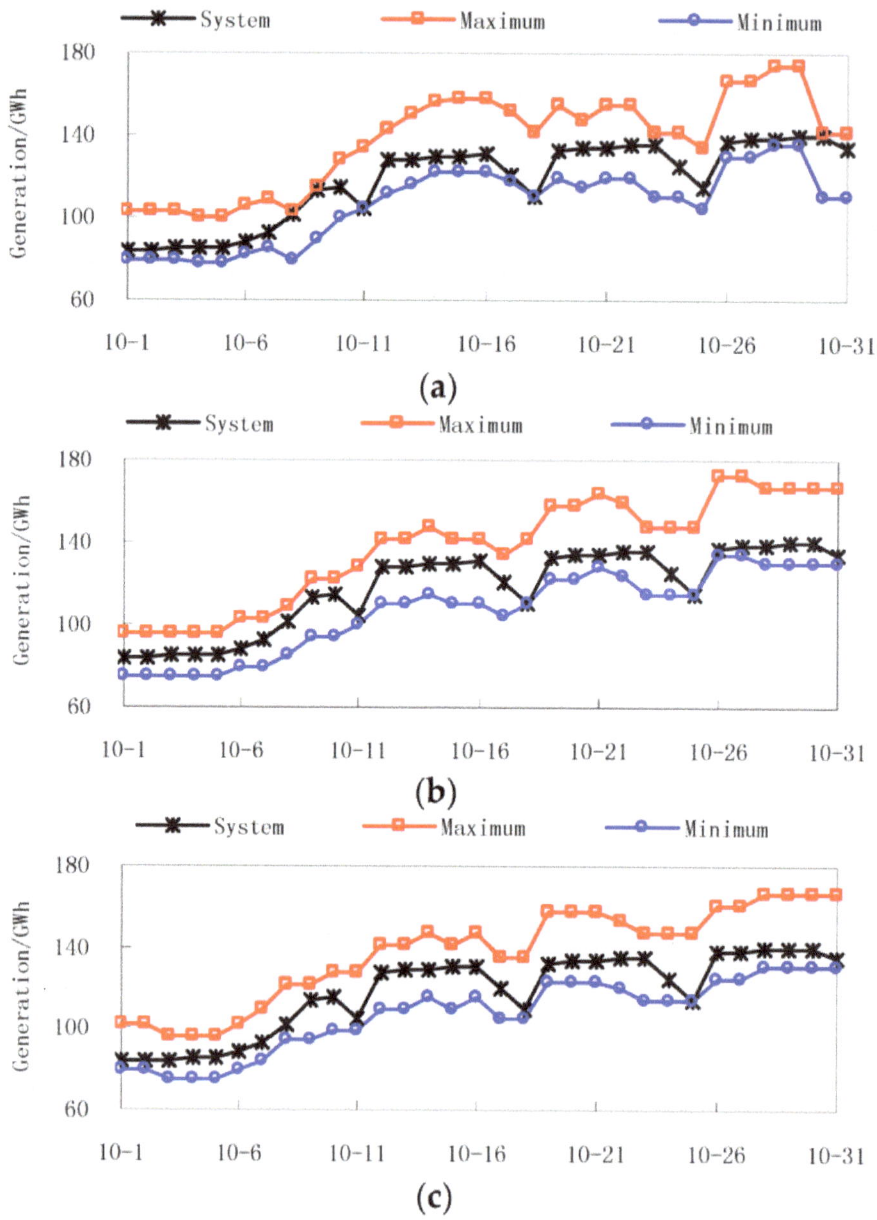

Figure 7: Energy balance maps by three methods: (**a**) POA; (**b**) first adjustment; (**c**) IPOA.

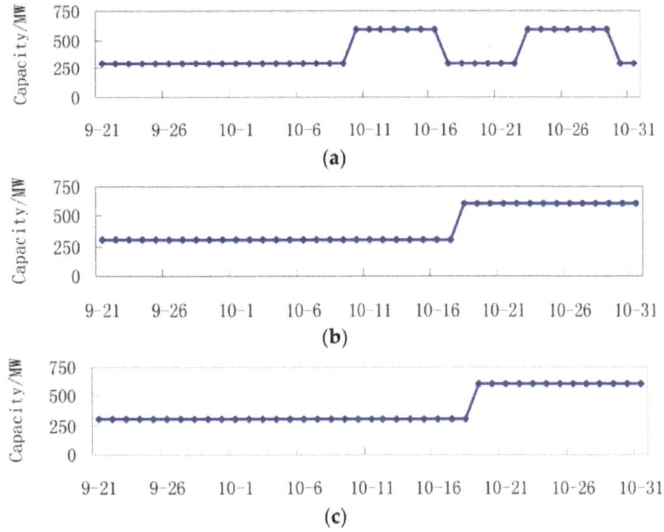

Figure 8: Boot capacity of Plant F for three methods: (**a**) POA; (**b**) first adjustment; (**c**) IPOA.

The calculation results on the YNPG are shown in Table 2. IPOA is compared with heuristics and POA. Max-min difference and objective value are taken as the measurement index. When the initial solution is obtained by heuristics, most plants' capacity utilization hours are different. As shown in Table 2, calculated by heuristic and POA in turn, the max-min differences are 307.2 h and 16 h, and the objective values are 10,283.0 h^2 and 23.0 h^2. Although the results have been significantly improved, the optimal solution is still not found. The reason is that the results obtained from POA run into local optima. The results by POA in Figure 7a illustrate that it has reached the system's power constraint boundary in many periods (10-08, 10-09, 10-30) and Figure 8a shows that two peak values of Plant F's boot mode reach the duration periods of peak in the operating capacity constraint, and it cannot be optimized any further by POA. Table 2, Figure 7b and Figure 8b show that the equalization operation and combination operation can change the structure of solutions and make the solution leave the constraint boundary under the condition of the unchanged objective function value (23.0 h^2). After the first combination operation, the boot process of Plant Fcombines two peaks (Figure 8a) into one peak (Figure 8b) and makes the duration periods of the peak change from 7 d (T^{up} = 7) to 14 d, leaving the constraint boundary. After applying IPOA, the second unit's boot time of Plant F is adjusted from 18 October (Figure 8b) to 19 October (Figure 8c), and the capacity utilization hours change from 432 h to 422.4 h.

At the same time, Table 2 shows that it has found the optimal results of the problem, which indicates that the strategies of equalization operation and combination operation can provide a new round of the search for the optimization space and finding the global optimum. At this moment, the optimal solution is obtained, as shown in Table 2, which manifests that the equalization operation and combination operation provide the search space for a new iteration to find the optimal solution. The generation requirements for the nine plants in October are shown in Figure 9 and the details of the results are listed in Table 3.

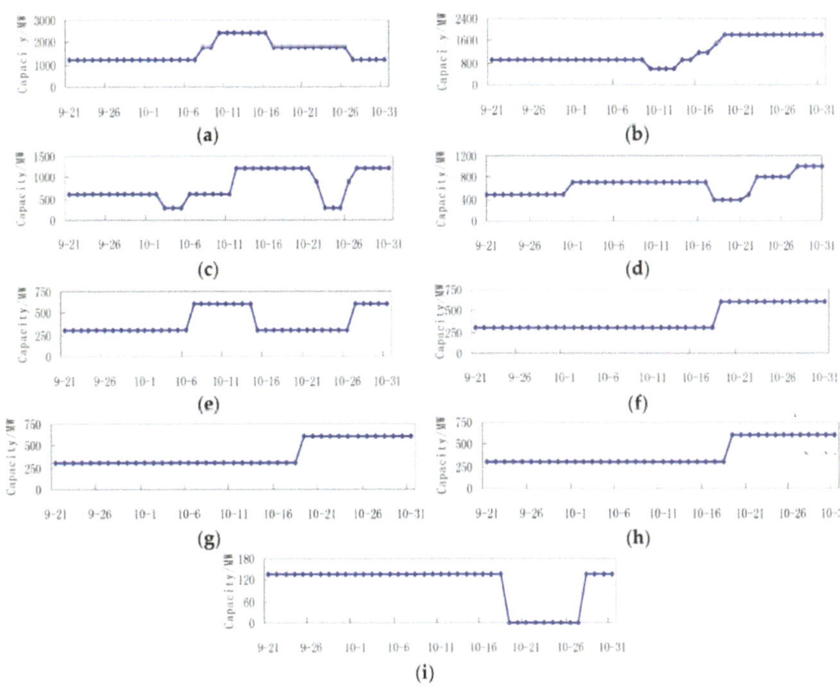

Figure 9: Results of thermal plants: (**a**) Plant A; (**b**) Plant B; (**c**) Plant C; (**d**) Plant D; (**e**) Plant E; (**f**) Plant F; (**g**) Plant G; (**h**) Plant H; (**i**) Plant I.

From Table 3, it can be seen that the boot capacity result of each plant can meet all constraints. In Plant B, for example, the days of the first valley in plant output are 4 (from 10th to 13th) which is larger than the minimum duration of valley in plant output $T_i^{\text{down}} = 3d$, while the days of first peak in plant output are 13 (from 19th to 31st) which is larger than minimum duration of peak in plant output $T_i^{\text{up}} = 7d$. Figure 9 lists the boot processes for each plant achieved by IPOA and all of the boot processes are satisfied Constraint Groups 3 and 4, which demonstrate the effectiveness of the proposed algorithm.

For comparison and illustration the impact of extra award hours, namely h_i^G, attach extra award hours for Plant C, D and F, and the award hours are 30, 20 and 10, respectively. The other conditions are unchanged. The optimization results are shown in Table 4. Because values of installed capacity utilization hours of each plant is discrete in calculation periods, the objective function value corresponding to optimal solutions is not always zero, and as a result, the objective value becomes 0.057 h^2. Moreover, installed capacity utilization hours of Plants C, D and F are more 29.52, 19.36 and 9.68 h than other plants, respectively, and the calculation precision meets the demand of practical engineering.

Table 3: Results of generation requirements for plants in October (MW)

Date	Plant A	Plant B	Plant C	Plant D	Plant E	Plant F	Plant G	Plant H	Plant I
1st	1200	900	600	700	300	300	300	300	135
2nd	1200	900	600	700	300	300	300	300	135
3rd	1200	900	300	700	300	300	300	300	135
4th	1200	900	300	700	300	300	300	300	135
5th	1200	900	300	700	300	300	300	300	135
6th	1200	900	600	700	300	300	300	300	135
7th	1200	900	600	700	600	300	300	300	135
8th	1800	900	600	700	600	300	300	300	135
9th	1800	900	600	700	600	300	300	300	135
10th	2400	600	600	700	600	300	300	300	135
11th	2400	600	600	700	600	300	300	300	135
12th	2400	600	1200	700	600	300	300	300	135
13th	2400	600	1200	700	600	300	300	300	135
14th	2400	900	1200	700	600	300	300	300	135
15th	2400	900	1200	700	300	300	300	300	135
16th	2400	1200	1200	700	300	300	300	300	135
17th	1800	1200	1200	700	300	300	300	300	135
18th	1800	1500	1200	400	300	300	300	300	135
19th	1800	1800	1200	400	300	600	600	600	0
20th	1800	1800	1200	400	300	600	600	600	0
21st	1800	1800	1200	400	300	600	600	600	0
22nd	1800	1800	900	500	300	600	600	600	0
23rd	1800	1800	300	800	300	600	600	600	0
24th	1800	1800	300	800	300	600	600	600	0
25th	1800	1800	300	800	300	600	600	600	0
26th	1800	1800	900	800	300	600	600	600	0
27th	1200	1800	1200	800	600	600	600	600	0
28th	1200	1800	1200	1000	600	600	600	600	135
29th	1200	1800	1200	1000	600	600	600	600	135
30th	1200	1800	1200	1000	600	600	600	600	135
31st	1200	1800	1200	1000	600	600	600	600	135

Table 4: Comparison of capacity utilization hours of each plant (h)

Items	h_i^G	Heuristic Search		IPOA	
		$h_i^p + h_i^T$	h_i	$h_i^p + h_i^T$	h_i
Plant A	0	240.00	240.00	422.40	422.40
Plant B	0	297.60	297.60	422.40	422.40
Plant C	30	399.12	369.12	451.92	421.92
Plant D	20	508.96	488.96	441.76	421.76
Plant E	0	422.40	422.40	422.40	422.40
Plant F	10	499.28	489.28	432.08	422.08
Plant G	0	547.20	547.20	422.40	422.40
Plant H	0	499.20	499.20	422.40	422.40
Plant I	0	528.00	528.00	422.40	422.40
Average value	-	-	431.50	-	422.24
Max-min difference	-	-	307.20	-	0.64
Objective value (h^2)	-	-	10,277.489	-	0.057

CONCLUSIONS

According to the real and practical demands of power dispatch centers of China, a simulation model for medium-term thermal UC is presented and solved by using an efficient solution named IPOA. As a new contribution to thermal power UC for power grids, the simulation model gives plenty of considerations to load demands, plant characteristics, practical engineering needs and other important factors. IPOA, which is developed to improve the performance of POA, is another contribution. The major advantage of IPOA is that strategies of operating capacity equalization and single station operating peak combination are performed to move the feasible solution away from constraint boundaries, along target isopleths, into the feasible zone's interior, in order to provide the search space for a new iteration to overcome the drawback of easily converging to local optima of POA. Basically, the mid-term boot scheduling plan of thermal plants has been generated by operators in YNPG using the proposed simulation model and method since 20 May 2013. Two-year implementation and the results of thermal power plants with 27 units in YNPG demonstrate that the proposed model has made a closer step from theoretical developments to real-world implementations, and that IPOA is an effective method for solving the MOCTU problem.

ACKNOWLEDGMENTS

This study is supported by the National Natural Science Foundation of China (No. 51209029), the Fundamental Research Funds for the Central Universities (20120041120002) and the Fundamental Research Funds for the Central Universities (No. DUT14QY15).

AUTHOR CONTRIBUTIONS

Shengli Liao and Gang Li devised the experimental strategy and carried out this experiment. Shengli Liao and Zhifu Li wrote the manuscript and contributed to the revisions. Jiayang Wang and Xinyu Wu contributed to the experiment and analysis of the data.

CONFLICTS OF INTEREST

The authors declare no conflict of interest.

REFERENCES

1. China Electric Power Yearbook Editorial Board. China Electric Power Yearbook; China Electric Power Press: Beijing, China, 2011. .
2. China Electric Power Yearbook Editorial Board. China Electric Power Yearbook; China Electric Power Press: Beijing, China, 2012. .
3. China Electric Power Yearbook Editorial Board. China Electric Power Yearbook; China Electric Power Press: Beijing, China, 2013. .
4. Ministry of Planning and Statistics. National Power Industry Statistics Express; China Electricity Council: Beijing, China, 2013. .
5. Liu, L.; Zong, H.; Zhao, E.; Chen, C.; Wang, J. Can China realize its carbon emission reduction goal in 2020: From the perspective of thermal power development. Appl. Energy 2014, 124, 199–212. . .
6. Mario, C.; Antonio, M.; Giuseppina, N. A Dynamic Fuzzy Controller to Meet Thermal Comfort by Using Neural Network Forecasted Parameters as the Input. Energies 2014, 7, 4727–4756. . .
7. Padhy, N.P. Unit commitment—A bibliographical survey. IEEE Trans. Power Syst. 2004, 19, 1196–1205. .
8. Niknam, T.; Khodaei, A.; Fallahi, F. A new decomposition approach for the thermal unit commitment problem. Appl. Energy 2009, 86, 1667–1674. . .
9. Senjyu, T.; Shimabukuro, K.; Uezato, K.; Funabashi, T. A fast technique for unit commitment problem by extended priority list. IEEE Trans. Power Syst. 2003, 18, 882–888. . .
10. Baldwin, C.J.; Dale, K.M.; Dittrich, R.F. A study of the economic shutdown of generating units in daily dispatch. Trans. Am. Inst. Electr. Eng. 1959, 78, 1272–1284. . .
11. Iguchi, M.; Yamashiro, S. An efficient scheduling method for weekly hydro-thermal unit commitment. In Proceedings of the 2002 IEEE Region 10 Conference on Computers, Communications, Control and Power Engineering, Beijing, China, 28–31 October 2002; Volume 3, pp. 1772–1777.
12. Dudek, G. Genetic algorithm with integer representation of unit start-up and shut-down times for the unit commitment problem. Eur. Trans. Electr. Power 2007, 17, 500–511. . .
13. Tong, S.K.; Shahidehpour, S.M.; Ouyang, Z. A heuristic short-term unit commitment. IEEE Trans. Power Syst.1991, 6, 1210–1216. . .
14. Hyeon, G.P.; Jae, K.L.; Kang, Y.C.; Jong, K.P. Unit commitment considering interruptible load for power system operation with wind power. Energies 2014, 7, 4281–4299. . .
15. Liao, S.L.; Cheng, C.T.; Wang, J.; Feng, Z.K. A hybrid search algorithm for midterm optimal scheduling of thermal power plants. Math. Probl. Eng. 2015, 2015. . .
16. Venkata, S.P.; Istvan, E.; Kurt, R.; Jan, D. A stochastic model for the optimal operation of a wind-thermal power system. IEEE Trans. Power Syst. 2009, 24, 940–949. . .
17. Saadawi, M.M.; Tantawi, M.A.; Tawfik, E. A fuzzy optimization-based approach to large scale thermal unit commitment. Electr. Power Syst. Res. 2004, 72, 245–252. . .
18. Dang, C.Y.; Li, M.Q. A floating point genetic algorithm for solving the unit commitment problem. Eur. J. Oper. Res. 2007, 181, 1670–1395. . .
19. Simonovic, S.P. Reservoir systems analysis: Closing gap between theory and practice. J. Water Resour. Plan. Manag. 1992, 118, 262–280. . .
20. Yang, G.; James, M.C.; Ming, N.; Rui, B. Economic modeling of compressed air energy storage. Energies 2013, 6, 2221–2241. . .
21. Rogelio, P.M.; Juan, A.M.; Miguel, J.P.; Lourdes, A.B.; Juan, M.M.S. A novel modeling of molten-

salt heat storage systems in thermal solar power plants. Energies 2014, 7, 6721–6740. . .

22. Wei, S.Y.; Xu, F.; Min, Y. Study and modeling on maintenance strategy for a thermal power plant in the new market environment. In Proceedings of the IEEE International Conference on Electric Utility Deregulation, Restructuring and Power Technologies, Hong Kong, China, 5–8 April 2004; pp. 200–204.
23. Janusz, B.; Andrzej, O. Modelling of thermal power plants reliability. In Proceedings of the International Conference on Power Engineering, Energy and Electrical Drives, Istanbul, Turkey, 13–17 May 2013; pp. 1038–1043.
24. Turgeon, A. Optimal short-term hydro scheduling from the principle of progressive optimality. Water Resour. Res. 1981, 17, 481–486. . .
25. Nanda, J.; Bijwe, P.R.; Kothari, D.P. Application of progressive optimality algorithm to optimal hydrothermal scheduling considering deterministic and stochastic data. Int. J. Electr. Power Energy Syst. 1986, 8, 61–64. . .
26. Nanda, J.; Bijwe, P.R. Optimal hydrothermal scheduling with cascaded plants using progressive optimality algorithm. IEEE Trans. Power Syst. 1981, 100, 2093–2099. . .
27. Lucas, N.J.D.; Perera, P.J. Short-term hydroelectric scheduling using the progressive optimality algorithm. Water Resour. Res. 1985, 21, 1456–1458. . .
28. Cheng, C.T.; Shen, J.J.; Wu, X.Y.; Chau, K.W. Short-term hydro scheduling with discrepant objectives using multi-step progressive optimality algorithm. J. Am. Water Resour. Assoc. 2012, 48, 464–479. . .
29. Labadie, J.W. Optimal operation of multireservoir systems: State-of-the-art review. J. Water Resour. Plan. Manag.2004, 130, 93–111. . .
30. Howson, H.R.; Sancho, N.G. A new algorithm for the solution of multi-stage dynamic programming problems. Math. Program. 1975, 8, 104–116. . .
31. Marino, M.A.; Loaiciga, H.A. Dynamic model for multireservoir operation. Water Resour. Res. 1985, 21, 619–630. . .
32. Marino, M.A.; Loaiciga, H.A. Quadratic model for reservoir management: Application to the Central Valley Project. Water Resour. Res. 1985, 21, 631–641.

CHAPTER 10 THERMAL TRACKING IN MOBILE ROBOTS FOR LEAK INSPECTION ACTIVITIES

Aitor Ibarguren , Jorge Molina, Loreto Susperregi and Iñaki Maurtua

IK4-Tekniker, Calle Iñaki Goenaga 5, Eibar 20600, Spain

ABSTRACT

Maintenance tasks are crucial for all kind of industries, especially in extensive industrial plants, like solar thermal power plants. The incorporation of robots is a key issue for automating inspection activities, as it will allow a constant and regular control over the whole plant. This paper presents an autonomous robotic system to perform pipeline inspection for early detection and prevention of leakages in thermal power plants, based on the work developed within the MAINBOT (http://www.mainbot.eu) European project. Based on the information provided by a thermographic camera, the system is able to detect leakages in the collectors and pipelines. Beside the leakage detection algorithms, the system includes a particle filter-based tracking algorithm to keep the target in the field of view of the camera and to avoid the irregularities of the terrain while the robot patrols the plant. The information provided by the particle filter is further used to command a robot arm, which handles the camera and ensures that the target is always within the image. The obtained results show the suitability of the proposed approach, adding a tracking algorithm to improve the performance of the leakage detection system.

INTRODUCTION

Efficient and effective maintenance is crucial for all kinds of industries. In the case of capital intensive investment industries, such as petrochemicals, the steel industry or power generation plants, it is even more relevant and has an important impact on the operation costs during the long lifecycle of their production means.

Citation: Ibarguren, Aitor; Molina, Jorge; Susperregi, Loreto; Maurtua, Iñaki. 2013. "Thermal Tracking in Mobile Robots for Leak Inspection Activities." Sensors 13, no. 10: 13560-13574. DOI: 10.3390/s131013560

Automating inspection activities in industrial plants, especially in extensive plants, poses strong requirements from different points of view: a huge number of elements to inspect (pipes, valves, switches, pumps, vessels, motors, vibrating machinery, chillers, ovens, etc.), handling multiple sensors or special non-destructive testing equipment to be used (visual, ultrasonic, vibration, radiography, thermography, eddy current, noise analysis, gas sensors, etc.) and extensive production facilities that spread out for thousands of square meters, either in the vertical or horizontal, and risky working conditions for maintenance personnel, due to the presence of hazardous materials.

This paper presents part of the work performed in the MAINBOT European project. This project aims at developing service robot applications to autonomously execute inspection tasks in extensive industrial plants. The objective is to develop a surveillance robotic system able to detect the leakage of fluids using a vision system in the thermal and visible ranges. To validate the proposed solution, a solar plant of cylindrical-parabolic collectors is used, testing the approach in a very demanding environment from a mobile manipulation point of view.

The paper is organized as follows. Section 2 gives information about related works. Section 3 introduces the inspection task to be performed by the system. Section 4 presents the proposed approach, the architecture for leak inspection in solar thermal plants and autonomous navigation. Sections 5 and 6 are devoted to the leakage detection algorithm and the tracking system, respectively. In Section 7, the experimental results of the system are shown. Finally, Section 8 poses the obtained results, as well as the future work to be done.

RELATED WORK

Robots have been used for maintenance tasks in a wide range of applications and environments. From preventive maintenance of high-voltage transmission power lines [1] to inspection of cables [2] and nuclear reactor pressure vessels [3] in the nuclear industry, autonomous systems are used in many industries in an attempt to improve maintenance tasks. Focusing on pipeline inspection, Suzuki et al. [4] propose an autonomous robot for industrial pipeline inspection by means of ultrasonic diagnosis equipment. In the same way, Camerini et al. [5] present an underwater inspection robot for offshore pipeline inspection, using the pipeline itself for guidance purposes. Even so, a few works add some tracking tools as particle filters to improve the inspection task.

Several works have also been performed using thermal images for tracking and detection. Jiping et al. [6] propose a target detection and tracking system based on different morphological operations. Senthil Kumar et al. [7] pose the fusion of thermal images with 2D images for tracking Unmanned Aerial Vehicles (UAVs), adding optical flow techniques, such as LucasKanade and HornSchunck methods. Finally, Padole and Alexandre [8] propose the use of particle filters for human tracking with thermal images, using motion information to feed the particle filter. The presented work also proposes the use of particle filtering, although it uses the whole image information and dynamics to feed the particle filter, not only the motion information.

TASK SPECIFICATION

Valle 1 and 2 are two solar thermal power plants, whose promoter and owner is Torresol Energy Investments, S.A., and which are located in San José del Valle (Spain); see Figure 1. Valle 1 and 2 are two adjacent solar thermal power plants that generate electricity by means of cylindrical-parabolic collectors.

Figure 1: Valle 1 and 2 solar thermal power plants.

The solar field is composed of nearly 7,500 parabolic cylinder collectors. These collectors transport a Heat Transfer Fluid (HTF), which absorbs the solar energy. HTF circulates at a high temperature (around 390 °C) inside the absorber tubes, which is used after to heat the molten salts to generate steam in the Steam Generation System (SGS).

Swivel joints are critical points where leakages may happen (this is the point where the collector tube connects with the infrastructure of pipes that are deployed all over the plant). HTF leakages are not desirable, as oil losses may be unsafe, due to the high temperatures reached in the solar power plant. Early detection and prevention of leakages is a key issue for the maintenance of those kinds of facilities. Even so, the huge area of solar power plants makes it difficult to perform proper maintenance, due to the large amount of kilometers of collectors and pipelines, as well as the hazardous environment with the really high temperatures reached in its elements.

Nowadays, the inspection is performed by human operators using a thermographic camera while they travel along different parts of the solar plant by car through poorly asphalted and dirt roads. The inspection is carried out while the vehicle is moving, looking at the screen to analyze the thermal image and detecting leakages. Even so, the operator must inspect the image for long time periods (around 2 h) while correcting the the pose of the camera to keep the pipeline in the field of view when irregularities in the terrain appear, which may lead to leakages being left undetected.

PROPOSED APPROACH

Based on the task specifications posed in the previous section, this paper proposes an autonomous robotic system to perform the pipeline inspection for early detection and prevention of leakages. The autonomous robot patrols along the power plant while inspecting the collectors using thermo graphic images to identify the leakages. The inspection system continuously analyzes the images in order to monitor the status of the elements, the tube in this case, and highlight anomalies.

Initially, a path is defined for the robotic platform based on a hybrid map approach mixing topological graphs with local occupancy grids. Using the information provided by a GPS/IMU sensor, the robot executes the planned path. During this execution, the system takes advantage of a local metric planner to avoid obstacles and follows the initial path as close as possible.

Even so, irregularities in the terrain (poorly asphalted road and dirt road) make it difficult to perform the inspection using a fixed camera mounted on the robot, as the pipeline can be out of range when slopes and bumps are found on the road. Taking this into account, the addition of a robotic arm is proposed to allow the manipulation of the thermal camera and to track the pipeline through the inspection task. The aim is to establish a coordination between the thermographic inspection and the robot arm movements in order to keep the objective in the field of view of the camera.

The next sections will give information about the used hardware and system architecture, including the different software units defined within the system and the navigation system.

ARCHITECTURE

The architecture presented in this paper is based on the specifications and work performed in the MAINBOT project. The main robotic platform used is a RobucarTT developed by Robosoft (http://www.robosoft.com/), designed for outdoor environments and with the Ackermann steering geometry. The robot includes a GPS/IMU sensor for localization purposes, which provides the robot position with an accuracy of 0.20 m. The platform also has a robotic arm attached to it, which is employed in this case to handle a thermographic camera, specifically, an FLIRThermoVision® A20. This thermographic camera is used for leakage detection, as well as for tracking purposes.Figure 2 shows the physical elements of the architecture.

Figure 2: Robotic platform and thermographic camera.

There are two main tasks to be performed by the robotic platform while the predefined path is being executed by the robotic platform: on the one hand, to inspect the pipelines to search for leakages; on the other hand, to track the pipeline in order to send movements to the robotic arm and to maintain the pipe in the field of view of the camera. To this end, as shown in Figure 3, two different modules have been defined in the software architecture:

Figure 3: Software architecture: leak inspection unit and thermal tracking unit.

- Leak inspection unit: the unit in charge of receiving images from the thermographic camera and analyzing them in order to detect leakages in the collectors. Based on different morphological operations in the image, the unit is able to detect the leakages in the collectors in an accurate way.
- Thermal tracking unit: the unit in charge of tracking the pipeline and commanding the robotic arm to keep the pipe in the field of view of the camera. A particle filter-based tracking system is proposed, because to its capacity to accurately model the underlying dynamics and its rapid adaptation to changing signal features.

Based on this architecture, the system is able to (a) inspect the pipelines and detect leakages and (b) maintain stable the detection process by means of the tracking process, overcoming the problems derived from the irregularities of the terrain, while the mobile robot executes the planned path for inspection.

Autonomous Navigation

The autonomous navigation approach of the MAINBOT project is based on the use of a hybrid map consisting of a topological graph overlaid with local occupancy grids. Since the workspace is a large area, the overall plan is formed on a topological graph, as planning in a large metric map quickly becomes unwieldy. However, local metric information is used for achieving precise localization (needed for some operations) and obstacle avoidance. Hence, the path planning is performed in two steps:

- The overall plan is created in the topological graph, using Dijkstra's algorithm. This is the basis for the low level planning.
- The robot navigates locally using local metric maps and a search-based planning algorithm.
  - The global metric planner integrated in MAINBOT generates a path from the current position to a desired goal by combining a series of short, kinematically feasible "motion primitives". Planning is done in x, y and theta dimensions, resulting in smooth paths that take robot orientation into account. This is especially important for a RobucarTT robot, as it has nonholonomic constraints (i.e., due to the Ackermann configuration).
  - The local metric planner can be seen as a controller that drives a mobile base in the plane.
  - An execution component called "move base" links the global and local planners to achieve the metric navigation.

This navigation approach allows for creating paths for pipeline inspection in two steps, using the "move base" component to execute the defined path as accurately as possible. The autonomous navigation module has been developed using ROS (http://www.ros.org) (Robot Operating System) libraries and packages.

LEAK INSPECTION UNIT

The aim of this unit is the detection of leakages in the collectors based on information provided by a thermographic camera. To this end, initially, (a) the image is analyzed, searching for a pipeline section. Once the pipeline has been found; (b) the section is inspected to detect abrupt changes in the temperature, which indicate that there are leakages in the collectors.

In this process, as the first step, the object parameters of the thermal camera must be fixed. To extract the temperature information from an image, the output data of the thermal camera must be interpreted based on the correct fixing of parameters, such as emissivity, object distance or reflected temperature.

The emissivity is a surface property that states the ability to emit energy; it is expressed as the ratio of the radiation emitted by a surface to the radiation emitted by a blackbody. Emissivity is a unitless quantity and spans from zero to one.

Following the energy conservation law, all energy exchange is compensated mutually: the flux incident, Φ_i, is equal to the flux reflected, Φ'_r, absorbed, Φ_a, and transmitted, Φ_t:

$$\Phi_i = \Phi'_r + \Phi_a + \Phi_t \tag{1}$$

In general cases, terms on the right side of Equation (1) are specifically weighted following particular radiative properties related to reflection (P), absorption (α) and transmission (τ). These properties are linked together considering the flux exchanges on a semitransparent object in its environment for which:

$$\rho+\alpha+\tau=1 \tag{2}$$

The general form of Kirchhoffs law provides a link between the absorption and emission processes and, thus, between emissivity and absorbance, since:

$$\varepsilon(\lambda, \theta', \phi') = \alpha(\lambda, \theta, \phi) \tag{3}$$

The object distance is defined as the distance from the camera to the surface. Finally, reflected temperature indicates the temperature reflected by the surface of an object.

In the case of the glass that covers the collector, it has a high transmissivity, around 94% (0.94), so the emissivity is fixed to 0.04. The estimated reflected temperature is fixed empirically to 10 °C and the object distance to 10 m, based on the features of the environment and the solar plant.

Once the settings are established, the camera processes the thermal information and provides an image where pixels give information about the temperature. These values of temperature are normalized to gray values performing a thermal adjustment between the minimum and maximum values defined (in our case, between 100 °C and 300 °C), as shown in Figure 4A.

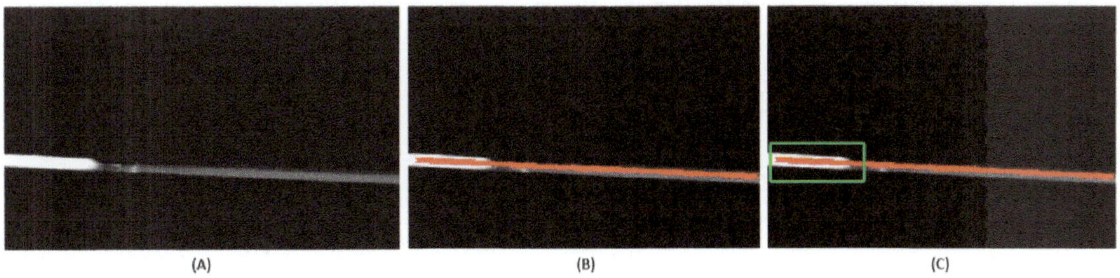

Figure 4: (**A**) Thermographic image; (**B**) Pipeline detection; and (**C**) Leak detection.

This image is then thresholded, highlighting the pixels with temperatures above 100 °C, as they are related with the pipelines. Skeletonization/Medial Axis Transform [9] is applied to this image, detecting the longest straight section in the image, as illustrated in Figure 4B. This straight section is then analyzed to detect temperature changes along the pipeline.

As the first step to detect the temperature changes along the collector, the temperatures of the previously obtained section are stored, as shown in Figure 5A, where the temperature along the collector is plotted. Based on these data, the absolute value of the first derivative is calculated as:

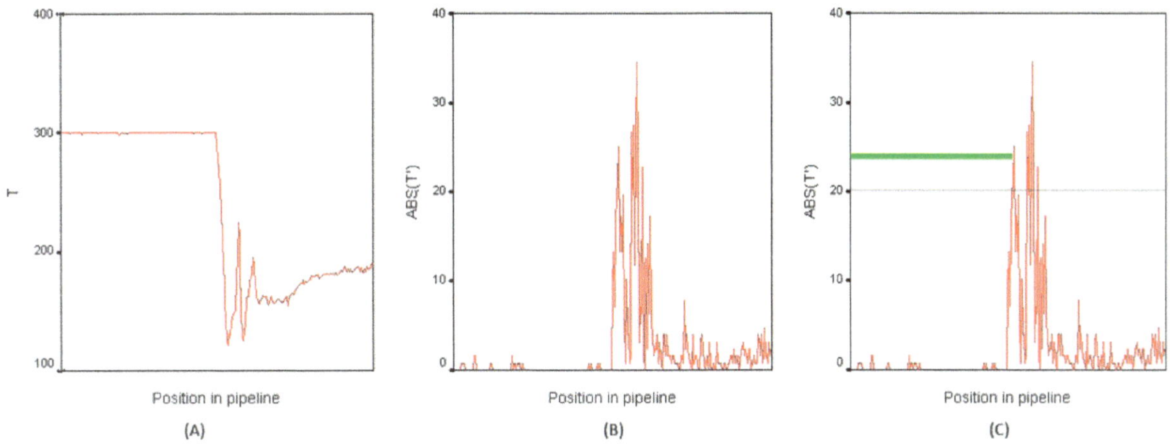

Figure 5: (**A**) Temperature along the pipeline; (**B**) Absolute value of the first derivative; and (**C**) Detection on the plot.

$$ABS(T') = |T_{i+1} - T_i|, \quad {}_{i=[1..N-1]} \tag{4}$$

where T_i is the temperature of pixel i of the pipeline section.

The information about the absolute value of the derivative is used to divide the section into different parts, which have similar temperature, defining a ΔT threshold (minimum change) to perform this division. An important consideration is the fact that the joints of different sections of the collector are metallic and have a high reflectivity, so the obtained temperature is unstable at this point and must be filtered. In order to filter them, an approximate resolution of the image is estimated and short parts with abrupt change of temperature are removed, defining a minimum width of the pipeline part. This allows us to filter short pipeline parts with abrupt temperature changes (joints), detecting real leakages, which fill a wider space on the image. Once the pipeline is divided, the mean temperature of each subsection is computed. The pipeline parts above the maximum temperature are labeled as leakages, as highlighted in green in Figures 4C and 5C.

Based on this algorithm, the Leak Detection Unit is able to raise alerts when leakages are detected while the robot is patrolling along the collectors.

THERMAL TRACKING UNIT

Irregularities in the terrain can make it difficult to perform a correct leak inspection, as pipelines can be out of the field of view of the camera when slopes and bumps are found along the road. To overcome this problem, a tracking system is proposed based on particle filtering. This tracking system follows the target pipeline through thermo graphic image sequences, and it is able to generate movement commands when the target is reaching the edge of the image. The next lines will give information about all the elements of the particle filter for thermal tracking.

Particle Filter

Particle filters [10,11], also known as Sequential Monte Carlo methods (SMC), are sequential estimation techniques that allow estimating unknown states, x_t, from a collection of observations $z_{1:t} = \{z_1, \ldots, z_t\}$. The state-space model is usually described by state transition and measurement equations:

$$x_t = f_t(x_{t-1}, v_{t-1}) \tag{5}$$

$$z_t = g_t(x_t, u_t) \tag{6}$$

where f and g are the state evolution and observation model functions, respectively, and v_t and u_t denote the process and observation noise, respectively.

Based on the previous equations, particle filters allow for approximating the posterior density (PDF) by means of a set of particles, $\{x_t^{(i)}\}_{i=1,\dots,n}$, using equation:

$$p(x_t|z_{1:t}) = \sum_{i=1}^{N} \omega_t^{(i)} \delta(x_t - x_t^{(i)}) \tag{7}$$

where each particle, $x_t^{(i)}$, has an importance weight, $\omega_t^{(i)}$, associated with it and δ is the Kronecker delta. These weights are computed following equation:

$$\omega_t^{(i)} = \omega_{t-1}^{(i)} \frac{p(z_t|x_t^{(i)})p(x_t^{(i)}|x_{t-1}^{(i)})}{q(x_t^{(i)}|x_{0:t-1}^{(i)}, z_{0:t})} \tag{8}$$

where $p(z_t|x_t^{(i)})$ is the likelihood function of the measurements, z_t, and, finally, $q(x_t^{(i)}|x_{0:t-1}^{(i)}, z_{0:t})$ is the proposal density function.

Based on the previously presented equations, the particle set evolves along time, changing the weights of the particles and resampling them in terms of the observations.

Particle filtering provides a robust tracking framework when dealing with non-linear and non-Gaussian state and observation functions, as it considers multiple state hypotheses simultaneously.

Particle Filtering for Thermal Tracking

In the first step, a tracking process has been defined in an attempt to maintain the object to be analyzed in the field of view of the thermal camera. A particle filter-based tracking is proposed, allowing us to correct the position of the robotic arm to maintain the pipe in the center of the image for a further analysis. In an attempt to develop a thermal tracking system for multiple detection tasks, the particle filter has been generalized to allow the tracking of objects with different shapes and configurable for each tracking process (easy reconfiguration for similar scenarios), although the system modeling explained in the next lines is tuned for the presented environment and posed problem.

System Modeling

In the presented scenario, the tracking process is able to follow a pipeline in successive frames. The state of the process is defined as:

$$X_t = [x_t, y_t, l_t, \alpha_t]^T \tag{9}$$

where x_t and y_t are the x and y pixel coordinates of the center of the pipeline in the image, l_t is the length of the pipeline and α_t is the orientation of the pipeline in time t.

Additionally, the state transition is defined as:

$$X_{t+1} = X_t + \Delta_t \dot{X}_t + V_t \tag{10}$$

$$\dot{X}_t = [\dot{x}_t, \dot{y}_t, \dot{l}_t, \dot{\alpha}_t]^T \tag{11}$$

where Δ_t is the time step, $\dot{X}_t$ is the dynamic part describing the variation of the state elements and V_t is an additive, zero mean Gaussian noise.

Likelihood Evaluation

For the likelihood evaluation, initially, the thermal image is analyzed to highlight the parts with a predefined temperature range, in this case, the temperature of the pipe to be tracked. To this end, thresholding and Skeletonization/Medial Axis Transform algorithms have been used, as in the leak inspection unit. From this step, a set of N connected regions are extracted; N possible pipe sections. Those regions form the observation, Z_t, where each region, z_i^t, is defined by their center in pixel coordinates, length and angle:

$$Z_t = z_{i=1..N}^t \tag{12}$$

$$z_i^t = [x_i^t, y_i^t, l_i^t, \alpha_i^t]^T \tag{13}$$

For the likelihood evaluation, initially, the distance between the particle and each of the observed regions is calculated using the pixel coordinates of the center, length and angle as:

$$dist_i = \lambda \sqrt{(x_t - x_i^t)^2 + (y_t - y_i^t)^2} + \beta\,|l_t - l_i^t|^2 + \gamma\,|\alpha_t - \alpha_i^t|^2 \tag{14}$$

where λ, β and γ are coefficients to weigh the importance of the pixel coordinates, length and angle, respectively. The distance between the particle and the observation is then calculated as the minimum distance between the particle and the N regions found:

$$dist(X_t, Z_t) = min(dist_i), \quad i=[1..N] \tag{15}$$

Finally, the likelihood is calculated as the exponential of the distance, as shown in the next equation:

$$P(Z_t|X_t) = e^{-dist(X_t, Z_t)} \tag{16}$$

Based on the presented likelihood evaluation, the particle filter estimates iteratively the process state as presented in the next paragraphs.

Particle Filtering Procedure

To initialize the process, when the first pipe is detected, a set of N random particles is drawn around its position and with its scale and orientation. Afterwards, the procedure of the particle filter is given as:

- Find the object in the initial thermal image and initialize N particles, $X_0^{(i)}$ with random samples around it, where $w_0^{(i)} = 1/N$;
- If ESS < threshold (effective sample size), draw N samples with selection with replacement;
- Predict $x_t^{(i)} = x_{t-1}^{(i)} + v_{t-1}$;
- Update importance weights $w_t^{(i)} = w_{t-1}^{(i)} P(Z_t|X_t)$; 5: Normalize weights $w_t^{(i')} = w_t^{(i)} / \sum_{j=1}^{N} w_t^{(j)}$;
- Set t = t + 1, go to Step 2.

In this procedure, ESS [12] (effective sample size) is calculated as:

$$cv_t^2 = \frac{var(w_t^{(i)})}{E^2(w_t^{(i)})} = \frac{1}{N}\sum_{i=1}^{N}(Nw_t^{(i)} - 1)^2 \quad (17)$$

$$ESS_t = \frac{N}{1 + cv_t^2} \quad (18)$$

where N is the number of particles and $w_t^{(i)}$ is the weight of particle i in time t.

Based on this discrete approximation of the posterior probability, the object is tracked along the inspection process.

EXPERIMENTAL RESULTS

To test the suitability of the proposed approach, a set of experiments have been designed trying to asses both the leakage detection algorithm and the tracking system. To this end, a database of image sequences was created in Valle facilities (see Figure 6) using actual production means and replicating the behavior of the maintenance robot:

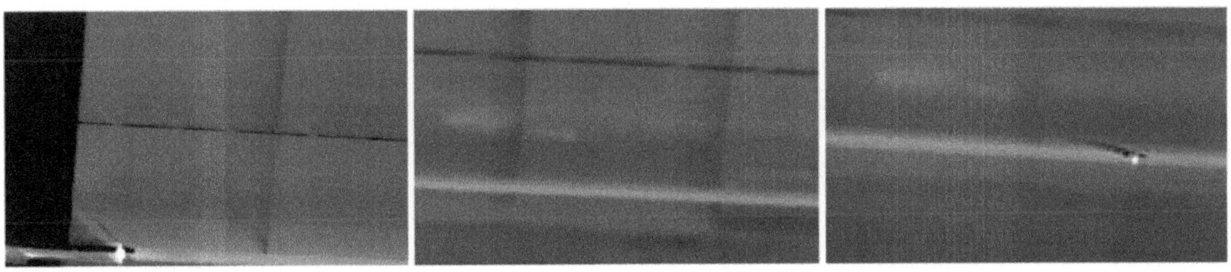

Figure 6: Sequence of thermographic images on Valle facilities.

- Recording of collectors and pipelines using a thermal camera 10 km at night, as performed now by human operators, divided in sequences of 150 m (half loop of collectors).
- Camera placed on a vehicle circulating at a speed of 20 km/h through a terrain with irregularities.
- Real leakages appearing in the images.

Based on this real data, two different experiments have been performed. The next lines give information about each experiment and the obtained results.

Results of the Leak Detection

An efficiency of the leak detection has been tested using 60 different thermal sequences. In those 60 sequences, human operators found a total of seven leakages during the recording session, which were labeled in the database. Those 60 sequences were analyzed by the previously presented algorithm, searching for leakages. A threshold, ΔT, of 70 °C was established for the detection algorithm empirically based on the gathered data. In order to have numeric data, a mean value for each stretch of the tube with temperature change (filtering joints) was also saved.

Figure 7 shows an example of the output of the algorithm while performing the validation test, where each column is related with a sequence. The first row indicates the image sequence ID, while the next rows show the mean temperature of the pipe section along the collectors where abrupt changes of temperature can be observed. Pipe parts with a high temperature (leakages) are marked in red.

P1A2	P1A4	P1A5	P1A7	P1A8	P1A10	P1A11	P1A12	P1A16	P1A24	P1A25	P1A28	P1A35
136,07	136,07	143,13	135,29	142,35	160,39	156,47	147,84	134,5	142,35	129,8	128,23	149,41
132,15	132,15	132,15	158,82	132,15	142,35	137,64	143,92	142,35	138,43	158,03	119,6	133,72
127,45	127,45	132,15	299,21	144,7	174,5	189,41	153,33	172,15	246,66	207,45	271,76	138,43
299,21	299,21	192,54	298,43	193,33	260,78	202,74	292,94	237,25	217,64	189,41	138,43	205,88
293,72	293,72	299,21	300	292,94	249,8	272,54	252,15	238,03	132,94	122,74	138,43	209,01
290,58	290,58	296,07	298,43	246,66	152,54	152,54	260	170,58	129,01	122,74		139,21
292,94	292,94	141,56	296,07	217,64	148,62	154,11	141,56	146,27				125,09
150,19	150,19	135,29	299,21	252,15			138,43	152,54				128,23
140	140	136,07	299,21	136,07			139,21					
139,21	139,21		148,62	125,88								
	...		145,49									
	...											
	148,62											
	143,13											
	179,21											
	258,43											
	263,92											
	263,13											
	243,52											
	143,13											
	144,7											

Figure 7: Example of changes in temperature in recorded image sequences.

The algorithm found all the leakages labeled by human operators, obtaining 100% sensitivity. Besides, six more leakages were found along the sequences, leakages that match exactly with the previous leakage patterns. Based on the similarity and after an analysis of the new leakages, they could be considered as real leaks that were missed by the human operators during visual inspection.

Results of the Tracking Process

To test the efficiency of the particle filter for tracking, 10 different image sequences have been used, each of them with 200 images approximately (a total of 2,000 images). For each sequence, the particle filter has been initialized using the first pipeline appearance, and the pipeline has been tracked through the rest of the images. In this experiment, the manipulation of the robotic arm has been left out of the scope of this manuscript, as the sequences have been recorded by a human operator from a car and it is not possible to simulate the arm movements. For each frame, the error between the output of the particle filter and the labeled images has been computed, as shown inFigure 8. Images show peaks in the error plots, derived from the image noise and irregularities in the terrain.

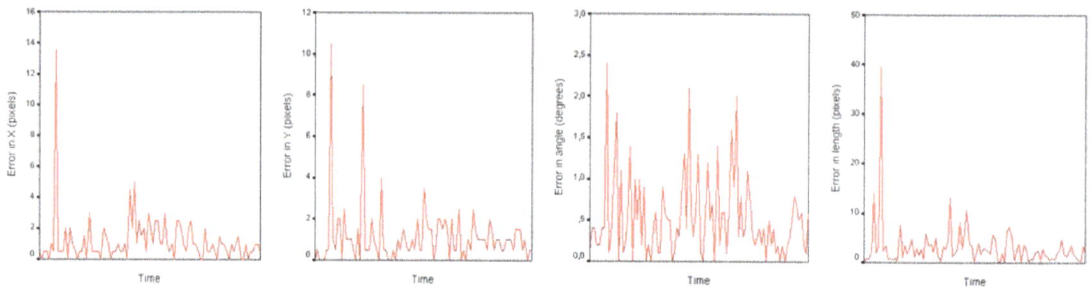

Figure 8: Tracking error during a recorded sequence in position in X and Y orientation and length.

Specifically, the experiment measures the tracking error in position, angle and length of the pipeline, using different parameters for the particle filter. Those are the specifications of the experiment:

- Six different particle filter configurations have been set up, mixing different state estimation methods and numbers of particles. Specifically, the state estimation methods are:
  - Best particle (the one with maximum weight);
  - Robust mean with the 10 particles with the maximum weight;
  - Weighted mean using the whole particle set;
- For each configuration, the previously cited image sequences have been used, tracking the pipeline through around 2,000 images; the mean error and standard deviation of the position, angle and pipe length has been measured for all the configurations;
- In the likelihood evaluation step, the same values for coefficients λ, β and γ have been used, applying values that give a similar weight to position, angle and length.

Table 1 shows the obtained results. The first column describes the number of particles and the estimation method, the second and third columns, the mean (μ) and the standard deviation (σ) of the error in the position and the fourth and fifth columns, the mean (μ) and the standard deviation (σ) of the error in the angle, and finally, the last columns show the mean (μ) and the standard deviation (σ) of the error in the length.

Table 1: Results of the Tracking Process

	Pos. Error (pix)		Angle Error (°)		Length Error (pix)	
	μ	σ	μ	σ	μ	σ
500 - Best particle	1.80	1.27	0.96	0.86	3.16	2.58
1,000 - Best particle	**1.42**	**0.95**	0.72	0.63	2.94	**2.51**
500 - Robust mean 10	1.82	2.11	0.51	0.41	2.86	5.35
1,000 - Robust mean 10	1.57	1.83	**0.41**	**0.34**	**2.65**	5.66
500 - Weighted mean	6.34	5.39	0.43	0.36	3.67	7.68
1,000 - Weighted mean	6.11	4.67	0.35	0.38	3.45	6.64

Results show a better performance of the particle filter with a set of 1,000 particles, improving the mean error and standard deviation in almost all the configurations. The addition of a bigger set of particles could decrease the error rates, although it would be necessary to optimize the computational time as much as possible to ensure a suitable frame rate. In the case of the state estimation methods, the best particle and robust mean methods show very similar results, overcoming in any case the errors of the weighted mean.

CONCLUSIONS AND FUTURE WORK

This paper presents an autonomous leakage detection system for maintenance tasks in extensive industrial plants, including a leakage detection system. Based on the information provided by a thermographic camera, the system is able to detect leakages in pipelines and collectors of thermal power plants. This system is enhanced by a particle filter-based tracking system, used to maintain the target in the field of view of the camera and stabilize the detection process when irregularities are found along the road.

The results show a high success rate of the leakage detection unit, reaching 100% sensitivity (on data labeled by operators) and detecting even more leakages on the recorded sequences. The use of thermographic information allows for detecting the fluids leaked from the collectors, taking advantage of morphological operations to highlight the leaks in the thermal images.

Besides, a tracking system has been added to manage the thermographic camera and to avoid the loss of the target. The particle filtering for tracking has shown a position error of less than 1.5 pixels and less than a 0.5° error in the angle. It also allows for modeling the system in a simple and effective way, adapting rapidly to the changing image features and data noise.

As further work, there are two main paths to follow: on the one hand, to add the robot arm movements and to mix the information extracted by the leak inspection unit and thermal tracking unit with the navigation module to create a complete system that takes into account the navigation information in the inspection task and vice versa. In the same way, it would also be interesting to test the tracking system in similar inspection tasks where different elements of the solar plant are analyzed (valves, vessels, etc.) and observe its suitability.

ACKNOWLEDGMENTS

This work has been performed within the scope of the project "MAINBOT: Mobile robots for inspection and maintenance activities in extensive industrial plants". MAINBOT is a small or medium-scale focused research project supported by the European Commission in the 7th Framework Program (285245). For further information, seehttp://www.mainbot.eu.

CONFLICTS OF INTEREST

The authors declare no conflict of interest.

REFERENCES

1. Debenest, P.; Guarnieri, M.; Takita, K.; Fukushima, E.F.; Hirose, S.; Tamura, K.; Kimura, A.; Kubokawa, H.; Iwama, N.; Shiga, F.; et al. Expliner—Toward a Practical Robot for Inspection of High-Voltage Lines. Proceedings of the Field and Service Robotics (FSR), Cambridge, MA, USA, 14–16 July 2009; pp. 45–55.
2. Lee, J.K.; Cho, B.H.; Jang, K.N.; Jung, S.C.; Oh, K.Y.; Park, J.Y.; Kim, J.S. Development of Autonomous Cable Inspection Robot for Nuclear Power Plant. Proceedings of World Academy of Science, Engineering and Technology, Rome, Italy, 28–30 April 2010; pp. 314–318.
3. Luk, B.L.; Collie, A.A.; Cooke, D.S.; Chen, S. Walking and Climbing Service Robots for Safety Inspection of Nuclear Reactor Pressure Vessels. Proceedings of the Asia Pacific Conference on Risk Management and Safety, Hong Kong, 1–2 December 2005; pp. 432–438.
4. Suzuki, M.; Yukawa, T.; Satoh, Y.; Okano, H. Mechanisms of Autonomous Pipe-Surface Inspection Robot with Magnetic Elements. Proceedings of the IEEE International Conference on Systems, Man and Cybernetics, Taipei, 8–11 October 2006; pp. 3286–3291.
5. Camerini, C.; Freitas, M.; Langer, R.A.; von Der Weid, J.P.; Marnet, R.; Kubrusly, A.C. A Robot for Offshore Pipeline Inspection. Proceedings of the 9th IEEE / IAS International Conference on Industry Applications (INDUSCON), Sao Paulo, Brazil, 8–10 November 2010; pp. 8–10.
6. Xu, J.; Ikram-ul-haq; Chen, J.; Dou, L.; Liu, Z. Moving Target Detection and Tracking in FLIR Image Sequences Based on Thermal Target Modeling. Proceedings of the International Conference on Measuring Technology and Mechatronics Automation (ICMTMA), Changsha, China, 13–14 March 2010; pp. 715–720.
7. Senthil Kumar, K.; Kavitha, G.; Subramanian, R.; Ramesh, G. Visual and Thermal Image Fusion for UAV

Based Target Tracking. In MATLAB—A Ubiquitous Tool for the Practical Engineer; Ionescu, C.M., Ed.; InTech: Rijeka, Croatia, 2001; pp. 307–326. .

8. Padole, C.N.; Alexandre, L.A. Motion Based Particle Filter for Human Tracking with Thermal Imaging. Proceedings of the 3rd International Conference on Emerging Trends in Engineering and Technology (ICETET 2010), Goa, India, 19–21 November 2010; pp. 158–162.

9. Aichholzer, O.; Aurenhammer, F. Straight Skeletons for General Polygonal Figures in the Plane. Proceedings of the 2nd Annual International Conference on Computing and Combinatorics, Hong Kong, 17–19 June 1996; pp. 117–126.

10. Doucet, A.; de Freitas, N.; Gordon, N. Sequential Monte Carlo Methods in Practice; Springer: New York, NY, USA, 2001. .

11. Kotecha, J.H.; Djuric, P.M. Gaussian Particle Filtering. IEEE Trans. Signal Process **2003**, 51, 2592–2601. .

12. Liu, J.S.; Chen, R.; Logvinenko, T. A theoretical Framework for Sequential Importance Sampling and Resampling. In Sequential Monte Carlo Methods in Practice; Springer: New York, NY, USA, 2001; pp. 225–246.

CHAPTER 11

DYNAMIC MODELING OF THE SOLAR FIELD IN PARABOLIC TROUGH SOLAR POWER PLANTS

Lourdes A. Barcia [1], Rogelio Peón Menéndez [2], Juan Á. Martínez Esteban [3], Miguel A. José Prieto [3], Juan A. Martín Ramos [3], F. Javier de Cos Juez [4] and Antonio Nevado Reviriego [5]

[1]Normagrup Technology S.A. Llanera 33420, Asturias, Spain

[2]Grupo TSK, Gijón 33203, Asturias, Spain

[3]Department of Electrical Engineering, University of Oviedo, Gijón 33203, Asturias, Spain

[4]Department of Explotation and Prospection of Mining, University of Oviedo, Asturias 33004, Spain

[5]Department of Electrical, Electronics and Control Engineering, National Distance Education University, Madrid 28040, Spain

ABSTRACT

Parabolic trough solar power plants use a thermal fluid to transfer thermal energy from solar radiation to a water-steam Rankine cycle in order to drive a turbine that, coupled to an electrical generator, produces electricity. These plants have a heat transfer fluid (HTF) system with the necessary elements to transform solar radiation into heat and to transfer that thermal energy to the water-steam exchangers. In order to get the best possible performance in the Rankine cycle and, hence, in the thermal plant, it is necessary that the thermal fluid reach its maximum temperature when leaving the solar field (SF). Also, it is mandatory that the thermal fluid does not exceed the maximum operating temperature of the HTF, above which it degrades. It must be noted that the optimal temperature of the thermal fluid is difficult to obtain, since solar radiation can change abruptly from one moment to another. The aim of this document is to provide a model of an HTF system that can be used to optimize the control of the temperature of the fluid without interfering with the normal operation of the plant. The results obtained with this model will be contrasted with those obtained in a real plant.

INTRODUCTION

Solar radiation can be exploited in several ways. While photovoltaic (PV) systems use luminous solar radiation to produce electricity, thermoelectric solar plants use thermal radiation. There are different technologies that take advantage of the solar thermal energy. They can be classified depending on the temperatures they can reach. In [1], differences between these technologies are shown. Parabolic trough collector (PTC) technology is one of the most mature and, for that reason, the most implanted one worldwide. They use a heat transfer fluid (HTF) that flows inside several absorber pipes placed in the focal axis of a set of mirrors. These mirrors focus the solar radiation over the pipes in such a way that the HTF absorbs the thermal energy of the sun.

Although some of the older concentrated solar power (CSP) plants do not have any sort of thermal energy storage, the fact is that during the last years several installations with a thermal energy storage (TES) system have appeared, as well as papers dealing with different methods to store thermal energy and with how to control such storage [1,2,3,4,5,6].

Citation: Lourdes A. Barcia, Rogelio Peón Menéndez, Juan Á. Martínez Esteban, Miguel A. José Prieto, Juan A. Martín Ramos, F. Javier de Cos Juez and Antonio Nevado Reviriego, Dynamic Modeling of the Solar Field in Parabolic Trough Solar Power Plants, doi:10.3390/en81212373

Figure 1 shows a schematic of a solar thermal power plant of PTCs with a TES system. This thermoelectric solar plant works basically like a conventional thermal one. It consists of a generator steam train that, thanks to the contribution of thermal energy, turns liquid water into steam. This steam drives a turbine coupled to an electric generator that produces electricity.

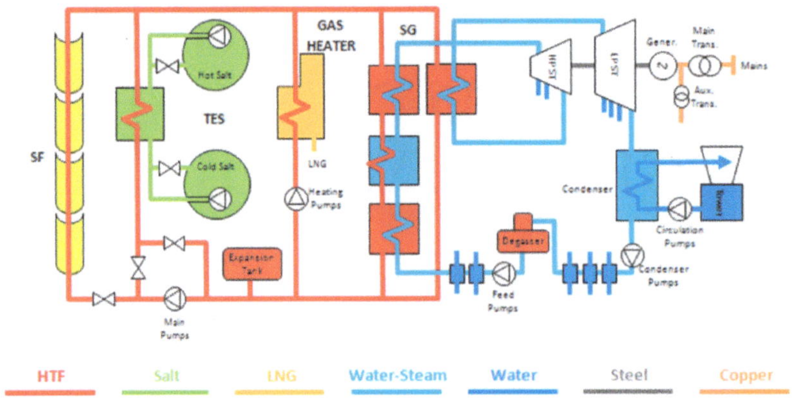

Figure 1: Schematic of a parabolic trough collector (PTC) solar thermoelectric plant.

In order for these plants to be able to use solar thermal radiation, they have a subsystem named solar field (SF). This system absorbs the solar radiation, extracts its thermal energy and transports it to the rest of the subsystems, such as the steam generator (SG) and the TES. A HTF crosses the SF absorbing thermal energy from the sun and transferring it to the rest of the plant.

The main objective is to achieve as high a fluid temperature as possible at the output of the SF, but without exceeding the maximum operating temperature of the fluid. The purpose of the control of the heating process is to define a set point temperature and guarantee that, if the temperature at the output of the SF tends to rise above that set point, the control causes the mass flow of the fluid to rise and *vice versa*. In order to optimize this control system, an accurate model of the process to be controlled is required, since the power plant will not stop production so that different control strategies can be tested. The goal of this paper is to produce a simple yet accurate version of such a model.

RESULTS AND DISCUSSION

Description of the Solar Field (SF)

In order to adequately model the thermal-fluid heating process, it is mandatory to understand the main elements of the HTF system. This system contains several solar collector assemblies (SCA), Figure 2 formed by solar collector elements (SCE), Figure 3. Each SCE has mirrors (28 mirrors in Figure 3) as well as absorber pipes arranged in such a way that the SCE is parabolic-trough shaped and the pipes are placed in its focal axis.

Depending on the solar thermal plant, the SCA may have six, eight or 12 SCEs. The dynamic model developed in this paper is based on the HTF system of the PTC solar plant "La Africana", located in Cordoba, Spain. In the SF of "La Africana", each SCE has three absorber tubes arranged in series and each SCA is formed by 12 SCEs.

SCAs are arranged in series forming loops. These loops can be U-shaped or W-shaped. In the model described, U-shaped loops with four SCAs each are considered.

Figure 2: Solar collector assembly (SCA).

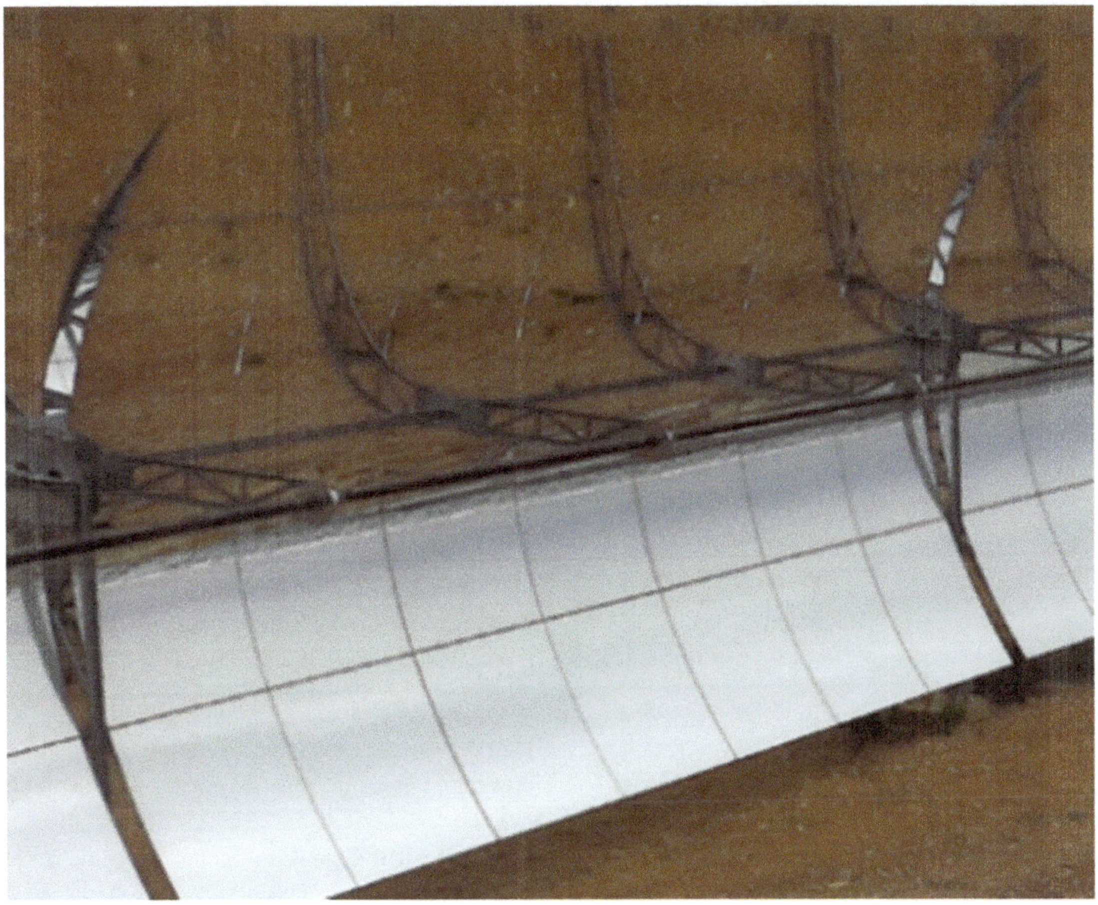

Figure 3: Solar collector element (SCE) with 28 mirrors and three absorber pipes.

The beginning of the loop is linked to a large diameter pipe (cold header) that contains cold HTF (HTF that has not been heated by the solar radiation yet). The end of the loop is linked to another large diameter pipe (hot header) in such a way that, once the fluid coming out of the loops has been heated, it is transported to the steam generator system (SGS) and/or the TES system, depending on the operation mode. Figure 4 shows one of these loops.

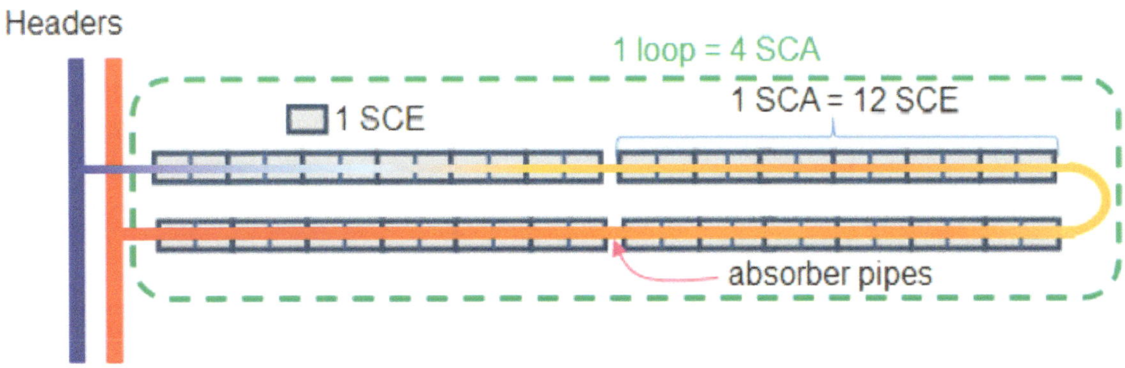

Figure 4: U-Shaped loop in PTC solar plants.

The elements included in the HTF system, also referred to as SF, modeled in this paper has 168 loops with 4 SCA per loop. Each SCA consists on 12 SCE that contains 3 absorber pipes. Therefore, this HTF systems has 24.192 absorber pipes transporting the fluid.

The most important elements to take into account in the modeling of this process are the absorber pipes, shown in Figure 5 They are in charge of absorbing the solar radiation and turning it into thermal energy. Thanks to the fluid that flows through them, they also transport that thermal energy to the rest of the plant where needed.

Figure 5: Headers.

They consist of two concentric pipes. The inner one is metallic with a blue film to minimize radiation thermal losses, whereas the outer pipe is made of glass. Vacuum is created between them in order to avoid convention and conduction thermal losses. Figure 6 shows the main parts of an absorber pipe.

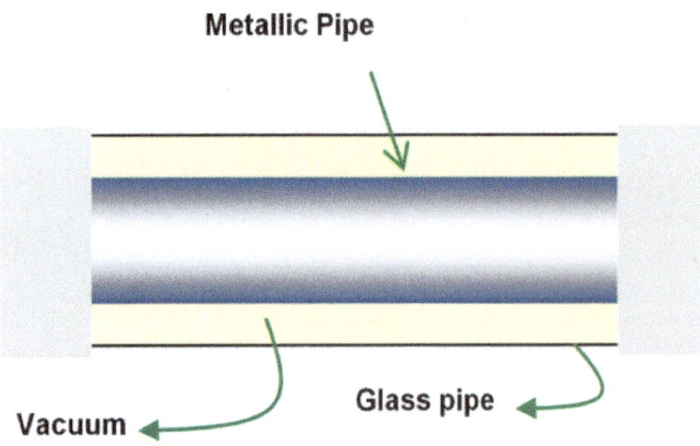

Figure 6: Absorber pipe.

Description of the Heat Transfer Fluid (HTF)-Heating Process

The solar radiation is reflected by the mirrors over the absorber pipes, thus heating the HTF. A comparative study of the different thermal fluids most currently used in PTC power plants (oil, molten salt and water steam) is presented in [7]. The authors conclude that direct steam generation is more efficient than the other two. In spite of that, and although other alternatives are being studied [8,9,10,11,12,13,14], most of the PTC plants use some sort of synthetic oil, mainly due to its lower freezing point (around 12 °C) when compared to other HTFs.

According to the literature, there are different kinds of fluids that have been or are being analyzed to be used as HTF. One of them is molten salt. Taking into account that they are used to store thermal energy in the TES, it would be relatively easy to use it also in the SF. The main problem is its high freezing point (around 140 °C). Due to the small internal diameter of the absorber pipes, having part of the molten salts freeze would cause blockages that could cause the pipes to break. Other installations use direct steam generators (DGS) instead, in which water acts as HTF. The main advantages of this technology are its lower thermal losses and, therefore, its higher cycle efficiency. Its drawbacks are the higher pressures required and the more complex control needed. These drawbacks are also present in installations that use gas and air as HTF. Two new research lines are emerging nowadays: Nano fluids and ionic fluids. The first one consists of adding metallic nanoparticles or carbon nanotubes in order to improve its optical and thermal properties. Ionic fluids are very promising. They are formed by joining anions and cations so that they can result in billions of different ionic fluids for multiple applications.

Using synthetic oil as HTF also has drawbacks, the main one being its maximum operation temperature, which is around only 400 °C in the best-case scenario. This critical point can be easily exceeded when direct solar irradiation (DNI) is high. Given that solar radiation is a phenomenon that can vary widely in a short space of time, a good regulation method is necessary. This method should allow the oil to leave the SF with the highest possible temperature while, at the same time, preventing the oil from exceeding its maximum operating point.

To meet this goal, the control method varies the mass flowof the fluid in such a way that, if the oil temperature exceeds a given set point (393 °C in the case considered), the mass flow will increase. Thus, the time the fluid stays in the SF will be shorter and its temperature will drop. Similarly, if the oil temperature is below the set point, the mass flow must be reduced. Table 1 and Table 2 show all the variables (perturbations, input, output and control variables) involved in the HTF-heating process:

Table 1: Input, output and control variables of the HTF-heating process.

Variable	Acronym	Description
Input	THTFc	Heat transfer fluid (HTF) temperature at the input of the solar field (SF) (°C)
Output	THTFh	HTF temperature at the output of the SF (°C)
Control	$\dot{m}_{HTF}$	HTF mass flow (kg/s)

Table 2: Perturbations of the HTF-heating process.

Perturbations	Description
Direct solar irradiation (DNI)	Direct normal irradiation (W/m^2)
ϕ	Angle of incidence (°)
Tamb	Ambient temperature (°C)

Figure 7 shows the block diagram that represents the process described above.

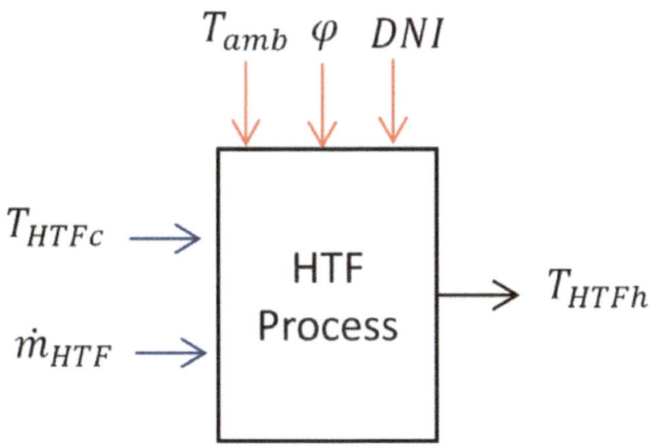

Figure 7: Heat transfer fluid (HTF)-heating process model.

The goal of this work is to obtain a simple, accurate and dynamic model that allows designers to optimize the control of this process.

Modeling of the HTF-Heating Process

Several models of different parts of the PTC thermoelectric plants have been developed in the last years. In [15], a mathematical model is described that compares the quasi-dynamic test (QDT) and the steady-state test (SST) in five different types of collectors. In [16], the authors present a dynamic model for the collector field and a steady-state model for the power plant; this paper is based on a real PTC solar plant and the whole plant has been mathematically modeled. In [17] models for different geometries and insulation materials with different thicknesses have been carried out in order to select the best one for high temperature processes; in this paper, all the models are numerically compared and, hence, absolute values are not provided. A paper written by one of the main manufacturers of solar absorber pipes [18] explains all the thermal exchanges between the different parts or elements that form a pipe; this paper is a good starting point to develop the dynamic model of the HTF-heating process. In reference [19], the authors established a computational fluid dynamics (CFD) model to calculate the temperature profile on the wall of the absorber tubes of DSG plants. The authors of [20] compare three different models of absorber pipes: a one- and a two-dimensional analytical model and a three-dimensional one using finite-element techniques.

It must be said that most of these references model the absorber pipes in order to determine the gradient of temperature between their elements, but they are not used as a dynamic model susceptible of being used in the control of the HTF-heating process.

To model the HTF-heating process, some simplifications of the SF must be carried out, namely, the 168 loops have been modeled as four serial-connected unitary blocks. Each unitary block contains 168 parallel-connected SCAs. This configuration is shown in Figure 8.

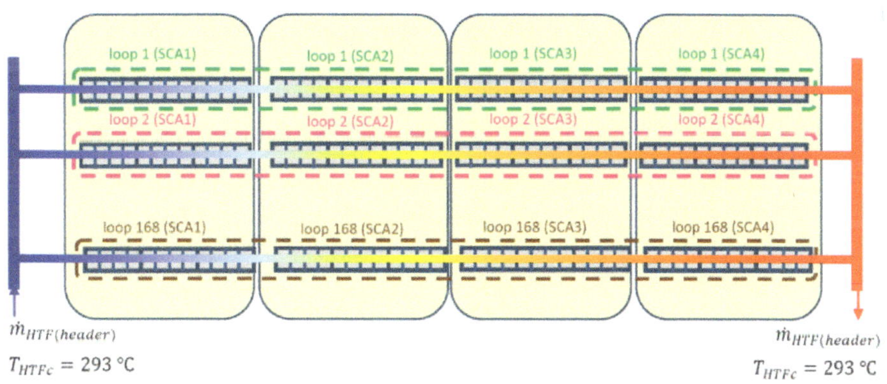

Figure 8: Loop simplification.

Similarly, the headers are modeled as four serial-connected blocks, each containing a quarter of the total amount of HTF. Figure 9 shows the block diagram of the SF to be used in the modeling of the HTF-heating process, and Figure 10 shows the blocks used to model the loops described above.

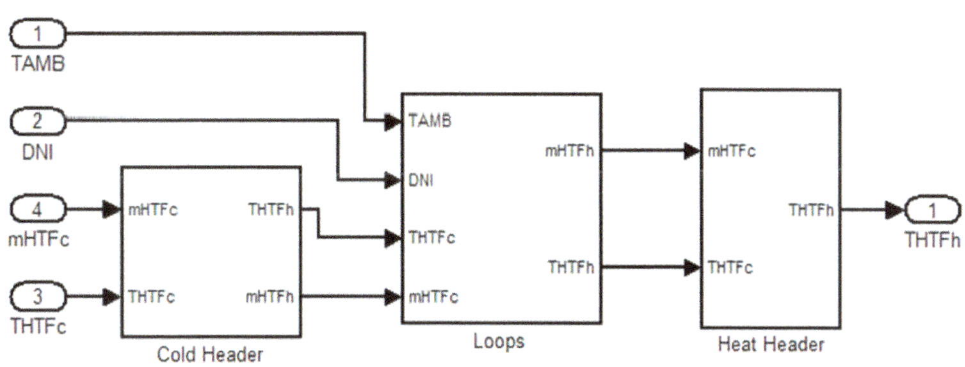

Figure 9: Solar field (SF) model.

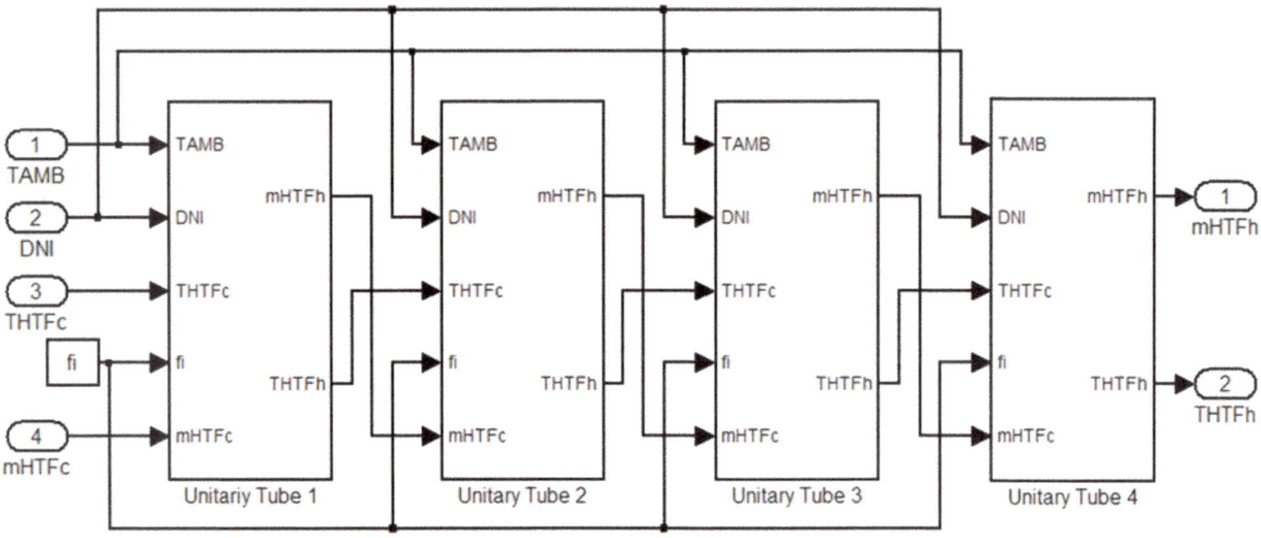

Figure 10: Loops model.

These Simulink blocks can be easily used for modeling larger or smaller SFs by respectively adding or removing unitary tubes.

Energy transfer from solar radiation to the fluid takes place inside the unitary tubes as well as inside the headers. For that reason, all the equations related to mass and energy conservation principles should be included in those blocks.

The conservation energy principle provides the following equations:

$$\dot{E}_{\text{input}} = \dot{E}_{\text{output}} \tag{1}$$

Energy can be transferred as work, heat and mass:

$$\dot{Q}_{\text{in}} + \dot{W}_{\text{in}} + (\dot{m} \cdot e)_{\text{in}} = \dot{Q}_{\text{out}} + \dot{W}_{\text{out}} + (\dot{m} \cdot e)_{\text{out}} \tag{2}$$

where Q˙ is the absorbed or released heat, W˙ is the work done by the HTF and e is the energy transported per unit mass of a moving fluid, whose expression is given by Equation (3):

$$e = h + \frac{1}{2}V^2 + gz \tag{3}$$

Finally, assuming that in the absorber pipes kinetic and gravitational energies are negligible and that, due to the mass conservation principle in the control volume considered in the process, input and output mass flows are similar, Equation (2) is simplified as indicated in Equation (4):

$$h(t) = \int_0^t \frac{\dot{m}_{\text{HTF}} \cdot (h_i - h_o) + \dot{Q}_{\text{gain}}}{\rho_{\text{HTF}} \cdot v_c} \cdot dt \tag{4}$$

where $\dot{Q}_{\text{gain}}$ is the heat that the HTF absorbs from the solar radiation once all the losses due to thermal transfer from the fluid to the environment have been taken into account.

The net heat transferred from the HTF to other subsystems of the solar plant is the result of the total heat absorbed by the fluid minus the thermal losses due to radiation, convection and conduction from the fluid to the air around it.

The heat absorbed by the fluid depends on geometric, optical and thermal factors. Firstly, not all the solar radiation that reaches the mirrors $(\dot{Q}_{\text{capt}})$ is reflected over the absorber tubes and then absorbed by the fluid, due to phenomena such as shadows, angle of incidence far from the optimal 90°, dirty mirrors, non-optimal reflectivity and transmissivity, etc. This results in the heat that reaches the fluid being reduced by the efficiency of the SCA (η_{SCA}). Additionally, of all the heat reaching the fluid, some is lost in unwanted thermal transfers $(\dot{Q}_{\text{thloss}})$:

$$\dot{Q}_{\text{gain}} = \dot{Q}_{\text{capt}} \cdot \eta_{\text{SCA}} - \dot{Q}_{\text{thloss}} \tag{5}$$

All these phenomena have been modeled in a Simulink block (Figure 11) called $\dot{Q}_{\text{gain}}$.

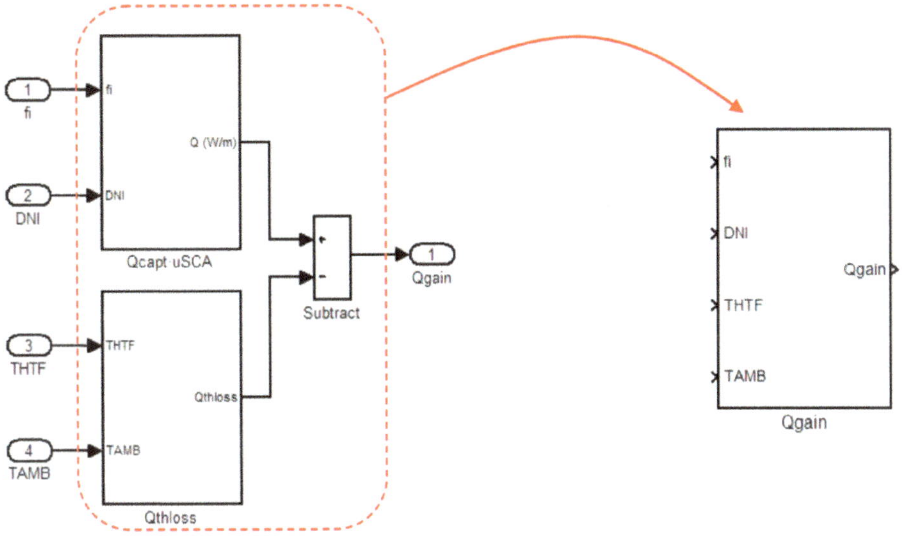

Figure 11: Simulink block model to calculate $\dot{Q}_{gain}$

Figure 12 shows all the thermal exchanges involved in the different parts of the SCE (mirrors, absorber tube, HTF) with the environment. Some of those losses are negligible when compared to others [15,16,17]. This is the case of conduction and convection losses between the outer wall of the metallic tube and the inner wall of the glass cover (thanks to the vacuum created between them), or of conduction thermal losses produced in the walls of the tubes to mention a few.

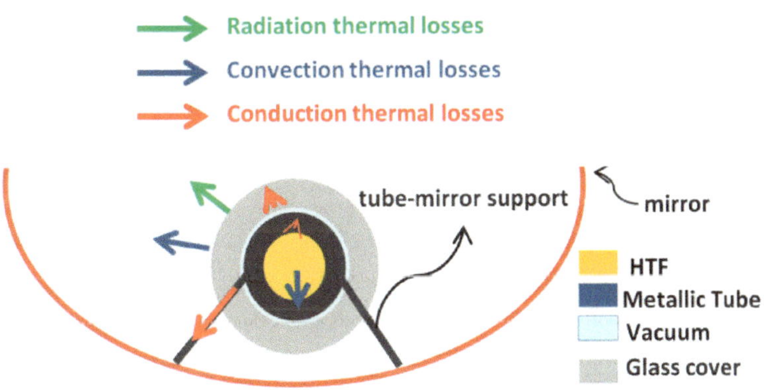

Figure 12: Thermal exchanges between solar collector elements (SCE) and surround ambient.

Figure 13 shows the Simulink blocks used to model all those thermal losses, which are included in the ($\dot{Q}_{thloss}$) block of Figure 11.

Several equations are included in these blocks:

$$\dot{Q}_{radGLASS-AMB} = \varepsilon_{CUB} \cdot \sigma \cdot Ae_{GLASS} \cdot \left(T_{GLASS}^4 - T_{AMB}^4\right) \tag{6}$$

where $\dot{Q}_{radGLASS-AMB}$ represents the radiation thermal losses per meter (W/m) from the outer wall of the absorber pipe glass cover to the environment, ε_{GLASS} is the emissivity coefficient of the external surface of the glass cover, σ is the Stefan-Boltzmann constant ($W/m^2 \cdot C^4$), Ae_{GLASS} is the external surface of the absorber pipe glass cover per unit length (m^2/m) and TGLASS and TAMB are, respectively, the glass cover and the ambient temperatures (°C):

$$\dot{Q}_{convGLASS-AMB} = h_{GLASS-AMB} \cdot Ae_{GLASS} \cdot \left(T_{GLASS} - T_{AMB}\right) \tag{7}$$

where $\dot{Q}_{convGLASS-AMB}$ represents the convection thermal losses per meter (W/m) from the outer wall of the absorber pipe glass cover to the surround ambientand $h_{GLASS-AMB}$ is the convection coefficient between the external surface of the glass cover and the surround ambient. This coefficient is highly dependent on wind speed and can be calculated as:

$$h_{GLASS-AMB} = \frac{Nu_{AMB}}{De_{GLASS}} \cdot K_{AMB} \tag{8}$$

In Equation (8), Nu_{AMB} is the dimensionless Nusselt number, De_{GLASS} is the outer diameter of the glass cover and K_{AMB} is the thermal conductivity of the air.

To calculate the Nusselt number, Hilpert equation for forced convection flow is used:

$$Nu_{AMB} = C \cdot Re^{m} \cdot Pr^{1/3} \quad (9)$$

where *Re* is the dimensionless Reynolds number and parameters, *C* and *m* depend on the value of the Reynolds number. Pris the dimensionless Prandtl number of the air, the value of which is around 0.7.

Finally, the dimensionless Reynolds number of the air over the outer wall of the glass cover is calculated from Equation (10):

$$Re = \frac{V_{air} \cdot De_{GLASS} \cdot \rho_{air}}{\mu_{air}} \quad (10)$$

where V_{air}, ρ_{air}, μ_{air} are the air speed, density and dynamic viscosity respectively.

Equation (11) is used to calculate thermal radiation transference from the outer surface of the metallic pipe to the inner wall of the glass cover $\dot{Q}_{radMET-CUB}$:

$$\dot{Q}_{radMET-GLASS} = \frac{\sigma \cdot Ae_{MET} \cdot (T_{MET}^4 - T_{GLASS}^4)}{\frac{1}{\varepsilon_{MET}} + \frac{1-\varepsilon_{GLASS}}{\varepsilon_{GLASS}} \cdot \left(\frac{De_{MET}}{Di_{GLASS}}\right)} \quad (11)$$

where Ae_{MET} is the external surface of the metalic pipe per unit length (m^2/m) and TMET and TGLASS are, respectively, the metallic pipe and the glass cover temperatures (°C).

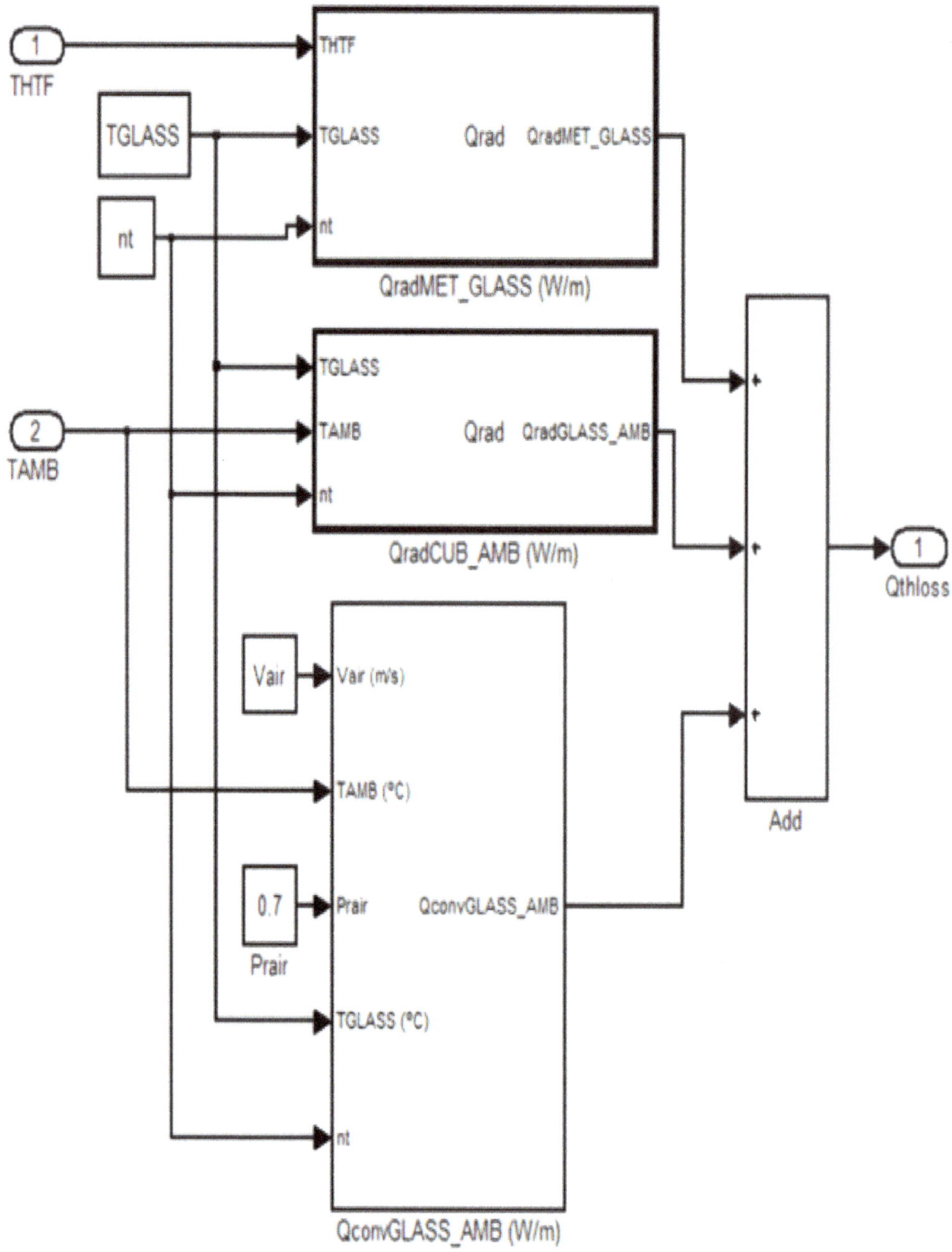

Figure 13: Thermal losses blocks.

Once all the blocks have been defined and all the equations have been included, the dynamic model of the HTF heating process is obtained. Figure 14 shows the Simulink diagram that includes all the blocks and equations described above. Inside that block there are the four unitary tubes that contain the time-dependent Equation (4) as well as the block $\dot{Q}_{gain}$ described in Figure 11. In this way, the fluid temperature varies along the path as do all the temperature-dependent variables (enthalpy, fluid density, *etc.*).

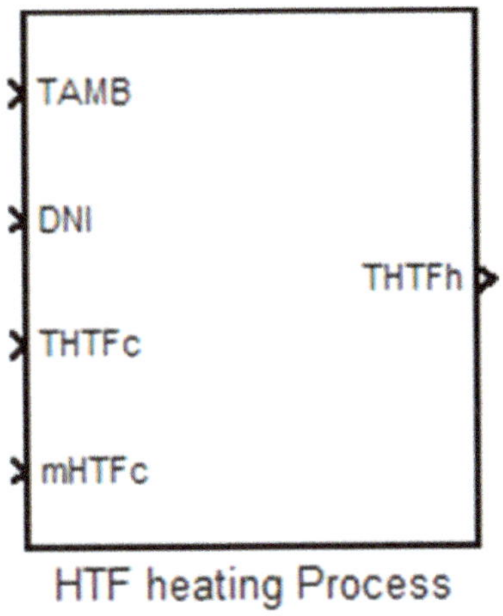

Figure 14: HTF heating process.

EXPERIMENTAL SECTION

To validate the already developed model, the control strategy usually applied in thermoelectric solar plants was implemented. Due to the complexity of a PTC solar plant and the amount of subsystems included in it, there are many controllers to regulate different outputs: mirrors position (in order to have the lowest possible optical losses), temperature and flow rate in the TES system (in order to have as much storage as possible during the molten salt charge process and the highest energy transference to the fluid in the discharging mode [1]) and so on. To cope with all this, solar plants must have a distributed control system (DCS) [21] that receives field signals (such as ambient temperature, HTF temperature, incident angle, the DNI, *etc.*) and sends them to the appropriate controller, which generates the adequate actions in each case.

The control of the HTF heating is typically performed in either of two different ways. Some plants have a semiautomatic proportional-integral-derivative (PID) regulator that relies on an operator observing the output variables of the process, mainly the HTF temperature, and changing the mass flow input set point to make this output temperature be constant despite the perturbations. Once the mass flow input set point has been fixed by the operator, internal PID regulators will operate the pumps to achieve that set point. This semiautomatic method has two drawbacks: in addition to its accuracy depending on the operator's experience, it is carried out by steps. Figure 15 shows a diagram of the regulator.

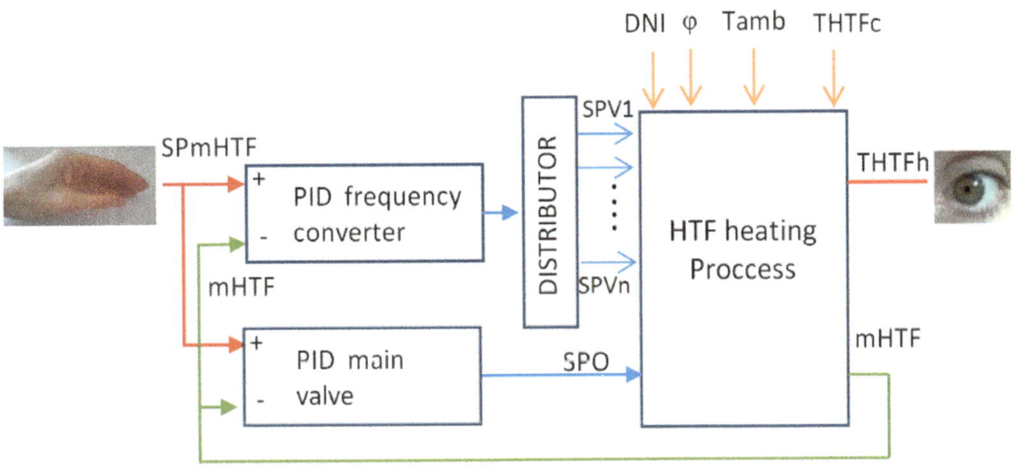

Figure 15: Semiautomatic control of the HTF-heating process.

More modern plants have substituted the operator by an external PID regulator in such a way that this new controller is the one that determines the HTF mass flow to achieve as a result of a comparison between the real temperature of the HTF at the output of the SF (THTFh) and the set point of the fluid temperature at the output of the SF, previously fixed (SPTHTFh). Figure 16 shows the PID automatic control.

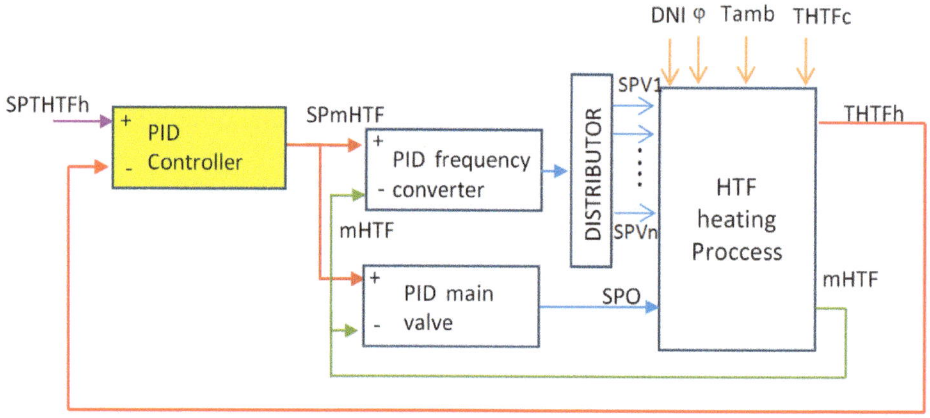

Figure 16: Proportional-integral-derivative (PID) automatic control of the HTF-heating process.

Figure 17 shows how the dynamic model of the heating process was linked to a model of a PID regulator using also Matlab and Simulink. Different perturbations where introduced in the input variables (Tamb, DNI and THTFc) and the performance of the system was analyzed.

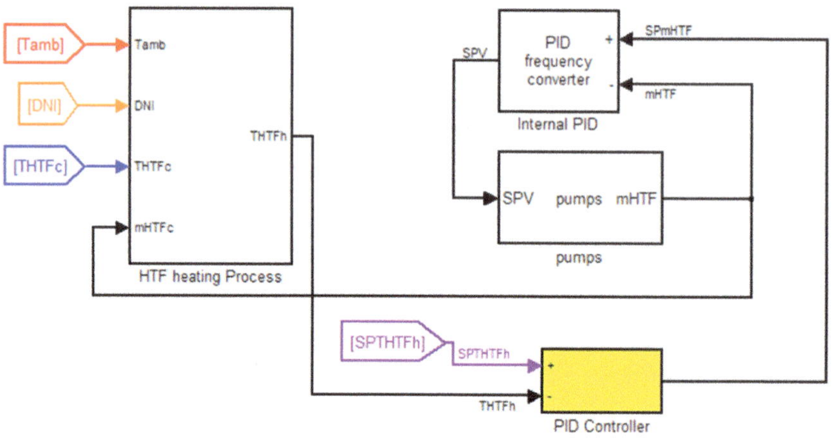

Figure 17: MatLab-Simulink model of the PID control for the HTF-heating process.

Figure 18 shows the response of the process to a sharp step perturbation in the DNI. This sort of perturbations may occur, for example, when multiple clouds suddenly disappear.

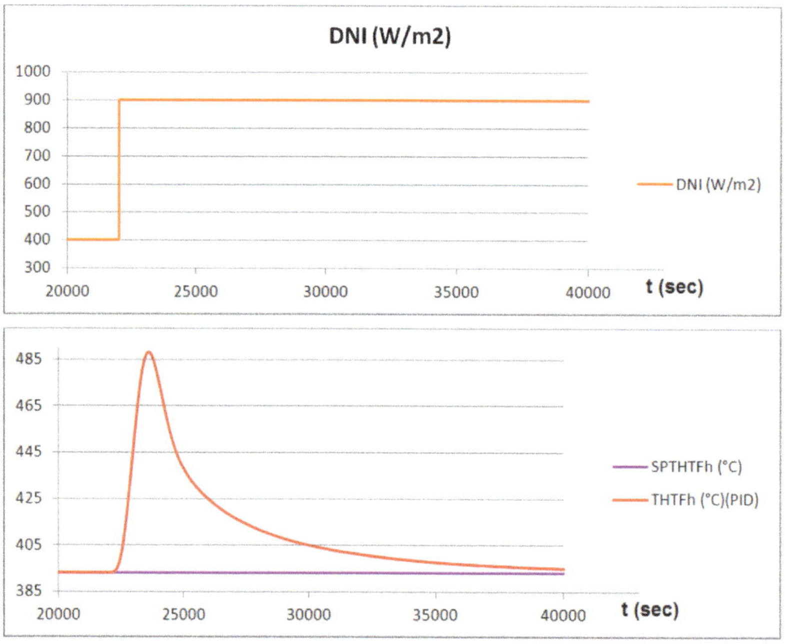

Figure 18: PID controller response to a sharp step in direct solar irradiation (DNI).

It is possible to see that the behavior of the PID when there is a sharp variation in the solar radiation is far from optimal, and a very high temperature can be reached, which would result in faster fluid degradation. Besides, the settling time is also high so any intent to reduce the high peak, will give rise to an even slower settling time. The slow response to a rapid and sharp change in an input variable as the DNI can be explained by the amount of time the oil stays in the SF, which gives rise to thermal inertia. One of the effects of this inertia is transport delay; a variation of temperature at the inlet of the pipe is not reflected at the output until sometime later. The other effect of thermal inertia is similar to a filtering. At a constant flow rate, a step of radiation produces a variation in the temperature at the loop output that looks like a pure delay stage with respect to the variation of radiation.

In the collectors that gather and mix the oil from the loops, the HTF of each loop has a different delay depending on their distance to the power block (also called power island), which is normally located in the center or north of the SF. The overall effect is that a step in the solar radiation results in a delay plus an additional filtering effect in such a way that the global response to that step is a pure delay followed by a second-order overdamped response.

Once the steady state has been established, the average error between the set point and the real temperature is below 1 °C.

The model was also subjected to values of the input variables extracted from an actual plant called "La Africana" situated in Cordoba, Spain. These values were taken in different years and seasons: September 2010, April 2011 and July 2013 and they were extracted from the DCS. The response of the model developed in this paper, to those input variables can be seen in Figure 19, Figure 20 and Figure 21.

In these figures, horizontal axis represents the time in seconds and the vertical axis shows the true representation of the absolute values of the different variables.

Figure 19 shows that, when variations in the input variables are not sharp, the PID response does not involve high peak temperatures. Table 3 shows the error average value (μ) and the error standard deviation (σ) calculated for all the values obtained during 5 h.

Table 3: Numerical results of simulation with PID (September 2010)

Control Strategy	Input Values	$e = SPTHTF_h - THTF_h$	
		μ	σ
Proportional-integral-derivative (PID) controller	September 2010	−1.92	3.27

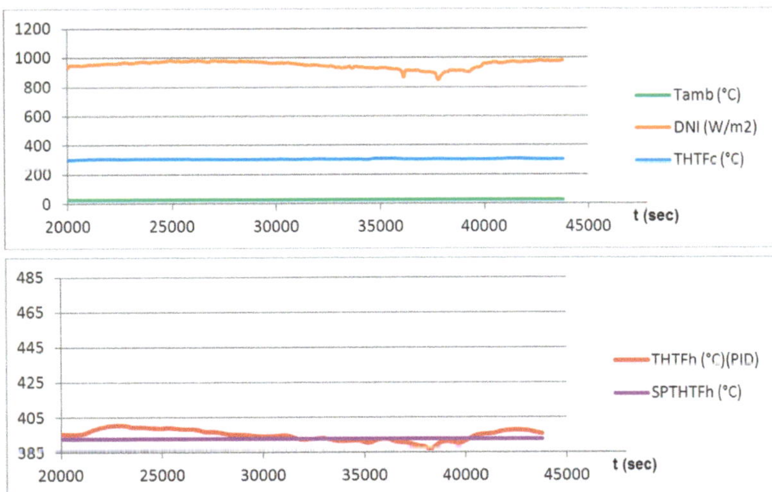

Figure 19: PID controller response to September 2010 values.

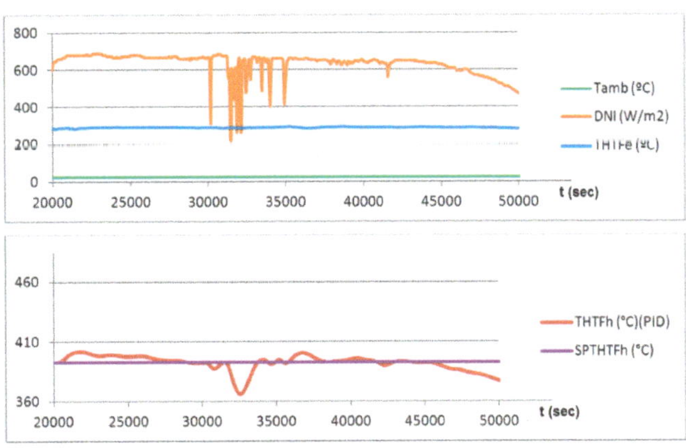

Figure 20: PID controller response to April 2011 values.

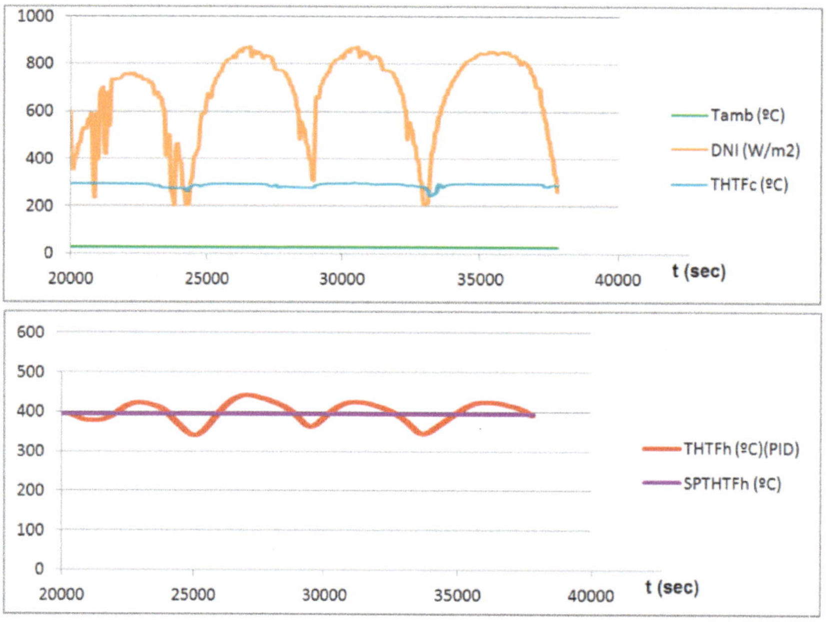

Figure 21: PID controller response to July 2013 values.

When solar radiation falls below a certain value, however, the PID controller is not able to provide rapid response and temperature deviates greatly from its set point, as can be seen in Figure 20. This is also confirmed by the value of the standard deviation in Table 4.

Table 4: Numerical results of simulation with PID (April 2011).

Control Strategy	Input Values	$e = SPTHTF_h - THTF_h$	
		μ	σ
PID controller	April 2011	6.35	15.11

If the set of values corresponding to a DNI under 500 W/m^2 are removed, the error and standard deviation values are shown in Table 5.

Table 5: Numerical results of simulation with PID (April 2011).

Control Strategy	Input Values	$e = SPTHTF_h - THTF_h$	
		μ	σ
PID controller	April 2011	−3.20	12.04

Figure 21 shows the PID response when there are quick and sharp changes in the solar direct normal irradiation. Even though the PID succeeds in following these variations, the oscillation is high and the standard deviation is large, as indicated in Table 6.

Table 6: Numerical results of simulation with PID (July 2013).

Control Strategy	Input Values	$e = SPTHTF_h - THTF_h$	
		μ	σ
PID controller	July 2013	−4.06	25.90

Looking at the PID response for different values of the input variables (some of which have been obtained from an actual thermal power plant), it can be shown that this controller has an adequate behavior when those input variables do not vary suddenly. However, it has a very slow response to quick variations; therefore this control method should be improved if the global efficiency of these plants has to be optimized.

The goal of this paper, however, is not determining what the best control strategy is, but rather defining a model of the SF that can be later used to optimize such strategies. In this regard, it must be said the results obtained with the model presented match those measured in the actual plant under similar conditions.

CONCLUSIONS

A dynamic model of the HTF heating process as it flows through the SF has been developed. The use of Simulink has allowed a model of the process to be obtained that is very visual and easy to follow. Besides, since it has been built from the simplest blocks to the more complicated ones, it can also be used to model the fluid heating process of other solar plants with different sizes by just adding or removing unitary tubes (see Figure 10).

Some of the fluid properties such as density, thermal conductivity, specific heat, *etc.* have been included as temperature dependent equations that can be easily changed so that the model can be also used in solar plants where the HTF is not synthetic oil (as in the case considered) but water steam, molten-salt, *etc.*

Looking at Figure 18, Figure 19, Figure 20 and Figure 21 it is possible to see that, even though a PID controller might be suitable when there are no rapid changes in environment conditions, its response is slow when a sharp change is produced. The developed model is able to identify this situation and can therefore be used to test other algorithms of control, such as predictive ones, adaptative, and so on, to optimize the behavior of the HTF temperature controller without interfering in the operation of the plant, *i.e.* using just simulations. Once the regulator has been selected, it can be implemented in the real plant.

ACKNOWLEDGMENTS

This work has been subsidized through the Plan of Science, Technology and Innovation of the Principality of Asturias, (Ref: FC-15-GRUPIN14-122).

CONFLICTS OF INTEREST

The authors declare no conflict of interest.

REFERENCES

1. Peón Menéndez, R.; Martínez, J.A.; Prieto, M.J.; Barcia, L.Á.; Martín Sánchez, J.M. A Novel Modeling of Molten-Salt Heat Storage Systems in Thermal Solar Power Plants. *Energies* 2014, *7*, 6721–6740.
2. Barlev, D.; Vidu, R.; Stroeve, P. Innovation in concentrated solar power. *Sol. Energy Mater. Sol. Cells* 2011, *95*, 2703–2725.
3. Kuravi, S.; Trahan, J.; Goswami, D.Y.; Rahman, M.M.; Stefanakos, E.K. Thermal energy storage technologies and systems for concentrating solar power plants. *Prog. Energy Combust. Sci.* 2013, *39*, 285–319.
4. Kenisarin, M.M. High-temperature phase change materials for thermal energy storage. *Renew. Sustain. Energy Rev.*2010, *14*, 955–970.
5. Gil, A.; Medrano, M.; Martorell, I.; Lazaro, A.; Dolado, P.; Zalba, B.; Cabeza, L.F. State of the art on high temperature thermal energy storage for power generation. Part 1—Concepts, materials and modellization. *Renew. Sustain. Energy Rev.* 2010, *14*, 31–55.
6. Medrano, M.; Gil, A.; Martorell, I.; Potau, X.; Cabeza, L.F. State of the art on high-temperature thermal energy storage for power generation. Part 2—Case studies. *Renew. Sustain. Energy Rev.* 2010, *14*, 56–72.
7. Montes, M.J.; Abánades, A.; Martínez-Val, J.M. Thermofluidynamic Model and Comparative Analysis of Parabolic Trough Collectors Using Oil, Water/Steam, or Molten Salt as Heat Transfer Fluids. *J. Sol. Energy Eng.* 2010, *132*.
8. Montes, M.J.; Abanades, A.; Martinez-Val, J.M.; Valdes, M. Solar multiple optimization for a solar-only thermal power plant, using oil as heat transfer fluid in the parabolic trough collectors. *Sol. Energy* 2009, *83*, 2165–2176.
9. Lopez-Gonzalez, D.; Valverde, J.L.; Sanchez, P.; Sanchez-Silva, L. Characterization of different heat transfer fluids and degradation study by using a pilot plant device operating at real conditions. *Energy* 2013, *54*, 240–250.
10. Feldhoff, J.F.; Schimtz, K.; Eck, M.; Schnatbaum-Laumann, L.; Ortiz-Vives, F.; Laing, D.; Schulte-Fischedick, J. Comparative system analysis of direct steam generation and synthetic oil parabolic trough power plants with integrated thermal storage. *Sol. Energy* 2012, *86*, 520–530.
11. Haddad, Z.; Abid, C.; Oztop, H.F.; Mataoui, A. A review on how the researchers prepare their nanofluids. *Int. J. Therm. Sci.* 2014, *76*, 168–189.
12. Flamant, G.; Gauthier, D.; Benoit, H.; Sans, J.L.; Boissiere, B.; Ansart, R.; Hemati, M. A new heat transfer fluid for concentrating solar systems: Particle flow in tubes. *Energy Procedia* 2014, *49*, 617–626.
13. Ghadimi, A.; Saidur, R.; Metselaar, H.S.C. A review of nanofluid stability properties and characterization in stationary conditions. *Int. J. Heat Mass Transf.* 2011, *54*, 4051–4068.
14. Alguacil, M.; Prieto, C.; Rodriguez, A.; Lohr, J. Direct steam generation in parabolic trough collectors. *Energy Procedia*2014, *49*, 21–29.
15. Osório, T.; Carvalho, M.J. Testing of solar thermal collectors under transient conditions. *Energy Procedia* 2012, *30*, 1344–1353.
16. Thorsen, S.; Blair, N.; Mitchell, J.A.; Beckman, W.A. Automatic Control of a 30 MWe SEGS VI parabolic trough plant.*Sol. Energy* 2004, *76*, 187–193.
17. Roldán, M.; Cañadas, I.; Casas, J.L.; Zarza, E. Thermal analysis and design of a solar prototype for high-temperature processes. *Int. J. Heat Mass Transf.* 2013, *56*, 309–318.
18. Burkholder, F.; Kutscher, C. *Heat Loss Testing of Schott's 2008 PTR70 Parabolic Trough Receiver*; National Renewable Energy Laboratory: Golden, CO, USA, 2009.
19. Roldán, M.; Valenzuela, L.; Zarza, E. Thermal analysis of solar receiver pipes with superheated steam. *Appl. Energy*2013, *103*, 73–84.
20. Eck, M.; Feldhoff, J.F.; Uhlig, R. Thermal Modelling and Simulation of Parabolic Trough Receiver Tubes. In Proceedings of the 4th International Conference of Energy Sustainability, Phoenix, AZ, USA, 17–22 May 2010; pp. 659–666.
21. Bong-Kuk, L.; Yong-Hak, S. The integrated monitoring and control system for the combined cycle power plant. In Proceedings of the International Conference on Control, Automation and Systems, Seoul, Korea, 14–17 October 2008; pp. 1479–1483.

CHAPTER 12 COST EFFECTIVE OPERATING STRATEGY FOR UNIT COMMITMENT AND ECONOMIC DISPATCH OF THERMAL POWER PLANTS WITH CUBIC COST FUNCTIONS USING TLBO ALGORITHM

E. B. Elanchezhian, S. Subramanian, S. Ganesan

Department of Electrical Engineering, Annamalai University, Chidambaram, India

ABSTRACT

This paper deals with a Unit Commitment (UC) problem of a power plant aimed to find the optimal scheduling of the generating units involving cubic cost functions. The problem has non-convex generator characteristics, which makes it very hard to handle the corresponding mathematical models. However, Teaching Learning Based Optimization (TLBO) has reached a high efficiency, in terms of solution accuracy and computing time for such non-convex problems. Hence, TLBO is applied for scheduling of generators with higher order cost characteristics, and turns out to be computationally solvable. In particular, we represent a model that takes into account the accurate higher order generator cost functions along with ramp limits, and turns to be more general and efficient than those available in the literature. The behavior of the model is analyzed through proposed technique on modified IEEE-24 bus system.

INTRODUCTION

The non-storable nature of electrical energy calls for permanent adjustment of production to consumption. Since consumption varies constantly throughout a day, any utility should forecast these variations, and each day, draw up a schedule for its production facilities for the next day. These processing functions are an essential stage in preparation of "real time" control of an electrical network. One of these processing functions is the determination of thermal Unit Commitment (UC). The UC problem is a large-scale, non-convex, non-linear, and mixed integer optimization problem. The optimal solution of the problem can be obtained by complete enumeration, which is prohibitive in practice owing to its excessive computational resource requirements. So, attempts are being continuously made to solve this problem by reliable iterative and heuristic methods.

A bibliographical survey on UC reveals that a good amount of numerical optimization techniques [1] -[4] and meta-heuristic [5] -[13] methods have been applied to achieve efficient and near optimal solutions. Traditionally to solve the UC effectively, conventional techniques need the incremental fuel cost curves to be featured of monotonically increasing and continuous. But practically, the generating units actually have non-monotonic incremental fuel cost curves. For simplicity

Citation: Elanchezhian, E. , Subramanian, S. and Ganesan, S. (2015) Cost Effective Operating Strategy for Unit Commitment and Economic Dispatch of Thermal Power Plants with Cubic Cost Functions Using TLBO Algorithm. Journal of Power and Energy Engineering, 3, 20-30. doi: 10.4236/jpee.2015.36003.

and easy solving purposes, the input-output characteristics of thermal generators are usually approximated by quadratic functions or piecewise linear functions [1] - [13]. However, many approaches have been addressed that the Economic Dispatch (ED) solution accuracy can be improved or the exact cost can be obtained with higher order functions or Cubic Cost Functions (CCF) [14] - [17]. Earlier papers in the literature have not considered the third order cost functions because they focus on their algorithms and lost the accuracy by approximating the realistic cost curve. In our previous work, we proposed an economic dispatch model using CCF [17]. Even though the UC research is solved for past years, less work is carried out for UC with Cubic Cost Functions (UC-CCF).

The evolutionary and population based techniques in the literature survey are probabilistic techniques and necessitate some common controlling parameters like population size, number of generations and elite size. Moreover, the algorithms also require finding of some of their own control parameters. Genetic Algorithm (GA) needs crossover rate and mutation rate, Particle Swarm Optimization (PSO) has its own parameters like inertia weight, and social and cognitive parameters. The global solution of any function is only achieved with the proper tuning of these algorithm-specific parameters. Improper tuning may lead to local optimal solution or increase in convergence time.

In this paper, a recently developed heuristic algorithm named teaching learning based optimization (TLBO) algorithm based on the effect of the influence of a teacher on the output of learners in a class, introduced by Rao et al. [18] , is utilized for the solution of UC problem. This TLBO algorithm has been implemented in various problem domains of engineering and technology. Most recently, it has been utilized in solving a few areas in power system [17] [19] - [21]. Unlike other population based techniques, TLBO requires only determination of common controlling parameters like population size and number of generations for its functionality. In the present work, TLBO based UC (TLBO-UC) with CCF (TLBO-UC-CCF) is carried on 26-unit test system considering ramp rate constraints for a time horizon of 24 hours to prove the scalability of the algorithm.

UNIT COMMITMENT PROBLEM FORMULATION

Mathematically, the unit commitment problem can be formulated as a mixed integer non-linear problem. The objective function and constraints are formulated as given below.

Objective Function

The objective function of UC problem is the minimization of the total cost (F_{Tuc}) (Equation (1)) which is the sum of the fuel cost (F) and the startup cost (SUC) of individual units (N) for the given period (T) subject to various constraints. Mathematically, the UC problem model can be formulated as

$$\text{Minimize } F_{Tuc} = \sum_{t=1}^{T}\sum_{i=1}^{N} U_{it}\left[F_i\left(P_i(t)\right) + SUC_i\left(1-U_{i(t-1)}\right)\right] \tag{1}$$

where F_{Tuc} is the total operating cost in \$, F_i (P_i (t)) is the fuel cost of unit i at hour t, P_i (t) is the output power of ith unit at hour t, U_{it} is the on/off status of ith unit at hour t.

The major component of the operating cost, for thermal units, is the power production cost of the committed units that is modelled as third order cubic function as shown in Equation (2)

$$F_i(P_i) = a_i + b_i P_i + c_i p_i^2 + d_i P_i^3 \,(\$/\text{h}) \tag{2}$$

where a_i, b_i, c_i, d_i are cost function coefficients of unit i in \$, \$/MW, \$/MW^2, \$/MW^3 respectively.

Constraints

System Power Balance

$$\sum_{i=1}^{N} U_i P_i(t) = P_d(t), \; t = 1,2,\cdots,T \tag{3}$$

where P_d (t) is the power demand at tth interval.

Spinning Reserve Constraints

The sum of the maximum power generating capacities of all the committed units at a time instant should be at least equal to the sum of the known power demand and minimum spinning reserve requirement at that time instant, i.e.

$$\sum_{i=1}^{N} U_{it} P_{i(\max)} \geq P_d(t) + R_t, \; t = 1,2,\cdots,T \tag{4}$$

where $P_{i\,(max)}$ is the known maximum power that can be generated by unit i at any time instant and R_t is the minimum spinning reserve requirement at time t.

Generation Limit Constraints

$$P_{i(\min)} \leq P(t)_i \leq P_{i(\max)} \quad \text{when } U_{it} = 1$$

$$P_i(t) = 0 \quad \text{when } U_{it} = 0 \tag{5}$$

where $P_{i\,(min)}$ and $P_{i\,(max)}$ represents the minimum and maximum generation limits of thermal units.

Unit Minimum up/down Time Constraints

$$\left[X_{i(t-1)}^{on} - T_i^{on}\right] * \left[U_{i(t-1)} - U_{i(t)}\right] \geq 0 \tag{6}$$

$$\left[X_{i(t-1)}^{off} - T_i^{off}\right] * \left[U_{i(t)} - U_{i(t-1)}\right] \geq 0 \tag{7}$$

where $X_{i(t)}^{on}$ and $X_{i(t)}^{off}$ is the time duration for which unit i has been on and off respectively at hour t.

Ramp Rate Limits

$$P_i(t) - P_i(t-1) \leq UR_i \quad \text{As generation increases} \tag{8}$$

$$P_i(t-1) - P_i(t) \leq DR_i \quad \text{As generation decreases} \tag{9}$$

where UR and DR represents the generator ramp up and ramp down limits

TEACHING LEARNING BASED OPTIMIZATION

By inspiring a teaching-learning process, Rao et al. proposed relatively a new population based meta-heuristic optimization algorithm called, Teaching Learning Based Optimization

(TLBO). It works on the effect of influence of a teacher on the output of learners in a class. The key advantage of this algorithm is it has no user-de- fined parameter which makes it better than other optimization algorithms. The TLBO method works on the philosophy of teaching and learning. Teacher and learners are the two vital components of the algorithm. The working of TLBO is divided into two parts, "teacher phase" and "learner phase". In this algorithm, a group of learners is considered as population and design variables are considered as different subjects offered to the learners and the learner's result is analogous to the "fitness" value of the optimization problem. In the entire population, the best solution is considered as the teacher. The UC problem is a highly combinatorial optimization problem can be solved in two folds: optimal unit selection and dispatch among online generating units. Usually, unit selection is expressed as binary variables and real numbers for dispatch. Thus, the problem contains both types of variables. The applicability of TLBO for solving such mixed integer problems is explored. The steps involved in the search procedure of the TLBO algorithm for the proposed UC-CCF problem is illustrated by flowchart in Figure 1 and are summarized as follows.

Initialization of TLBO-UC-CCF problem

Step 1: Define the UC-CCF optimization problem as minimization problem.

Step 2: Population size (Ps), number of design variables (Nd) which represents number of generating units, minimum up/down time, initial status, maximum and minimum generation limits (limits of design variables) and stopping criteria (maximum number of iterations) are defined in this step.

Teacher phase

Step 3: Evaluate the difference between existing mean and best mean result by utilizing Teaching factor (T_f).

Learner phase

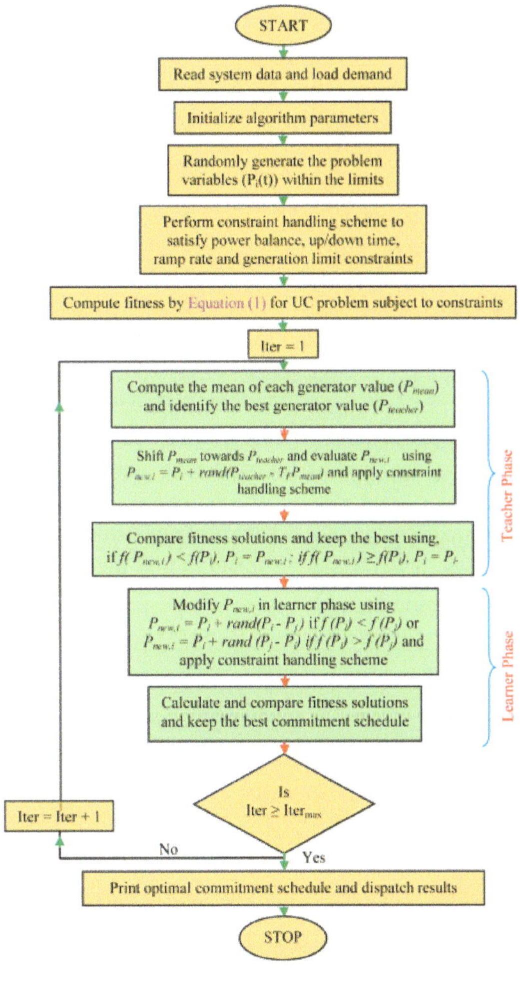

Figure 1: Flowchart of TLBO based UC-CCF.

Step 4: Update the learner's generation solution with the help of teacher's generation.

Step 5: Update the learner's generating solution by utilizing the generating solution of some other learner.

Termination criteria

Step 6: Repeat the procedure from Step 2 to 5 till the maximum number of iterations is met.

Constraint Handling Techniques

A key factor in the application of heuristic algorithms to UC problem is how the algorithm handles the constraints relating to the problem. In the proposed solution method, the constraints of the UC are implemented using a combination of preservation and penalty function methods. The generating capacity limits and ramp up/down rate constraints are also handled by the preservation method. The production of the initial population and in the two phases of TLBO, the output of all committed units minus the reference unit are chosen arbitrary within their respective generating capacity limits, whereas the output of the reference unit is constrained by the system power balance constraint. So, for the dispatchable units other than reference unit, the generating capacity limit constraint is automatically satisfied in both initial population and two phases of TLBO. Therefore, the violation of the capacity limit constraint should be only considered as the penalty term for the reference unit. The minimum up-time/down-time duration constraints of the UC problem are checked for each unit over the scheduling horizon in each interval. If there is any violation in the minimum up or minimum down time constraint then repair mechanism [11] is used to overcome the violation.

NUMERICAL SIMULATION RESULTS AND DISCUSSIONS

In this section, the numerical results aimed at showing the efficiency and the effectiveness of the proposed approach is presented. The proposed TLBO-UC method is tested on modified IEEE-24 bus system having 26 units over a 24-hour scheduling horizon. In order to avoid misleading results due to the stochastic nature of the TLBO, 100 runs were averaged with each run starting with random initial populations. The performance of the proposed problem is implemented using MATLAB-7.9 and realized in 2.40 GHz, i3 processor with 4 GB RAM. The obtained results are compared with other methods in the literature.

This test system consists of 26 units involving cubic cost equations. System particulars involving system data, load demand, ramp rate and capacity limits of the aforementioned units are obtained from literature [2] [3] [14] [15] . In order to determine the best parameters of the proposed TLBO, a number of simulations are carried out. After a number of careful experimentation, following optimal values of TLBO-UC parameters have finally been settled and most appropriate to adopt: Population size (Ps) = 30 and Number of iterations = 100. These parameters are obtained from fine tuning process which is problem dependant task. Simulations are carried out for a 24 hour horizon. The generation cost, % reserve and start-up cost for 24 hour horizon is shown in Table 1. The optimal commitment schedule and generation output derived by the proposed algorithm considering the ramp rate and cubic cost equation is shown in Table 2 for 24 hour horizon.

In this combination for unit scheduling and generation output, the total generation cost obtained by the proposed approach is \$795488.869. Table 3 shows the comparison of the total generation cost and execution time achieved by TLBO method with the popular methods, e.g. Piecewise Linear Iterative (PLI) [15] and Dynamic Programming-Sequential and Truncated Combination (DP-STC) [15] . The corresponding generation cost using the PLI and DP-STC is \$795698.33 and \$795489.09 respectively. The total operation cost considering start-up cost obtained by TLBO is \$800158.869.

For the 26 unit system, the number of units committed for the entire scheduling period is shown inFigure 2. Figure 3 illustrates variation of load demand and the summation of the maximum capacity of the committed units for each hour in the entire time horizon. Figure 4 exemplify the generating units ON/OFF status of the 26 unit system for 24 hour horizon. Execution time complexity of each optimization method is very important for its application to real systems. Table 3 shows the execution time of 0.32 seconds which is less than the other methods in literature. The performance analysis of the test system involving CCF as considered by the authors of the present work being totally a new one, no comparison could be done with regard to the performance analysis results. The best, average and worst fuel costs among the 100 runs of algorithm satisfying the system constraints are listed in Table 4.

In this case, two criteria of goodness are considered. First, the popular "Mean" and "Standard Deviation" that has been used approximately in all of optimization algorithms are computed and listed in Table 4. The small values of the standard deviation confirm the capability and reliability of the TLBO to find the best compromise solution.

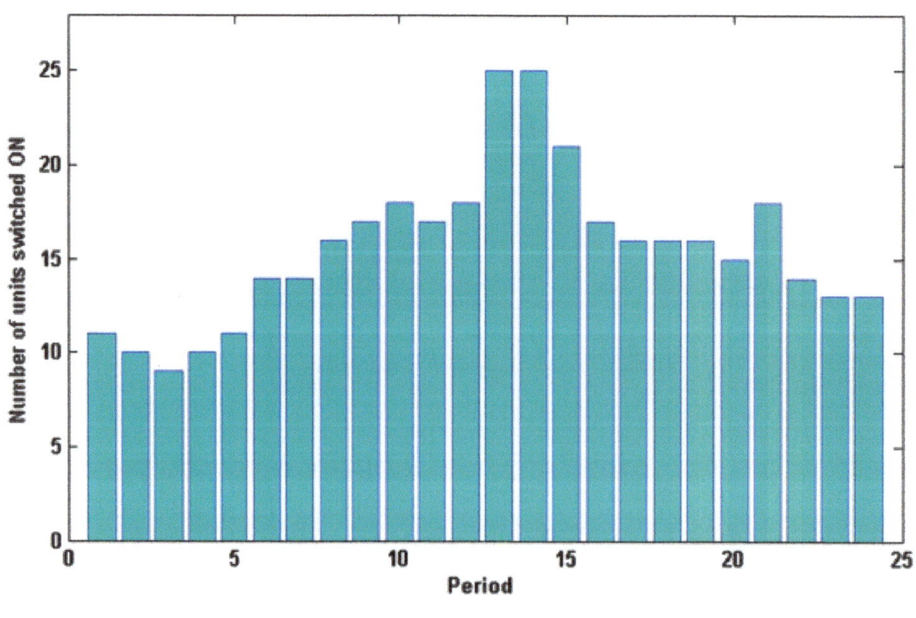

Figure 2: Number of units switched ON for 24 hours.

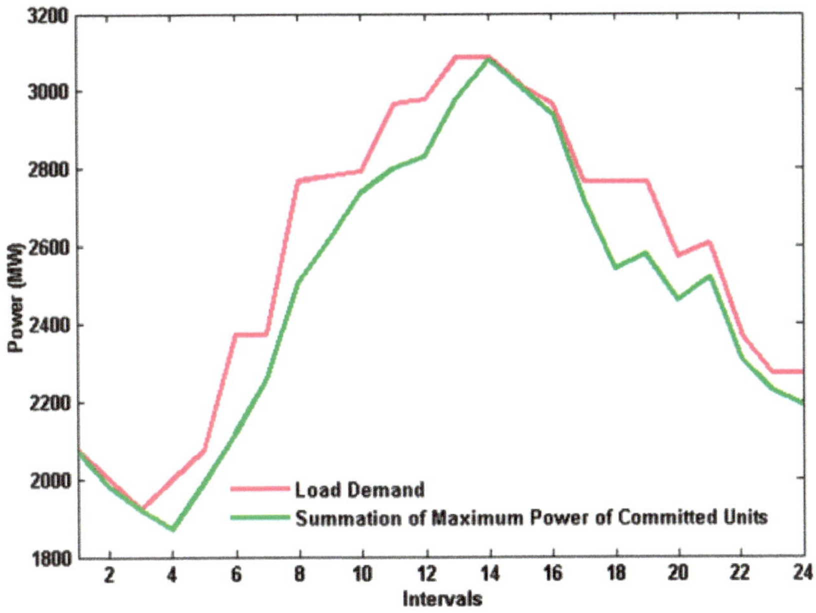

Figure 3: Power demand and sum of maximum capacity of committed units for 24 hours.

The success rate is defined as ($Run_{Success}/Run_{Tot}$) x 100, where $Run_{Success}$ is the number of successful experiments which converge to the best solution within the range and Run_{Tot} is the total number of runs performed. Results of the success rate for 26-unit with cubic cost functions is 63% and is listed in Table 4 which demonstrates that TLBO has satisfactory success rate. From the results, it is clear that the proposed method is robust and it is applicable to practical systems.

Like other evolutionary algorithms, TLBO uses the stochastic techniques, thus randomness is an intrinsic feature of these techniques. Several runs with different initial learners have been conducted to test the performance and consistency of the proposed algorithm. The spread of best costs for 100 runs are calculated and graphically displayed in Figure 5 to illustrate the robustness of the proposed algorithm. To gain further insights into the solution quality of the test case, the obtained statistical results are reported in Table 4 by means of best, worst and average values. From the statistical results it is evident that the generation costs obtained by different trails are close to the best solution, thereby validating that the proposed method has the higher probability of attaining reliable and quality solution.

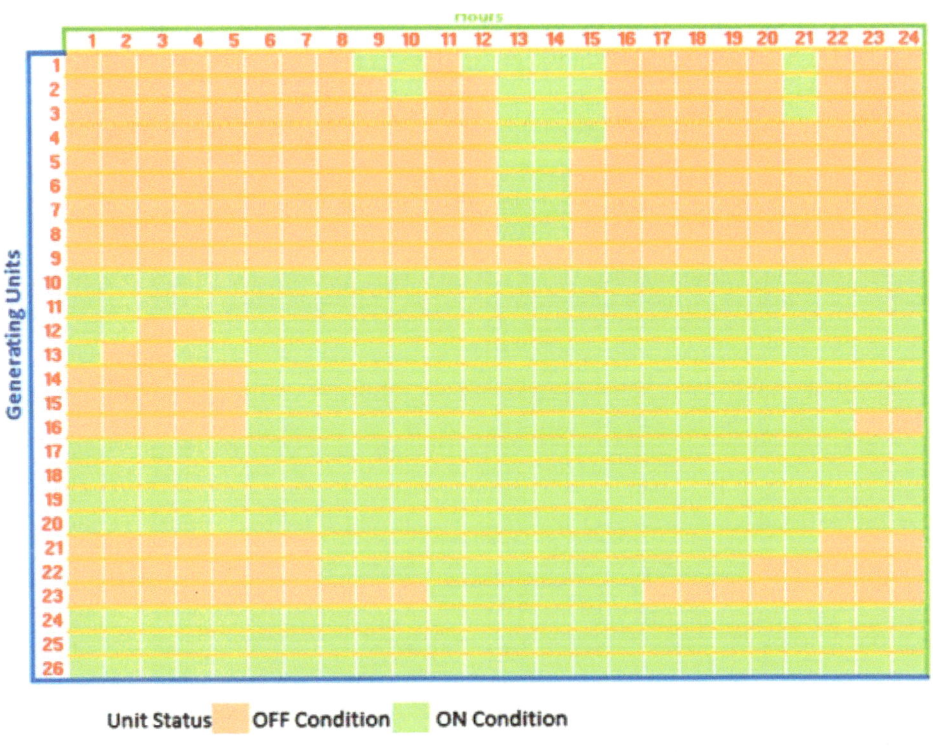

Figure 4: Units ON/OFF representation of 26 unit test system for 24 hours.

Table 1: Operation costs and available reserve for 24 hours.

Hour (h)	Demand (MW)	Reserve %	Generation cost ($)	Start-up cost ($)
1	2070	0.19	23627.42	3220
2	1980	0.90	22268.16	0
3	1920	0.10	21351.82	0
4	1870	6.41	20701.82	80
5	1990	4.05	22482.68	80
6	2120	10.7	25286.92	300
7	2260	4.80	27778.05	0
8	2510	9.32	33640.50	600
9	2620	5.76	36251.28	0
10	2740	1.86	39154.75	0
11	2800	5.56	40740.86	300
12	2830	4.94	41474.64	0
13	2980	3.40	45727.71	90
14	3080	0.16	48818.60	0
15	3010	0.10	45959.10	0
16	2940	0.84	44085.14	0
17	2720	1.73	38603.78	0
18	2540	8.24	34335.21	0
19	2580	6.79	35276.52	0
20	2460	4.32	32211.16	0
21	2520	3.34	33715.07	0
22	2310	2.70	28726.08	0
23	2230	1.93	27013.60	0
24	2190	3.69	26257.99	0
		Total	**795488.8690**	**4670.0**

Table 2: Optimal scheduling of modified IEEE-24 bus system.

Units → / Hours ↓	1	2	3	4	5	6	7	8	9	10	11	12	13
1	0.00	0.00	0.00	0.00	0.00	0.00	0.00	0.00	0.00	76.0	76.00	75.75	72.25
2	0.00	0.00	0.00	0.00	0.00	0.00	0.00	0.00	0.00	72.91	69.90	67.20	DN
3	0.00	0.00	0.00	0.00	0.00	0.00	0.00	0.00	0.00	76.00	74.00	DN	UP
4	0.00	0.00	0.00	0.00	0.00	0.00	0.00	0.00	0.00	43.71	41.10	UP	15.19
5	0.00	0.00	0.00	0.00	0.00	0.00	0.00	0.00	0.00	76.00	75.10	15.20	53.65
6	0.00	0.00	0.00	0.00	0.00	0.00	0.00	0.00	0.00	76.00	75.62	53.70	69.68
7	0.00	0.00	0.00	0.00	0.00	0.00	0.00	0.00	0.00	76.00	76.00	76.00	76.00
8	0.00	0.00	0.00	0.00	0.00	0.00	0.00	0.00	0.00	76.00	76.00	76.00	76.00
9	2.40	0.00	0.00	0.00	0.00	0.00	0.00	0.00	0.00	76.00	76.00	76.00	76.00
10	10.0	2.40	0.00	0.00	0.00	0.00	0.00	0.00	0.00	76.00	76.00	76.00	76.00
11	0.00	0.00	0.00	0.00	0.00	0.00	0.00	0.00	0.00	76.00	76.00	76.00	76.00
12	2.40	0.00	0.00	0.00	0.00	0.00	0.00	0.00	0.00	76.00	76.00	76.00	76.00
13	11.5	2.40	2.40	2.40	2.40	4.00	4.00	4.00	0.00	76.00	76.00	76.00	76.00
14	12.0	12.0	12.0	12.0	12.0	20.0	20.0	15.0	0.00	76.00	76.00	76.00	76.00
15	12.0	12.0	11.7	9.30	0.00	0.00	0.00	0.00	0.00	76.00	76.00	76.00	76.00
16	0.00	0.00	0.00	0.00	0.00	0.00	0.00	0.00	0.00	76.00	76.00	76.00	76.00
17	0.00	0.00	0.00	0.00	0.00	0.00	0.00	0.00	0.00	76.00	76.00	76.00	76.00
18	0.00	0.00	0.00	0.00	0.00	0.00	0.00	0.00	0.00	76.00	76.00	76.00	76.00
19	0.00	0.00	0.00	0.00	0.00	0.00	0.00	0.00	0.00	76.00	76.00	76.00	76.00
20	0.00	0.00	0.00	0.00	0.00	0.00	0.00	0.00	0.00	76.00	76.00	76.00	76.00
21	2.40	2.40	2.40	0.00	0.00	0.00	0.00	0.00	0.00	76.00	76.00	76.00	76.00
22	0.00	0.00	0.00	0.00	0.00	0.00	0.00	0.00	0.00	76.00	76.00	76.00	76.00
23	0.00	0.00	0.00	0.00	0.00	0.00	0.00	0.00	0.00	76.00	76.00	76.00	76.00
24	0.00	0.00	0.00	0.00	0.00	0.00	0.00	0.00	0.00	76.00	76.00	76.00	76.00

Units → / Hours ↓	14	15	16	17	18	19	20	21	22	23	24	25	26
1	0.00	0.00	0.00	155	155	155	155	0.00	0.00	0.00	350	400	400
2	0.00	0.00	0.00	155	155	155	155	0.00	0.00	0.00	350	400	400
3	0.00	0.00	0.00	155	155	155	155	0.00	0.00	0.00	350	400	400
4	0.00	0.00	0.00	155	155	155	155	0.00	0.00	0.00	350	400	400
5	UP	UP	UP	155	155	155	155	0.00	0.00	0.00	350	400	400
6	25.0	25.0	25.0	155	155	155	155	UP	UP	0.00	350	400	400
7	68.9	62.0	55.1	155	155	155	155	UP	UP	0.00	350	400	400

8	100	100	98.0	155	155	155	155	69.0	69.0	0.00	350	400	400
9	100	100	100	155	155	155	155	124	119.6	UP	350	400	400
10	100	100	100	155	155	155	155	179	174.6	UP	350	400	400
11	100	100	100	155	155	155	155	188.50	168.5	69	350	400	400
12	100	100	100	155	155	155	155	174.8	154.8	124	350	400	400
13	100	100	100	155	155	155	155	197	197	178.9	350	400	400
14	100	100	100	155	155	155	155	197	197	197	350	400	400
15	100	100	100	155	155	155	155	197	197	197	350	400	400
16	100	100	100	155	155	155	155	197	195.1	173.9	350	400	400
17	100	100	100	155	155	155	155	183	163	DN	350	400	400
18	100	100	100	155	155	155	155	92.8	73.2	DN	350	400	400
19	100	100	100	155	155	155	155	112.8	93.2	0.00	350	400	400
20	100	100	100	155	155	155	155	86	DN	0.00	350	400	400
21	100	100	100	155	155	155	155	138.8	DN	0.00	350	400	400
22	85.30	78.6	72.1	155	155	155	155	DN	0.00	0.00	350	400	400
23	81.40	74.6	DN	155	155	155	155	DN	0.00	0.00	350	400	400
24	61.50	54.5	0.00	155	155	155	155	0.00	0.00	0.00	350	400	400

Table 3: Comparison of best results obtained by different methods.

Method	Total start-up cost ($)	Total generation cost ($)	Total operation cost ($)	Execution time (s)
PLI	-	795698.3300	-	0.33
DP-STC	-	795489.0900	-	0.34
TLBO	4670	795488.8690	800158.869	0.32

Table 4: Performance analysis of best feasible solution for 26 unit test system by TLBO.

Test system	26-unit system
Population size	30
Best cost ($)	795488.8690
Average cost ($)	795509.9318
Worst cost ($)	795553.5564
Standard deviation	22.515
Success rate 795,488 ($) - 795,510 ($)	63%
Average CPU time (s)	0.32

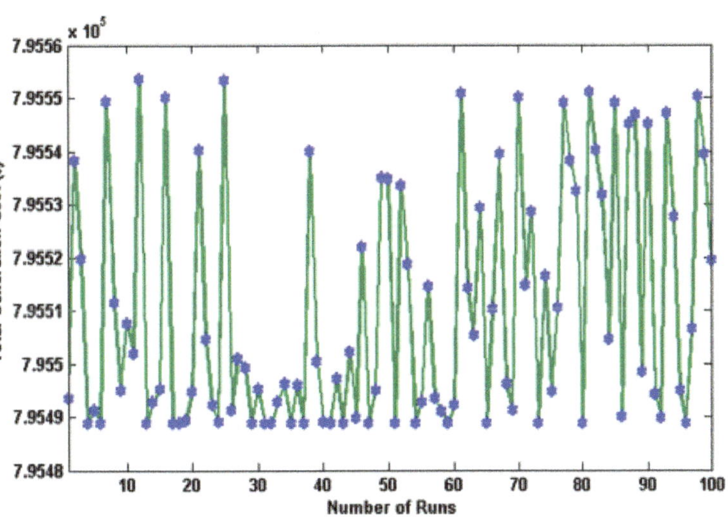

Figure 5: Distribution of the best generation cost of each simulation run for 24 hours.

CONCLUSIONS

In this paper, a nature inspired TLBO technique has been proposed for solving unit commitment problem with ramp constraints on the thermal generating units involving cubic cost functions. The keys of the effectiveness of the approach are the efficient algorithm for ED problems with cubic cost equations recently proposed that exactly solves the UC problem involving non-convex higher order cost functions without any form of approximation. The sophisticated heuristics for producing a ramp feasible and demand feasible solution is developed. Solving ramp constrained UC problem turns out to be efficient with high provable accuracy on large scale realistic instances in reasonable computational time. The proposed TLBO-UC-CCF method remarkably reduces the generation cost and computation time, and yields more accurate generation scheduling, which shows its adaptability to any higher order generation cost functions.

Incorporating reserve constraints in the UC problem is usually difficult; indeed, they are often used in literature. However, extending our approach to reserve constrained UC with higher order cost functions is not straightforward, and it will be the subject of a future work.

ACKNOWLEDGEMENTS

E. B. Elanchezhian, S. Subramanian, S. Ganesan The authors gratefully acknowledge the authorities of Annamalai University, Annamalai Nagar, Tamilnadu, India for the facilities provided to carry out this research work.

REFERENCES

1. Wood, A.J. and Woolenberg, B.F. (1996) Power Generation, Operation and Control. Wiley, New York.
2. Tong, S.K., Shahidehpour, S.M. and Ouyang, Z. (1991) A Heuristic Short-Term Unit Commitment. IEEE Transactions on Power Systems, 6, 1210-1216.http://dx.doi.org/10.1109/59.119268
3. Wang, C. and Shahidehpour, S.M. (1993) Effects of Ramp Rate Limits on Unit Commitment and Economic Dispatch. IEEE Transactions on Power Systems, 8, 1341-1350. http://dx.doi.org/10.1109/59.260859
4. Wang, S.J., Shahidehpour, S.M., Kirschen, D.S., Mokhtari, S. and Irisarri, G.D. (1995) Short Term Generation Scheduling with Transmission and Environmental Constraints Using an Augmented Lagrangian Relaxation. IEEE Transactions on Power Systems, 10, 1294-1301. http://dx.doi.org/10.1109/59.466524
5. Dudek, G. (2004) Unit Commitment by Genetic Algorithm with Specialized Search Operators. Electric Power Systems Research, 72, 299-308.http://dx.doi.org/10.1016/j.epsr.2004.04.014
6. Wang, L.F. and Singh, C. (2009) Unit Commitment Considering Generator Outages through a Mixed-Integer Particle Swarm Optimization. Applied Soft Computing, 9, 947-953. http://dx.doi.org/10.1016/j.asoc.2008.11.010
7. Eslamian, M., Hosseinian, S.H. and Vahidi, B. (2009) Bacterial Foraging-Based Solution to the Unit Commitment Problem. IEEE Transactions on Power Systems, 24, 1478-1488.http://dx.doi.org/10.1109/TPWRS.2009.2021216
8. Ebrahimi, J., Hosseinian, S.H. and Gharehpetian, G.B. (2011) Unit Commitment Problem Solution using Shuffled Frog Leaping Algorithm. IEEE Transactions on Power Systems, 26, 573-581. http://dx.doi.org/10.1109/TPWRS.2010.2052639
9. Abookazemi, K., Ahmad, H., Tavakolpour, A. and Hassan, M.Y. (2011) Unit Commitment Solution Using an Optimized Genetic System. International Journal of Electrical Power & Energy Systems, 33, 969-975. http://dx.doi.org/10.1016/j.ijepes.2011.01.009
10. Vaisakh, K. and Srinivas, L.R. (2011) Evolving Ant Colony Optimization Based Unit Commitment. Applied Soft Computing, 11, 2863-2870.http://dx.doi.org/10.1016/j.asoc.2010.11.019
11. Chandrasekaran, K., Hemamalini, S., Simon, S.P. and Padhy, N.P. (2012) Thermal Unit Commitment Using Binary/ Real Coded Artificial Bee Colony Algorithm. Electric Power Systems Research, 84, 109-119. http://dx.doi.org/10.1016/j.epsr.2011.09.022
12. Datta, D. and Dutta, S. (2012) A Binary-Real-Coded Differential Evolution for Unit Commitment Problem. International Journal of Electrical Power & Energy Systems, 42, 517-524. http://dx.doi.org/10.1016/j.ijepes.2012.04.048
13. Roy, P.K. and Sarkar, R. (2014) Solution of Unit Commitment Problem Using Quasi-Oppositional Teaching Learning Based Algorithm. International Journal of Electrical Power & Energy Systems, 60, 96-106. http://dx.doi.org/10.1016/j.ijepes.2014.02.008
14. Moon, Y.H., Park, J.K., Kook, H.J. and Lee, Y.H. (2001) A New Economic Dispatch Algorithm Considering Any Higher Order Generation Cost Functions. International Journal of Electrical Power & Energy Systems, 23, 113-118. http://dx.doi.org/10.1016/S0142-0615(00)00043-0
15. Lin, W.-M., Gow, H.-J. and Tsay, M.-T. (2007) A Partition Approach Algorithm for Non-Convex Economic Dispatch. International Journal of Electrical Power & Energy Systems, 29, 432-438. http://dx.doi.org/10.1016/j.ijepes.2006.11.002
16. Saber, A.Y., Chakraborthy, S., Abdur Razzak, S.M. and Senjyu, T. (2009) Optimization of Economic Load Dispatch of Higher Order General Cost Polynomials and Its Sensitivity Using Modified Particle Swarm Optimization. Electric Power Systems Research, 79, 98-106.http://dx.doi.org/10.1016/j.epsr.2008.05.017
17. Elanchezhian, E.B., Subramanian, S. and Ganesan, S. (2014) Economic Power Dispatch with Cubic Cost Models Using Teaching Learning Algorithm. IET Generation, Transmission & Distribution, 8, 1187-1202. http://dx.doi.org/10.1049/iet-gtd.2013.0603
18. Rao, R.V., Savsani, V.J. and Vakharia, D.P. (2012) Teaching-Learning-Based Optimization: An Optimization Method for Continuous Non-Linear Large Scale Problems. Information Sciences, 183, 1-15. http://dx.doi.org/10.1016/j.ins.2011.08.006
19. Sultana, S. and Roy, P.K. (2014) Optimal Capacitor Placement in Radial Distribution Systems Using Teaching Learning Based Optimization. International

Journal of Electrical Power & Energy Systems, 54, 387-398. http://dx.doi.org/10.1016/j.ijepes.2013.07.011

20. Singh, M., Panigrahi, B.K. and Abhyankar, A.R. (2013) Optimal Coordination of Directional Over-Current Relays Using Teaching Learning-Based Optimization (TLBO) Algorithm. International Journal of Electrical Power & Energy Systems, 50, 33-41.http://dx.doi.org/10.1016/j.ijepes.2013.02.011

21. Patel, S.J., Panchal, A.K. and Kheraj, V. (2014) Extraction of Solar Cell Parameters from a Single Current-Voltage Characteristic Using Teaching Learning Based Optimization Algorithm. Applied Energy, 119, 384-393. http://dx.doi.org/10.1016/j.apenergy.2014.01.027

CHAPTER 13

OPTIMISATION OF ELECTRICAL DISTRIBUTION SYSTEM BY USING SOLAR THERMAL POWERED SYSTEMS AND ITS IMPACT ON ELECTRICAL DISTRIBUTION FEEDERS

Punnaiah Veeraboina[1], G. Yesuratnam[2]

[1]Department of Electrical Engineering, Centre for DNA Fingerprinting and Diagnostics (CDFD), Hyderabad, India

[2]Department of Electrical Engineering, Osmania University, Hyderabad, India

ABSTRACT

In this research, the performance of the solar thermal powered systems (STPS) is analyzed with different models (without inserts, with inserts and with Nano fluids with different concentrations) and its impact on the Electric load in a residential/ Institutional Electrical Distribution system. For this purpose, the electrical and solar thermal water heater is tested and validated. Solar thermal powered systems and its impact on the Institutional electrical distribution feeders are tested and compared with the energy efficiency (EE) and cost optimization. The goal of this paper is to analyze the impact of solar thermal energy on electrical energy consumption in the electrical distribution feeder level. The electrical system cost and energy consumptions are tabulated and observed that there is a considerable savings.

INTRODUCTION

As the cost of energy continues to rise and research budgets grow ever tighter, lab's autoclave is an excellent place to create savings without compromising on usability or results. Autoclaves kill different pathogens by subjecting the equipment and tools to steam 121˚C for 30 minutes. Institutional electrical water heating is considered the second most energy-intensive activity which accounts for approximately 15% of a home's energy consumption, after HVAC [1]. This fraction could be as high as 30% for residents in winter dominated parts of the world [2] . Solar thermal water heating systems which utilize solar energy to produce hot water can be easily retrofitted into existing electric/gas water heating systems. Solar domestic hot water systems have been widely adopted since they use solar radiation which is free of cost, despite the changes that need to be incorporated to existing systems—such as installation of solar collectors, preheat tanks, internal or external heat exchangers, and circulation control and piping systems [3] [4] . B. S. Swanson and J. H. Fletcher [5] investigate the effects of solar thermal water heaters (includes the integral collector storage and active drain-back open loop systems) on peak demand.

Generally, a residential electric water heater with a tank for hot water storage is set to heat water at the outlet to a setting point of value T_{set}, with a T_{delta} dead band [6] [7] . As soon as hot water is drawn, an equal amount of cold water replenishes the

Citation: Veeraboina, P. and Yesuratnam, G. (2016) Optimisation of Electrical Distribution System by Using Solar Thermal Powered Systems and Its Impact on Electrical Distribution Feeders. Energy and Power Engineering, 8, 219-229. doi: 10.4236/epe.2016.84020.

tank at the same flow rate. When the mixed water temperature at the outlet drops below the lower bound, i.e. $T_{tank} < T_{set} + T_{delta}$, the heating elements are turned on to work at rated power till the outlet hot water temperature attains the set point value, i.e. $T_{set} + T_{delta}$. Water heaters with storage tanks are not free of the standby loss, unlike the tank less water heaters. The heating elements are turned off unless the outlet temperature is not within the range of the dead band, i.e. $T_{set} - T_{delta} \leq T_{tank} \leq T_{set} + T_{delta}$.

Solar water heating systems are in high demand. Typically, 30% - 40% of a family's electricity bill is devoted to devoted water. Sun Heat's system can save the individual family from 70% to 90% of the total amount spent on the Electricity used for heating water. The system generally meets all of the summer time heating needs. During times of decreased sunlight, the system will preheat the water then bring it up to temperature by the conventional water heating system already in place.

For the flat plate collectors, glass panes play an important role in auxiliary electric energy consumption. Flat plate collectors with two glass panes save about 10% - 15% auxiliary electrical energy compared to those with no glass panes and about 3% - 5% energy saving compared to collectors with one glass pane. This is because there are reduced wind convective losses with glass panes. However, there are also transmittance losses from glass panes and there are upper limits on how many glass panes can be used.

The present research is focused on optimization of electrical distribution system (i.e., to reduce the electrical materials used in the system) by using the enhanced solar water heating system (SWHS) and its impact on the electrical distribution feeders. The main purpose is to reduce the Electrical energy consumption and system ratings (i.e., transformer size, cable sizes, MCB's and feeder sizes).

AUTOCLAVES

Autoclaving is one of the techniques used in moist heat sterilisation. All R&D organisations and other hospitals requiring their tools to be sterilized need autoclaves. Inlet water to the Autoclaves is normal raw water which is having a temperature between 20°C - 26°C depending upon the season. The operational temperature for the Autoclave is 121°C. There is huge gap between the inlet temperature to the Autoclaves and required for operation. By using the different technics, the experiments are carryout on the Solar flat plate collectors with different models (with inserts, without inserts & by using Nano-fluids) to enhance the output temperature at 90°C. Considerable energy is saved in this system by using solar heated water than the conventional Electrical water heaters. The savings are not only the Electrical Energy savings but also the savings on the SITC (Supply, Installation, Testing and Commissioning) of Electrical Installations (Electrical Distribution System). The principle behind the working of an Autoclave is that-The higher the pressure created inside the autoclave, the higher would be the attainable temperature inside it.

The boiling point of water at normal atmospheric pressure is 100°C. when the free-flowing steam at a temperature of 100°C is subjected under pressure of 1 atmosphere above the sea-level pressure i.e., 15 lbs pressure per sq. inch, the temperature inside the autoclave happens to rise up to 121°C, which is an usual and common parameters employed in the moist heat sterilisation.

Steam enters the chamber jacket, passes through an operating valve and enters the rear of the chamber behind a baffle plate. It flows forward and down through the chamber and the load, exiting at the front bottom. A pressure regulator maintains jacket and chamber pressure at a minimum of 15 psi, the pressure required for steam to reach 121°C. Overpressure protection is provided by a safety valve. The average American home would take 45 years to consume that much energy. This extreme example serves as a reminder of how resource-intensive high- volume, high-throughput medical-grade autoclaves can be.

Mode of Action/Principle of Moist Heat Sterilization

Moist heat destroys microorganisms by the irreversible denaturation of enzymes and structural proteins. The temperature at which denaturation occurs varies inversely with the amount of water present. Sterilization in saturated steam thus requires precise control of time, temperature, and pressure.

Pressure serves as a means to obtain the high temperatures necessary to quickly kill microorganisms. Specific temperatures must be obtained to ensure the microbicidal activity. Minimum sterilization time should be measured from the moment when all the materials to be sterilized have reached the required temperature throughout.

Most of the manufactures recommends the sterilization process in an autoclave are 15 min. at 121°C and pressure at 2 atm. Table 1 represents the different combinations of temperature, pressure and time. Control and monitor the process is purely based on the temperature and pressure settings to obtain the required steam temperature.

Sterilization may be carried out some time even below at 121°C shown in Figure 1, it's mostly depends upon the pressure and temperature during the sterilisation.

Steam held at elevated temperature and pressure for time is used to transfer moist heat. Steam has much greater heat transfer than boiling water, −80 calories to boil water, 540 calories to produce steam. Heat acts to denature proteins, effectively killing all cells present.

The steam of the autoclave shown above has been provided from a solar flat plate collector which was stored in a thermal storage system. The required temperature for the Autoclave system is 121°C. Normally the solar water heating system is designed for 60°C. But by using the different technics (with inserts and with Nano-fluids) the temperature of the solar flat plat collector is enhanced from 60°C - 90°C.

Hospitals, R&D labs and other regional health centres can greatly benefit from using solar Autoclaves. Rural Primary Health Clinics where Electricity is not available to generate steam could benefit from solar Autoclaves.

Impact on Solar thermal Autoclaves on Electrical Distribution Load

Impact on Solar thermal Autoclaves on Electrical Distribution Load in Institutional Building @60°C and its cost analysis at 500 kVA level is shown in Table 2.

Table 1: Different combinations of temperature, pressure and time.

S No.	Temperature in °C	Approx. corresponding pressure (atm)	Mini. sterilization time (minutes)
1	126 - 129	2.50	10
2	129 - 134	2.75	7.5
3	134 - 138	3.0	5

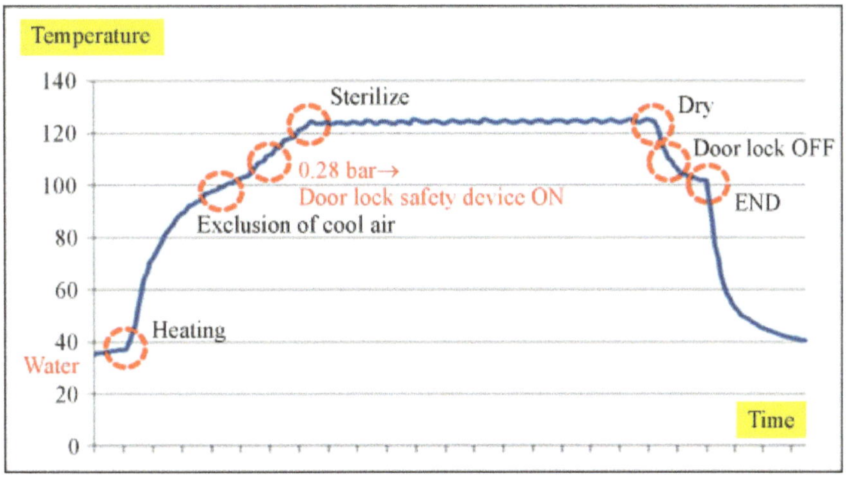

Figure 1: Time vs temperature graph for sterilisation.

Table 2: Electrical load and the cost analysis of solar autoclaves.

Description	No of heaters	kW	Units/day	No. of units/ month	Unit rate (Rs)	Expenditure/ month (Rs)	Expenditure/ annum (Rs)
Autoclaves on electric heaters	04 × 2 systems (8 Nos)	18 kW × 04 × 02 = 144 kW	432 (72 kW for 30 Min × 6 loads × 2 systems	8640 (20 day month)	7.00	60,480	725,760
Autoclaves on electric heaters	04 × 1 system (4 Nos)	18 kW × 04 × 01 = 72 kW	216 (72 kW for 30 Min × 6 loads × 01 systems	4320 (20 day month)	7.00	30,240	362,880
Autoclaves on SWHS	Nil	Nil	08 flat plate collectors	Nil	Nil	Nil	Nil

Capacity of the Autoclave is 72 kW. Each Electrical Element is 18 kW.

Number of Electrical elements is 04 Nos.

So, total capacity of the Autoclave is 18 kW × 04 Nos = 72 kW.

Electrical Power Consumption/day:

18 kW × 04 Nos × 30 min (0.5 Hrs) = 36 kWh.

Daily number of loads is 06 times, Number of Autoclaves is 02 Nos.

So, the total Electrical Energy consumption/day = 36 kW × 06 Nors × 02 systems = 432 kWh.

Number of units per month:

Normally R&D labs will work only 05 day week, so, the number days per month = 20 days.

Number of units consumed per month = 432 kWh × 20 days = 8640 units.

As per Cat-II tariff, per unit charge is Rs 7 per unit.

So, the Electrical Power consumption per year is 8640 units × Rs 7 = Rs 60,480.

Electrical Power Consumption per year:

Rs 60,480 × 12 Months = Rs 725,760.

Figure 2 represents the SLD (Single Line Diagram) and Bill of Materials (BoQ) of 500 kVA Electrical Distribution system at feeder level.

Designed capacity of the Electrical Distribution system is 500 kVA:

Without using the solar water heating system, the required temperature of the Autoclaves is around 121°C. To increase the temperature at 121°C, 144 kW (18 kW × 8 Nos) of Electrical Heaters are used. The capacity of the Electrical Load is 144 kW at 0.98 P.F (say 150 kVA).

The Electrical Distribution system, one of the feeder (250 Amps, TPN MCCB-Power Load) is connected to the domestic water heating load, the total Electrical load of 40 rooms of each 2.5 kW capacity Electrical heaters are connected.

There for the total capacity of the domestic heating Electrical load = 2.5 kW × 40 rooms = 100 kW = 102 kVA (100 kW at 0.98 P.F). The load reduction is 150 kVA + 102 kVA = 252 kVA, the other Distribution loads are 248 kVA.

Based on the above Electrical loads the recommended size of Transformer is 500 kVA and its cost analysis is shown in Table 3.

Automatic Power Factor Correction (With Thyristor Switched)

Techniques can be applied to the industries, power systems and also households to make them stable and due to that the system becomes stable and efficiency of the system as well as the apparatus increases. The use of microcontroller reduces the costs. Automatic Power factor correction (APFC) device reads the power factor from line voltage and line current, calculating the compensation requirement switch on different banks. Improving the power factor of an installation requires a bank of capacitors which acts as a source of reactive energy. This arrangement is said to provide reactive energy compensation.

An inductive load having a low power factor requires the generators and transmission/distribution systems to pass reactive current with associated power losses and exaggerated voltage drops. If a bank of shunt capacitors is added to the load, its (capacitive) reactive current will take the same path through the power system as that of the load reactive current. Since, as pointed out, this capacitive current Ic is in direct phase opposition to the load reactive current (IL).

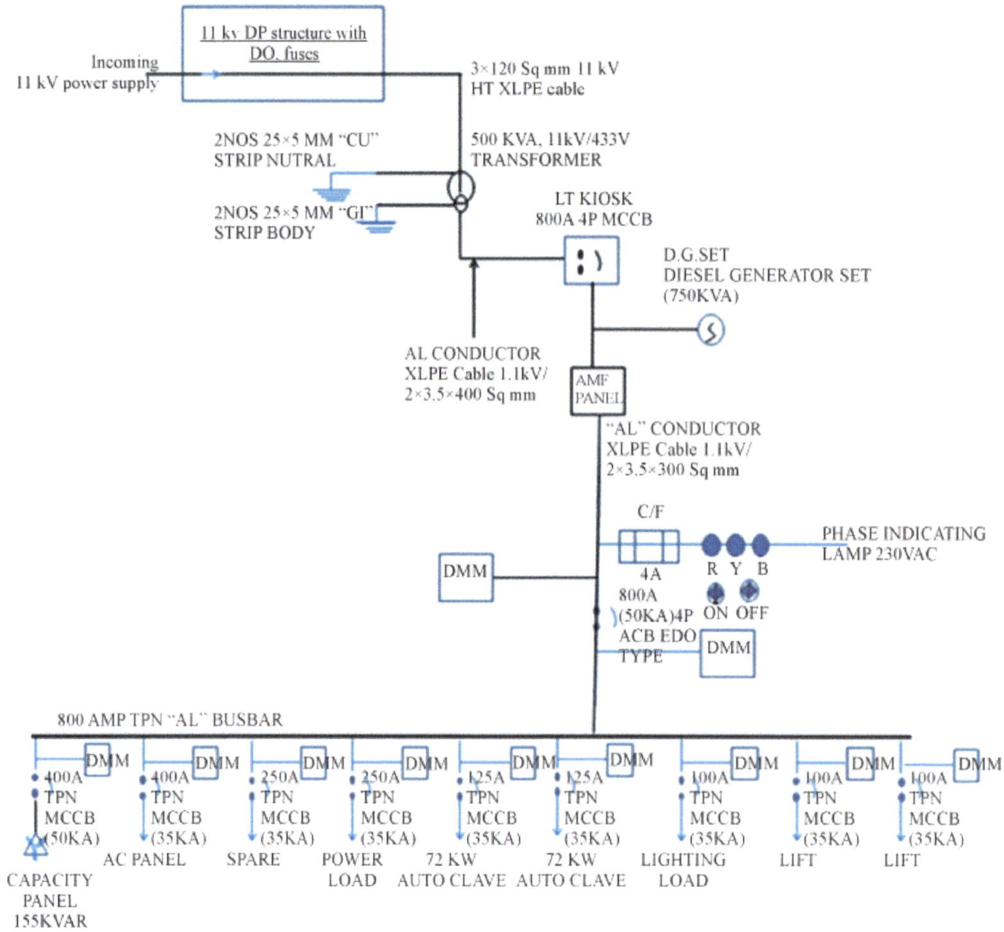

Figure 2: Single line diagram of the electrical distribution system for 500 kVA transformer level.

Table 3: Cost analysis of Electrical Distribution systems at 500 kVA Transformer level.

S No.	Description	Qty	UoM	Rate (Rs)	Total system cost (Rs)
1	HT cable: supply of 11 kV 3C × 120 Sq. mm HT XLPE armoured Cable.	50	Mtr	1100	55,000
2	LT cables: supply of 3.5c × 400 Sq. mm XLPE insulated, 1.1 kV grade armoured aluminium cable.	100	Mtr	1800	180,000
3	HT metering panel (11kV)	1	Nos	332,000	332,000
4	500 KVA O/D TRANSFORMER (OIL FILLED)	1	each	720,000	720,000
5	APFC PANEL (Thyristor Switched)	1	No	459,400	459,400
6	LT KIOSK 800A TPN 4P 50KA EOD type ACB	1	No	288,000	288,000
7	Main LT Panel: I/C: 800A 4 pole EDO type ACB (50 KA)-1 No. O/G (400 A TPN MCCB-2 Nos, 250 A-2 Nos, 125 A-2 Nos, 100 A-3 Nos)	1	No	644,000	644,000
8	AMF Panel suitable for 750 kVA	1	Set	1,027,500	107,500

The two components flowing through the same path will cancel each other, such that if the capacitor bank is sufficiently large and Ic = IL, there will be no reactive current flow in the system upstream of the capacitors. This is indicated in Figure 3 which shows the flow of the reactive components of current only.

In Figure 3: R is active-power elements of the load, L is (inductive) reactive-power elements of the load, and C is (capacitive) reactive-power elements of the power-factor correction equipment (i.e. capacitors).

kVAR required for transformer compensation:

Transformer	Required Kva
≤315 kVA T.C	= 5% of KVA
315 kVA to 1000 kVA	= 6% of KVA
≥1000 kVA	= 8% of KVA

Calculation of required capacitor:

Fixed compensation of the 500 kVA Transformer is (500 kVA × 6%) = 30 kVAR.

Suppose actual P.F is 0.8, required P.F is 0.98 and total connected load is 354 KVA.

kW = 354 × 0.8 = 283.20. Required capacitance = 283.20 × Multiplying Factor = 283.20 × 0.547 (from Table 4 to find Value according to P.F 0.8 to P.F of 0.98) = 155.00 kVAR.

Total capacity of the APFC panel = Fixed compensation + Variable compensation i.e., 155 kVAR + 30 kVAR = 185 kVAR (available rating).

The capacity of the APFC panel in Figure 4 shows 155 kVAR capacity rest is fixed compensation. Load Reduction by STWH at 11 kV Electrical Distribution System is represented in Table 5.

Impact on Solar Thermal Autoclaves on Electrical Distribution Load at 315 kVA

Impact on Solar Thermal Autoclaves on Electrical Distribution Load in Institutional Building @60°C and its cost analysis at 315 kVA level.

Designed capacity of the electrical distribution system is 315 kVA:

With using the solar water heating system, the hot water output is at 60°C, but required temperature is around 121°C. To increase the temperature from 60°C - 121°C, 18 kW × 4 Nos of Electrical Heater are used. The capacity of the Electrical Load is 75 kVA (72 kW at P.F 0.95).

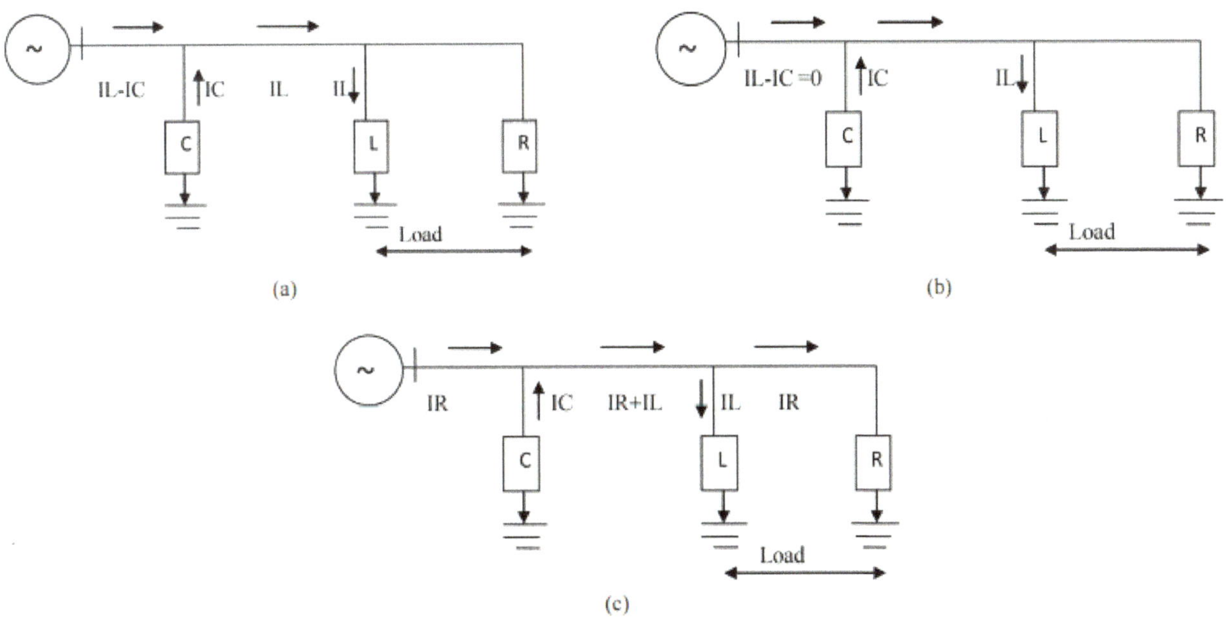

Figure 3: Showing the essential features of power-factor correction. (a) Reactive current components only flow pattern; (b) When IC = IL, all reactive power is supplied from the capacitor bank; (c) With load current added to case (b).

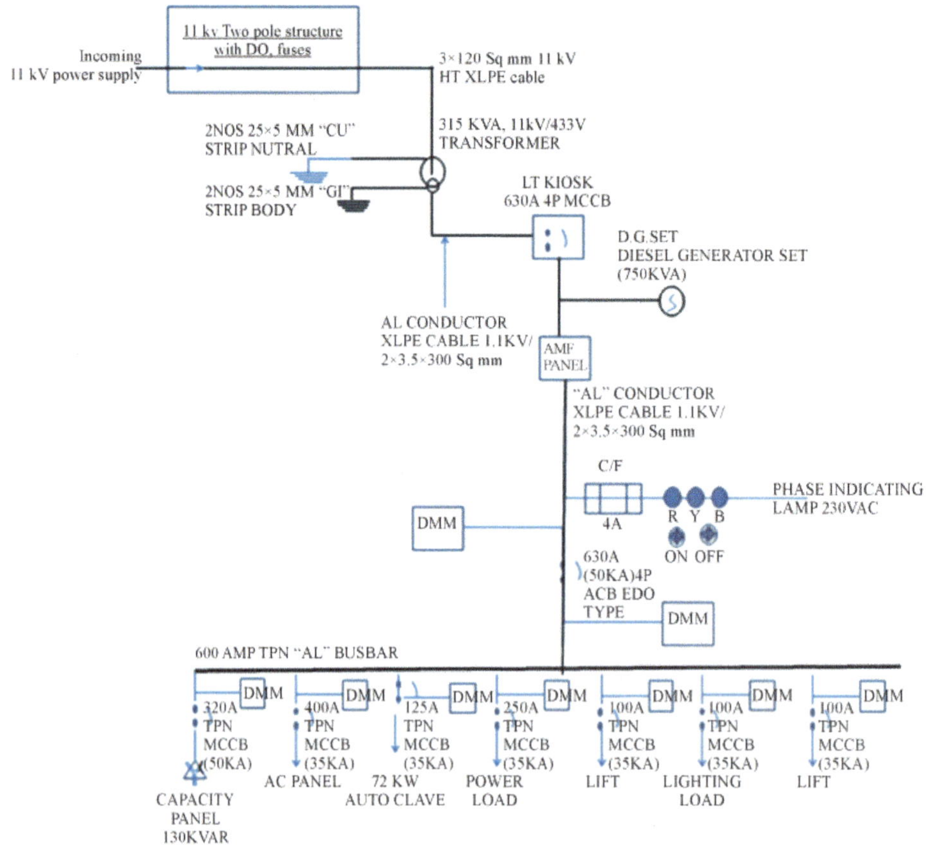

Figure 4: Single line diagram of the electrical distribution system for 315 kVA transformer level.

Table 4: Table for finding the required multiplication factor (P.F from 0.6 to unity) [8].

	0.9	0.91	0.92	0.93	0.94	0.95	0.96	0.97	0.98	0.99	1
0.6	0.849	0.878	0.907	0.938	0.970	1.005	1.042	1.083	1.130	1.191	1.333
0.7	0.536	0.565	0.594	0.625	0.657	0.692	0.729	0.770	0.817	0.878	1.020
0.8	0.266	0.294	0.324	0.355	0.387	0.421	0.458	0.499	0.547	0.608	0.750
0.85	0.135	0.164	0.194	0.225	0.257	0.291	0.328	0.369	0.417	0.477	0.620
0.9		0.029	0.058	0.089	0.121	0.156	0.193	0.234	0.281	0.342	0.484
0.91			0.030	0.060	0.093	0.127	0.164	0.205	0.253	0.313	0.456
0.92				0.031	0.063	0.097	0.134	0.175	0.223	0.284	0.426
0.93					0.032	0.067	0.104	0.145	0.192	0.253	0.395
0.94						0.034	0.071	0.112	0.160	0.220	0.363
0.95							0.037	0.078	0.126	0.186	0.329

The Electrical Distribution system, one of the feeder (250 Amps, TPN MCCB-Power Load feeder) is connected to the domestic water heating load, the total Electrical load of 40 rooms of each 2.5 kW capacity Electrical heaters are connected,

There for the total capacity of the domestic heating Electrical load is 102 kVA (100kW at 0.98 P.F).

The load reduction is 177 kVA (say 180 kVA).

Required size of Transformer after load reduction i.e., 500 kVA - 180 kVA = 320 kVA

Recommended size of Transformer is 315 kVA and its cost analysis is shown in Table 6.

Calculation of Required Capacitor

Fixed compensation of the 315 kVA Transformer is (315 kVA × 5%) =15.75 kVAR.

Suppose actual P.F is 0.8, required P.F is 0.98 and total load is 282 KVA.

kW = 282 × 0.8 = 225.60, Required capacitance = 123.40 kVAR.

Total capacity of the APFC panel is 139 kVAR (i.e., fixed compensation + variable compensation). The capacity of the APFC panel in Figure 5 shows 139 kVAR capacity rest is fixed compensation.

Impact on Solar Thermal Autoclaves on Electrical Distribution Load in Institutional Building @60°C and Its Cost Analysis Shown in Table 7 at 250 kVA Level

Designed capacity of the Electrical Distribution system is 500 kVA. With using the solar water heating system with inserts (twisted inserts with Nano fluids), the hot water output is at 90°C, but required temperature is around 121°C.

Table 5: Autoclave and its load reduction by STWH systems.

Description	Electrical load (Auto claves) kVA	Electrical load (DWHS) kVA	Load reduction kVA	Recommended TFR sizes in kVA	@60°C	@90°C
Conventional Electrical System.	150	105	Nil	500	Nil	Nil
Load reduction by STWH's-with inserts @90°C	75	105	180	315	Up to 60°C for DWHS	With twisted inserts
Load reduction by STWH's with Nano fluids and twisted tape inserts@90°C	19	105	236	250	Up to 60°C for DWHS	With inserts and Nano fluids

Table 6: Cost analysis of electrical distribution systems at 315 kVA TFR level.

S No.	Description	Qty	UoM	Rate (Rs)	Amount (Rs) for 315 kVA
1	HT Cable: Supply of 3 Core 11 KV 120 Sqmm HT XLPE (E) armoured Cable.	50	Mtr	1100	55,000
2	LT cables: Supply of 400 Sqmm 3.5 Core XLPE insulated, 1100V grade armoured aluminium cable.	100	Mtr	1800	180,000
3	HT Metering Panel (11kV)	1	No	332,000	332,000
4	315 kVA O/D Transformer (Oil Filled)	1	each	600,000	600,000
5	APFC PANEL(Thyristor Switched) 100 kVAR	1	No	368,000	368,000
6	LT KIOSK 630A TPN 4P 50KA EOD type ACB	1	Set	260,000	260,000
7	Main LT Panel: I/C: (800A 4 pole EDO type ACB-50 KA-1 No) and Outgoings (400 A TPN MCCB (35 KA)-2 Nos, 250 A-2 Nos, 200 A-2 Nos, 100 A-2 Nos.	1	Set	540,500	540,500
8	AMF Panel suitable for 500 kVA	1	Set	860,000	860,000
	Total				3,195,500

To increase the temperature from 90°C - 121°C, 18 kW × 1 No of Electrical Heater are used. The capacity of the Electrical Load is 19 kVA (18 kW at P.F 0.95).

The load reduction is 126 kW (133 kVA, P.F as 0.95).

The Electrical Distribution system, one of the feeder (250 Amps, TPN MCCB-Power Load feeder) is connected to the domestic water heating load, the total Electrical load of 40 rooms of each 2.5 kW capacity Electrical heaters are connected,

There for the total capacity of the domestic heating Electrical load = 102 kVA (100 kW at 0.98 P.F). Load reduction is 235 kVA.

The required size of Transformer after load reduction i.e., 500 kVA − 235 kVA= 265 kVA. So, recommended size of Transformer is 250 kVA, the cost analysis of the system is represents in Table 8.

Calculation of required capacitor:

Fixed compensation of the 500 kVA Transformer is (250 kVA × 5%) = 12.5 kVAR.

Actual P.F is 0.8, required P.F is 0.98 and total load is 210 KVA.

kW = 210 × 0.8 = 168,

Required capacitor = kW × Multiplying Factor = 168 × Multiplying Factor= 168 × 0.547 (from table to find Value according to P.F 0.8 to P.F of 0.98) = 92 kVAR.

Total capacity of the APFC panel is 105 kVAR (i.e., fixed compensation + variable compensation).

The capacity of the APFC panel in Figure 5 shows 80 kVAR capacity rest is fixed compensation.

Methodology

In the first stage of the study, the hot water temperature is enhanced to 90°C by using inserts into the solar flat plate collector fins. The enhanced water is fed to the Autoclaves for generation of the steam at 121°C. Information such as current (Amps), voltage (Voltage) and power (kW) was collected directly from the nameplate affixed on the Autoclave.

Table 7: Impact of solar thermal systems on electrical distribution system at 90°C.

Description	No of Geysers	kW	Units/day/ geyser	No. of units	Unit rate (Rs)	Cost/day (Rs)	Cost/month (Rs)	Cost/annum (Rs)
Electric Geyser (2.5 kW)	40	100	11	440	7.00	3080	92,400	1,108,800
SWHS	Nil	Nil	08 FPC	Nil	Nil	Nil	Nil	Nil

Table 8: Cost analysis of electrical distribution systems at 250 kVA TFR level.

S No.	Description	Qty	UoM	Rate (Rs)	Amount (Rs) at 250 kVA level
1	HT Cable: Supply of 11 kV, 3c × 120 Sqmm HT XLPE armoured Cable.	50	Mtr	1100	55,000
2	LT cables: Supply of 400 Sqmm 3.5 Core XLPE insulated, 1100V grade armoured aluminium cable.	100	Mtr	1800	180,000
3	HT Metering Panel 11 kV	1	No	332000	332,000
4	250 kVA O/D Transformer (Oil Filled)	1	each	520000	520,000
5	APFC Panel (Thyristor Switched)	No	1	294400	294,400
6	LT KIOSK 800A TPN 4P 50KA EOD type ACB	1	Set	260000	260,000
7	Main LT Panel: I/C (800A 4 pole EDO type ACB-50 kA-1 No, Outgoing (400 A TPN MCCB-35 kA-1 Nos, 250 A-3 Nos, 100 A-3 Nos.	1	Set	455,000	455,000
8	AMF Panel suitable for 320 kVA	1	Set	660,000	660,000
	Total				2,756,400

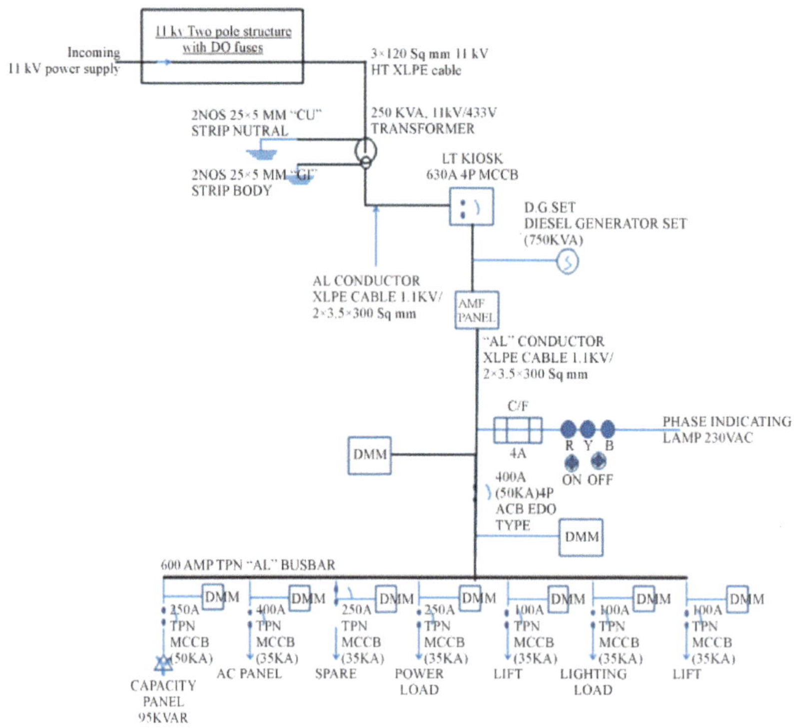

Figure 5: Single Line Diagram (SLD) of the electrical distribution system for 250 kVA transformer level.

After measuring the power consumption of the Autoclaves without solar hot water, the capacity of the Transformer was estimated. Then by using the solar hot water at 90°C was estimated. The cost savings of the system was calculated with Autoclave Electrical Heaters and without Electrical heaters for 60°C and 90°C. For the equipment which was tested, the weekly power consumption (kWh/week) can be obtained using the same equation. But instead of using the nameplate power rating, the actual measured power rating was used. According to the calculations, the cost of the Transformer changes from rating to rating. In this context, it is important to note that this rate of cost of higher capacity Transformer and its associated feeders is less than smaller capacities and also considerable Electrical Energy savings are noted with and without Solar Hot water systems.

CONCLUSIONS

The electrical energy consumption and its cost analysis have been calculated for systems and observed that there is a considerable savings by using the solar hot water at 90°C. The use of Al_2O_3nanoparticles as the dispersed phase in water can significantly enhance the convective heat transfer in the transition flow, and the enhancement increases with increase in particle concentration and further can be increased by using the different types of inserts. The temperature of the outlet get at 90°C and the same has been connected to the Autoclave.

For laboratory equipment sterilisation, commonly autoclaves are used. These equipments are consumes highest electrical power consumption in terms of kWh. The use of solar energy, the feed water to the autoclave is fed at 90°C by using different methods to enhance the temperature. Average weekly power consumption, capacities and cost of operation were analysed. By using the enhanced methods, the cost of the electrical distribution network and its capacities are optimised at feeder level.

REFERENCES

1. 1. Paull, L., Li, H. and Chang, L.C. (2010) A Novel Domestic Electric Water Heater Model for a Multi-Objective Demand Side Management Program. Electric Power Systems Research, 80, 1446-1451. http://dx.doi.org/10.1016/j.epsr.2010.06.013 [Paper reference:1]
2. Knudsen, S. (2002) Consumers' Influence on the Thermal Performance of Small SDHW Systems-Theoretical Investigations. Solar Energy, 73, 33-42. http://dx.doi.org/10.1016/S0038-092X(02)00018-X [Paper reference:1]
3. 3. Kalogirou, S.A. (2004) Solar Thermal Collectors and Applications. Progress in Energy and Combustion Science, 30, 231-295. http://dx.doi.org/10.1016/j.pecs.2004.02.001 [Paper reference:1]
4. Zhang, R., Zhu, H., Wen, H. and Li, M. (1993) Performance Analysis of a Solar Glass Tube Collector. Renewable Energy, 3, 877-883. [Paper reference:1]
5. Swanson, B.S. and Fletcher, J.H. (2008) Analysis of Solar Thermal Water Heaters on Peak Demand. Proceedings of ECTC 2008, 2008 ASME Technical Conference, 3-4 October 2008. [Paper reference:1]
6. 6. http://www.energysavers.gov/tips/water heating.cfm [Paper reference:1]
7. Taylor, Z.T., Gowri, K. and Katipamula, S (2008) GridLAB-D Technical Support Document: Residential End-Use Module Version 1.0. [Paper reference:1]
8. https://electricalnotes.wordpress.com/2011/03/20/automatic-power-factor-correction [Paper reference:1]

CHAPTER 14 A HYBRID SEARCH ALGORITHM FOR MIDTERM OPTIMAL SCHEDULING OF THERMAL POWER PLANTS

Shengli Liao,[1] Chuntian Cheng,[1] Jing Wang,[2] and Zhongkai Feng[1]

[1]Institute of Hydropower and Hydroinformatics, Dalian University of Technology, Dalian 116024, China
[2]Huadian Electric Power Research Institute, Hangzhou 310030, China

ABSTRACT

A hybrid search algorithm consisting of three stages is presented to solve the midterm schedule for thermal power plants (MTSFTPP) problem, where the primary objective is to achieve equal accumulated operating hours of installed capacity (EAOHIC) for all thermal power plants during the selected period. First, feasible spaces are produced and narrowed based on constraints on the number of units and power load factors. Second, an initial feasible solution is obtained by a heuristic method that considers operating times and boundary conditions. Finally, the progressive optimality algorithm (POA), which we refer to as the vertical search algorithm (VSA), is used to solve the MTSFTPP problem. A method for avoiding convergence to a local minimum, called the lateral search algorithm (LSA), is presented. The LSA provides an updated solution that is used as a new feasible starting point for the next search in the VSA. The combination of the LSA and the VSA is referred to as the hybrid search algorithm (HSA), which is simple and converges quickly to the global minimum. The results of two case studies show that the algorithm is very effective in solving the MTSFTPP problem accurately and in real time.

INTRODUCTION

Electric power generation in China mainly consists of coal-fired thermal power plants and hydroelectric power plants [1, 2]. At the end of 2013, China's total installed generation capacity reached 12473.8 GW, with approximately 64.2% derived from coal-fired thermal power plants, 5% from thermal power using other fuels (e.g., oil and gas), 22.5% from hydropower, 6.1% from wind power, 1.2% from nuclear power, and the remainder from other sources [2]. However, coal-fired units are difficult to start up and shut down, and they respond slowly to system demand because preheating and cooling of the stream turbines is required. Moreover, startup and shutdown of a coal-fired unit require considerable amounts of fuel [3]. Therefore, thermal plants are usually scheduled to meet the base demand to ensure that the generation process proceeds smoothly and to minimize the startup and shutdown times of thermal units.

Over the past decade, extreme weather events such as storms, droughts, high temperatures, and damaging hail have occurred frequently in China, which has caused the demand on

Citation: Shengli Liao, Chuntian Cheng, Jing Wang, and Zhongkai Feng, "A Hybrid Search Algorithm for Midterm Optimal Scheduling of Thermal Power Plants," Mathematical Problems in Engineering, vol. 2015, Article ID 710876, 13 pages, 2015. DOI:10.1155/2015/710876.

power systems to change substantially from day to day. These large deviations in demand over short periods cause difficulties in the scheduling of the power systems, a problem that is exacerbated by the large number of coal-fired thermal units in use. Consequently, the power dispatch centers in China must determine the optimal midterm schedule for thermal power plants (MTSFTPP), where "midterm" is a period of one to six months, to determine which units to start up and shut down or the boot capacity of every thermal plant each day before determining the short-term and daily schedules.

The midterm schedule is important for the economical operation of power systems. The objective of the scheduling problem is to find the optimal set of thermal generating units in a power system that will satisfy the system demand, operational restrictions, reliability constraints, and security requirements for each period considered. Unlike the short-term unit commitment problem associated with power systems, which involves determining a startup and shutdown schedule of units to meet the demand over a period of one day to one week, loads must be dispatched to each plant in the midterm to allow for maintenance and coordination with hydropower plants. Optimal midterm scheduling smoothes the demand on thermal plants, allowing them to operate more efficiently, and makes the best use of hydropower plants. In contrast with the short-term optimal scheduling of thermal power plants, there has been little research on the midterm problem.

The MTSFTPP problem is a variation of the unit commitment (UC) problem, which is a large-scale, nonlinear, nonconvex, mixed-integer combinatorial optimization problem. Various techniques [4–8] have been used to solve this computationally expensive problem [5]. Among these techniques are dynamic programming (DP) [9–12], branch-and-bound (BAB), Lagrangian relaxation (LR) [13–15], integer programming (IP) [16, 17], and metaheuristic algorithms such as genetic algorithms (GA) [18–24], particle swarm optimization (PSO), and neural networks [25–27].

Many methods such as DP, BAB, and IP suffer from the curse of dimensionality; that is, the solution space increases exponentially with the number of generating units. As a result, the computation time becomes unacceptable. LR, which decomposes the primal problem into a set of single plant or unit optimization subproblems that are easier to solve independently, has shown potential in solving the UC problem. However, this approach requires the conversion of the optimal dual solution into a feasible solution of the primal problem because of the duality gap, which represents the primary difficulty associated with LR. Metaheuristic algorithms have been widely used to solve the UC problem in recent years, but the quality of the solution depends on the choice of the control parameters, and choosing the control parameters of these algorithms is a very difficult task.

Essentially, the MTSFTPP problem is a multistage (in this context, the term "stage" refers to time) decision problem with unit constraints (type 1), thermal power load factor constraints (type 2), operating time constraints (type 3), and boundary condition constraints (type 4). Considering that types 1 and 2 are single-stage constraints and independent between any two stages, whereas type 4 constraints can be met by the initial solution, the optimization process mainly involves type 3 constraints, which are time constraints. This study investigates a hybrid search algorithm (HSA) to solve the MTSFTPP problem. The primary objective is to achieve equal accumulated operating hours of installed capacity [25] (EAOHIC) for all thermal power plants over a given period, with extra hours that are given to the plants with lower emissions, higher efficiency, and occasional unit maintenance. A more detailed description of the problem is provided in Section 2. Section 3presents a detailed explanation of the HSA. A case study is presented in Section 4, and conclusions are presented in Section 5.

PROBLEM FORMULATION

Objective Function

Equality, impartiality, and transparency are important in electric power dispatching. Based on power system security, stability, and economics, units should be treated equally in problems involving operations management and information disclosure. In many western countries, the driving force behind deregulation is the belief that competitive markets make the industry more efficient, leading to lower costs. In the case of China, similar to other developing countries, the incentive for restructuring is derived from the desire to provide equal opportunities to all participants. In China, the firms established by the government are responsible for power system operations, and the power systems are operated in an integrated fashion. Therefore, equality is usually the main concern in long-term and midterm scheduling in China. In this study, EAOHIC is the primary consideration in the MTSFTPP problem. Operating hours are accumulated from the beginning of the respective warm-up periods to the end of the computational period (Figure 1). The EAOHIC constraint for the MTSFTPP problem is formulated as follows:

$$H_i = H_j \quad (i \neq j), \tag{1}$$

Where H_i, H_j are the accumulated operating hours of installed capacity for plants i, j, respectively, in the selected period.

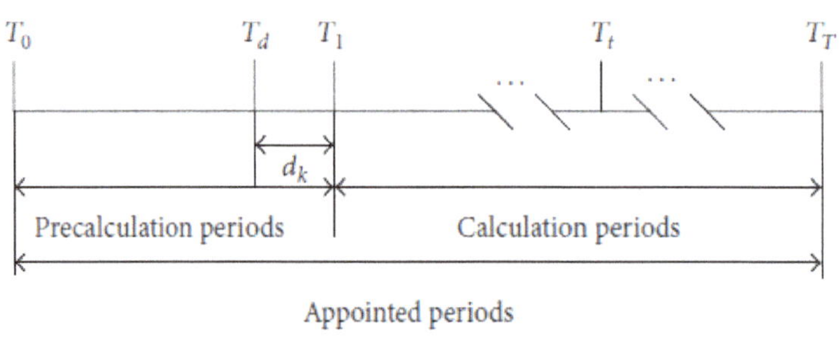

Figure 1: Accumulated operating hours defined from the start of the warm-up period to the end of the computational period (horizon).

Although the concept captured by (1) is unambiguous and easily understood, the problem is very difficult to solve directly because the formulation is related to all plants and the objective cannot be evaluated by the computation of a single mathematical expression. Considering that the variance can be used to describe the degree of dispersion of data, let $H = (H_1, H_2, \ldots, H_k, \ldots, H_m)$ be the set of accumulated operating hours of the installed capacity for the plants, and let D(H) be the variance of the set H. Equation (1) can be replaced by the following problem:

$$\begin{aligned} f &= \min D(H), \\ D(H) &= \frac{1}{m}\sum_{k=1}^{m}\left(H_k - \overline{H}\right)^2, \\ \overline{H} &= \frac{1}{m}\sum_{k=1}^{m} H_k, \\ H_k &= H_{kp} + H_{kT} - H_{ke}, \\ H_{kT} &= \frac{1}{N_k} * \overline{C_k} * T * 24 * r, \\ \overline{C_k} &= \frac{1}{T}\sum_{t=1}^{T} C_{kt}, \end{aligned} \quad (2)$$

where the variables are defined as follows: H is the set of H_k, the accumulated operating hours of the installed capacity of plant k during the selected period, in hours (h); m is the total number of plants; $D(H)$ is the variance of set H, in h^2; H is the average of set H, in h; H_{kp} is the operating hours of installed capacity of plant k accumulated during the warm-up period, in h; H_{kT} is the operating hours of installed capacity of plant k accumulated during the computational period, in h; H_{ke} is the extra hours assigned to plant k, in h; T is the total number of steps t in the computational period, where each step is one day;N_k is the installed capacity of plant k, in MW; $\overline{C_k}$ is the average of the operating capacity of plant k during the computational period, in MW; r is the load factor; and C_{kt} is the operating capacity of plant k in period t, in MW.

Constraints

Due to the operational requirements, the minimization of the objective function is subject to the following types of constraints.

(1) Unit Constraints. Consider

$$u_{t,k}^{\min} \le u_{t,k} \le u_{t,k}^{\max}, \quad (3)$$

Where $u_{k,t}^{\min}$ and $u_{k,t}^{\max}$ are the minimum and maximum numbers of units, respectively, in plant k in period t and $u_{k,}$ is the number of active units in plant k in period t.

(2) Thermal Power Load Factor Constraints. Consider

$$C_{st} * r^{\min} \le P_t \le C_{st} * r^{\max}, \quad (4)$$

where C_{st} is the operating capacity of the system in period t, in MW, and $r^{\min}$ and $r^{\max}$ are the minimum and maximum load factors of thermal power, respectively, which are calculated from the long-term real-world operation of power grid. P_t is the demand of system in period t, in MW.

(3) Time Constraints. Midterm optimal thermal power plant scheduling only dispatches the system load to the plants and does not distribute the load among the units in each plant. Therefore, we transform the constraints on the response of a plant into minimum-time constraints at minimum or maximum output. Consider

$$\begin{aligned} T_{k,j}^{\text{peak}} &\ge T_k^{\text{peak,min}}, \\ T_{k,j}^{\text{valley}} &\ge T_k^{\text{valley,min}}, \end{aligned} \quad (5)$$

Where $T_{k,j}^{\text{peak}}$ is the duration of peak j in the output of plant k, $T_k^{\text{peak,min}}$ is the minimum duration of the peak in the output of plant k, $T_{k,j}^{\text{valley}}$ is the duration of valley j in the output of plant k, and $T_k^{\text{valley,min}}$ is the minimum duration of the valley in the output of plant k.

(4) Boundary Condition Constraints. The planning for the current period depends on the previous states of the units.

For the plant k, considering the previous period d_k as the maximum in $T_k^{\text{peak,min}}$ and $T_k^{\text{valley,min}}$, the states of the units before the starting point affect the operation of the units through the minimum peak and valley time restrictions, as shown in Figure 1. Consider

$$\begin{aligned} T_{k,0}^{\text{peak}} &\ge T_k^{\text{peak,min}}, \\ T_{k,0}^{\text{valley}} &\ge T_k^{\text{valley,min}}, \end{aligned} \quad (6)$$

where $T_{k,0}^{\text{peak}}$ and $T_{k,0}^{\text{valley}}$ are the first duration of peak and valley in the output of plant k, respectively. Note that the boundary condition constraints are the same as the time constraints (type 3), the time constraints considering the time horizon in $[T_1, T_T]$, while the boundary condition constraints considering in $[T_d, T_T]$.

SOLUTION METHODOLOGY

The MTSFTPP problem is a UC problem, and each approach to the problem has its disadvantages, as mentioned previously. Constraints of types 1 and 2 involve a single period and can be treated independently to reduce the search space, constraints of type 4 can be satisfied in the initial solution, and constraints of type 3 must be satisfied over multiple periods. This study investigates a novel method using three stages to solve the problem.

Producing a Feasible Solution Space Satisfying Constraint Types 1 and 2

Mathematically, the MTSFTPP problem, like the UC problem, suffers from the curse of dimensionality as the number of units and the computational period increases. Thus, it is difficult to solve this problem by considering a simple combination of units. For a power grid with five thermal plants, each having four units, the total number of unit combinations will reach 8.67×10^{41} for a horizon of one month, and the search will be computationally expensive. Fortunately, the actual constraints on the thermal power load factors and the number of units are beneficial because they reduce the search space when searching for feasible solutions. The specific procedure is as follows.

Step 1. For the thermal plants with different values of unit capacity, each plant will be represented by several "virtual plants" whose capacity depends on the number of units of a given type. For instance, for a thermal plant with 5 units, including 2 units with 300 MW of capacity and 3 units with 200 MW of capacity, the plant will be divided into 2 "virtual plants." By initializing the plants in this manner, the final number of virtual plants will be greater than the number of actual plants and will be denoted as M.

Step 2. Start from the first period; that is, let $t = 1$.

Step 3. Sort the combinations U_t that meet the unit constraints in period t; that is, $U_t = \{u_t^1, u_t^2, \dots, u_t^i, \dots, u_t^{N_{Ut}}\}$ $(i = 1 \sim N_{Ut})$, where N_{Ut} is the number of combinations that meet constraint 1 (3) in period t, $u_t^i = (u_{t,1}^i, u_{t,2}^i, \dots, u_{t,k}^i, \dots, u_{t,M}^i)$, $u_{k,t}^{\min} \le u_{k,t}^i \le u_{k,t}^{\max}$ $(k = 1 \sim M)$.

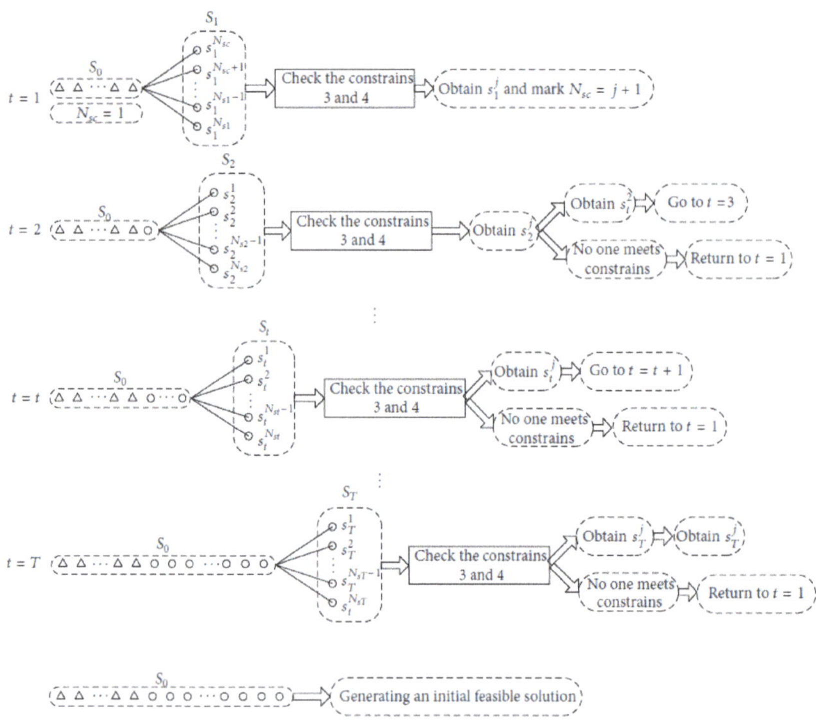

Figure 2: The procedure for generating an initial feasible solution.

Step 4. Select the combinations S_t that satisfy the thermal power load factor constraints from U_t in period t; that is, $S_t = \{s_t^1, s_t^2, \dots, s_t^j, \dots, s_t^{N_{st}}\}$ $(j = 1 \sim N_{st})$, where N_{st} is the number of combinations that satisfy constraints 1 and 2 ((3) and (4)) in period t.

Step 5. If $t < T$, set $t = t + 1$ and return to Step 3. Otherwise, stop and retain the feasible solution space $S = \{S_1, S_2, \dots, S_t, \dots, S_T\}$ satisfying constraints 1 and 2 ((3) and(4)).

Generating an Initial Feasible Solution Using a Heuristic Method

At this point, we have obtained the feasible space satisfying constraints of types 1 and 2. Finding the optimum in the next stage requires a solution that satisfies all of the operational and system constraints. A heuristic method is proposed for this goal. The procedure is as follows.

Step 1. Obtain the operating data for d days before the starting point. These data will be stored in an array and denoted as the scheme S_0. The midterm schedule must satisfy the boundary constraints (type 4) between actual operation and future scheduling.

Step 2. Start from the first period; that is, let $t = 1$.

Step 3. Obtain all combinations S_t for period t $(t = 1 \sim T)$ using the method described in Section 3.1.

Step 4. Choose a combination that satisfies constraint type 3 in (5) and that is suitable for the boundary constraints, type 4. For each subset s_t^j $(j = 1 \sim N_{st})$ in combination space S_t at period t, check the type 3 and type 4 constraints. Inparticular, the type 4 constraints use the historical data S_0 to satisfy the minimum-time constraints at the starting points. Clearly, S_0 includes the historical data when $t = 1$.The search is stopped and the current subset appended to S_0 when a feasible subset s_t^j $(j = 1 \sim N_{st})$. The subset number

in the first period is stored to reduce the search in the next cycle. In this case, continue to the next period and set $t = t+1$. Otherwise, return to the first period, start from the stored number, and repeat the previous procedure.An initial feasible solution will be generated for the optimization. A flow chart and pseudocode for generating an initial feasible solution are shown in Figure 2 and Algorithm 1, respectively.

Heuristic for generating an initial feasible solution
Input actual operation data for d days before the start point $S_0 = \{S_{01}, \ldots, S_{0k}, \ldots, S_{0M}\}$
feasible solution space $S = \{S_1, \ldots, S_t, \ldots, S_T\}$, $S_t = \{s_t^1, \ldots, s_t^j, \ldots, s_t^{N_{st}}\}$, $s_t^j = (u_{t1}^j, \ldots, u_{tk}^j, \ldots, u_{tM}^j)$
initial $N_{sc} = 1$
Output initial feasible solution I
for all periods from $t = 1$ to T
  for all the feasible states $s_t^j \subset S_t$ from $j = b$ to N_{st}
    if $t = 1$
      $b = N_{sc}$
    end if
    else
      $b = 1$
    end else
    Define num to count, num = 0
    for all plants states $u_{tk}^j \in s_t^j$ from $k = 1$ to M
      if $\tau_{k,j}^{\text{peak}}(S_{0k}, u_{tk}^j) \geq T_k^{\text{peak,min}}$ & $\tau_{k,j}^{\text{valley}}(S_{0k}, u_{tk}^j) \geq T_k^{\text{valley,min}}$
        num = num + 1
      end if
    end for
    if num = M
      Expand S_0' by appending s_t^j to S_0
        Store s_t^j in I
        if $t = 1$
          $N_{sc} = j + 1$
        end if
    end if
    if num < M & $j = N_{st}$
      $t = 1$, $j = N_{sc}$,
    end if
  end for
end for

Algorithm 1: Algorithm for generating an initial feasible solution.

Hybrid Search Algorithm

Numerous operational and system constraints are imposed in the unit commitment problem. The aforementioned initial feasible solution meets only some constraints, and the optimization for unit commitment still must be performed. Sections 3.3.1 and 3.3.2 present a two-step procedure for optimization.

Vertical Search Algorithm

Over the past few decades, the progressive optimality algorithm [28] (POA) has been shown to have great advantages over classical optimization methods and has been one of the most widely used techniques for hydroelectric generator scheduling and water resources problems [29–31]. The POA attempts to find the best solution in a given decision space based on Bellman's Principle and is free of a particular model structure. The advantages of the POA over other optimization techniques are that it can decompose a multistate decision problem into several nonlinear programming subproblems to reduce the dimensionality. In fact, the optimal scheduling of thermal power plants is a multistate and multistage problem, and hence it is suitable for the POA. The POA is defined as follows [27]:

$$f = \min \sum_{t=1}^{T} g(x_{t-1}, x_t),$$

$$G(x_{t-1}, x_{t+1}) = \min_{1 \leq j \leq N_t} \left[g(x_{t-1}, x_t^j) + g(x_t^j, x_{t+1}) \right], \tag{7}$$

where x_t^j is the state j in period t, $G(x_{t-1}, x_{t+1})$ is the objective function in period t, and x_{t-1} and x_{t+1} are the decision variables in periods $t - 1$ and $t + 1$, respectively. For the optimal scheduling of thermal power plants, the problemcan be mathematically stated as follows:

$$f = \min D(H),$$
$$G(s_t, s_{t+1}) = \min_{1\le j\le N_{st}} \left[D\left(H\left(s_t^j, s_{t+1}\right)\right)\right] \quad 1 \le t < T,$$
$$G(s_{t-1}, s_t) = \min_{1\le j\le N_{st}} \left[D\left(H\left(s_{t-1}, s_t^j\right)\right)\right] \quad t = T, \tag{8}$$

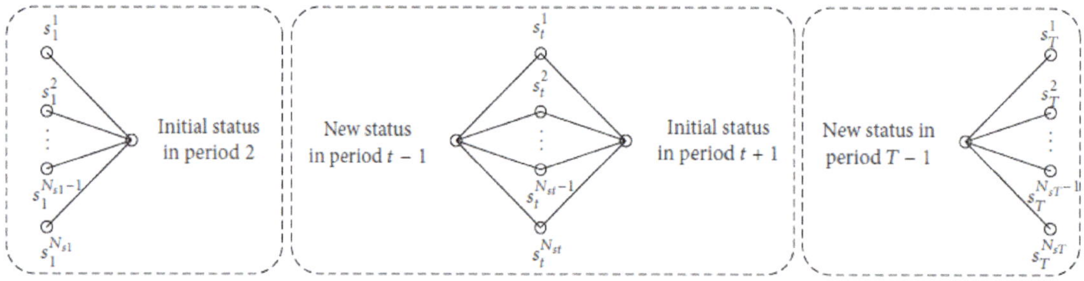

Figure 3: Search principle of VSA using POA.

VSA
Input initial feasible solution $I = \{I_t \mid I_t \subset S_t, 1 \le t \le T\}, I_t = (i_{t1}, \ldots, i_{tk}, \ldots, i_{tM})$
actual operation data for d days before the start point $S_0 = \{S_{01}, \ldots, S_{0k}, \ldots, S_{0M}\}$
feasible solution space $S = \{S_1, \ldots, S_t, \ldots, S_T\}, S_t = \{s_t^1, \ldots, s_t^j, \ldots, s_t^{N_{st}}\}, s_t^j = (u_{t1}^j, \ldots, u_{tk}^j, \ldots, u_{tM}^j)$
initial value V_0
Output optimized solution
for $n = 1$ to N
  for all periods from $t = 1$ to T
    for all the feasible states $s_t^j \subset S_t$ from $j = 1$ to N_{st}
      Define num to count, num = 0
      for all plants states $u_{tk}^j \in s_t^j$ from $k = 1$ to M
        if $\tau_{k,j}^{\text{peak}}(S_{0k}, u_{tk}^j) \ge T_k^{\text{peak,min}}$ & $\tau_{k,j}^{\text{valley}}(S_{0k}, u_{tk}^j) \ge T_k^{\text{valley,min}}$
          num = num + 1
        end if
      end for
      if num = M
        Calculate $V = D(H)$
        if $V < V_0$
          Replace i_{tk} using u_{tk}^j; $V_0 = V$
        end if
      end if
    end for
  end for
  Obtain new solution R
  If $I = \text{R}$
    Obtain optimized solution
  end if
end for

Algorithm 2: Algorithm for VSA.

Where $H(s_t^j, s_{t+1})$ and $H(s_{t-1}, s_t^j)$ are functions that give the operating hours of installed capacity when the status is s_t^j and $G(s_t, s_{t+1})$ and $G(s_{t-1}, s_t)$ are the objective functions for period t. With this formulation, we choose s_t^j from S_t with the status in the other periods fixed to obtain the optimal objective function value. A diagram of the operating of the POA is shown in Figure 3, and pseudocode for the algorithm is shown in Algorithm 2.

The aforementioned search is a sequential procedure, which we refer to as the vertical search algorithm (VSA), as distinguished from the following search algorithm.

Lateral Search Algorithm

The solution obtained using the VSA is clearly affected by the starting point (i.e., initial guess) and can converge to a local minimum. Hence, another search strategy is presented to improve the solution by enlarging the search space. The search addresses the objective that all plants have equal or approximately equal operating hours of installed capacity. To attain this goal, the plants whose operating hours are greater than or less than the average value will be adjusted. The plant with the greatest bias from the average is selected for adjustment. Generally, we choose the periods of maximum or minimum plant output to perform the adjustments. This state involves situations in which the system load allows for a change in the number of active units. Another state occurs in periods of maximum or minimum output during which a change in the number of active units in other periods is not possible. For the first case, the adjustment is started from the first period among the periods of maximum or minimum output. The increase or decrease in the number of units depends on the difference from the average and the constraints. If the plant's operating hours of installed capacity are greater than the average, we decrease the number of units for selected periods. This adjustment is a trial procedure because of the complex constraints and because the starting period for the adjustment is related to the constraints. For the second case, the changes in the maximum/minimum output periods will cause a change in the unit status in subsequent periods due to the minimum-time constraints. The adjustment features two aspects: satisfying all of the constraints in the maximum- or minimum-output periods and satisfying the constraints in subsequent periods. In either case, a new feasible solution will be produced. Consequently, an updated optimal solution can be obtained using the VSA. Herein, we refer to the search procedure as the lateral search algorithm (LSA). An optimal solution that satisfies the constraints can be obtained using the VSA and LSA alternately. A flowchart for the hybrid search algorithm that combines the VSA and LSA is shown in Figure 4.

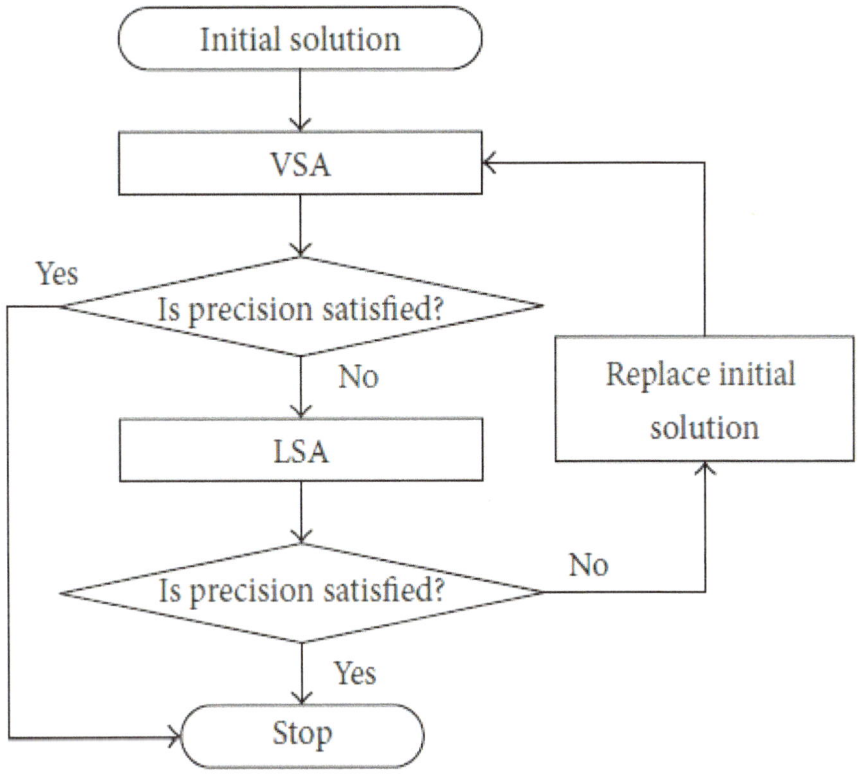

Figure 4: Search procedure for HSA using VSA and LSA.

CASE STUDY

The proposed method was applied to the MTSFTPP problem for an actual power system, the Yunnan Power Grid (YPG) in China. Currently, it is used as the primary tool to determine the midterm optimal scheduling of thermal power plants by the operators of YPG. Yunnan is among the richest in hydropower resources of all of the provinces in China. Its potential hydropower capacity is 103.6 GW, which contributes 15.3% of the national capacity and ranks third in the nation. The exploitable hydropower capacity is 90 GW, which contributes 17.9% and ranks second in the nation. By the end of 2013, the total installed capacity of the YPG had reached 26.1 GW, with hydropower capacity accounting for 16.7 GW and thermal capacity accounting for 8.935 GW. Proper scheduling of thermal power plants can make the best use of hydropower systems and improve system security.

The thermal power system in the YPG consists of 9 thermal power plants with a total installed capacity of 8.935 GW and 27 units. The dispatching center in the YPG controls the thermal power system. Table 1 lists the basic data for these plants. The HSA was implemented in Java and executed on a PC with an Intel Core 2 Duo 2.93 GHz CPU and 2 GB of memory. An actual schedule for 2013 in the YPG was used to test the validity and the computational efficiency of the proposed method.

Table 1: Thermal power system in the YPG

Plant number	Units (number * capacity, MW)	Capacity (MW)
1	4 * 600	2400
2	6 * 300	1800
3	4 * 300	1200
4	2 * 200 + 2 * 300	1000
5	2 * 300	600
6	2 * 300	600
7	2 * 300	600
8	2 * 300	600
9	1 * 135	135

Two typical months, May and September, were selected to demonstrate the practicality and the efficiency of the method. May represents the beginning of the flood season, when the hydropower plants can operate at their peak and the demand on the thermal system can be decreased. Conversely, September represents the beginning of the dry season, when the hydropower system has the lowest output and the demand on the thermal system increases.

According to the real-world operating experience and users' actual demands, the basic parameters for the HSA are set as follows:

(i) minimum load factor $r^{\min} = 0.7$;

(ii) maximum load factor $r^{\max} = 0.9$;

(iii) initial load factor $r = 0.8$;

(iv) actual operating hours of installed capacity in the warm-up period $H_{kp} = 0$ $(k = 1 \sim 9)$;

(v) extra hours $H_{ke} = 0$ $(k = 1 \sim 9)$;

(vi) minimum number of units $u_{kt}^{\min} = 1$ $(k = 1 \sim 8,$ $t = 1 \sim T)$, and $u_{kt}^{\min} = 0$ for $k = 9$;

(vii) maximum number of units $u_{k,t}^{\max} = UA_{k,t}^{\max}$ $(k = 1 \sim 9,$ $t = 1 \sim T)$, where $UA_{k,t}^{\max}$ is the number of available units for plant k in period t;

viii) minimum duration of peak in plant output $T_k^{\text{peak,min}} =$ 7 days $(k = 1 \sim 9)$;

(ix) minimum duration of valley in plant output $T_k^{\text{valley,min}} = 3$ days $(k = 1 \sim 9)$;

(x) considering previous periods $d_k = 7$ days $(k = 1 \sim 9)$.

The average cost time of 10 runs for May is 9764 ms and for September is 10241 ms, and the calculation efficiency can be met the actual demand in midterm scheduling.

In a comparison of the results obtained by the HSA with those obtained by the heuristic, the VSA and the LSA are shown in Table 2. As indicated by the table, the results improve progressively in going from the heuristic to the HSA for most of the evaluation criteria, including the average, the max-min difference, and the objective value, for both May and September. It should be noted that the most improved solutions were achieved by the VSA, which demonstrated that the VSA has the capability to find the optimal solution.

The LSA attempts to improve the objective function value and avoid convergence to a local minimum. As shown in Table 2, Plant 1 was changed in May by the LSA. Although the max-min difference remained the same, the objective function was diminished from 50.06 to 36.4. There is little difference in the results obtained between September and May. Due to the adjustments to the generating units in the valley periods by the LSA, the max-min difference was greater, but the objective function decreased from 10.81 to 7.17. Equal values for all three evaluation criteria were attained for May and September using the HSA, which indicates that the HSA has the capability to find the global minimum.

Table 2: Comparison of three methods for two typical months

	May				September			
	Heuristic	VSA	LSA	HSA	Heuristic	VSA	LSA	HSA
Operating hours of installed capacity (h)								
Plant 1	475.2	484.8	499.2	499.2	264	288	288	288
Plant 2	499.2	499.2	499.2	499.2	304	284.8	284.8	288
Plant 3	475.2	499.2	499.2	499.2	244.8	278.4	292.8	288
Plant 4	499.2	499.2	499.2	499.2	268.8	282.2	282.2	288
Plant 5	489.6	499.2	499.2	499.2	288	288	288	288
Plant 6	499.2	499.2	499.2	499.2	288	288	288	288
Plant 7	499.2	499.2	499.2	499.2	288	288	288	288
Plant 8	489.6	499.2	499.2	499.2	288	288	288	288
Plant 9	480	480	480	499.2	268.8	288	288	288
Average (h)	489.6	495.5	497.1	499.2	278.04	285.9	287.5	288
Max-min difference (h)	24	19.2	19.2	0	59.2	9.6	10.6	0
Objective value (h2)	97.28	50.06	36.4	0	282.61	10.81	7.17	0

The generation requirements for the 9 plants in May are shown in Figure 5 and the details of results are listed in Table 3. Meanwhile, the generation requirements and details in September are shown in Figure 6 and Table 4. From Tables 3 and 4, it can be seen that the boot capacity result of each plant can meet all constraints. In Plant 2 in Table 3, for example, the days of first peak in plant output are 8 (from 7th to 14th) which is bigger than minimum duration of peak in plant output $T_k^{peak,min} = 7$ days, while the days of first valley in plant output are 4 (from 17th to 20th) which is bigger than minimum duration of valley in plant output $T_k^{valley,min} = 3$days. Figure5 shows that the outputs of most of the plants by the end of May were reduced to take advantage of the hydropower system. Plant 4 appears to have been an exception to this trend in May and could not satisfy the peak time constraint by the end of May, but this result is misleading. Table 1 shows that Plant 4 has two types of turbine units, 2 units of 200 MW capacity and 2 units of 300 MW capacity. As described in Section 3.1, Plant 4 was divided into two "virtual plants" for computational purposes. Table 3 and Figure 7 show the generation requirements for "virtual plants" 1 and 2 and demonstrate that these plants exhibit the same performance as the other plants. Conversely, the outputs of all of the plants increased from the beginning of September. Figures 5and 6 demonstrate that the MTSFTPP results for May and September are reasonable.

Table 3: Results of generation requirements for Plants 1 to 9 in May (MW)

Date	Plant 1	Plant 2	Plant 3	Plant 4			Plant 5	Plant 6	Plant 7	Plant 8	Plant 9
				Virtual plant 1	Virtual plant 2	Total					
1st	2400	1500	1200	400	600	1000	300	300	300	300	135
2nd	2400	1500	1200	400	600	1000	600	300	300	300	135
3rd	2400	1500	900	200	600	800	600	600	600	300	135
4th	2400	1500	900	200	600	800	600	600	600	600	135
5th	2400	1500	900	200	600	800	600	600	600	600	135
6th	2400	1500	1200	400	600	1000	600	600	600	600	0
7th	2400	**1800**	1200	400	600	1000	600	600	600	600	0
8th	2400	**1800**	1200	400	600	1000	600	600	600	600	0
9th	2400	**1800**	1200	400	600	1000	600	600	600	600	135
10th	2400	**1800**	1200	400	300	700	600	300	300	600	135
11th	2400	**1800**	1200	400	300	700	600	300	300	600	135
12th	2400	**1800**	1200	400	300	700	600	300	300	600	135
13th	1800	**1800**	1200	400	600	1000	600	600	600	600	135
14th	1800	**1800**	900	400	600	1000	600	600	600	600	135
15th	1800	1500	900	400	600	1000	600	600	600	600	135
16th	2400	1200	900	400	600	1000	300	600	600	600	135
17th	2400	**900**	1200	200	600	800	300	600	600	600	135
18th	2400	**900**	1200	200	600	800	300	600	600	300	135
19th	2400	**900**	1200	200	600	800	300	600	600	300	135
20th	2400	**900**	1200	400	300	700	600	300	300	300	135
21st	2400	1200	1200	400	300	700	600	300	300	300	135
22nd	2400	1200	1200	400	300	700	600	300	300	300	135
23rd	2400	1200	1200	400	300	700	600	300	300	600	135
24th	1200	1800	300	400	600	1000	600	600	600	600	135
25th	600	1800	300	400	600	1000	600	600	600	600	135
26th	600	1800	300	400	600	1000	600	600	600	600	135
27th	600	1800	900	200	600	800	300	600	600	600	135
28th	1200	1800	900	0	600	600	300	600	600	600	135
29th	1200	1800	900	0	600	600	300	600	600	600	135
30th	1200	1800	900	0	600	600	300	600	600	300	0
31st	2400	900	900	400	300	700	300	300	300	300	0

Table 4: Results of generation requirements for Plants 1 to 9 in September

Date	Plant 1	Plant 2	Plant 3	Plant 4	Plant 5	Plant 6	Plant 7	Plant 8	Plant 9
1st	600	300	300	300	300	300	300	300	135
2nd	600	300	300	600	300	300	300	300	0
3rd	600	300	600	600	300	300	300	300	0
4th	600	300	600	600	300	300	300	300	0
5th	600	600	600	600	300	300	300	300	0
6th	600	600	600	600	300	300	300	300	135
7th	1200	600	600	600	300	300	300	300	135
8th	1200	600	600	600	300	300	300	300	135
9th	1200	600	600	600	300	300	300	300	135
10th	1200	900	600	600	300	300	300	300	135
11th	1200	900	600	600	300	300	300	300	135
12th	1800	900	600	600	300	300	300	300	135
13th	1800	900	600	600	300	300	300	300	0
14th	1800	900	600	600	300	300	300	300	0
15th	1800	900	600	600	300	300	300	300	0
16th	1800	900	600	600	300	300	300	300	0
17th	1800	900	600	600	300	300	300	300	0
18th	1800	900	600	600	300	300	300	300	0
19th	1800	900	600	600	300	300	300	300	0
20th	1800	900	600	600	300	300	300	300	135
21st	1200	900	600	600	300	300	300	300	135
22nd	600	1200	600	600	300	300	300	300	135
23rd	600	1500	600	600	300	300	300	300	135
24th	600	1500	600	600	300	300	300	300	135
25th	1200	1500	600	600	300	300	300	300	135
26th	1200	1500	600	600	300	300	300	300	135
27th	1200	1500	600	600	300	300	300	300	0
28th	1200	1500	600	600	300	300	300	300	0
29th	1200	1500	600	600	300	300	300	300	0
30th	1200	300	1200	900	300	300	300	300	0

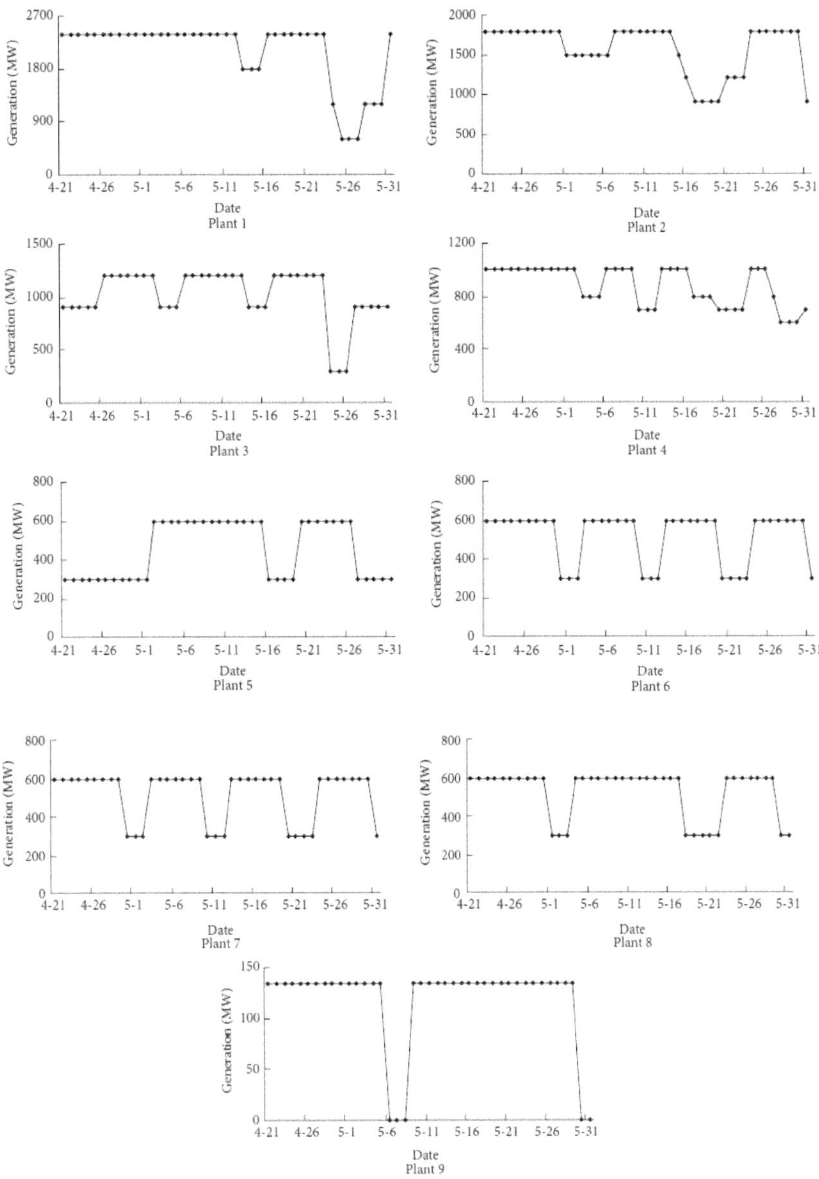

Figure 5: The process of generation requirements for Plants 1 to 9 in May.

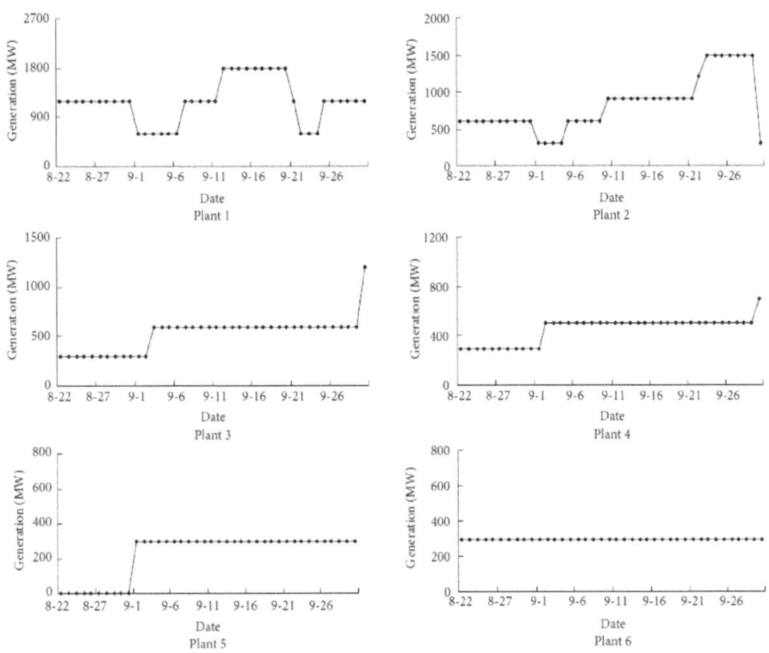

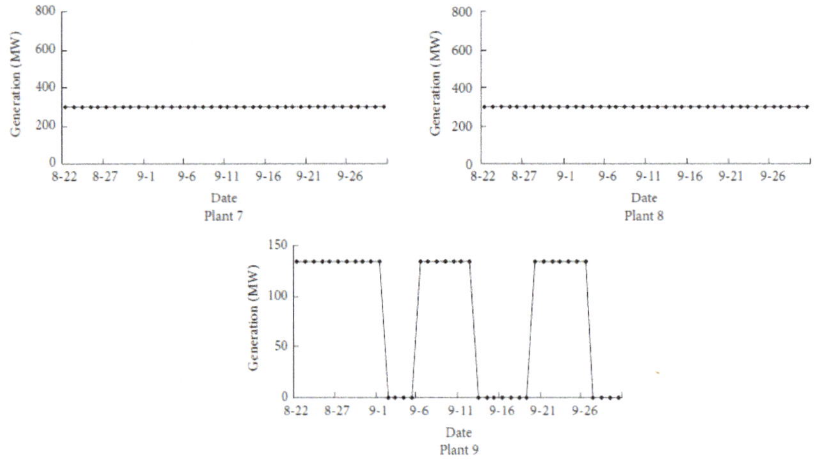

Figure 6: The process of generation requirements for Plants 1 to 9 in September.

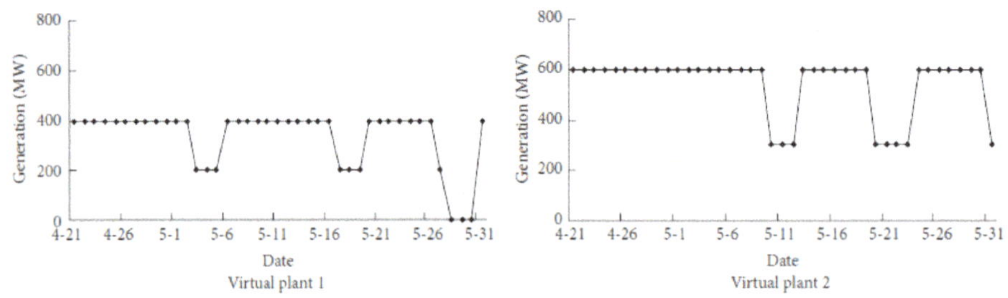

Figure 7: The process of generation requirements for "virtual plants" 1 and 2.

Figures 8 and 9 show the generation and capacity requirements in May and September, respectively. It can be observed that the generation requirements are between maximum and minimum capacity.

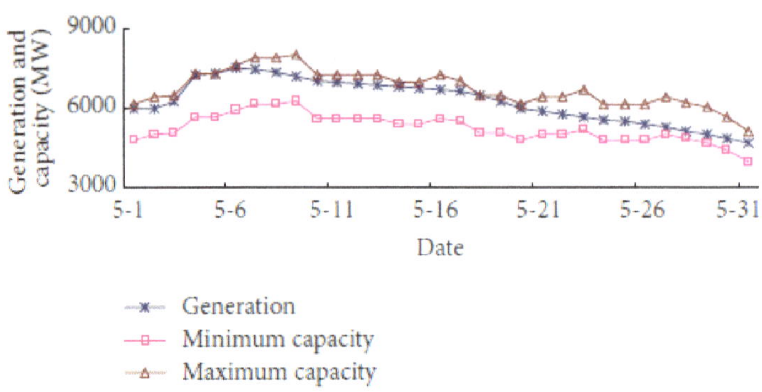

Figure 8: Generation and capacity requirements in May.

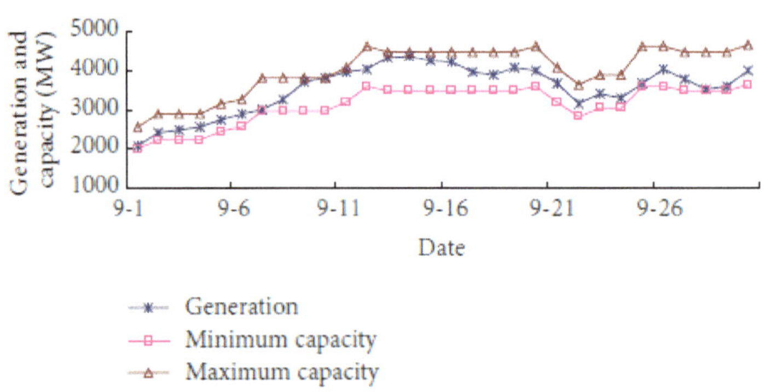

Figure 9: Generation and capacity requirements in September.

CONCLUSIONS

A model for the MTSFTPP problem was proposed. The model includes three stages. The feasible spaces for the MTSFTPP problem are first produced by narrowing the search space based on constraints 1 and 2. Next, an initial feasible solution for the MTSFTPP problem is obtained using virtual plants and considering constraints 3 and 4. The POA is used to solve the MTSFTPP problem, and the optimal search is referred to as the VSA. To avoid convergence to a local minimum, the LSA is used, and the updated solution is used as a new feasible starting point for the next search with the VSA. Thus, the HSA is produced by combining the LSA with the VSA. A complete framework for solving the MTSFTPP problem was established and demonstrated using a case study. The results showed that the proposed method can efficiently produce satisfactory solutions with practical requirements.

A novel method focusing on solving strategies for the MTSFTPP problem was proposed, in which the thermal system load was given after the other electric sources were optimized. However, the joint modeling of thermal power system with other source systems in midterm schedule can be considered as a research basis for future studies.

Conflict of Interests

The authors declare that there is no conflict of interests regarding the publication of this paper.

ACKNOWLEDGMENTS

This study is supported by the Major International Joint Research Project (no. 51210014), the National Natural Science Foundation of China (no. 51209029), and the Fundamental Research Funds for the Central Universities (no. DUT14QY15).

REFERENCES

1. China Electric Power Yearbook Editorial Board, China Electric Power Yearbook, China Electric Power Press, 2003 2013.
2. National Electric Power Industry Statistics Express, 2014, http://www.sxcoal.com/dl/3636879/articlenew.html.
3. L. Liu, H. Zong, E. Zhao, C. Chen, and J. Wang, "Can China realize its carbon emission reduction goal in 2020: from the perspective of thermal power development," Applied Energy, vol. 124, pp. 199–212, 2014.
4. N. P. Padhy, "Unit commitment—a bibliographical survey," IEEE Transactions on Power Systems, vol. 19, no. 2, pp. 1196–1205, 2004.
5. T. Niknam, A. Khodaei, and F. Fallahi, "A new decomposition approach for the thermal unit commitment problem," Applied Energy, vol. 86, no. 9, pp. 1667–1674, 2009.
6. S. K. Tong, S. M. Shahidehpour, and Z. Ouyang, "A heuristic short-term unit commitment," IEEE Transactions on Power Systems, vol. 6, no. 3, pp. 1210–1216, 1991.
7. T. Senjyu, K. Shimabukuro, K. Uezato, and T. Funabashi, "A fast technique for unit commitment problem by extended priority list," IEEE Transactions on Power Systems, vol. 18, no. 2, pp. 882–888, 2003.
8. A. Viana, J. P. de Sousa, and M. A. Matos, "Fast solutions for UC problems by a new metaheuristic approach," Electric Power Systems Research, vol. 78, no. 8, pp. 1385–1395, 2008.
9. S. Subramanian and S. Ganesan, "A novel hybrid approach using modified priority list and sequential approach for unit commitment," International Journal of Energy Sector Management, vol. 4, no. 4, pp. 555–576, 2010.
10. C. T. Cheng, S. S. Li, and G. A. Li, "A hybrid method of incorporating extended priority list into equal incremental principle for energy-saving generation dispatch of thermal power systems," Energy, vol. 64, pp. 688–696, 2014.
11. W. J. Hobbs, G. Hermon, S. Warner, and G. B. Shelbe, "An enhanced dynamic programming approach for unit commitment," IEEE Transactions on Power Systems, vol. 3, no. 3, pp. 1201–1205, 1988.
12. A. Rong, H. Hakonen, and R. Lahdelma, "A variant of the dynamic programming algorithm for unit commitment of combined heat and power systems," European Journal of Operational Research, vol. 190, no. 3, pp. 741–755, 2008.
13. W. L. Peterson and S. R. Brammer, "Capacity based Lagrangian relaxation unit commitment with ramp rate constraints," IEEE Transactions on Power Systems, vol. 10, no. 2, pp. 1077–1084, 1995.
14. Q. Z. Zhai, X. H. Guan, and J. Cui, "Unit commitment with identical units: successive subproblem solving method based on Lagrangian relaxation," IEEE Transactions on Power Systems, vol. 17, no. 4, pp. 1250–1257, 2002.
15. T. Seki, N. Yamashita, and K. Kawamoto, "New local search methods for improving the lagrangian-relaxation-based unit commitment solution," IEEE Transactions on Power Systems, vol. 25, no. 1, pp. 272–283, 2010.
16. A. Grey and A. Sekar, "Unified solution of security-constrained unit commitment problem using a linear programming methodology," IET Generation, Transmission & Distribution, vol. 2, no. 6, pp. 856–867, 2008.
17. B. Venkatesh, T. Jamtsho, and H. B. Gooi, "Unit commitment—a fuzzy mixed integer linear programming solution," IET Generation, Transmission and Distribution, vol. 1, no. 5, pp. 836–846, 2007.
18. N. Amjady and A. Shirzadi, "Unit commitment using a new integer coded genetic algorithm," European Transactions on Electrical Power, vol. 19, no. 8, pp. 1161–1176, 2009.
19. B. Yu, Z. Z. Yang, X. S. Sun, B. Yao, Q. Zeng, and E. Jeppesen, "Parallel genetic algorithm in bus route headway optimization," Applied Soft Computing Journal, vol. 11, no. 8, pp. 5081–5091, 2011.
20. C. Y. Dang and M. Q. Li, "A floating-point genetic algorithm for solving the unit commitment

problem,"European Journal of Operational Research, vol. 181, no. 3, pp. 1370–1395, 2007.

21. G. Dudek, "Genetic algorithm with integer representation of unit start-up and shut-down times for the unit commitment problem," European Transactions on Electrical Power, vol. 17, no. 5, pp. 500–511, 2007.

22. V. Senthil Kumar and M. R. Mohan, "Solution to security constrained unit commitment problem using genetic algorithm," International Journal of Electrical Power and Energy Systems, vol. 32, no. 2, pp. 117–125, 2010.

23. C. T. Cheng, C. P. Ou, and K. W. Chau, "Combining a fuzzy optimal model with a genetic algorithm to solve multi-objective rainfall-runoff model calibration," Journal of Hydrology, vol. 268, no. 1–4, pp. 72–86, 2002.

24. C.-T. Cheng, W.-C. Wang, D.-M. Xu, and K. W. Chau, "Optimizing hydropower reservoir operation using hybrid genetic algorithm and chaos," Water Resources Management, vol. 22, no. 7, pp. 895–909, 2008.

25. M. P. Walsh and M. J. O›Malley, "Augmented hopfield network for unit commitment and economic dispatch," IEEE Transactions on Power Systems, vol. 12, no. 4, pp. 1765–1774, 1997.

26. M. H. Wong, T. S. Chung, and Y. K. Wong, "Evolving neural network approach in unit commitment solution," Microprocessors and Microsystems, vol. 24, no. 5, pp. 251–262, 2000.

27. J.-X. Xie, C.-T. Cheng, K.-W. Chau, and Y.-Z. Pei, "A hybrid adaptive time-delay neural network model for multi-step-ahead prediction of sunspot activity," International Journal of Environment and Pollution, vol. 28, no. 3-4, pp. 364–381, 2006.

28. H. R. Howson and N. G. Sancho, "A new algorithm for the solution of multi-state dynamic programming problems," Mathematical Programming, vol. 8, pp. 104–116, 1975. . · . · View at Zentralblatt MATH · View at MathSciNet · .

29. A. Turgeon, "Optimal short-term hydro scheduling from the principle of progressive optimality," Water Resources Research, vol. 17, no. 3, pp. 481–486, 1981.

30. N. J. D. Lucas and P. J. Perera, "Short-term hydroelectric scheduling using the progressive optimality algorithm," Water Resources Research, vol. 21, no. 9, pp. 1456–1458, 1985.

31. J. Nanda, P. R. Bijwe, and D. P. Kothari, "Application of progressive optimality algorithm to optimal hydrothermal scheduling considering deterministic and stochastic data," International Journal of Electrical Power and Energy Systems, vol. 8, no. 1, pp. 61–64, 1986.

CHAPTER 15

ENHANCING ENERGY EFFICIENCY OF BOILER FEED PUMPS IN THERMAL POWER PLANTS THROUGH OPERATIONAL OPTIMIZATION AND ENERGY CONSERVATION

Rajashekar P Mandi, [1] S Seetharamu[1] and Udaykumar R Yaragatti[2]

[1]Energy Efficiency and Renewable Energy Division, Power Research Institute, Bangalore, India

[2]Dept. of Electrical and Electronic Engineering Central, National Institute of Technology, Surathakal, India

ABSTRACT

This paper describes the various methods for enhancing the energy efficiency of Boiler Feed Pumps (BFP) in thermal power plants based on the energy audit study conducted in 28 numbers of 210 MW coal fired thermal power plants in India. The auxiliary power used by BFPs is about 2.4 to 3.2 % of the total gross energy generation. BFPs are the largest (in rating) motor in a thermal power plant. The detailed energy audit of auxiliary equipment in various thermal power stations, operational optimization and appropriate control system had shown ample scope for improving the energy efficiency of BFPs. The energy efficiency improvement of BFPs by reducing the re-circulation flow, pressure drop across feed water circuit elements, enhancing overall efficiency of BFPs, etc., are discussed with case studies. The implementation of energy conservation measures reduce the average auxiliary power used by BFPs for 210 MW plant from 2.97 % to 2.32 % of gross energy generation and release an additional power of about 480 MW that reduce the CO_2 emission by about 3.9 million tonne per year

Keywords:

Energy efficiency, boiler feed pumps, auxiliary power, pump efficiency, sec, energy conservation

INTRODUCTION

The total installed power generation capacity in India is about 223.344 GW, out of which the generation from coal fired power plant is about 56 %. The average auxiliary power used of coal fired thermal power plant is 8.44 % of gross energy generation at an average plant load factor (PLF) of 73.3 % (CEA, 2013). Among the auxiliary equipment BFPs are the major energy consuming equipment that forms 30 % of total auxiliary power used by thermal power plants (Siddhartha Bhatt M et al., 1999). The estimated auxiliary power used BFPs of coal fired thermal power plants in India is about 4420 MW and average estimated CO_2 emission is about 36 million tonne per year on account of auxiliary power used by BFPs. The thermal power plant availability and reliability depends largely upon the operational reliability of auxiliary equipment like BFPs and also the

Citation: Mandi RP, Seetharamu S, Yaragatti UR (2014). Enhancing energy efficiency of boiler feed pumps in thermal power plants through operational optimization and energy conservation. International Research Journal of Power and Energy Engineering, 1(1): 002-011.

capability of the auxiliary system (Kiran Kumar et al., June 2013). The auxiliary power consumption is on higher side in Indian power plants as compared to other developed countries due to poor plant load factor, poor coal quality, excessive steam flow, internal leakage in equipment, inefficient drives, lack of operational optimization of equipment, ageing of equipment, hesitation in technology up-gradation, obsolete equipment, design deficiencies, oversizing of equipment, use of inefficient controls, etc. (Ashis Saha et al., 2005). The auxiliary power consumption of BFPs can be reduced by improving the plant load factor of the plants, by operational optimization, adoption of advanced control techniques and implementation of energy conservation measures (Namdeo D Adate et al., 2013). The operating overall efficiency of BFPs is in the range of 54.3 – 73.45 % as compared to design value of 80 %. By improving the overall efficiency of BFPs by average value of 7 – 10 % will reduce the average auxiliary power of BFPs by 480 MW and equivalent reduction of CO_2 emission is 3.9 million tonne per year.

Boiler Feed Pumps (BFP) are the major energy consuming equipment which are essential to increase the feed water pressure in a coal fired thermal power plants. In a 210 MW power plant BFP is supplied along with Booster pump which is mounted on the same shaft. The Booster pump increases the feed water pressure from 0.5 – 0.66 MPa (Deaerator pressure) to intermediate pressure of about 1.2 – 1.4 MPa and BFP main pump increases the FW pressure from Booster pump discharge to about 17 – 18 MPa (Drum pressure). In a typical 210 MW power plant, three numbers of multistage pumps with hydraulic scoop coupling to control the feed water flow. Booster pumps are of centrifugal pumps. Boiler feed pumps are axial split multistage, horizontal, barrel type, high capacity, high speed (about 5000 rpm), centrifugal pumps. There are three BFPs with HT induction motors of 6.6 kV and the motor rating will be of either 4.0 MW or 3.5 MW in a 210 MW power plants (Rajashekar P Mandi et al., 2010). Two pumps will be working continuously and third pump will be stand-by. The feed water flow will be regulated by scoop (fluid coupling) control and 3 element feed control valve station. There are two types of driving systems are used for Boiler Feed Pumps i.e., steam operated turbine drive or motor drive system. Generally in 500 MW above plant size, Turbine (steam operated) Driven BFP (TDBFP) are used because the motor size will be very big of the order of about two numbers of 10 MW size. Starting current of these motors will be very high and influence the voltage and other supply parameters in the network. The auxiliary steam at cold re-heat line will be used to run TDBFP. This steam is already taken part in producing partial output power in High Pressure Turbine (HPT). The overall efficiency of conversion from thermal energy (coal) to hydraulic output at BFP output is higher in case of TDBFP compared to motor driven BFP. The average conversion efficiency of coal to hydraulic power in TDBFP is about 62 % whereas at MDBFP is 26 %. But in 210 MW and lower size units adopt motor driven BFP due to lower operation and maintenance and also to optimize the space utilization.

Figure 1 is the energy used by different auxiliary equipment in a 210 MW power plant. The average specific auxiliary power used by BFP is 2.42 % of gross power generation at maximum continuation rating (MCR) and is high compared to design value of 2.28 % at MCR condition

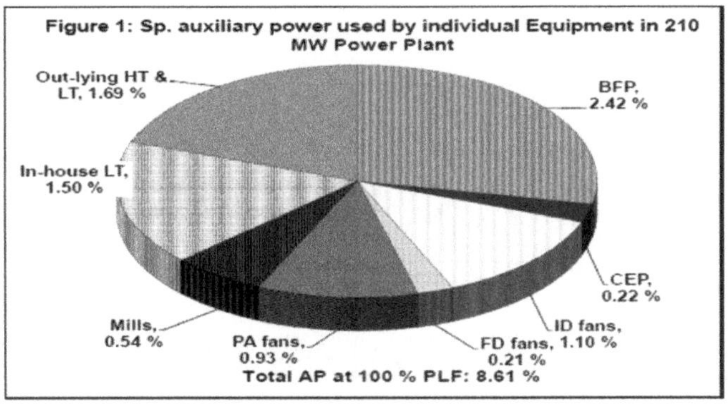

Figure 1: Sp auxiliary power used by individual Equipment in 210 MW power plant

Boiler feed pumps have to overcome the hydrodynamic resistances offered by High Pressure Heater (HPH) (regenerative feed water heaters), feed regulating station (FRS), economizer, water walls, superheaters, reheaters, etc. (Figure 2) and provide the high pressure steam at boiler outlet and then to turbines

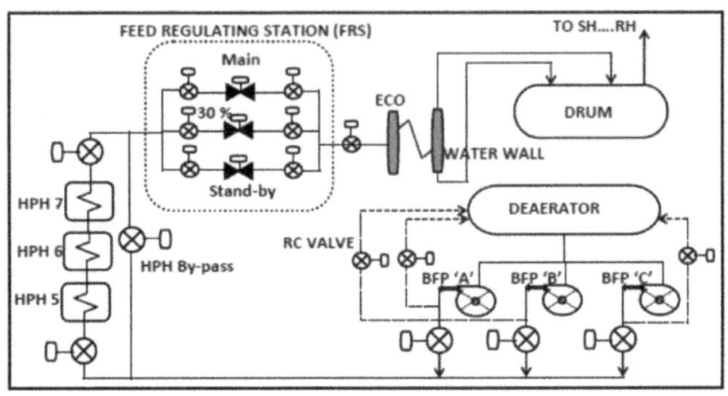

Figure 2: Schematic of feed water circuit

The performance test of BFPs is carried out in 28 units of 210 MW units with similar design of BFPs in India to evaluate the energy losses in BFPs and to reduce these losses thereby reducing the environmental burden of CO_2 by enhancing the energy efficiency of BFPs through implementation of energy conservation measures and operational optimization. During this performance test, the plant load is maintained nearly constant (load variation < 3 %) for a period of 120 minutes continuously; coal from the same source is used; no change-over of auxiliary equipment; no intermittent bottom ashing; and no soot-blowing operation during the test

Performance Evaluation of BFPS

The input power to pump-motor depends on the efficiency of pumps, motors, pressure gain (net head i.e., dynamic head and velocity head) across pump and flow (John S Hsu et al., 1998). The feed water (FW) circuit consists of hydrodynamic resistive elements like HP heaters, Feed regulating station (FRS) where the FW pressure will be regulated to obtain final main steam pressure at turbine inlet, Economizer coils, Water Walls, Superheaters and Reheaters (Boiler circuit). BFPs have to overcome these pressure drop offered by all these hydrodynamic resistive elements. The pressure drop across all these elements influences the pressure gain across BFP. The above variables are directly dependent on the plant load factor (PLF). All these variables (array) are plotted with variation in plant load factor (array). The Pearson product moment correlation method is used for finding the correlation coefficient (R^2) between x-axis array (i.e., PLF independent variable) and y-axis array (i.e., P_{in} ΔPR, ΔP_{HPH}, ΔP_{FRS}, ΔP_{ECO}, ΔP_{Boiler}, $\dot{m}_{FW}$ dependent variables)

Generally, coal fired thermal power stations are termed as base load power plants and are bound to operate at full load condition always. The predicted performance values for all the parameters of all auxiliary equipment are provided by manufacturers' for operating the plant at

100 % PLF (MCR condition) and 60 % PLF. The design values are extrapolated to 70 % of PLF because the plant operating below 70 % PLF with poor coal quality will be unstable and sometimes the plant may run even with oil support which is not economically advisable. Therefore, in this study the performance of BFP is considered for the plant load variation between 70 % to 100 % PLF.

Pressure gain across pump

As the plant load on the unit increases, the discharge pressure at pump increases to provide the necessary steam pressure at turbine inlet (Becnel CL et al., 1987). Figure 3 is the pressure gain across pump with plant load. The correlation coefficient for the variation of pressure gain with plant load factor is 0.9539 and the noise level in data is less. At MCR condition, the average measured pressure gain is 16.85 MPa and is lower than the design value of 17.55 MPa at Maximum Continuous Rating (MCR) condition i.e., 210 MW plant load. The pressure gain at full load capacity of BFP (max. efficiency point) is 20.07 MPa. The margin provided for operation of BFPs at MCR condition is about 13 % but the actual average measured operating point of BFP is about 16 % of full load capacity of BFPs. Operation of these pumps at non-optimal operating point cause reduction in efficiency of pumps that increase the power loss. The power loss operating at pressure gain at MCR condition is 0.05 % of gross generation and the power loss compared at actual measured operating point at MCR condition is 0.27 %. The lower pressure gain at BFPs is mainly because of higher margin is provided for BFPs for higher operational reliability during single cycle operation. The measured average pressure gain at 70 % PLF is 16.30 MPa and is lower than the design value of 17.52 MPa. The deviation in pressure gain measured at different plant loading is computed using the MATLAB software average value got from the regression analysis with respect to operating the plant load at MCR condition and is:

$$\partial P_r = \left(1-\frac{PR_T}{PR_{MCR}}\right)\times 100 \quad (1)$$

Where ∂P_r the deviation in pressure gain (%), PR_{MCR} is the pressure gain (MPa) at MCR condition (100 % PLF) and PR_T is the pressure gain at tested plant load (MPa)

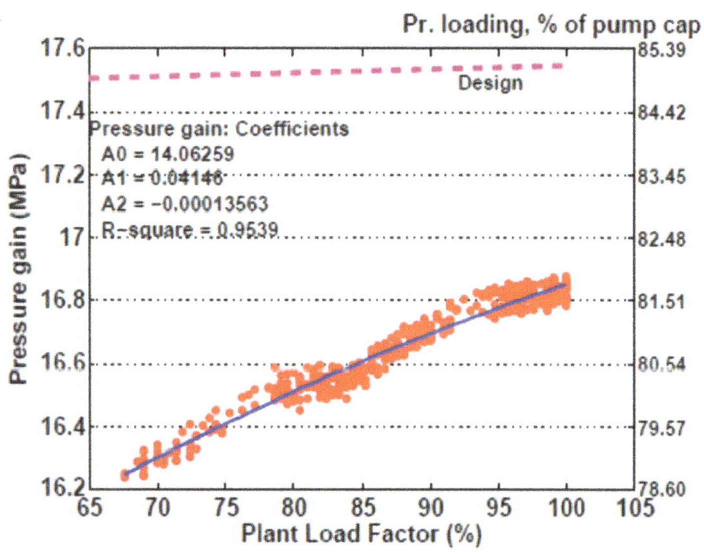

Figure 3: Variation of pressure gain across BFP

The deviation in pressure gain for operating the plant at 70 % of MCR is 3.3 % (Figure 4). The variation is less because all the turbines operate on a constant pressure mode operation where the main steam pressure at turbine inlet will be same for different plant load but marginal variation will be there.

Pressure drop across hydrodynamic resistive elements

The pressure drop across each hydrodynamic resistive elements are also analyzed and the power loss due to these elements are computed. The pressure drop across HPH is varying in the range of 0.53 MPa at MCR and 0.50 MPa at 70 % PLF. The deviation between MCR to 70 % PLF is 5.9 %. The power loss in HP heaters is computed in the range of 108.2 kW at MCR and 82.3 kW at 70 % PLF. The power loss in HP heater forms about 1.87 % to 2.07 % of total BFP power input. The pressure drop across FRS is varying in the range of 0.60 MPa at MCR and 0.56 MPa at 70 % PLF. The deviation between MCR to 70 % PLF is 6.5 %. The power loss in FRS is computed in the range of 122.1 kW at MCR and 92.2 kW at 70 % PLF. The power loss in FRS forms about 2.09 % to 2.34 % of total BFP power input. The pressure drop across Economizer is varying in the range of 0.189 MPa at MCR and 0.178 MPa at 70 % PLF. The deviation between MCR to 70 % PLF is 5.8 %. The power loss in Economizer is computed in the range of 38.41 kW at MCR and 29.23 kW at 70 % PLF. The power loss in Economizer forms about 0.66 % to 0.74 % of total BFP power input. The pressure drop across Boiler is varying in the range of 1.42 MPa at MCR and 1.34 MPa at 70 % PLF. The deviation between MCR to 70 % PLF is 5.8 %. The power loss in Boiler is computed in the range of 289.26 kW at MCR and 220.0 kW at 70 % PLF. The power loss in Boiler forms about 4.99 % to 5.55 % of total BFP power input.

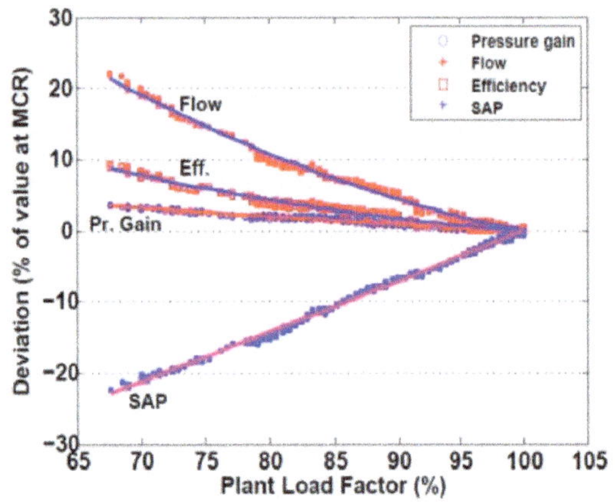

Figure 4: Deviation in parameter with PLF.

Feed water flow

Similarly, the feed water flow for both BFPs is plotted with PLF (Figure 5) and the correlation coefficient for the variation between feed water flows with PLF is 0.9726. The variation of flow is more sensitive to change in plant load and the noise level in data is very less. At MCR condition the average measured FW flow is 732 m ^{3} /h and is on par with the design value of 708.3 m ^{3}/h at MCR condition. The FW flow at full load capacity of BFP (max. efficiency point) is 715.6 m 3 /h (357.8 m ^{3} /h per pump). The margin provided for operation of BFPs at MCR condition is about 1 % but the actual average measured

FW flow is higher than the design value and operating point of BFPs is about – 2.3 % of full load capacity of BFPs. The higher FW flow may be due to operating BFPs at lower pressure, increased specific steam consumption of turbines, higher DM make up water, use of higher auxiliary steam for tracing lines, soot blowers, etc. The measured average FW flow at 70 % PLF is 591.5 m ^{3} /h and is higher than the design value of 444.2 m ^{3} /h. The deviation in FW flow measured at different plant loading is computed with respect to operating the plant load at MCR condition and the deviation (%) is computed as:

$$\partial \dot{m} = \left(1 - \frac{\dot{m}_T}{\dot{m}_{MCR}}\right) \times 100 \qquad (2)$$

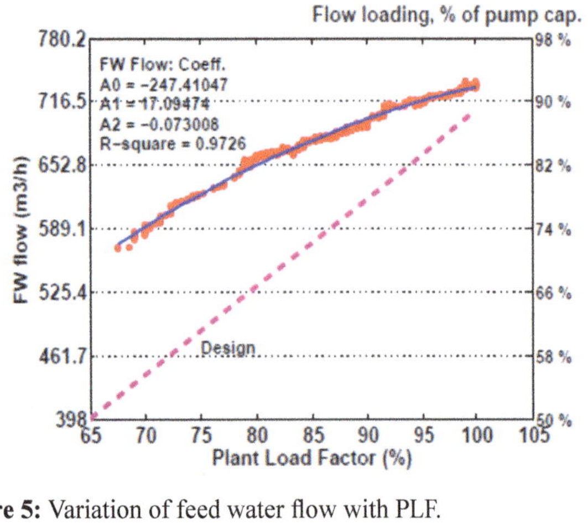

Figure 5: Variation of feed water flow with PLF.

Where $\partial \dot{m}$ the deviation (%) in FW flow is, $\dot{m}_{MCR}$ is the FW flow (m ^{3} /h) at MCR condition (100 % PLF) and $\dot{m}_T$ is the FW flow (m ^{3} /h) at tested plant load.

The deviation in feed water flow for operating the plant at 70 % of MCR is 19.2 %.

Efficiency

Similarly, the combined motor and pump (i.e., overall efficiency) is also plotted with PLF (Figure 6). The correlation coefficient of second order polynomial curve fit for the overall efficiency with PLF is 0.9551 and the noise level in data is less compared to flow and is on par with pressure gain. At MCR condition the average overall efficiency is 66.78 % and is lower than the design value of 72.1 % at MCR condition. The overall efficiency at full load capacity of BFP (max. efficiency point) is 73.3 %. The power loss for operating the plant at design MCR condition is 0.05 % of gross generation and the power loss compared at actual measured operating point at MCR condition is 0.27 %. The overall efficiency is low because of lower discharge pressure, higher FW flow, passing in re-circulation flow through RC valve, problem in pump like higher clearance between impeller and casing, pitting and erosion of pump impeller, etc., (El-Wakil MM, 1984). The deviation in overall efficiency at different plant loading is computed with respect to operating the plant load at MCR condition and the deviation (%) is computed as:

$$\partial \eta_O = \left(1 - \frac{\eta_{OT}}{\eta_{OMCR}}\right) \times 100 \qquad (3)$$

Where $\partial \eta_O$ the deviation in overall efficiency (%) is, η_{OMCR} is the overall efficiency at MCR condition (%) and η_{OT} is the overall efficiency at tested plant load (%).

The deviation in overall efficiency for operating the plant at 70 % of MCR is 7.9 %

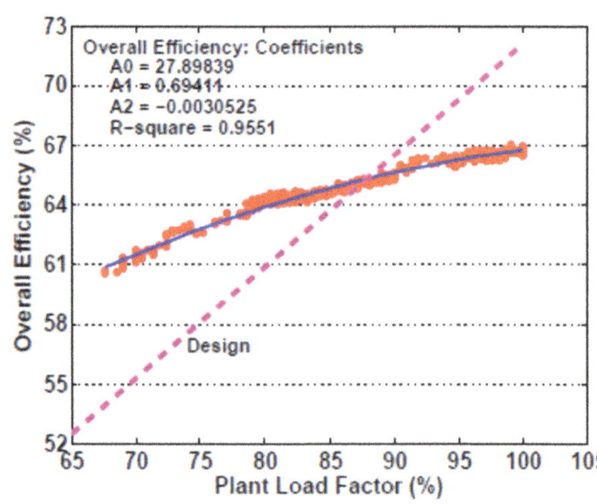

Figure 6: Variation of overall efficiency with PLF.

Power input

The variation of power input is plotted with the variation in plant load (Figure 7). It can be seen from the Figure 7 that as the plant load increases the auxiliary power of BFP increases. The correlation coefficient for the variation between measured power inputs with PLF is 0.9776. The correlation coefficient for power input is slightly better compared to pressure gain and FW flow and the noise level in data is less. At MCR condition the average measured power input is 5129.07 kW and is higher than the design value of 4788 kW at MCR condition. The power input at full load capacity of BFP (max. efficiency point) is 5894 kW. The actual measured power input is high compared to MCR condition because BFPs operating point is shifted thereby lower efficiency of pumps, higher FW flow and lower BFP discharge pressure. The average power input at 70 % PLF is 4352.66 kW and is higher than the design value of 3795.75 kW.

The deviation in power input (%) is computed by using average value through best curve fit of second order polynomial:

$$\partial P = \left(1 - \frac{P_T}{P_{MCR}}\right) \times 100 \tag{4}$$

Where ∂P is the deviation in power input (%), P_{MCR} is the power input (kW) at MCR condition (100 % PLF) and P_T is the power input (kW) at tested plant load

The deviation in power input for operating the plant at 70 % of MCR is 15.1 %.

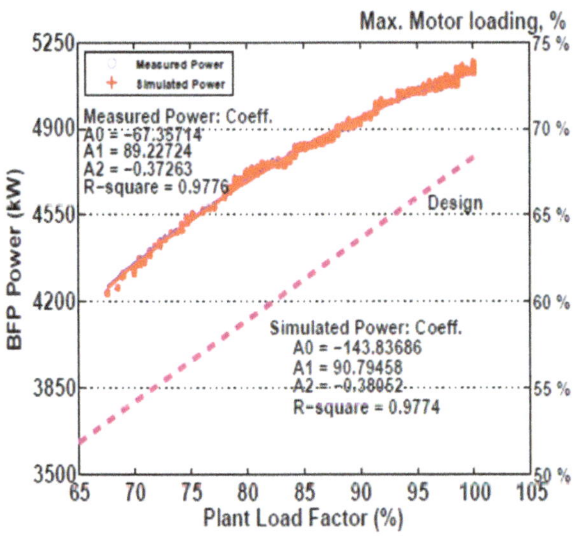

Figure 7: Variation of Power input with PLF.

Specific auxiliary power

In order to evaluate the auxiliary power used by BFPs, the specific auxiliary power is computed and is the ratio of Power input to plant load. The variation of specific auxiliary power for both pumps is plotted with variation in plant load (Figure 8). The specific auxiliary power (%) for BFP motors is computed as

$$AP = \frac{\sum_{i=1}^{i=n} P_i \times 100}{PL \times 10^3} \tag{5}$$

Where P is measured power input (kW), n is the number of equipment and PL is the plant load (MW) at generator output.

The correlation coefficient for the variation between specific auxiliary powers with PLF is 0.9966 and the noise level in data is less. At MCR condition the average SAP is 2.44 % and is higher than the design value of 2.28 % at MCR condition because of higher losses in pumps, higher FW flow, etc. The average SAP at 70 % PLF is 2.96 % and is higher than the design value of 2.63 %. The correlation coefficient for specific auxiliary power is better than the power input. The deviation in specific auxiliary power (% of gross gen.) is computed by using average value through best curve fit of second order polynomial.

$$\partial AP = \left(1 - \frac{AP_T}{AP_{MCR}}\right) \times 100 \tag{6}$$

Where ∂AP is the deviation in specific auxiliary power input (%), AP_{MCR} is the specific auxiliary power (% of gross gen.) at MCR condition and AP_T is the specific auxiliary power (% of gross gen.) at tested plant load

The specific auxiliary power at MCR condition is 2.44 % whereas at 70 % PLF the specific auxiliary power is 2.96 %. The deviation in specific auxiliary power for operating the plant at 70 % of MCR is 21.3 %.

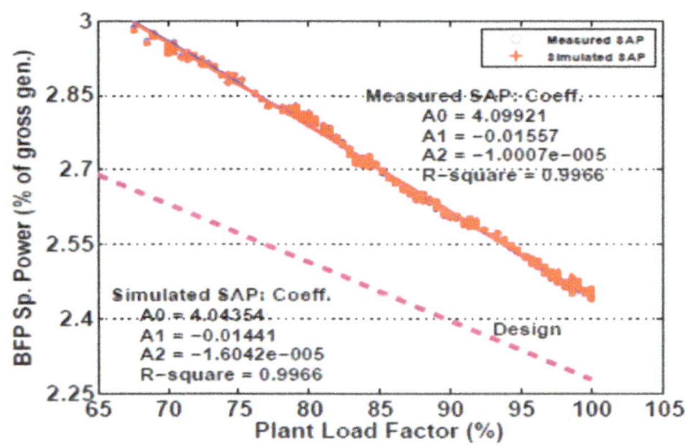

Figure 8: Variation of Specific Auxiliary Power with PLF.

SIMULATION STUDIES

The mechanical power output of pump is directly related with pressure gain and feed water flow. The electrical power input to BFP motor terminals is also directly related with mechanical power output along with motor and pump efficiency (i.e., overall efficiency). But all these parameters will not have the similar kind of variation trend for different plant load conditions. The simulation of variation of auxiliary power with plant load is carried out with respect to variation in pressure gain, pressure drop across hydrodynamic resistive elements (HPH, FRS, ECO, and Boiler), feed water flow and overall efficiency. The Artificial Neural Network (ANN) feed forward technique is adopted to simulate the variation of power input. In this technique, three layer model is adopted i.e., input layer, hidden layer and output layer.

ANNs are computational models which simulate the function of biological networks that composed of neurons (Anderson JA, 2003). The unique concept of ANN is the multi layered feed forward neural networks. Figure 9 is the ANN architecture. In this case three layer concept is adopted i.e., input layer, hidden layer and output layer. A node in one layer is connected to all nodes in the next layer i.e., feed forward architecture (Parisi DR et al., 2001). The input layer takes all the input parameters, the information is transmitted to hidden layer where they will be processed and output is computed in output layer (Singh KK et al., 2004). In this study, the input layers are chosen as plant load factor, pressure gain, feed water flow, overall efficiency, measured electrical power input and Pressure drop across HPH, FRS, ECO and Boiler (Mustafa Golcu et al., 2010). Back propagation training algorithm which is a gradient descent technique to minimize the sum of square errors is used (Engin T, 2007). The output layer is the simulated power input to BFP motor.

$$E = \sum_{i=1}^{N}\sum_{j=1}^{Q}\left(P_{PR} - P_{BFP}\right)^2 \quad (7)$$

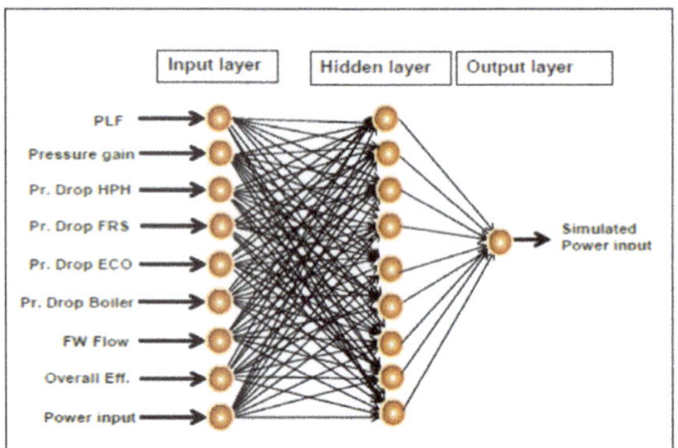

Figure 9: ANN Architecture for BFP

Figure 9: ANN Architecture for BFP

Where P_{PR} is the simulated power input (%), i is input data set value from 1 to N and j is the output data set value from 1 to Q. The simulation will try to minimize the error near to zero. In ANN, the data have to be trained by assigning appropriate weights and biases. The process of training a neural network involves tuning the values of the weights and biases of the network to optimize network performance. There are two different ways in which training can be implemented i.e., incremental mode (adapt command) or batch mode (train command). In incremental mode, the gradient is computed and the weights are updated after each input is applied to the network. In batch mode, all the inputs in the training set are applied to the network before the weights are updated. In this study we have adopted batch mode of training process. In this batch mode training process, after each cycle, the error between simulated value by the ANN and input value is propagated backward to adjust the weight in a manner mathematically to converge. Adjustments of weights are done using the following relation:

$$\Delta W_{ij} = -\alpha\frac{\partial E}{\partial W_{ij}} + \beta\Delta W_{ij}(A-1) \quad (8)$$

Where α is the learning rate, β is the momentum coefficient and A is the current step

Batch training can be done using train which is the best option for training because it typically has access to more efficient training algorithms. For small and medium size networks where enough memory is available, Levenberg-Marquardt training algorithms are suitable. Therefore, in this case, Levenberg-Marquardt algorithm is used. In this multilayer architecture tan-sigmoid transfer function ' tansig ' is used for hidden layer and linear transfer function ' purelin ' is used for output layer.

An input data set of 422 for all input variables taken for simulating the output parameter. All these input data set are normalized in the range of 0.1 to 0.9 by using the following technique:

$$D_n = \frac{\left(D_T - D_{T\text{-min}}\right)*0.8}{\left(D_{T\text{-max}} - D_{T\text{-min}}\right)} + 0.1 \quad (9)$$

Where D_n is the normalized value, D_T is the measured data, D_T-max is the maximum value of measured data and $D_{T\text{-min}}$ is the minimum value of measured data.

RESULTS AND DISCUSSIONS

The simulated power at 100 % PLF is 5070 kW and at 70 % PLF is 4246 kW % as compared to measured value of 5083 kW at 100 % PLF and 4241 at 70 % PLF. The simulated specific auxiliary power at 100 % PLF is 2.42 % and at 70 % PLF is 2.88 % as compared to measured value of 2.41 % at 100 % PLF and 2.89 % at 70 % PLF. The Pearson product moment correlation coefficient (R^2) value for measured power input is 0.9776 and is for simulated power input is 0.9774. The error between actual measured values (% of measured input power) is varying between -7.65635 to 0.83244 %. The error is slightly negative as the plant load factor increases. The correlation coefficient between the measured input power and predicted power is plotted in Figure 10. The correlation coefficient computed (as per Pearson product moment correlation method) (R^2) is 0.99995 and Root Means Square Error (RMSE) for the correlation between predicted power input to measured power input is 0.00019 (Verma AK et al., 2010). The error is negative for lower plant load and is slightly positive at higher PLF. At lower PLF operation of the plant will be less stable compared to plant operating above 80 % PLF. The noise level of data will be less at a PLF more than 80 %. At present Indian power plants average PLF of 210 MW power plants is about 82 % and is lower may be due to use of poor coal quality and ageing of the plants (Shanmugam KR et al., 2005 and CEA, 2012)

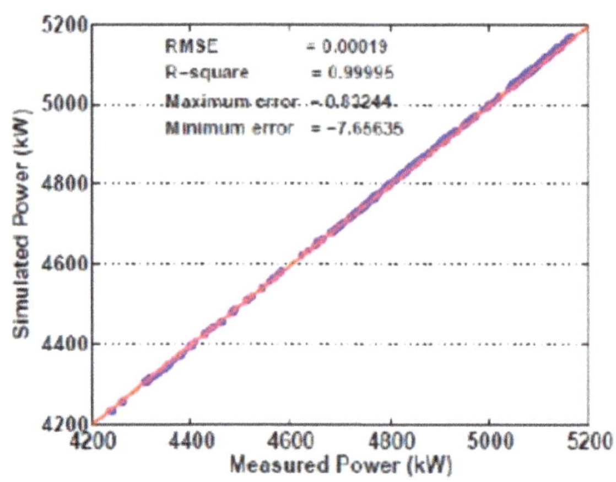

Figure 10: Correlation coefficient between measured and simulated power.

The MATLAB Simulink programming is done to evaluate the performance of BFPs. This Simulink program input the simulated and curve fit values from the ANN program. The performance tests are conducted for 210 MW power plants at different plant load conditions. The measured data are compared with the simulated data. The variations in power due to variation of different parameters are discussed.

Table 1 gives the performance results of BFPs at Unit 1 at Raichur Super Thermal Power Station (RTPS), Raichur, Karnataka, India (CPRI, 2011).

Table 1: Performance Results of BFPS at RSTPS.

Sl. No.	Particulars	Unit	Predicted value at 80.95 % PLF	BFP 1B	BFP 1C
01	Plant load	MW	-	170.0	
02	BFP Motor rating	MW	4.00	4.00	4.00
02	Booster pump suction pressure	MPa	-	0.56	0.56
03	BFP discharge Pressure	MPa	-	14.20	15.84
04	Pressure gain	MPa	**16.47**	13.64	15.28
05	FW flow	m^3/h	**325.4**	393.78	363.63
06	Electrical power	kW	**2330**	2775.92	2854.24
07	Load factor of motor	%	**54.17**	64.10	65.94
08	Motor efficiency	%	**93.00**	92.36	92.42
09	Electrical power input to pump	kW	**2166.90**	2563.84	2637.89
10	Mechanical power output	kW	**1460.92**	1464.10	1514.57
11	Main pump and Booster pump efficiency	%	**67.42**	57.09	57.40
12	Overall efficiency	%	**62.70**	52.73	53.05
13	Specific Energy Consumption	$kWh/m^3/h$ of FW flow	**7.16**	7.76	8.64
14	Specific Auxiliary Power	% of plant load	**2.74**	3.31	
15	Increased power due to increased FW flow	kW	-	481.95	309.28
16	Reduction in power due to reduced pressure gain	kW	-	-476.91	-190.22
17	Increased power due to poor overall (motor+pump) efficiency	kW	-	440.88	405.18
18	Net increase in power	kW	-	445.92	524.24

The performance test is conducted at an average plant load of 170 MW (PLF: 80.95 %). The observations from the study are as follows:

a) the total power input for both BFPs is 5630.16 kW (3.31 % plant load) and is higher than the

b) average predicted Power input i.e., 2.74 % at 80.95 % PLF. The power input is even above the economical operating band i.e., 2.71 – 2.74 %.

c) The main reason for increase of FW flow is mainly due to passing in re-circulation valve and higher specific steam consumption of turbines.

d) Since BFP is the multistage high pressure pump during start-up of the pump, the FW flow is bypassed to deaerator through re-circulation valve. During normal operation of the plant, the re-circulation valve will be closed. But due to passing in these valves, the FW flow is increased in BFP. The additional FW flow increases the power of BFP. The replacement of valve seat of re-circulation valve for both pumps reduced the power of BFP by 0.30 MU/month. The investment for replacing the valve seat of re-circulation valve is $ 60,000 and payback period is 4 months. The reduction in specific auxiliary power is 0.25 % of plant load.

e) The pump efficiencies are in the range of 57.09 to 57.40 % and are lower compared to predicted pump efficiency at 80.95 % PLF of 67.42 %. The pump efficiency is low due to more clearances inside the pump. The pump impeller set (cartridge set) is replaced in both pumps. The replacement of BFP cartridge had enhanced the pump efficiency by average of 7 % (Mukherjee S et al., 2008) and had reduced the power of BFP by 0.40 MU/month. The investment for replacing the pump cartridge is $ 75,000 and payback period is 5 months. The reduction in specific auxiliary power is 0.32 % of plant load

Table 2: Performance Results of BFPS at RGGSTPS.

Sl. No.	Particulars	Unit	Predicted value at 100 % PLF	BFP 3A	BFP 3B
01	Plant load	MW	-	210.0	
02	BFP Motor rating	MW	4.00	4.00	4.00
02	BFP Pump (Booster) suction pressure	MPa	-	0.56	0.56
03	BFP discharge Pressure	MPa	-	17.90	18.60
04	Pressure gain	MPa	17.16	17.34	18.04
05	Pressure drop across HP heaters	MPa	0.25	0.28	0.28
06	Pressure drop across FRS	MPa	0.10	0.35	0.35
07	Pressure drop across Economizer	MPa	0.15	0.16	0.16
08	Pressure drop across water wall and SH	MPa	1.25	1.31	1.31
09	Pump discharge FW flow	m^3/h	391.03	394.56	390.63
10	Re-circulation flow	m^3/h	-	1.29	38.18
11	Total FW flow in pump	m^3/h	391.03	395.85	428.81
12	Electrical power	kW	2578.45	3022.19	3216.86
13	Load factor of motor	%	59.95	70.05	65.94
14	Motor efficiency	%	93.00	92.72	92.91
15	Electrical power input to pump	kW	2397.96	2802.17	2988.78
16	Mechanical power output	kW	1828.50	1464.10	2107.99
17	Main pump and Booster pump efficiency	%	76.25	66.75	70.53
18	Overall efficiency	%	70.91	61.89	65.53
19	Specific Energy Consumption	$kWh/m^3/h$ of FW flow	6.59	7.63	7.50
20	Specific Auxiliary Power	% of plant load	2.46	2.97	
21	Increased power due to higher Pr. Drop across HP heaters	kW	-	4.56	5.89
22	Increased power due to higher Pr. Drop across FRS	kW	-	38.01	49.10
23	Increased power due to higher Pr. Drop across Economizer	kW	-	1.52	1.96
24	Increased power due to higher Pr. Drop across Water walls and SH	kW	-	9.13	11.78
25	Reduction in power due to reduced net pressure gain at boiler outlet	kW		-25.85	-33.38
26	Increased power due to passing in re-circulation valve	kW	-	8.59	315.47
27	Increased power due to increase in FW flow (higher SSC)	kW		23.51	-3.30
28	Increased power due to poor overall (motor+pump) efficiency	kW	-	384.27	290.89
29	Net increase in power	kW	-	443.74	638.41

Table 2 gives the performance results of BFPs at Unit 3 at Guru Gobind Singh Super Thermal Power Station (RGGSTPS), Rupnagar, Punjab, India (CPRI, 2011).

The performance test is conducted at an average plant load of 210 MW (PLF: 100 %). The observations from the study are as follows:

a) The total power input for both BFPs is 6239.05 kW (2.97 % plant load) and is higher than the average predicted Power input i.e., 2.46 % at 100 % PLF. The power input is even above the economical operating band i.e., 2.43 – 2.49 %.

b) The increased power due to higher pressure drop across High Pressure (HP) Heaters is 10.45 kW, across Feed Regulating Station (FRS) is 87.11 kW, across Economizer is 3.48 kW and across Water walls and super-heaters (SH) is 20.91 kW. The power change due to higher pressure available at Boiler outlet is 59.23 kW. The net power increased due to change in pressure drop

c) across feed water circuit elements is 62.72 kW that forms 0.03 % of plant load.

d) Reducing the pressure drop across HP heaters by acid cleaning of HP heater tubes reduced the pressure drop from average value of 0.28 MPa to 0.24 MPa. This had reduced the auxiliary power of BFP by 6.5 MWh/month. The investment is $ 3,500 and the payback period is 11 months.

e) Reducing the pressure drop across FRS from an average value of 0.35 MPa to 0.10 MPa by operational optimization i.e., reducing the pressure drop across feed control valve, will reduce the auxiliary power of BFP by 50.2 MWh/month.

f) The replacement of valve seat of re-circulation valve for BFP 3B reduced the power of BFP 3B by 0.18 MU/month. The investment for replacing the valve seat of re-circulation valve is $ 30,000 and payback period is 4 months. The reduction in specific auxiliary power is 0.12 % of plant load.

g) The pump efficiencies are in the range of 66.75 to 70.53 % and are lower compared to predicted pump efficiency at 100 % PLF of 76.25 %. The pump efficiency is low at BPF 3A due to more clearances inside the pump. The pump impeller set (cartridge set) is replaced for BFP 3A. The replacement of BFP cartridge had enhanced the pump efficiency by average of 7 % and had reduced the power of BFP by 0.14 MU/month. The investment for replacing the pump cartridge is $ 37,500 and payback period is 6 months. The reduction in specific auxiliary power is 0.12 % of plant load.

CONCLUSIONS

The specific auxiliary power of BFP at 100 % PLF is about 2.46 % of gross energy generation and is increased for operating the plant at reduced PLF of 70 % which is 2.87 % of gross energy generation. Reducing the passing in re-circulation valve will reduce the auxiliary power of BFP in the range of 10 – 15 % of BFP power and 0.2 – 0.4 % of gross energy generation (CO_2 reduction of about 3,400 – 7,200 t/year for one 210 MW plant). Improvement of pump efficiency by changing the impeller will enhance the BFP efficiency by about 7 – 10 % that will reduce the auxiliary power of BFP by 0.40 MU/month for one 210 MW plant. Operational optimization of BFPs and implementation of energy conservation measures for BFPs will reduce the auxiliary power of BFPs in Indian Thermal Power Generating Sector by about 480 MW and equivalent reduction of CO_2 emission is 3.9 million tonne per year

ACKNOWLEDGEMENTS

The authors are thankful to the management of CPRI and NITK for providing opportunity to carry out the study and allowing us to publish the work in the conference.

REFERENCES

1. Anderson JA (2003), An Introduction to neural networks, Prentice Hall.
2. Becnel CL, Kilgrore JW, Merrill EF (1987), Determining motor efficiency by field testing, IEEE Transactions on Industrial Applications, Vol. IA-23, pp. 440 –443.
3. Central Electricity Authority (2012), Performance Review of Thermal Power Stations 1999-00, 2006-07, 2007-2008, 2008-09, 2009-10, 2010-11 and 2011-12',website: http//:www.cea.nic.in, Accessed on June 2013.
4. Central Electricity Authority (2013), Growth of Electricity Sector in India from 1947 to 2013, website: http//:www.cea.nic.in, Accessed on June 2013.
5. Central Power Research Institute (2011), Instrumented and Diagnostic Energy audit at Raichur Super Thermal Power Station, Raichur, Karnataka, India, Report No: CPRI/ERED/EA/83/11.
6. Central Power Research Institute (2011), Instrumented and Diagnostic Energy audit at Guru Gobind Super Thermal Power Station, Rupnagar, Punjab, India, Report No: CPRI/ERED/EA/95/11.
7. El-Wakil M.M. (1984), Power plant Technology, International Edition 1984, McGraw Hill Book Company, pp. 219-250.
8. Engin T (2007), Prediction of relative efficiency reduction of centrifugal slurry pumps: empirical- and artificial-neural network-based methods, Proceeding of IMechE Vol. 221 Part A: J. Power and Energy, JPE224 # IMechE 2007.
9. John S Hsu, Kueck JD, Olszewski M, Casada DA, Otaduy PJ, Tolbert LM (1998), Comparison of Induction Motor Field Efficiency Evaluation Methods, IEEE transactions on industry applications, Vol. 34, No. 1, pp. 117-125.
10. Kiran Kumar, Srinivasa Rao G (2013), Performance Analysis from the Energy Audit of A thermal Power Plant, International Journal of Engineering trends and Technology (IJETT), ISSN:2231-5381, Vol. 4, Issue 6, pp.2485-2490.
11. Mukherjee S, Sengupta AK, Srivastava SK (2008), Boiler feed Pumps: Efficiency Improvement initiatives at Vindhyachal, Proceedings of National Conference 2008 on Indian Power Stations Competitive Edge beyond best practices in Power Generation, at PMC, NTPC, Noida, New Delhi, during Feb. 13 - 15, 2008.
12. Mustafa Golcu, Yasar Pancar, Sevil Ergur H, Goral O (2010), Prediction of Head, efficiency and Power Characteristics in a semi-open impeller, International Journal of Mathematical and Computational Applications, Vol. 15, Issue No.1, pp. 137 – 147.
13. Namdeo D Adate, Awale RN (2013), Energy Conservation through Energy Efficient Technologies at Thermal Power Plant, International Journal of Power System Operation and Energy Management, ISSN (PRINT): 2231 – 4407, Vol. 3, Issue 1, pp. 95 – 100.
14. Parisi DR, Laborde MA (2001), Modeling steady-state heterogeneous gas –solid reactors using feed forwardneural networks', International Journal ofComputers and Chemical Engineering, 25,1241.
15. Rajashekar P Mandi, Seetharamu S, Udaykumar R Yaragatti (2010), Enhancing energy efficiency of auxiliary power system in a 210 MW coal fired power plant through energy efficiency, Published in IEEE Explorer, Proc. of Fifth International Conference on Industrial and Information Systems-2010, NITK Surathkal, during July 29 – August 1, 2010 pp. 549 - 554.
16. Saha A, Chatterjee S (2005). Achievements in Thermal and Electrical Energy Conservation at Budge Budge Generating Station, CSES Limited, Proceedings of National workshop on Energy Conservation for Power Engineers, at PMC, NTPC, Noida, New Delhi, during Oct. 20-21, 2005, pp. 32-42.
17. Siddhartha Bhatt M, Mandi RP (1999), Performance enhancement in coal fired thermal power plants part III: Auxiliary power, International Journal of Energy Research, Vol. No.23, pp.779-804.
18. Singh KK, Shenoy KT, Mahendra AK, Ghosh SK (2004), Artificial neural network based modeling of head and power characteristics of pump-mixer, International Journal of Chemical Engineering Science, 59:2937-2945
19. Shanmugam KR, Praveen Kulshreshtha (2005), Efficiency analysis of coal based thermal power generation in India during post-reform era, International Journal Global Energy Issues, Vol. 23, Issue No. 1, 2005, pp. 15-28.
20. Verma AK, Singh TN, Monjez M (2010). Intelligent prediction of heating value of coal, Iranian Journal of Earth Sciences 2:32-38

CHAPTER 16

MATHEMATICAL SAFETY ASSESSMENT APPROACHES FOR THERMAL POWER PLANTS

Zong-Xiao Yang, Lei Song, Chun-Yang Zhang, Chong Li, and Xiao-Bo Yuan

Institute of Systems Science and Engineering, Henan University of Science and Technology, Luoyang 471003, China

ABSTRACT

How to use system analysis methods to identify the hazards in the industrialized process, working environment, and production management for complex industrial processes, such as thermal power plants, is one of the challenges in the systems engineering. A mathematical system safety assessment model is proposed for thermal power plants in this paper by integrating fuzzy analytical hierarchy process, set pair analysis, and system functionality analysis. In the basis of those, the key factors influencing the thermal power plant safety are analyzed. The influence factors are determined based on fuzzy analytical hierarchy process. The connection degree among the factors is obtained by set pair analysis. The system safety preponderant function is constructed through system functionality analysis for inherence properties and nonlinear influence. The decision analysis system is developed by using active server page technology, web resource integration, and cross-platform capabilities for applications to the industrialized process. The availability of proposed safety assessment approach is verified by using an actual thermal power plant, which has improved the enforceability and predictability in enterprise safety assessment.

INTRODUCTION

Prevention and mitigation of accidents for over a century have had great impact on process industries such as power industries and chemical industries [1]. Many scientists have been working on methodologies and technologies to resolve safety problems in process plants that can diagnose primary faults in process systems [2–8]. The methodology research, such as fault tree analysis (FTA) and cause consequence analysis (CCA), of safety assessment for thermal power plants is very important, considering management safety, environment protection, and safety in equipment operation, which can change safety factors from uncontrolled status to controlled status [9]. Experts are still searching for ideal mathematical models for safety assessment that include fuzzy mathematics, cluster analysis, and osculating value method of multiobjective decision. Some pertinent mathematical models of system safety assessments are necessary for analyzing accident causes on process industries.

Citation: Zong-Xiao Yang, Lei Song, Chun-Yang Zhang, Chong Li, and Xiao-Bo Yuan, "Mathematical Safety Assessment Approaches for Thermal Power Plants," Mathematical Problems in Engineering, vol. 2014, Article ID 864682, 13 pages, 2014. DOI:10.1155/2014/864682

Preponderant system functions are used to analyze structure effect and cooperation mechanisms [10–12]. These functions have been used for safety assessment of thermal power plants, but there are concerns when constructing the Delphi weight judgment matrix of combined index awkwardly, because it is strenuous processing when the judgment matrix is exceptionally large. There will be problems of a lack of data sources and a defect of high level subjectivity, so it is difficult to compute the influences among the factors in a system accurately. The analytic hierarchy process (AHP) is widely used in systems of thermal power plants and other fields [2]. It has been combined with entropy weighting, artificial neural networks, and other methods and often combined with fuzzy mathematics to create the fuzzy analytical hierarchy process (FAHP) [13–15]. FAHP is replacing AHP in several academic fields including machinery evaluation and risk evaluation, because it better represents the fuzzy features of systems. Otherwise, the set pair analysis (SPA) is used to analyze systematic relations, systematic forecasts, systematic evolutions, and other uncertain problems [16–18].

An integrated mathematical modeling approach for thermal power plant safety assessment is proposed in this paper. The safety status can be obtained by constructing preponderant functions of the system safety, and the nonlinear structural layers will be constructed by FAHP and SPA, respectively. An AHP-based safety assessment decision analysis system (SADAS) will be developed by using ASP.NET technology for identifying the systemic hazard evaluation and real time performance evaluation in the safety engineering.

System structure and cooperation functions utilize index weight to judge the matrix by Delphi method at first and then use AHP to quantitative analysis. But when there are many indexes, relationship judgment matrix of combination factors becomes so very large that it is not easy to be obtained in reality. So, it is difficult to implement.

Every kind of factor, which influencing the safety of thermal power plant, is analyzed by using the perspective of system analysis approaches. According to present situation of safety assessment for power plants, a mathematical model and approach for them are proposed, in which the key factors influencing the power plants safety are considered; on the basis of establishing multihierarchy valuation structure for power plants, A synthetic safety assessment model for power plant was constructed by combining fuzzy analytical hierarchy process (FAHP), set pair analysis (SPA), and system functionality, in which the key factors influencing the power plants safety were considered, and interaction among the factors influencing the power plants safety was analyzed quantitatively; a preponderant function of system safety was constructed by use of the result to solve the problem that the system safety function preponderates over the weight sum of safety functions of subsystems. Based on this, the UML and C#.NET is used to design and develop assessment components. A simulation system for safety assessment platform is developed. The instances show that the systemic question and real time assessment problem in the safety analyses assessment are resolved; maneuverability and wide adoption are improved. The model can integrate immanence relation and nonlinearity influence of factors in power plant, which make safety assessment more reasonably, and it can also offer scientific foundation for decision-making in safety management.

SAFETY ASSESSMENT APPROACHES

System Functionality

System functionality reflects the emergence of objective systems, and is used to solve linear and nonlinear contribution problems of each subsystem for a complex objective system. The system function can be defined in active scope of the system, optimized as the structure evolves. The holistic behavior at the system is not equal to the algebraic sum of subsystems or parts usually. Various energy and attribute integration functions can emerge from the combined resources (e.g., human beings, materials, equipment, environment, management, software, data, etc.) in a complex industrial system. Assume X describes the system safety of thermal power plants, describes the important factors in the system, that is, $x = (x_1, x_2. . . , xn) \in X$and W describes the system holistic behavior. The preponderant function of system safety can be defined as a mathematical formula as follows:

$$\begin{aligned} w &= \varphi(x_1, x_2, \ldots, x_n) \\ &= \sum_{i=1}^{n} \alpha_i x_i + \sum_{i>j=1}^{n} \alpha_{ij} x_i x_j + \sum_{i>j>k=1}^{n} \alpha_{ijk} x_i x_j x_k \\ &\quad + \cdots + \alpha_{12\cdots n} \prod_{i=1}^{n} x_i, \end{aligned} \tag{1}$$

where the sum of $\alpha_i x_i$ describes the linear contribution to system safety and the high order terms represent the nonlinear contributions to the system safety. The preponderant function of system safety is the emergence of the system mathematical model. This is the system of linear and nonlinear problems and also is the system structure function of cooperation and cooperative function problem.

In the formula (1), the first summation term in the right of equation is described the linear contribution of important factors, and the rest of summation terms means the nonlinear cooperative relationship of the factors. The effect of cooperative relationship is not only associated with their values (x_1, x_2) but also associated with the match appropriateness. That is to say, management injects energy into the system, so-called management energy or other functions, of course. The functional coefficient α describes a corresponding cooperation quantity in the objective system. The factors interact with each other in the system and the effect produced is related to respective activities and mutual cooperation, which is embodied by α. Evaluating security state of the system can be obtained, comparing the preponderant function value of system and the system function optimal value. The W value measures system security value. Observers determine the minimum security value of the system by experience.

Fuzzy Analytical Hierarchy Process

FAHP provides a basis for choosing key factors through calculating their weights for a given objective and ameliorates some disadvantages (such as difficulty in estimation consistency of judgment matrix, distinction that contrasts the judgment matrix with the human thinking) of AHP. FAHP has been used in many domains [2]. Some basic analysis steps of FAHP are summarized as follows [13].

Step 1 (construction of hierarchy). The first step of FAHP is to determine all the important criteria and their relationship in the form of a hierarchy, which is very crucial because the selected criteria can influence the safety assessment. The hierarchy structure can be determined from the target layer through the intermediate layer to the bottom layer.

Step 2 (definition of the new fuzzy comparison matrix). The criterion on the same layer of the hierarchy is compared with each of the criteria from the upper layers. A pairwise comparison is performed by using a triangular fuzzy number that employs 0.1–0.9 scales. The scale ranges from 0.1 for "absolutely less than" to 0.5 for "equal" to 0.9 for "absolutely more than" covering the entire spectrum of the comparison.

A triangular fuzzy number regarding a given value can be denoted by (L,M,U), which represent the minimum value, most-likely value, and maximum value of the fuzzy number, respectively. The reverse judgment is treated as complementary and reversed order of the fuzzy number of the corresponding judgment. For example, suppose that criterion B_1 compared with criterion B_2 is "more important" and denoted by fuzzy number (0.6, 0.75, 0.9) and the reverse judgment B_2 compared with criterion B_1 is "less important" and described by $(1-0.9, 1-0.75, 1-0.6)$, namely, (0.1, 0.25, 0.4). After the upper triangular matrix is completed, the lower triangular matrix is filled with the element of its reverse judgment. The upper level factor A dominates the lower level factors $B_1, B_2 \cdots B_n$. The b_{ij} denotes the relative importance of criterion B_i compared with criterion B_j in the same hierarchy. Suppose a matrix A of order n; when $b_{ij} = (l_{ij}, m_{ij}, u_{ij})$ and $b_{ji} = (l_{ji}, m_{ji}, u_{ji})$ are elements of matrix A of order N, they possess the following properties:

$$A = \begin{bmatrix} b_{11} & b_{12} & \cdots & b_{1n} \\ b_{21} & b_{22} & \cdots & b_{2n} \\ \cdots & \cdots & \cdots & \\ b_{n1} & b_{n2} & \cdots & b_{nn} \end{bmatrix},$$

$$l_{ij} + U_{ji} = 1, \quad m_{ij} + m_{ji} = 1, \quad u_{ij} + u_{ji} = 1,$$

$$i \neq j, \quad 1 \leq i, \quad j \leq N,$$

$$l_{ii} = 0.5, \quad m_{ii} = 0.5, \quad u_{ii} = 0.5, \quad 1 \leq i \leq N. \tag{2}$$

The judgment matrix A is triangular fuzzy number complementary judgment matrix.

Step 3 (calculation of local weight of order vector). Triangular fuzzy weight vector q_i is defined in (3), using fuzzy numbers $b_{ij} = (l_{ij}, m_{ij}, u_{ij})$ and $b_{ji} = (l_{ji}, m_{ji}, u_{ji})$ of matrix A. Vector q_i represents the fuzzy relative weight of each factor. The formula to compute relative weight vector q_i of every couple criteria denoted by triangular fuzzy number is given by

$$q_i = \left(\frac{\sum_{j=1}^{n} l_{ij}}{\sum_{i=1}^{n}\sum_{j=1}^{n} u_{ij}}, \frac{\sum_{j=1}^{n} m_{ij}}{\sum_{i=1}^{n}\sum_{j=1}^{n} u_{ij}}, \frac{\sum_{j=1}^{n} u_{ij}}{\sum_{i=1}^{n}\sum_{j=1}^{n} u_{ij}} \right)$$

$$i, j \in N. \tag{3}$$

There are two relative weight vectors of random element, represented as $q_i = (x_i, y_i, z_i)$ and $q_j = (x_j, y_j, z_j)$, respectively, and the element of possibility degree matrix P_{ij} that represents possibility degree of $q_i \geq q_j$ can be calculated by formula (4). The possibility degree matrix P is still a triangular fuzzy number complementary matrix:

$$P_{ij} = \lambda \max\left\{1 - \max\left\{\frac{y_j - x_i}{y_i - x_i + y_j - x_j}, 0\right\}, 0\right\}$$
$$+ (1-\lambda) \max\left\{1 - \max\left\{\frac{z_j - y_i}{z_i - y_i + z_j - y_j}, 0\right\}, 0\right\},$$
$$i, j \in N, \tag{4}$$

where λ specifies the risk attitude of decision-makers. When $\lambda > 0.5$, the decision-maker is optimistic (tolerant of risk); when $\lambda = 0.5$, the decision-maker is neutral to risk; when $\lambda < 0.5$, the decision-maker is pessimistic (averse to risk). Lastly, the local weight of order vector can be computed by getting a priority vector of possibility degreematrix $P(p_{ij})$.The element of the order vector is given by

$$w_i = \frac{\sum_{j=1}^{n} (p_{ij}) + (n/2) - 1}{n(n-1)}, \quad i \in N, \tag{5}$$

where w_i is the ith criterion weight, where $w_i > 0$ and $\sum w_i = 1, 1 \leq i \leq N$.

Step 4 (calculation of general weight of element). The general weight of the criterion regarding the ith subcriterion is denoted by R. This safety evaluation hierarchy is a tree structure, so R can be computed by the following recursion formula:

$$R_i = \begin{cases} 1, & i \in \text{RootNode} \\ w_i \times R_{F(i)}, & i \notin \text{RootNode}, \end{cases} \tag{6}$$

where R_i is the general weight of the ith criterion, $R_{F(i)}$ is the general weight of the father criterion, $F(i)$ is the father criterion of the ith criterion, w_i is the local weight of ith criterion, and i is the node that represents risk factor.

Set Pair Analysis

Set pair analysis (SPA) is used to analyze systematic relations, systematic forecasts, systematic evolutions, and other uncertain problems [17]. When applied to connection degree and combined with system functionality, it can be a simple way to construct the preponderant function of system safety avoiding subjective quantify.

Connection Degree

A set pair is composed of two sets A and B, namely, $H = (A, B)$. The connection degree is the key to SPA, which describes the differences between A and B. It can be applied to the study of the relationship between certainty and uncertainty in a macroscopic system. Connection degre μ between A and B is denoted as follows:

$$\mu = a + b_i + c_j, \tag{7}$$

where $i, j \in [-1, 1]$ is a stabilization coefficient for a set pair and a, b, and c are the degree of identity, the degree of discrepancy, the degree of opposition, respectively, for the set pair. Moreover, $a, b, c \in [0, 1]$, and $a + b + c = 1$.

The Features of SPA

Through the connection degree μ, SPA can be used to analyze and process various uncertain problems such as those caused by fuzzy, stochastic, and incomplete information. The main feature of SPA is that SPA accepts all kinds of uncertainties objectively in the real world and analyzes and processes various uncertain problems as well as the relationship between definitiveness and changefulness through three aspects: identity, discrepancy, and opposition. A method of obtaining the connection degree of safety factors in thermal power plants from routine inspection data is proposed in this paper, which represents the application of SPA in the field of safety assessment.

Integrated Analysis Approaches

A mathematical integrated safety assessment approach is presented to design and develop a safety assessment decision support system to cope with the increasingly important industrial system safety, combining with the theories of FAHP, SPA, system functionality, and so forth.

System Hierarchy Description for Safety Assessment

The enterprise safety assessment for thermal power plants is the decision-making goal, the hierarchical structure of which is composed of three levels, the target layer A, the criteria layer B, and the subcriteria layer C. The factors affecting the system safety are considered as criteria and the architecture is as shown in Figure 1. As shown in Figure 1, invited experts use the 0.1–0.9 scale for comparisons of every index to get the original triangular fuzzy number judgment matrix, each factor weight vector q denoted by triangular fuzzy number, and possibility degree matrix P.

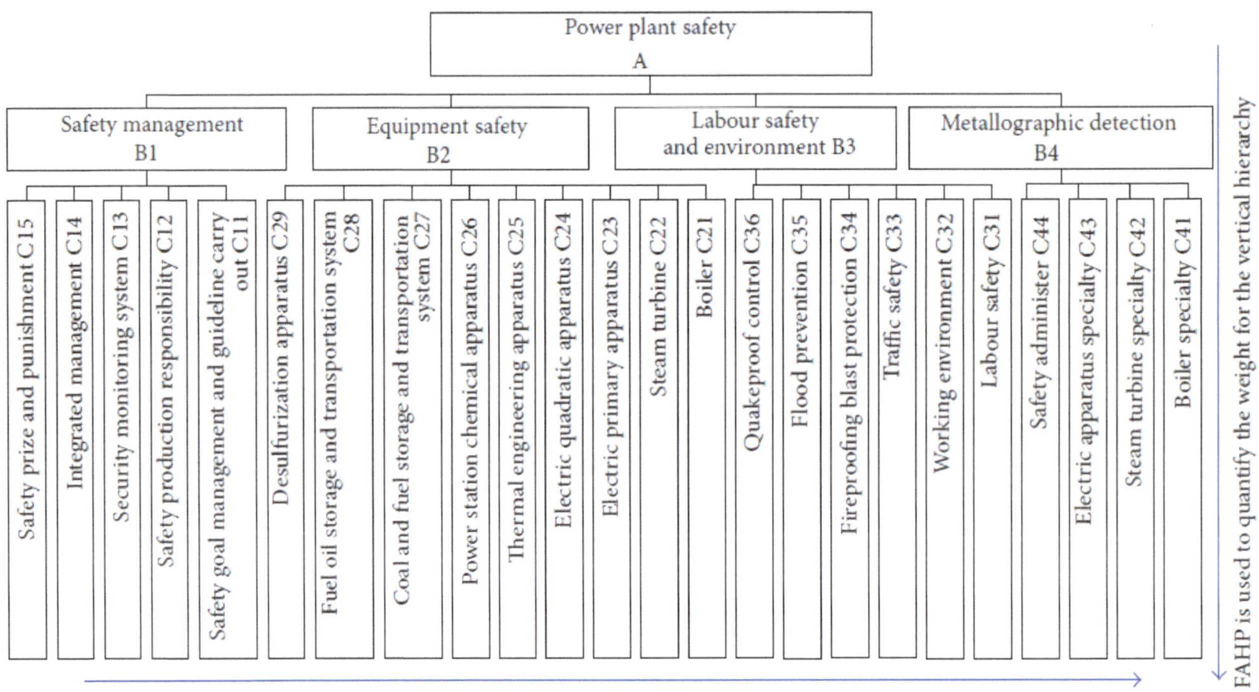

Figure 1: The architecture of safety assessment for thermal power plants.

The hierarchy of the system includes two contexts. The first one is a physical or vertical hierarchy where the subordinate relationship exists between lower layer and upper one, and subcriteria possess a weight to their parent criteria. The second one is a multiple order function hierarchy derived from horizontal interaction among factors where there is an expanded functional structure for the linear energy and nonlinear energy among the criteria in all layers. For example, layer B is composed of four parts, namely, the safety management B1, the equipment safety B2, the labor safety and working environment B3, and metallographic detection B4. The linear hierarchy function of level B originates from the four independent parts, and the nonlinear expanded structure functional derives from interaction among two, three, or all of them. The vertical hierarchy pertains to the research object of FAHP, while the horizontal interaction pertains to system functionality. The analysis process of integrated safety assessment structure is shown in Figure 2.

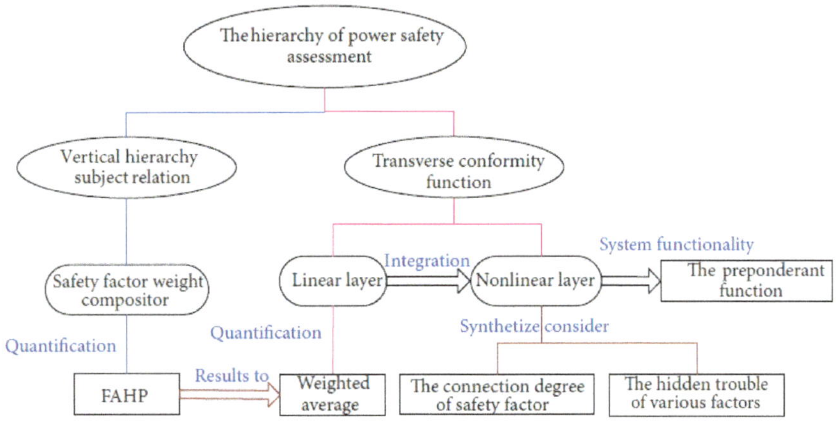

Figure 2: The process of the integrated safety assessment.

Calculation of Connection Degrees

Suppose a thermal power plant has been safely inspected m times to n criteria; the matrix $A_{ij} = (d_{ij})_{m\times n}$ can be constructed, whose number of rows and columns is m and n, and whose element d_{ij} denotes the safety state of jth criteria in ith safety inspection. The calculation process to obtain the connection degree for safety assessment criteria is shown as follows.

Step 1 (calculation of correlation coefficient between the factors). d_i is a column vector of the matrix A. The correlation coefficient r between two factors is given by

$$r_{ij} = \left|\frac{\operatorname{cov}(d_i, d_j)}{\sqrt{V_x}\sqrt{V_y}}\right|, \quad i, j \in N, \tag{8}$$

where x or y indicates the discrete random variable, respectively, while V_x or V_y is the variance of x or y, respectively.

Step 2 (calculate connection degree between the factors). For the two same evaluation factors, the higher the identity degree of the factors is, the more similarly the trend changes, the higher the corresponding department coordination will appear. The identity degree can be described as follows:

$$h_{ijk} = \left(d_{(k+1)i} * \zeta_{ki(k+1)} - d_{ki}\right) * \left(d_{(k+1)j} * \zeta_{kj(k+1)} - d_{kj}\right), \quad i, j, k \in N$$

$$a_{ij} = \frac{\text{Countif}\left(h_{ijk} > 0\right)}{m-1} \times r_{ij}, \quad i, j \in N, \tag{9}$$

where aij is degree of identity between ith and jth factors."Countif " indicates the number satisfying certain conditions. $\zeta_{ki(k+1)}$ indicates the score percentage of the parent factor of ith difference between kth and $(k+1)$th inspection. The degree of discrepancy can be computed by

$$b_{ij} = 1 - r_{ij}, \quad i, j \in N. \tag{10}$$

For the two same evaluation factors, the higher the degree of opposition of the factors is, the more the trend reverses, the worse the contradiction of corresponding department coordination will appear. c_{ij} denotes degree of opposition between ith and jth factors, which should satisfy the following:

$$c_{ij} = \frac{\text{Countif}\left(h_{ijk} < 0\right)}{m-1} \times r_{ij}, \quad i, j \in N. \tag{11}$$

From formula (7), the connection degree between ith and jth factors can be given by

$$\mu_{ij} = a_{ij} + b_{ij} i + c_{ij} j. \tag{12}$$

The matrix whose elements are μ_{ij} satisfies the following properties. Firstly, μ_{ij} have symmetry relationship, id est, and a randomμij, satisfying $\mu_{ij} = \mu_{ji}$. It shows that the connection degree of ith factor and jth factor is equal to the connection degree of jth factor and ith factor. Secondly, the diagonal value in the matrix is constant; namely, $\mu_{ij} = 1$. It shows that the connection degree of two same factors is 1 and degree of discrepancy and opposition is 0.

Calculation of Functional Coefficient α of Factors

The cooperation harmony of departments is revealed by $(a_{ij} - c_{ij})$ with reliability $(a_{ij} + c_{ij})$.Harmony of each department should satisfy $\Sigma(a_{ij} - c_{ij})(a_{ij} + c_{ij})$. The variance of ith factor's record score is marked as var(d_i). The functional coefficient α of ith factor is given by

$$\alpha_i = \frac{\sum_{j=1}^{n}\left(a_{ij} - c_{ij}\right)\left(a_{ij} + c_{ij}\right) - \operatorname{var}(d_i)}{\sum_{i=1}^{n}\left(\sum_{j=1}^{n}\left(a_{ij} - c_{ij}\right)\left(a_{ij} + c_{ij}\right) - \operatorname{var}(d_i)\right)}, \quad \lambda \in 2n, \quad n, i, j \in N. \tag{13}$$

Creation of the Preponderant Function of System Safety

Safety assessment consists of linear and nonlinear components, and the functional coefficient relates not only to factor's independent safety state but also to connection degrees. The linear part can be quantified by means of weighted averages from FAHP. The fluctuating range of a factor's score reveals its safety state which is denoted by their variance. Formula (14) is the mathematical model for the safety assessment of thermal power plants based on FAHP, SPA, and system functionality:

$$F(x) = f_1 + f_2 + \cdots + f_n$$

$$= R_i \sum_{i=1}^{n} x_i + \alpha_i \alpha_j \sum_{i>j=1}^{n} x_i x_j + \alpha_i \alpha_j \alpha_k \sum_{i>j>k=1}^{n} x_i x_j x_k$$

$$+ \cdots + \prod_{i=1}^{n} \alpha_i \prod_{i=1}^{n} x_i. \tag{14}$$

Decision Analysis System

Once introduced into the area of safety assessment for thermal power plants, decision support is the kind of system that combines information technology and decision theories in management science. The internet technology-based decision support system is a very good technical solution to safety assessment, which can help appraisers of thermal power plants make objective, scientific, and speedy decisions. However, decision support systems have made few inroads in practice because the safety assessment process for thermal power plants is a very complicated project that involves knowledge of many fields, so there is a problem as to how to combine expert judgments and mathematical models, transfer expert experiences into a language which computer can identify, manage these expert knowledge scientifically and orderly, and make inferences based on offered knowledge.

We will construct a special safety assessment website using a browser/server model to synchronously integrate different expert safety assessments for thermal power plants. This method of constructing SADAS is based on FAHP, SPA, system functionality, and information technology for thermal power plants and is used to improve the level of safety management. Knowledge and date capturing from safety assessments are completed by knowledge transformation, that is to say, collecting former assessment data, expert experience, and evaluations standard of safety assessment of thermal power plants; identifying the relationships among knowledge points in the complex information; determining the relationships among standards; and transforming the formal knowledge according to production rule and hierarchical network. The concept of knowledge restructuring is very important to restructure the knowledge of knowledge base and form new information when new relationship is clarified during the process of inference.

SYSTEM DESIGN AND IMPLEMENTATION

The web-based SADAS is used to provide a platform for evaluation experts to build a multiple criteria evaluation index system and for administrators to get decision-making advice so that the policy maker can make an easy decision. In this study, we construct a system architecture made up of five functional parts:(1)user management module (UMM), (2) system evaluative model (SEM),(3) matrix parameter identification (MPI), (4) report module (RM), and (5) help module (HM), as shown in Figure 3.

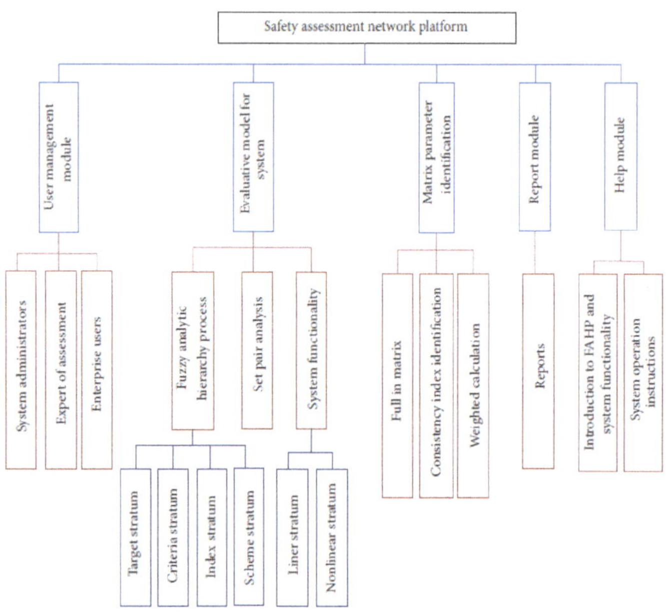

Figure 3: The system architecture of SADAS.

The end users are classified into database administrators, whose responsibility is keeping the system working normally, avoiding database breakdown, and servicing the database system; the evaluation experts who establish and modify the evaluation system architecture and evaluation index system; and the enterprise users who are to make daily rectification decisions on the enterprise safety in UMM functional part. All users can access SADAS functions within their authorization.

The AHP and system functionality module (SFM) will be embodied in SEM functional part. AHP is a multiple criteria decision-making tool that has found applications in almost all the fields related to decision making [2]. SFM can be accessed from the system, optimized through structure evolvement. The unitary function behavior of the system is not equal to the algebraic sum of all parts [21]. The module is used to avoid too many judgment matrixes. In addition, SFM can be considered as an interface (connection tool) to obtain effective data through mining the thermal power plant safety assessment information system which is introduced in another paper [22]. MPI can automatically process and calculate the judgment matrix by using AHP and SFM; whether the calculation results satisfy the condition or not is called consistency in AHP. The matrix and result will be stored in the database for enterprise user query. The RM module provides graphic or character-based results to help managers make decisions according to the decisions of experts; HM module supplies the users who first access the system with the results of AHP and SFM.

The data flow diagram, shown in Figure 4, describes the data through the flow of information through the system and the system implementation. When the user inputs an accepted name and password, he will log into the system and do some operations according to the user identity; the data flows into the data dictionary along with the user's operation. The data flows in the process of transforming the input to output based on the figure mode of data transmission and processing. A complex system is decomposed step by step according to the hierarchical structure of the system, reflecting the main function of the new system, that is, input-output of the external environment, internal processing of the system, data communication, data storage, and so on. An arrow is used to denote the data flow direction. An entity relationship diagram (ERD) is a visual diagram based on the cognition that objective realities in the world are composed of entity of fundamental objects and the relations of such objects. ERD is used to construct a database in the global design process, which includes authenticity, simplicity, avoidance redundancy, and aptitude. The design technology of ERD is an important foundation for database design and database application development, and ERD of the web-based decision analysis system is designed according to the principles mentioned here, as shown in Figure 5. When ERD is complete, the data dictionary can be defined expediently. In the paper, there are six data dictionaries to store data from web pages or other information, the effect and structure of which will be described here.

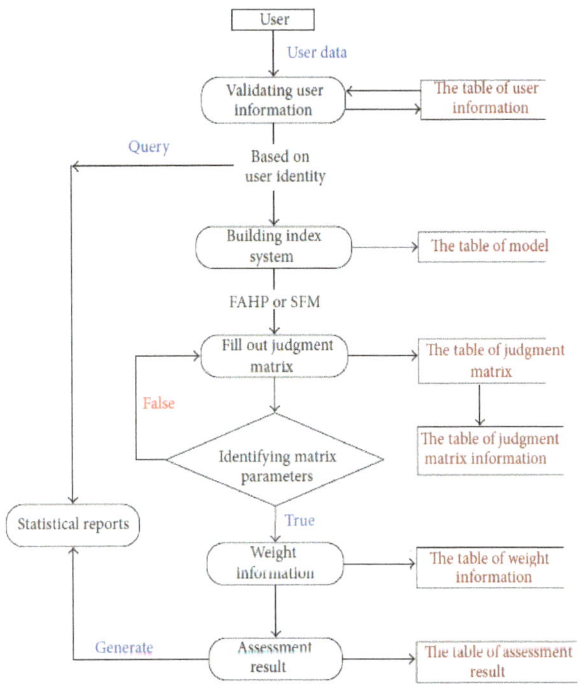

Figure 4: The flow chart about FAHP.

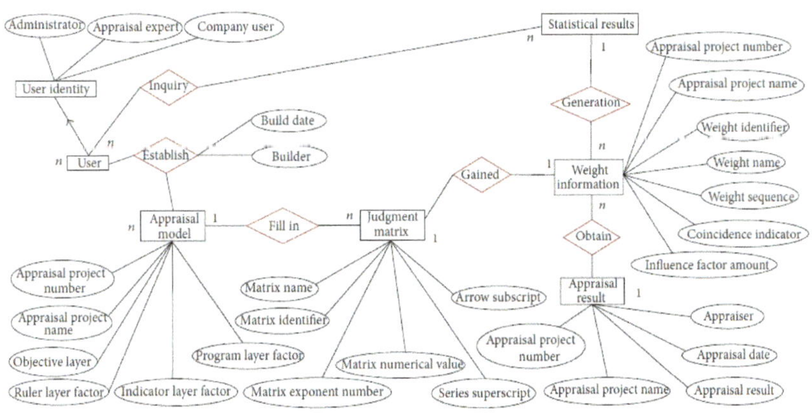

Figure 5: The entity relationship diagram for SADAS.

The function of user information is to store basic user information in the four columns of the table so that users can browse through database information according to their permissions. The table of evaluation established by evaluation experts is designed to store the index system of decision-making project and the table structure is meant especially to provide universality of the decision-making object. The factor of the judgment matrix is stored into the table named judgment matrix. An example of table structure can be seen in Figure 6. The data of judgment matrix is stored into the table of judgment matrix through an eigenvalue approach to the pairwise comparisons, and the judgment matrix information table can store nine rank matrixes; but if the length of field name is modified, the table can store another nine rank matrixes. It is difficult to use Oracle9i to store hierarchical relational tree, because it is relational database. Here, the hierarchical relational tree is decomposed into various father-son relationships which, in turn, can be integrated to form a whole model structure of AHP through combination of the judgment matrix table and judgment matrix information table to achieve the desired function. The weight of judgment matrix is stored in the table of weight information. Assessment result table is used to store the result of decision-making using evaluation expert reports.

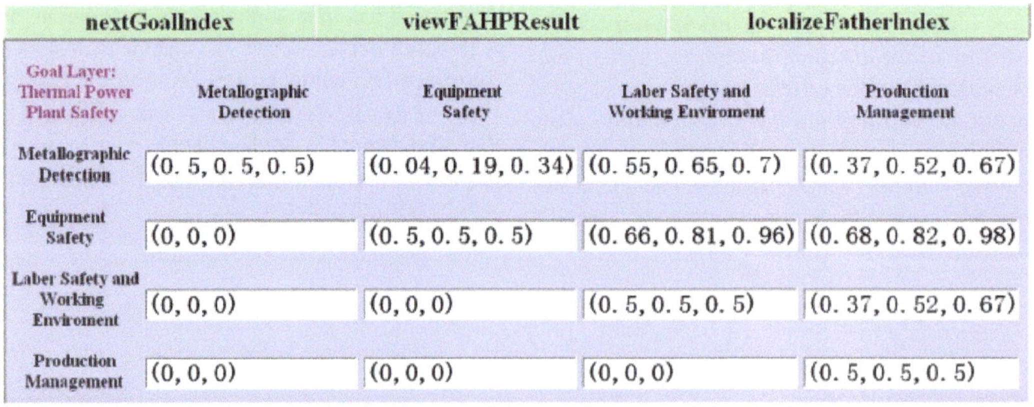

nextGoalIndex		viewFAHPResult		localizeFatherIndex
Goal Layer: Thermal Power Plant Safety	Metallographic Detection	Equipment Safety	Laber Safety and Working Enviroment	Production Management
Metallographic Detection	(0. 5, 0. 5, 0. 5)	(0. 04, 0. 19, 0. 34)	(0. 55, 0. 65, 0. 7)	(0. 37, 0. 52, 0. 67)
Equipment Safety	(0, 0, 0)	(0. 5, 0. 5, 0. 5)	(0. 66, 0. 81, 0. 96)	(0. 68, 0. 82, 0. 98)
Laber Safety and Working Enviroment	(0, 0, 0)	(0, 0, 0)	(0. 5, 0. 5, 0. 5)	(0. 37, 0. 52, 0. 67)
Production Management	(0, 0, 0)	(0, 0, 0)	(0, 0, 0)	(0. 5, 0. 5, 0. 5)

Figure 6: A part of input interface for the judgment matrix.

Remote decision support of safety assessment is developed by adopting concepts described in this paper. The system effectively separates the inference engine and knowledge base and mainly concentrates on the design of the whole system. The dynamic web page techniques ASP and .NET technology are used to develop B/S structure and IIS6.0 is utilized to design an open inference engine. Production rule-based knowledge is constructed in the form of database which is developed according to expert experience and standard manual of safety assessment. The data table is constructed from the former eight tables.

All of HTML and ASP is put into the package to format the enforceability file with the profession installation program making which can be set up at the platform of windows. In addition, data dictionary can be established through a file in the package. The original data are input into SADAS after SADAS is deployed into the web server, and the user will be able to browse and use the system to carry out safety assessments. Figure 6shows the input interface to the judgment matrix, and the data that are input via the interface is obtained by using the network on the basis of the subject offer experience of the experts. Figure 7 shows the result of the judgment matrix by using FAHP and combining it with information and computer technology.

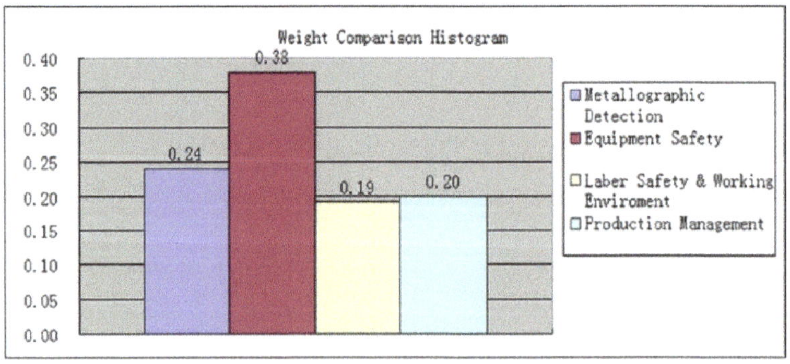

The Judgment Matrix

Goal Layer: ALL	Metallographic Detection	Equipment Safety	Laber Safety & Working Enviroment	Production Management
Metallographic Detection	(0.50, 0.50, 0.50)	(0.04, 0.19, 0.34)	(0.55, 0.65, 0.70)	(0.37, 0.52, 0.67)
Equipment Safty	(0.66, 0.81, 0.96)	(0.50, 0.50, 0.50)	(0.66, 0.81, 0.96)	(0.68, 0.82, 0.98)
Laber Safety & Working Enviroment	(0.30, 0.35, 0.45)	(0.10, 0.19, 0.34)	(0.50, 0.50, 0.50)	(0.37, 0.52, 0.67)
Production Management	(0.40, 0.48, 0.63)	(0.10, 0.18, 0.32)	(0.33, 0.48, 0.63)	(0.50, 0.50, 0.50)

Figure 7: The interface of FAHP result.

Figure 8 shows the system functionality for the safety assessment of thermal power plants based on FAHP and SPA. The data from the safety assessment manage information system has been deployed into the several power plants since 2007 for inputting the x_i to obtain the result about F(x)and it can be measured by using FAHP method as dimensionless variable. The assessment value of the thermal power plant safety can be acquired through calculating F(x) and the safety degree of the thermal power plant can be confirmed by comparing the optimal safety status (the experts set the values) obtained through calculating x in formula (17). The safety assessment model is issued only in the criteria level here while the other levels can also be disposed in the same way.

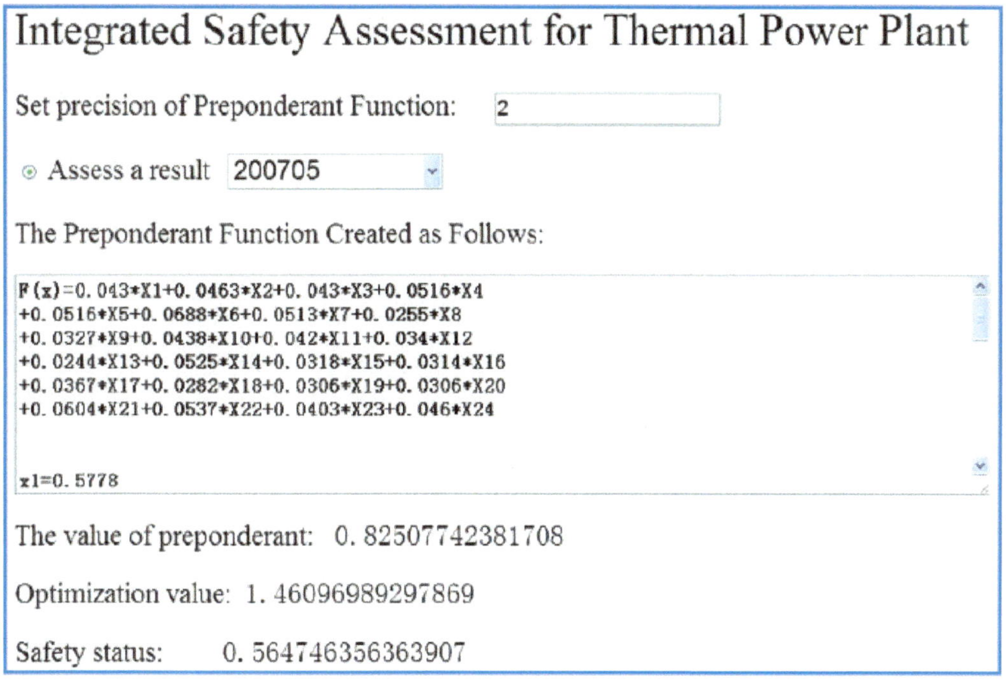

Figure 8: The interface of creating preponderant function of system safety.

As the key report function of the system, the function of SADAS is to enable the enterprise user to make decisions. In order to accelerate the development progress, we chose ASP and JS to generate reports on the basis of the weight of judgment matrix from FAHP. Moreover, the calculation result which is very small must be amplified to generate the figure report that can be queried expediently in the web pages. Figure 9 shows the result of the safety assessment of thermal power plant obtained by using the technology described here.

Set Time Range

2007-05-01 to 2008-06-03

Date	200705	200707	200803	200804	200806
Optimization	1.0002	1.0002	1.0002	1.0002	1.0002
F(x)	0.6258116	0.8861147	0.9139098	0.7878595	0.7622265
Safety Status	0.6256865	0.8859375	0.913727	0.7877019	0.7620741

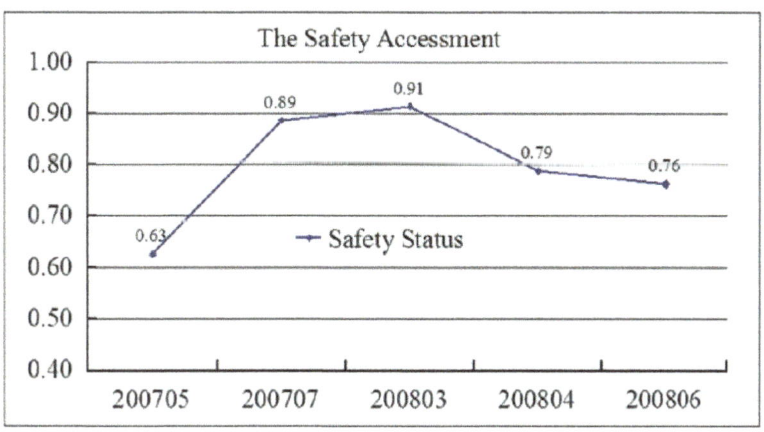

Figure 9: The interface of creating preponderant function of system safety.

The evaluation index system established by SADAS for an actual thermal power plant that is used in daily work and the effectiveness of SADAS will be validated in the future.

SAFETY ASSESSMENT FOR THE THERMAL POWER PLANTS

The enterprise safety assessment for thermal power plants is the decision-making goal, the hierarchical structure of which is composed of three levels, the target level A, the criteria level B, and the subcriteria level C. Hazards will be analyzed quantitatively and qualitatively to confirm the probability of risk and work out the necessary measures to attain the lowest accident rate and superior safety investment performance. The factors affecting safety analysis and the architecture are shown in Figure 1.

Once the hierarchy is established, expert knowledge is elicited through interviews and questionnaires whose results will be used to generate the weighting factors for the upper level; triangular fuzzy values in 0.1–0.9 scales are used to obtain the upper triangular values of judgment matrix. The corresponding lower triangular matrix represents negative judgment. An example of the matrix A established on 0.1–0.9 scales is shown as follows:

$$A_{4\times4} = \begin{bmatrix} (0.50,0.50,0.50) & (0.04,0.19,0.34) & (0.55,0.65,0.70) & (0.37,0.52,0.67) \\ (0.66,0.81,0.96) & (0.50,0.50,0.50) & (0.66,0.51,0.96) & (0.68,0.82,0.98) \\ (0.30,0.35,0.45) & (0.04,0.19,0.34) & (0.50,0.50,0.50) & (0.37,0.52,0.67) \\ (0.33,0.48,0.63) & (0.02,0.18,0.32) & (0.33,0.48,0.63) & (0.50,0.50,0.50) \end{bmatrix}. \tag{15}$$

The relative weight vector Q for every set pair denoted a by triangular fuzzy number is given as Q_1: (0.1513, 0.2325, 0.3480), Q_2: (0.2591, 0.3675, 0.5354), Q_3: (0.1254, 0.1950, 0.3087), and Q_4: (0.1223, 0.2050, 0.3276). The possibility degree matrix P, the weight vector W, and the general weight vector R are calculated from formulas (3)–(6), which are given as follows:

$$P = \begin{bmatrix} 0.5 & 0 & 0.69 & 0.64 \\ 1 & 0.5 & 1 & 1 \\ 0.31 & 0 & 0.5 & 0.46 \\ 0.36 & 0 & 0.54 & 0.5 \end{bmatrix},$$

$$W = [0.2355, 0.3750, 0.2004],$$

$$R = [0.2355, 0.3750, 0.1891, 0.2004]. \tag{16}$$

Next, the weight sequence about the subcriteria level C relative to the upper criteria level B can be calculated similarly and the impact factors about thermal power plant safety can be analyzed from the calculated results. The ranking sequence for thermal power plant safety is shown in Table 1.

Table 1: Local and general weight of index in B and C layer.

C	$R_{B1} = 0.2355$	$R_{B2} = 0.3750$	$R_{B3} = 0.1891$	$R_{B4} = 0.2004$	R	Ranking order
W_{C11}	0.1825	—	—	—	0.0430	11
W_{C12}	0.1968	—	—	—	0.0463	8
W_{C13}	0.1825	—	—	—	0.0430	12
W_{C14}	0.2191	—	—	—	0.0516	5
W_{C15}	0.2191	—	—	—	0.0516	6
W_{C21}	—	0.1834	—	—	0.0688	1
W_{C22}	—	0.1368	—	—	0.0513	7
W_{C23}	—	0.0679	—	—	0.0255	23
W_{C24}	—	0.0872	—	—	0.0327	17
W_{C25}	—	0.1169	—	—	0.0438	10
W_{C26}	—	0.1121	—	—	0.0420	13
W_{C27}	—	0.0906	—	—	0.0340	16
W_{C28}	—	0.0651	—	—	0.0244	24
W_{C29}	—	0.1401	—	—	0.0525	4
W_{C31}	—	—	0.1680	—	0.0318	18
W_{C32}	—	—	0.1659	—	0.0314	19
W_{C33}	—	—	0.1939	—	0.0367	15
W_{C34}	—	—	0.1492	—	0.0282	22
W_{C35}	—	—	0.1616	—	0.0306	20
W_{C36}	—	—	0.1616	—	0.0306	21
W_{C41}	—	—	—	0.3015	0.0604	2
W_{C42}	—	—	—	0.2681	0.0537	3
W_{C43}	—	—	—	0.2010	0.0403	14
W_{C44}	—	—	—	0.2294	0.0460	9

The 5-time scoring rates for all the criteria come from routine inspection (the score is a 0-1 scale) and the functional coefficient α can be calculated by using (8)–(13) as shown in Tables 2 and 3.

Table 2: Rate of scoring and functional coefficient of index in B layer

	Rate of scoring (%)						
A	1st	2nd	3rd	4th	5th	α	Sequence
B1	55.39	87.04	91.83	70.72	83.74	0.3722	2
B2	62.86	89.96	93.22	64.37	80.77	0.3985	1
B3	47.67	88.79	93.97	68.10	70.77	0.1053	4
B4	45.38	86.50	94.59	73.48	86.50	0.1240	3

Table 3: Recording score and horizontal functional coefficient of index in C layer

	Rate of scoring (%)						
	1st	2nd	3rd	4th	5th	α	Sequence
C11	57.78	83.33	88.89	96.67	70.00	0.1022	1
C12	51.00	86.71	94.14	79.14	65.71	0.0120	20
C13	67.50	88.00	91.00	93.00	78.00	0.1002	3
C14	58.33	86.67	83.33	84.71	90.59	0.0318	14
C15	58.00	91.00	85.00	85.00	75.00	0.0102	21
C21	60.33	88.33	91.08	77.42	68.00	0.0536	8
C22	57.08	89.91	92.30	73.53	71.68	−0.0159	24
C23	67.23	91.58	92.87	70.68	69.13	0.1012	2
C24	66.22	92.13	95.43	91.74	82.99	0.0840	4
C25	64.53	89.16	95.42	99.91	78.53	0.0138	19
C26	63.21	89.78	90.95	83.22	77.34	0.0484	9
C27	61.76	90.98	93.53	49.06	64.53	0.0236	16
C28	64.33	86.00	96.00	71.67	71.67	0.0142	18
C29	61.67	96.46	86.01	90.50	76.00	0.0652	6
C31	35.16	88.39	91.13	74.68	65.97	0.0474	10
C32	68.33	91.11	94.44	51.67	75.00	0.0658	5
C33	73.64	89.09	100.00	81.82	70.91	0.0460	11
C34	23.33	80.00	96.67	62.00	57.33	0.0071	22
C35	92.00	100.00	100.00	84.00	84.00	0.0455	12
C36	100.00	100.00	100.00	80.00	80.00	0.0571	7
C41	59.61	87.30	90.01	76.53	77.16	0.0317	15
C42	54.96	86.90	89.17	70.94	79.239	0.0156	17
C43	59.09	80.00	80.81	60.72	69.71	0.0327	13
C44	59.71	86.20	91.80	70.00	82.50	0.0066	23

The preponderant function of system safety on the basis of the criteria in B level can be expressed as

$$F_B(X) = 0.292x_1 + 0.208x_2 + 0.125x_3 + 0.375x_4$$
$$+ 0.148x_1x_2 + 0.039x_1x_3 + 0.046x_1x_4 + 0.042x_2x_3$$
$$+ 0.049x_2x_4 + 0.013x_3x_4 + 0.016x_1x_2x_3$$
$$+ 0.018x_1x_2x_4 + 0.005x_1x_3x_4 + 0.005x_2x_3x_4. \quad (17)$$

The preponderant function of system safety for the other layers can also be obtained in the same way. The assessment value for thermal power plant safety can be acquired through calculating the $F(x)$ in which x_i would be assigned by factors safety status value and the unitary safety status of the thermal power plant can be assessed by comparing the optimal safety status value of $F(x)$ in which x_i would be assigned by 1 (generally 1 for 0.1–0.9 scale).

In practical application, the x_1–x_4 of the formula (17) are assigned and calculated and were compared with the best security state

(i.e., security experts set value) to determine the safety level of thermal power plant. Table 4 displays an example result for the thermal power plant safety.

Table 4: The safety assessment result for the thermal power plants

	1st	2nd	3rd	4th	5th	Optimal value
$F(x)$ basis of B	0.6321	1.1700	1.2628	0.8741	1.0721	1.3822
Safety status	0.4573	0.8465	0.9136	0.6324	0.7756	1
$F(x)$ basis of C	0.8619	1.3447	1.4168	1.1921	1.0858	1.5907
Safety status	0.5418	0.8454	0.8907	0.7494	0.6826	1

The sequence of factors sorted by general weight is B2 > B1 > B4 > B3 and the key factors are *equipment safety* and *metallographic detection*, where B3 has the most hidden dangers in the B level that cause at least functional coefficient in level B and B2 is crucial since its expanded function is the strongest. The first three in order of arrangement by increasing α in level C are C11 > C23 > C13, which should be emphasized because they have the strongest expanded tendency. It indicates that strict management and surveillance are effective ways to ensure safety in production. The *electric primary apparatus* and *electric quadratic apparatus* are the first two in order of increasing by α in B2, indicating that there is strong dependence of other equipment on them.

CONCLUSIONS

It is practicable to establish the preponderant function for system safety using FAHP, system functionality, and SPA, which avoid problems of data sources because all the data come from the routine work of safety assessment experts. In addition, this provides a reference value for enterprise safety; however, it could lead to a great deal of calculations. Therefore, the development of a computer system is useful to facilitate the safety assessment process, which will in turn promote rationalization of safety assessment. Besides, with the web-based decision support system developed here to implement FAHP and SPA, the evaluation index system can be easily established and decision-making advice can be obtained through the steps involved in the FAHP methodology as realized by the ASP.NET technology. Furthermore, the system can be updated and expanded to face the developments, for some interfaces reserved in the system are available to integrate other data acquisition systems and the advice obtained will guide the management to cope with emergent events and reduce life and property loss. It is not only a promising methodology to help resolve industrial safety problem, but also a great help to lay the foundation for the next study of safety assessment and decision-making. Finally, it has values for the design and development of web-based decision analysis system corresponding to the methodology and technology presented in this paper.

In the future, we want to integrate the safety assessment management information system which has been deployed into several power plants combining with the data mining or neural network technology to obtain a more ideal safety assessment result to guide safety management activities. The real time results for this system need to be improved.

Conflict of Interests

The authors declare that there is no conflict of interests regarding the publication of this paper.

ACKNOWLEDGMENTS

The financial support from the National Natural Science Foundation of China (NSFC) under the Grants no. 71071078 and no. 70671035 is acknowledged gratefully. The authors also gratefully acknowledge the helpful comments and suggestions of the reviewers, which have improved the paper.

REFERENCES

1. S. Banerjee, Industrial Hazards and Plant Safety, Taylor & Francis, New York, NY, USA, 2002.
2. O. S. Vaidya and S. Kumar, "Analytic hierarchy process: an overview of applications," European Journal of Operational Research, vol. 169, no. 1, pp. 1–29, 2006. .
3. M. Deng, N. Bu, and A. Inoue, "Output tracking of nonlinear feedback systems with perturbation based on robust right coprime factorization," International Journal of Innovative Computing, Information and Control, vol. 5, no. 10, pp. 3359–3366, 2009.
4. Z. Yang and Z. Zhang, "Hierarchical network-based safety assessment decision support system for thermal power plants," International Journal of Innovative Computing, Information and Control, vol. 6, no. 7, pp. 2955–2964, 2010.
5. N. Bu and M. Deng, "System design for nonlinear plants using operator-based robust right coprime factorization and isomorphism," IEEE Transactions on Automatic Control, vol. 56, no. 4, pp. 952–957, 2011. .
6. O. Vayvay, Y. Ozcan, and M. M. Cruz-Cunha, "ERP consultant selection problem using AHP, fuzzy AHP and ANP: a case study in Turkey," Journal of Business Management and Economics, vol. 3, no. 3, pp. 106–117, 2012. View at Google Scholar
7. J. R. Yu and W. Y. Shing, "Fuzzy analytic hierarchy process and analytic network process: an integrated fuzzy logarithmic preference programming," Applied Soft Computing, vol. 13, no. 4, pp. 1792–1799, 2013. . View at Google Scholar
8. J. L. Coze, "Outlines of a sensitising model for industrial safety assessment," Safety Science, vol. 51, no. 1, pp. 187–201, 2013. . View at Google Scholar

9. S. S. Leu and C. M. Chang, "Bayesian-network-Based safety risk assessment for steel construction projects," Accident Analysis & Prevention, vol. 54, pp. 122–133, 2013. . View at Google Scholar

10. S. A. Umpleby and E. B. Dent, "The origins and purposes of several traditions in systems theory and cybernetics," Cybernetics and Systems, vol. 30, no. 2, pp. 79–103, 1999.

11. G. Paul and D. Harley, "Knowledge-system theory in society: charting the growth of knowledge-system models over a decade," Journal of the American Society For In Formation Science and Technology, vol. 58, no. 14, pp. 2372–2381, 2007. . View at Google Scholar

12. Srividya and H. N. Suresh, "Prioritizing feeders for maintenance activities using AHP in nuclear power plants," International Journal of Reliability, Quality and Safety Engineering, vol. 14, no. 3, pp. 275–282, 2007.

13. K. Zhou and S. Feng, "Evaluating the conceptual design schemes of complex products based on FAHP using fuzzy number," Journal of Systems Engineering and Electronics, vol. 16, no. 3, pp. 574–578, 2005.

14. X. Wang, H. K. Chan, R. W. Y. Yee, and I. Diaz-Rainey, "A two-stage fuzzy-AHP model for risk assessment of implementing green initiatives in the fashion supply chain," International Journal of Production Economics, vol. 135, no. 2, pp. 595–606, 2012. .

15. S. K. Lee, G. Mogi, and K. S. Hui, "A fuzzy analytic hierarchy process (AHP)/ data envelopment analysis (DEA) hybrid model for efficiently allocating energy R&D resources: in the case of energy technologies against high oil prices," Renewable and Sustainable Energy Reviews, vol. 21, pp. 347–355, 2013. View at Google Scholar

16. Z. Wang, H. Ma, C. Lai, and H. Song, "Set pair analysis model based on GIS to evaluation for flood damage risk," Procedia Engineering, vol. 28, pp. 196–201, 2012. . View at Google Scholar

17. F. F. Wu and X. Wang, "Eutrophication evaluation based on set pair analysis of baiyangdian lake, North China," Procedia Environmental Sciences, vol. 13, pp. 1030–1036, 2012. . View at Google Scholar

18. Y. Liu, "Study on environment evaluation and protection based on set pair analysis—a case study of chongqing," Energy Procedia, vol. 14, pp. 14–19, 2012. . View at Google Scholar

19. J. M. G. da Silva, M. Z. Oliveira, D. Coutinho, and S. Tarbouriech, "Static anti-windup design for a class of nonlinear systems," International Journal of Robust and Nonlinear Control, vol. 24, no. 5, pp. 793–810, 2014. . View at Google Scholar

20. S. Bellavia, B. Morini, and M. Porcelli, "New updates of incomplete LU factorizations and applications to large nonlinear systems," Optimization Methods & Software, vol. 29, no. 2, pp. 321–340, 2014. View at Google Scholar

21. F. Liu, Y. C. Ye, and Y. Huang, "Current and development foreground of safety evaluation methods research," China Water Transport, vol. 7, no. 1, pp. 179–181, 2007. View at Google Scholar

22. Z. Yang, X. Yuan, Z. Feng, K. Suzuki, and A. Inoue, "A fault prediction approach for process plants using fault tree analysis in sensor maiaolfunction," in Proceedings of the IEEE International Conference on Mechatronics and Automation (ICMA ‹06), pp. 2415–2420, June 2006.
.

CHAPTER 17

PREFERRED ECONOMIC DISPATCH OF THERMAL POWER UNITS

Selvaraj Durai, Srikrishna Subramanian, Sivarajan Ganesan

Department of Electrical Engineering, Annamalai University, Chidambaram, India

ABSTRACT

Economic Dispatch (ED) problem is one of the main concerns of the power generation operations which are basically solved to generate optimal amount of power from the generating units in the system by minimizing the fuel cost and by satisfying the system constraints. The accuracy of ED solutions is highly influenced by the fuel cost parameters of the generating units. Generally, the parameters are subjected to transform due to aging process and other external issues. Further the parameters associated with the transmission line modelling also change due to aforementioned issues. The loss coefficients which are the functions of transmission line parameters get altered from the original value over a time. Hence, the periodical estimation of these coefficients is highly essential in power system problems for obtaining ideal solutions for ED problem. Estimating the ideal parameters of the ED problem may be the best solution for this issuc. This paper presents the Teaching Learning Based Optimization (TLBO) algorithm to estimate the parameters associated with ED problem. The estimation problem is formulated as an error minimization problem. This work provides a frame work for the computation of coefficients for quadratic function, piece- wise quadratic cost function, emission function, transmission line parameters and loss coefficients. The effectiveness of TLBO is tested with 2 standard test systems and an Indian utility system.

Keywords:

Parameter Estimation, Cost Coefficients, Emission Coefficients, Error Estimation, Transmission Line Parameters, Teaching Learning Based Optimization

---

Citation: Durai, S, Subramanian, S. and Ganesan, S. (2015) Preferred Economic Dispatch of Thermal Power Units. Journal of Power and Energy Engineering, 3, 47-69. DOI: 10.4236/jpee.2015.311005.

INTRODUCTION

Ideal Economic Dispatch

The economical operation of power system needs more accurate representation of fuel cost coefficients and transmission loss coefficients. Due to changing weather conditions and other external issues, the aforesaid coefficients are not ideal and hence the existing parameters with ED solutions also change. Therefore, the objective of this article is to revise the existing parameters for the Preferred Economic Dispatch (PED) solutions.

Literature Review

Various optimization techniques have been proposed by many researchers to deal with the parameter estimation problems. To obtain the exact parameters for ED problems different parameter estimation rehearsal has been projected to solve estimation problems in power systems. State estimation based technique such as Least Error Square (LES) and least absolute value methods have been demonstrated. Among the two techniques, LES technique has been the most famous static estimation technique and in use for a long time as the preferred technique for optimum estimation of parameters. In general, some limitations and disadvantages are associated with this approach. El-Hawary and Mansour conducted performance analysis of LES, Bard algorithm, Marquardt algorithm and Powell regression algorithm for estimating coefficients [1] . The methods based on the least absolute value approximations and curve fitting techniques have been reported for fuel cost coefficients estimation [2] . Two polynomial curve fitting methods, Gram-Schmidt orthonormalization and least square are also applied to evaluate the fuel cost coefficients [3] . Further Henry Y. K. Chen and Charles E. Postel applied sequential regression technique to online parameter estimation of input-output curves for thermal units [4] .

Research endeavours indicate that the field of Transmission Line Parameters (TLP) estimation in power system studies is less focused. Traditionally, tower and conductor geometric parameters, conductor type, assumed ambient conditions etc., are utilized for estimating TLP [5] - [7] . The sending and receiving end voltage and current phasors are utilized to derive the TLP and propagation constant [8] . The synchronized phasor measurements at both ends of the transmission line emphasises the online parameter estimation [9] - [12] . Wagenaars et al., have proposed the measurement method, based on a pulse response measurement, to determine the transmission line parameters of the shield-to-phase and phase-to-phase modes [13] . Researchers have reported Newton-Raphson based method for estimating TLP parameters utilizing the measurements of voltage, current and power [14] .

A new category of classical optimization techniques has emerged to cope with some of the traditional algorithms in the field of power system estimation problems. The heuristic techniques such as Genetic Algorithm (GA) [15] , Particle Swarm Optimization (PSO) [16] [17] , Artificial Bee Colony (ABC) [18] , Artificial Neural Network (ANN) [19] [20] , Fuzzy Logic (FL) [21] , Ant Colony Optimization (ACO) [22] and Bacterial Foraging Algorithm (BFA) [23] - [25] have been used for solving various estimation problems such as load forecasting, state estimation and induction motor parameters.

Optimization Techniques

The heuristic techniques outperform the mathematical methods but their solution quality is sensitive to the algorithmic controlling parameters like population size and number of generations. Besides common control parameters, different algorithms require their own algorithm specific control parameters. Major disadvantage of ABC, PSO and GA are the presence of various parameters that need to be carefully tuned to reach acceptable estimation performance. Recently, TLBO, a nature inspired algorithm that is based on the effect of influence of a teacher on the output of learners in a class [26] [27] is proposed. It is a powerful evolutionary algorithm that maintains a population of students, where each student represents a potential solution to an optimisation problem. Each searching generation includes initializing of class, teacher phase, learner phase and termination criterion. The advantage of TLBO are simple, easy implementation and necessitates few control parameters for tuning that make it suitable to implement for power system parameter estimation problems. TLBO is an algorithm-specific, parameter-less algorithm that does not require any algorithm-specific parameters to be tuned [28] . This algorithm can find the global solution for nonlinear constrained optimization problems with less computation effort and high consistency [29] . The authors have proposed the TLBO algorithm for estimating the accurate parameters of the input output characteristics of thermal units [30] . The adaptable properties of this algorithm encourage the authors to use TLBO as a parameter estimator.

Research Gap and Motivation

The most important issue in ED problem is to have an accurate estimate of parameters. The exact ED solution in power generation and the estimation of parameters in transmission systems is an important task in power systems. The demonstration of literature review reveals that the estimation of accurate fuel cost coefficients and transmission line parameters are carried out independently. The impact of change in transmission line parameters on the loss coefficients of ED problem is seldom carried out. These points motivate us to contribute research in estimation of cost and emission coefficients, transmission line parameters and transmission loss coefficients for the exact ED solution. The TLBO algorithm is implemented for solving parameter and ED problems on different scale of test systems.

Highlights

- Accurate parameters of different generator cost functions and transmission lines are estimated.
- The preferred economic dispatch is carried out.
- A 19 unit practical Indian utility system is considered.
- A nature inspired TLBO is applied for both estimation of parameters and for solving ED problems.

Paper Organization

The rest of the paper is structured as follows: Problem formulation is explored in Section 2. Section 3 describes the implementation for preferred ED. The detailed discussions about numerical results achieved by various test systems are detailed in Section 4. Section 5 describes the potential verification of TLBO. Finally, Section 6 presents the conclusions of this article.

PROBLEM FORMULATION

State of the Art Model

The state of the art multi objective ED model is presented as follows.

$$FC_i^{Exi} = W_1\left(\sum_{i=1}^{N} a_i^{Exi}P_i^2 + b_i^{Exi}P_i + c_i^{Exi} + \left|e_i^{Exi}\sin\left(f_i^{Exi}\left(P_{i,\min} - P_i\right)\right)\right|\right) + W_2\left(\sum_{i=1}^{N} e_{0i}^{Exi}P_i^2 + e_{1i}^{Exi}P_i + e_{2i}^{Exi}\right) + r_i \quad i = 1,2,\cdots,N \tag{1}$$

FC_i^{Exi} is a existing fuel cost function, N is the number of generating units and P_i is the power generated by ith generating unit.

a_i^{Exi}, b_i^{Exi}, c_i^{Exi} are the existing cost coefficients, e_i^{Exi}, f_i^{Exi} are the existing valve point coefficients and e_{0i}^{Exi}, e_{1i}^{Exi}, e_{2i}^{Exi} are the existing emission coefficients .

W_1 and W_2 are the weights of the function.

r_i is the error associated with the ith equation.

Loss function:

$$P_L^{Exi} = \sum_{i=1}^{N}\sum_{j=1}^{N} P_i B_{ij}^{Exi} P_j + \sum_{i=1}^{N} B_{0i}^{Exi} P_i + B_{00}^{Exi} + r_i \quad i = 1,2,\cdots,N \tag{2}$$

P_L^{Exi} indicates existing transmission loss.

B_{ij}^{Exi}, B_{0i}^{Exi}, B_{00}^{Exi} are the existing transmission loss coefficients.

Proposed Parameter Estimation Problems

For preferred ED solution, the accurate parameters for fuel cost, emission, and transmission line and transmission loss are needed. Those parameters can be estimated by solving the following problems.

Cost coefficients Estimation, FC^{Est} = Estimate [a, b, c, e, f]

Emission coefficients Estimation, E^{Est} = Estimate [e_0, e_1, e_2]

Transmission line parameters Estimation, TLP^{Est} = Estimate [R, X, B]

Transmission loss coefficients Estimation, KC^{Est} = Estimate [B_{ij}, B_{0i}, B_{00}]

In general to estimate the accurate parameters, the following function are used

$$\text{Estimate}\left[FC^{Est}, E^{Est}, TLP^{Est}, KC^{Est}\right] \tag{3}$$

Accurate Model Using Estimated Parameters

The cost function of generating unit is expressed using estimated coefficients as follows,

$$FC_i^{Est} = W_1\left(\sum_{i=1}^{N} a_i^{Est}P_i^2 + b_i^{Est}P_i + c_i^{Est} + \left|e_i^{Est}\sin\left(f_i^{Est}\left(P_{i,\min} - P_i\right)\right)\right|\right) + W_2\left(\sum_{i=1}^{N} e_{0i}^{Est}P_i^2 + e_{1i}^{Est}P_i + e_{2i}^{Est}\right) + r_i \quad i = 1,2,\cdots,N \tag{4}$$

where FC_i^{Est} is an estimated fuel cost.

b_i^{Est} a_i^{Est}, c_i^{Est}, e_i^{Est}, f_i^{Est}, e_{0i}^{Est}, e_{1i}^{Est}, e_{2i}^{Est} are the estimated cost coefficients, valve point coefficients and emission coefficients respectively.

Loss function:

$$P_L^{est} = \sum_{i=1}^{N}\sum_{j=1}^{N} P_i B_{ij}^{Est} P_j + \sum_{i=1}^{N} B_{0i}^{Est} P_i + B_{00}^{Est} \quad i = 1,2,\cdots,N \tag{5}$$

B_{ij}^{Est}, B_{0i}^{Est}, B_{00}^{Est} are the estimated transmission loss coefficients and P_L^{Est} is the estimated transmission loss.

Error Minimization

As detailed in [30] , the error is estimated for fuel cost and emission coefficients as follows

The error at each step i can be calculated as,

$$\text{Error} = \left(FC^{Exi} - FC^{Est}\right) \text{ or } \left(E^{Exi} - E^{Est}\right) \tag{6}$$

$$\%\,\text{Error for Fuel cost} = \sum_{k=1}^{n}\left|\frac{FC_i^{Exi} - FC_i^{Est}}{FC_i^{Exi}}\right| * 100 \tag{7}$$

$$\%\,\text{Error for Emission} = \sum_{k=1}^{n}\left|\frac{E_i^{Exi} - E_i^{Est}}{E_i^{Exi}}\right| * 100 \tag{8}$$

The % error for each step k of transmission line parameter is expressed as,

$$\%\,\text{Error} = \sum_{k=1}^{n}\left|\frac{TLP_k^{Exi} - TLP_k^{Est}}{TLP_k^{Exi}}\right| * 100 \tag{9}$$

The objective function is minimised subjected to physical limits of each coefficients and TLP.

IMPLEMENTATION OF TLBO FOR PREFERRED ECONOMIC DISPATCH (PED)

This section details the computational flows for parameter estimation and PED problems.

Parameter Estimation Procedure

Step 1: Read the required data for estimating the parameters, population size, maximum number of iterations ($Iter_{max}$), maximum and minimum limits of parameters (PA^{min}, PA_i^{max}).

Step 2:Randomly generate parameters using the following equation,

$$PA_i^j = PA_i^{\min} + rand_z * \left(PA_i^{\max} - PA_i^{\min}\right) \quad i = 1,\cdots,CG,\ j = 1,\cdots,PS \tag{10}$$

Step 3: Check for the constraint violation.

Step 4: If the limit is violated, then upgrade the parameters of the units using the following equation

$$P_i^j = \begin{cases} PA_i^{\min}; \left(PA_i^j < PA_i^{\min}\right) \\ PA_i^{\max}; \left(PA_i^j > PA_i^{\min}\right) \\ PA_i^j; \left(PA_i^{\max} \geq PA_i^j \geq PA_i^{\min}\right) \end{cases} \quad (11)$$

Step 5: Calculate FC^{Est} by using Equations (4).

Step 6: The error and % error associated with each measurement can be calculated by using Equations (6), (7), (8) and (9).

Step 7: Teacher phase: Compute the difference between existing mean result and best mean by utilizing teaching factor (ψ_F).

Step 8: Learner phase: Evaluate the learners' parameter values with the help of teacher's parameter values.

Step 9: Update the parameters subject to constraints.

Step 10: Stopping Criterion: If Iter $\geq$ $Iter_{max}$ print the optimal results, otherwise go to step 7.

Computational Flow of PED

Hence the estimated parameters are suitably achieved. The ED solution is carried out and the computational flow is as follows in Figure 1.

NUMERICAL SIMULATION RESULTS AND DISCUSSION

This section presents three different test systems such as 10 unit, IEEE 30 bus and a practical Indian Utility system to demonstrate the efficacy of the proposed solution technique for perfect estimate of parameters and preferred ED solutions. The algorithm is implemented in Matlab package and simulations are carried out on a personal computer having the configuration of Intel(R) core i3 CPU with 2 GB RAM. The following algorithmic parameters are chosen for all cases in order to validate the performance of the algorithm: Population size = 10;

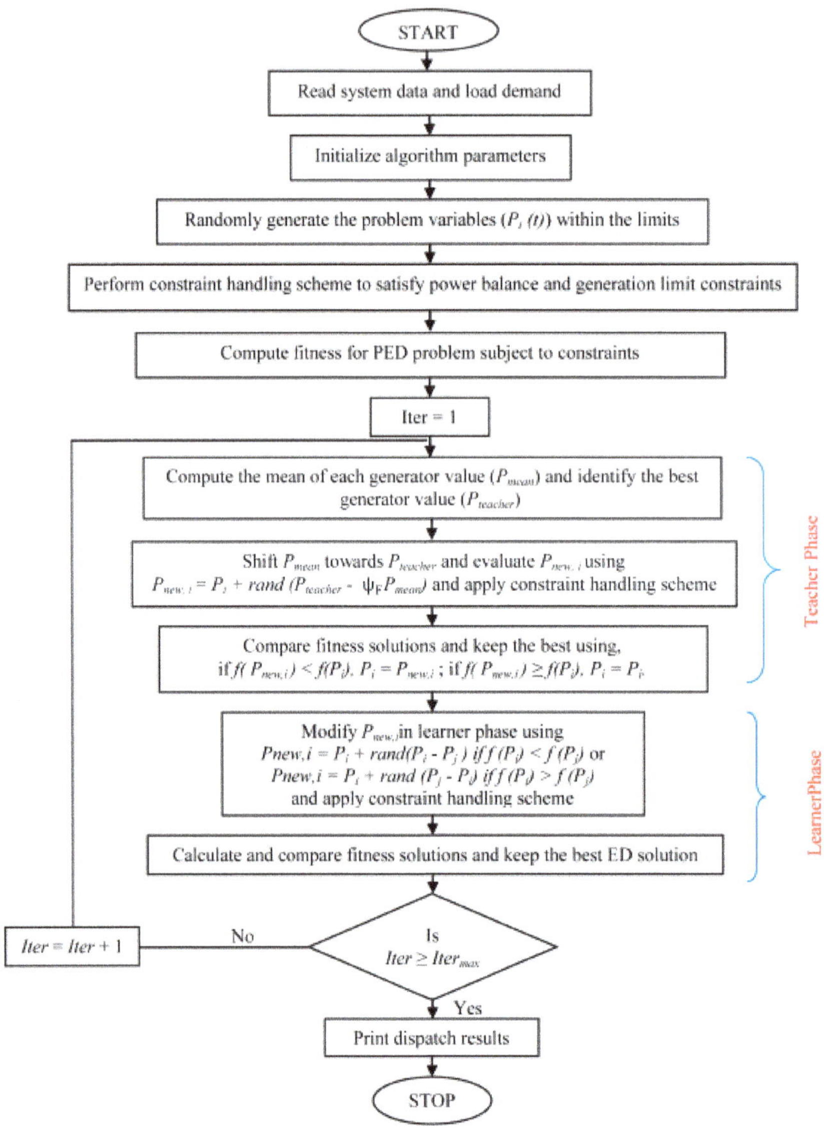

Figure 1: Flowchart for preferred economic dispatch using TLBO.

Maximum number of iterations = 120.

Test System 1: 10 Unit System with Piecewise Quadratic Functions

This test system consists of 10-generating units in which each unit has two or three fuel options like coal, oil and gas. As more than one fuel is used, the generator input-output characteristic is expressed as a piecewise quadratic function. The valve point effects are also considered. The test system data is available in [31] .

The generator characteristic is expressed as a piecewise quadratic function including valve point effects increases the number of variables to be estimated. In this test system, the generator 1 has two fuel options and the remaining units have three fuel options. It is specified that the fuel 2 is uneconomical for generator 9. The simulation is carried out using the proposed TLBO which acquires new values for cost coefficients and are presented in the Table 1. The estimation process has been performed for each generation level and the results of existing fuel cost (FC^{Exi}) and estimated fuel cost (FC^{Est}) are presented independently for each fuel type in the Table 1.

Table 1: Estimation of piecewise quadratic cost function coefficients using TLBO-10 unit system

Unit	P (MW)		Fuel type	Coefficients estimated by TLBO					FC^{Exi} (\$/h)		FC^{Est} (\$/h)		$\sum_{error}$	
	Min	Max		a_i	b_i	c_i	e_i	f_i	Min	Max	Min	Max	TLBO	ABC
1	100	196	1	25.19938	−0.380846	0.0021609	0.025324	−3.028257	8.980000	32.6800	8.723994	33.59283	1.168802	3.98151
	196	250	2	20.03000	−0.305900	0.0018897	0.021337	−3.009000	32.66577	60.9879	32.67157	61.67932	0.697156	2.81373
2	50	114	2	1.816721	−0.037780	0.0011015	0.001901	−0.410800	22.54490	48.4386	23.48145	49.14125	1.639115	3.25496
	114	157	3	12.94016	−0.198000	0.0016731	0.013159	−1.880000	2.716000	12.1088	2.681891	11.82574	0.317203	2.75457
	157	230	1	120.6888	−1.282684	0.0042228	0.112869	−11.94430	12.13152	22.4996	12.12238	23.09567	0.605175	2.93069
3	200	332	1	39.04620	−0.305817	0.0014527	0.039875	−3.115954	35.75000	96.9444	35.98904	97.64218	0.936732	1.59272
	332	388	3	−2.97500	0.032890	0.0008075	−0.002976	0.336952	131.3536	187.056	131.7986	187.3240	0.712539	1.27426
	388	500	2	−58.1400	0.484892	1.196E−05	−0.058140	4.764000	96.94140	131.236	96.94769	131.3470	0.116458	1.66413
4	99	138	1	1.990265	−0.031040	0.0010434	0.001982	−0.310400	9.181389	17.6636	9.143991	17.57876	0.122282	1.15842
	138	200	2	52.94079	−0.633800	0.0027499	0.053850	−6.248000	17.77095	36.2506	17.84586	36.22169	0.103847	0.43500
	200	265	3	258.6681	−2.290143	0.0058688	0.269957	−22.39430	36.60000	64.2124	35.39341	64.15721	1.261771	4.91182
5	190	338	1	13.80777	−0.088330	0.0010722	0.013720	−0.870300	35.80990	106.192	35.73286	106.449	0.333505	1.15977
	338	407	2	99.87032	−0.515692	0.0015799	0.099573	−5.182580	106.2448	152.504	106.0639	151.7490	0.936775	5.70216
	407	490	3	−52.9900	0.445057	0.0001478	−0.053890	4.560945	152.4276	200.634	152.6311	200.6286	0.208988	0.41948
6	138	200	1	52.95000	−0.631800	0.0027309	0.052750	−6.308000	17.77095	36.2506	17.76962	35.88033	0.371620	1.59502
	85	138	2	1.976247	−0.031040	0.0010540	0.001973	−0.310400	6.915125	17.6642	6.952709	17.76567	0.139008	0.91532
	200	265	3	265.4849	−2.333976	0.0059502	0.261282	−23.28925	36.60000	64.2124	36.69646	64.94455	0.828617	2.44408
7	200	331	1	18.88226	−0.132500	0.0011050	0.018850	−1.355000	36.71000	96.3699	36.58365	96.11233	0.383942	2.96959
	331	391	2	44.00000	−0.220000	0.0011464	0.043769	−2.211000	96.37086	133.271	96.77999	133.2703	0.410536	1.94797
	391	500	3	−43.0000	0.356873	0.0002422	−0.043000	3.552032	133.3239	195.993	133.5603	196.0080	0.251265	2.66318
8	99	138	1	1.984912	−0.031000	0.0010294	0.001992	−0.310000	9.181389	17.6636	9.005129	17.31185	0.528055	1.15768
	138	200	2	52.76360	−0.630000	0.0027294	0.052870	−6.375000	17.77095	36.2506	17.80167	35.96774	0.313608	1.13761
	200	265	3	266.5000	−2.330000	0.0059150	0.260000	−23.00000	36.60000	64.2124	37.10000	64.53166	0.819262	4.66804
9	213	370	1	88.56713	−0.566963	0.0015544	0.088509	−5.684923	38.15592	91.3812	38.32396	91.61140	0.398192	2.68979
	130	213	3	14.20019	−0.018100	0.0006124	0.014100	−0.181000	22.21239	38.1385	22.19758	38.14001	0.016320	1.07008
	370	440	3	14.10104	−0.018100	0.0006111	0.014280	−0.181000	91.30359	124.739	91.06478	124.4491	0.529615	5.65214
10	200	362	1	13.99000	−0.099100	0.0011054	0.013887	−0.992500	38.17400	122.414	38.38720	122.9832	0.781727	1.87694
	362	407	3	46.65000	−0.201400	0.0011360	0.046801	−2.014000	152.6864	198.035	152.7299	198.0353	0.043682	1.10165
	407	490	2	−61.0900	0.508437	4.157E−05	−0.061100	5.074000	122.4382	152.677	122.6075	152.8768	0.368904	1.08521

The error is also computed by comparing FC^{Exi} and FC^{Est} using Equation (6). The obtained error using the proposed algorithm is compared with ABC algorithm which clearly illustrates the superiority of TLBO over other algorithm.

The ED is carried out separately considering both existing and estimated fuel cost coefficients. The obtained results for various demands are presented in Table 2 along with % error, which clearly illustrates the impact of estimated coefficients on ED problem.

Test System 2: IEEE 30 Bus System

The TLBO algorithm has been executed on the test system consists of six generating units interconnected with 41 branches of transmission lines. Generator data, emission characteristics and line data are obtained from [32] . TLBO algorithm is applied to estimate the cost, emission coefficients and transmission line parameters. For each unit, the fuel cost and emission coefficients are computed. Table 3 and Table 4 show the TLBO outcomes of the estimated coefficients.

Table 2: Economic dispatch with improved parameters for 10 unit system

Demand (MW)	FC^{Exi} (\$/h)	FC^{Est} (\$/h)	% Error
2500	525.7539	532.2159	1.22090
2600	608.7376	615.3032	1.07856
2700	627.2440	634.3932	1.13978

Table 3: Estimation of quadratic cost function coefficients by TLBO-IEEE 30 bus system

Unit	P (MW)	Coefficients			FC^{Exi} (\$/h)	FC^{Est} (\$/h)	Error	Total Error
			Existing	TLBO				
	50	a_i	0.00375	0.003709	109.375	109.6930	0.318013	
1	125	bi	2.00000	2.008420	308.594	309.0026	0.408860	0.763171
	200	c_i	0.00000	0.000000	550.000	550.0363	0.036298	
	20	a_i	0.01750	0.017377	42.0000	41.95063	0.049366	
2	60	b_i	1.75000	1.750000	168.000	167.5557	0.444297	1.283524
	80	ci	0.00000	0.000000	252.000	251.2101	0.789861	
	15	a_i	0.06250	0.061676	29.0625	29.55164	0.489144	
3	30	b_i	1.00000	1.044963	86.2500	86.85768	0.607683	1.286063
	50	c_i	0.00000	0.000000	206.250	206.4392	0.189236	
	10	a_i	0.00834	0.008000	33.3340	33.30000	0.034000	
4	20	b_i	3.25000	3.250000	68.3360	68.20000	0.136000	0.586500
	35	c_i	0.00000	0.000000	123.967	123.5500	0.416500	
	10	a_i	0.02500	0.022322	32.5000	33.03565	0.535646	
5	20	b_i	3.00000	3.080341	70.0000	70.53576	0.535763	1.071758
	30	c_i	0.00000	0.000000	112.500	112.5003	0.000349	
	12	a_i	0.02500	0.023081	39.6000	40.22570	0.625698	
6	25	b_i	3.00000	3.075169	90.6250	91.30487	0.679869	1.369166
	40	c_i	0.00000	0.000000	160.000	159.9364	0.063598	

Table 4: Estimation of emission function coefficients by TLBO-IEEE 30 bus system

Unit	P(MW)	Coefficients			E^{Exi} (Kg/h)	E^{Est} (Kg/h)	Error	Total Error
			Existing	TLBO				
	50	e_{0i}	0.01260	0.012470	−0.5170	−0.01556	0.501437	
1	125	e_{1i}	−1.1000	−1.08064	82.3580	82.60794	0.249943	2.212806
	200	e_{2i}	22.9830	22.84106	306.983	305.5216	1.461426	
	20	e_{0i}	0.0200	0.019823	31.3130	31.64200	0.329003	
2	60	e_{1i}	−0.1000	−0.08000	91.3130	91.87402	0.561024	1.354070
	80	e_{2i}	25.3130	25.31300	145.313	145.7770	0.464043	
	15	e_{0i}	0.02700	0.026724	31.4300	31.29213	0.137871	
3	30	e_{1i}	−0.0100	−0.00900	49.5050	49.19559	0.309407	1.178847
	50	e_{2i}	25.5050	25.41431	92.5050	91.77343	0.731569	
	10	e_{0i}	0.02910	0.028212	27.7600	27.81122	0.051223	
4	20	e_{1i}	−0.0050	−0.00100	36.4400	36.26489	0.175107	1.073847
	35	e_{2i}	24.9000	25.00000	60.3725	59.52498	0.847516	
	10	e_{0i}	0.02900	0.030000	27.5600	26.67813	0.881871	
5	20	e_{1i}	−0.0040	−0.00100	36.2200	35.66813	0.551871	1.455612
	30	e_{2i}	24.7000	23.68813	50.6800	50.65813	0.021871	
	12	e_{0i}	0.02710	0.028000	29.1364	28.48005	0.656352	
6	25	e_{1i}	−0.0055	−0.00500	42.1000	41.88305	0.216952	1.541352
	40	e_{2i}	25.3000	24.50805	68.4400	69.10805	0.668048	

Existing fuel cost and existing emission for three different power generations are compared with its FC^{Est} and E^{Est} and it is found that those values are very close to each other. The new values of coefficients and their related total error are presented in the Table 3 and Table 4. From the results the error obtained by TLBO for all specified output of generating units is less which shows the accurate estimation of the system parameters.

TLP are obtained using Equations (A.2-A.5) for each line and the estimated TLP results are listed in Table 5. The estimated TLP values are compared with base case values and the percentage error is computed using Equation (9). The % error values for transmission lines are almost close to zero except for the lines 24 and 29 for which the TLP values are above 1. The efficiency of transmission lines 11-16, 37 are high as compared to other transmission lines, due to the presence of three winding transformer in the transmission lines. The estimated R, X, B values for each transmission line are presented in Table 5. The estimated transmission loss coefficients using Equations (A.10-A.11) are detailed in the Table 6.

By using existing parameters, economic emission dispatch is carried out and the results are listed in Table 7. The dispatch results are obtained for three different load demands of 275 MW, 284.4 MW and 300 MW. With the new estimated parameters for fuel cost function, emission and loss, the multi-objective ED problem is solved and the results are compared with the existing results. From the results, the percentage deviation of fuel cost, emission and network loss for estimated parameters are slightly elevated as compared to the existing parameters. The % error of all these values reveals the impact of accurate parameter estimation for ED problems.

Test System 3: 62 Bus Indian Utility System

In this case, a 62 bus Indian utility system is considered for estimating the improved parameters by using the proposed algorithm. The chosen test system consists of nineteen power producers, eighty nine (220KV) lines with eleven tap changing transformers.

Table 5: Estimation of transmission line parameters for IEEE 30 bus system (in p.u)

Bus		Estimated			% Error with respect to given data			%Efficiency	
nl	nr	R	X	B	R_e	X_e	B_e	Base Case	TLBO
1	2	0.019100	0.057400	0.02640	0.52	0.17	0.00	97.0336	97.014
1	3	0.045143	0.185000	0.02040	0.13	0.11	0.00	96.7376	96.6021
2	4	0.056500	0.173500	0.01830	0.88	0.12	0.54	97.6514	97.5072
3	4	0.013200	0.037700	0.00416	0.00	0.53	0.95	99.0193	99.0112
2	5	0.047150	0.198100	0.02080	0.11	0.10	0.48	96.5316	96.4118
2	6	0.057894	0.176000	0.01860	0.35	0.17	0.53	96.8020	96.7192
4	6	0.011900	0.041194	0.00446	0.00	0.50	0.89	99.1441	99.1147
5	7	0.045600	0.114000	0.01010	0.87	0.60	0.98	98.9923	98.6166
6	7	0.026672	0.080000	0.00840	0.10	0.73	0.47	99.0280	99.0081
6	8	0.012000	0.041600	0.00470	0.00	0.95	0.67	99.6538	99.3394
6	9	0.000000	0.208000	0.00000	0.00	0.00	0.00	100.000	100.000
6	10	0.000000	0.553000	0.00000	0.00	0.54	0.00	100.000	100.000
9	11	0.000000	0.206000	0.00000	0.00	0.96	0.00	100.000	100.000
9	10	0.000000	0.109000	0.00000	0.00	0.91	0.00	100.000	100.000
4	12	0.000000	0.253000	0.00000	0.00	0.55	0.00	100.000	100.000
12	13	0.000000	0.139800	0.00000	0.00	0.14	0.00	100.000	100.000
12	14	0.123100	0.255600	0.00000	0.00	0.12	0.00	99.0801	99.0584
12	15	0.066196	0.130100	0.00000	0.01	0.23	0.00	98.8170	98.7090
12	16	0.094500	0.198500	0.00000	0.00	0.10	0.00	99.2953	99.2099
14	15	0.220121	0.199500	0.00000	0.40	0.10	0.00	99.6200	99.5142
16	17	0.082396	0.192017	0.00000	0.00	0.15	0.00	99.6856	99.5790
15	18	0.107300	0.218300	0.00000	0.00	0.09	0.00	99.3638	99.2110
18	19	0.063900	0.129090	0.00000	0.00	0.09	0.00	99.8418	99.7114
19	20	0.031000	0.068000	0.00000	1.18	0.00	0.00	99.7729	99.2677
10	20	0.093526	0.207000	0.00000	0.08	0.96	0.00	99.0979	99.0061
10	17	0.032285	0.084200	0.00000	0.35	0.36	0.00	99.7364	99.4325

15	18	0.107300	0.218300	0.00000	0.00	0.09	0.00	99.3638	99.2110
18	19	0.063900	0.129090	0.00000	0.00	0.09	0.00	99.8418	99.7114
19	20	0.031000	0.068000	0.00000	1.18	0.00	0.00	99.7729	99.2677
10	20	0.093526	0.207000	0.00000	0.08	0.96	0.00	99.0979	99.0061
10	17	0.032285	0.084200	0.00000	0.35	0.36	0.00	99.7364	99.4325
10	21	0.034499	0.074500	0.00000	0.86	0.53	0.00	99.3010	99.1052
10	22	0.072300	0.149700	0.00000	0.55	0.13	0.00	99.3219	99.2125
21	22	0.011200	0.023600	0.00000	1.03	0.00	0.00	99.9496	99.8451
15	23	0.099858	0.200000	0.00000	0.14	0.99	0.00	99.3771	99.0043
22	24	0.114842	0.178100	0.00000	0.14	0.50	0.00	99.2247	99.1239
23	24	0.131505	0.268000	0.00000	0.37	0.74	0.00	99.6375	99.1248
24	25	0.188200	0.329000	0.00000	0.16	0.03	0.00	99.4132	99.0121
25	26	0.254154	0.378000	0.00000	0.10	0.53	0.00	98.7084	98.2095
25	27	0.109100	0.208700	0.00000	0.18	0.00	0.00	99.5007	99.4003
28	27	0.000000	0.393000	0.00000	0.00	0.76	0.00	100.000	100.000
27	29	0.219621	0.415000	0.00000	0.08	0.07	0.00	98.6190	98.6171
27	30	0.320200	0.602500	0.00000	0.00	0.03	0.00	97.7849	97.7771
29	30	0.239820	0.453000	0.00000	0.03	0.07	0.00	99.1302	99.0330
8	28	0.063600	0.197967	0.02140	0.00	1.02	0.00	100.000	99.9900
6	28	0.016900	0.059898	0.06480	0.00	0.00	0.31	99.6877	99.6425

nl: start bus of a line; nr: end bus of a line; Re, Xe and Be: Errors in TLP.

Table 6: Estimated B-Coefficients for IEEE 30 bus system

B_{ij} =	0.021757	0.010710	−0.00084	−0.00059	0.000804	0.003558
	0.010710	0.018204	−0.00076	−0.00093	0.000581	0.002856
	−0.00084	−0.00076	0.020189	−0.00920	−0.00927	−0.00636
	−0.00059	−0.00093	−0.00920	0.017105	0.006700	0.003504
	0.000804	0.000581	−0.00927	0.006700	0.016060	−0.00011
	0.003558	0.002856	−0.00636	0.003504	−0.00011	0.025026
B_{i0} =	1.65E−05	0.002078	−0.00351	0.002386	0.001136	0.003069
B_{00} =	0.001409					

Table 7: Economic emission dispatch with improved parameters for IEEE 30 unit system

Demand (MW)	FC^{Exi} (\$/h)	FC^{Est} (\$/h)	% Error	E^{Exi} (Kg/h)	E^{Est} (Kg/h)	% Error	P_L^{Exi} (MW)	P_L^{Est} (MW)	% Error
275.0	568.67	568.74	0.01231	331.71	337.88	1.86006	5.72	5.97	4.37063
283.4	591.51	591.94	0.07270	346.00	348.25	0.65029	5.74	6.06	5.57491
300.0	638.20	640.68	0.38859	390.59	394.16	0.91400	7.43	7.74	4.17227

The bus, line and generator data of the test system are taken from [32] .

The proposed method of determining the improved parameters is applied for the chosen test system and the obtained results are listed in Table 8. For each unit the fuel costs using improved parameters for three different real power outputs are also specified. In comparison with base case values the error is calculated for each generating unit. The obtained error is small which reveals that the estimation of the coefficient is accurate.

Further the estimation of TLP is carried out using A.2-A.5. The obtained TLP values by TLBO algorithm are listed in Table 9. The % error values for each transmission line are close to zero, except for few lines, which shows that the obtained TLP values are accurate. With the improved TLP values, the transmission line efficiency is calculated and is listed in Table 9. It is also inferred fromTable 9 that the efficiency of each transmission line obtained using TLBO is nearly the same as compared to the base case.

The PED is carried out with improved cost coefficients of generators and improved TLP for three different load demands of 2900 MW, 2967 MW, 3200 MW are obtained and the results are presented in Table 10. The estimated transmission loss coefficients using Equations (A.10-A.11) are presented in the Table 11. The results obtained are compared with the existing fuel cost and loss and the % deviation of fuel cost and transmission loss is computed and presented. The obtained results clearly illustrate the impact of improved parameters on ED problem.

POTENTIAL VERIFICATION OF TLBO

Solution Quality

Using the proposed algorithm several runs have been carried out by varying the population size. In this analysis there is no larger difference in the average fitness value above the population size of 10. Hence the population size of 10 is preferred for all estimation process. The simulations are performed to estimate the quadratic cost function coefficients, piecewise quadratic valve point coefficients, emission coefficients, transmission line parameters and transmission loss coefficients using three different test cases. For these cases the existing coefficients at the time of installation are only available, hence the obtained improved parameters are compared with only these values. From the obtained simulation results, the cost coefficients, emission coefficients and transmission line parameters are very close to the existing values and are presented in Tables 2-6 and Tables 8-10. The comparisons of total error in each case clearly indicate that the proposed method provides the best estimate of accurate coefficients.

Table 8: Estimation of quadratic cost function coefficients using TLBO?62 Bus systems

Unit	P (MW)	Coefficients			FC^{Exi} (Rs/h)	FC^{Est} (Rs/h)	Error	Total error
			Existing	TLBO				
	50	a_i	0.0070	0.007000	452.5000	451.9880	0.512035	
1	150	b_i	6.8000	6.805155	1272.500	1272.503	0.003454	1.292177
	300	c_i	95.000	94.23022	2765.000	2765.777	0.776688	
	50	a_i	0.0055	0.005473	243.7500	243.9042	0.154235	
2	200	b_i	4.0000	4.013154	1050.000	1051.133	1.133461	1.403032
	450	c_i	30.000	29.56281	2943.750	2943.865	0.115336	
	50	a_i	0.0055	0.005500	258.7500	258.2589	0.491087	
3	200	b_i	4.0000	4.002228	1065.000	1064.843	0.156884	1.048092
	450	c_i	45.000	44.39751	2958.750	2959.150	0.400120	
	0	a_i	0.0025	0.002500	10.00000	9.569746	0.430254	
4	50	b_i	0.8500	0.850000	58.75000	58.31975	0.430254	1.290761
	100	c_i	10.000	9.569746	120.0000	119.5697	0.430254	
	50	a_i	0.0060	0.006000	265.0000	264.6022	0.397802	
5	150	b_i	4.6000	4.600000	845.0000	844.6022	0.397802	1.193407
	300	c_i	20.000	19.60220	1940.000	1939.602	0.397802	
	50	a_i	0.0055	0.005500	303.7500	303.4052	0.344756	
6	200	b_i	4.0000	4.000000	1110.000	1109.655	0.344756	1.034268
	450	c_i	90.000	89.65524	3003.750	3003.405	0.344756	
	50	a_i	0.0065	0.006500	293.2500	292.8159	0.434064	
7	100	b_i	4.7000	4.700000	577.0000	576.5659	0.434064	1.302191
	200	c_i	42.000	41.56594	1242.000	1241.566	0.434064	
	50	a_i	0.0075	0.007503	314.7500	314.4751	0.274862	
8	250	b_i	5.0000	5.000000	1764.750	1764.653	0.096978	0.830749
	500	c_i	46.000	45.71773	4421.000	4421.459	0.458909	
	0	a_i	0.0085	0.008500	55.00000	54.58443	0.415572	
9	300	b_i	6.0000	6.000000	2620.000	2619.584	0.415572	1.246717
	600	c_i	55.000	54.58443	6715.000	6714.584	0.415572	
	0	a_i	0.0020	0.002000	58.00000	57.36662	0.633381	
10	50	b_i	0.5000	0.500000	88.00000	87.36662	0.633381	1.900144
	100	c_i	58.000	57.36662	128.0000	127.3666	0.633381	
	50	a_i	0.0045	0.004486	156.2500	155.9085	0.341495	
11	100	b_i	1.6000	1.600000	270.0000	269.5571	0.442877	1.396220
	150	c_i	65.000	64.69230	406.2500	405.6382	0.611848	
	0	a_i	0.0025	0.002259	78.00000	77.74735	0.252646	
12	25	b_i	0.8500	0.850000	100.8125	100.409	0.403531	1.512366
	50	c_i	78.000	77.74735	126.7500	125.8938	0.856189	

	50	a_i	0.0050	0.005000	177.5000	176.8509	0.649138	
13	150	b_i	1.8000	1.800000	457.5000	456.8509	0.649138	1.947415
	300	c_i	75.000	74.35086	1065.000	1064.351	0.649138	
	0	a_i	0.0045	0.004452	85.00000	83.87105	1.128948	
14	75	b_i	1.6000	1.617354	230.3125	230.2146	0.097931	1.618938
	150	c_i	85.000	83.87105	426.2500	426.6421	0.392060	
	0	a_i	0.0065	0.006494	80.00000	81.00000	1.000000	
15	250	b_i	4.7000	4.700862	1661.250	1662.108	0.858182	1.859955
	500	c_i	80.000	81.00000	4055.000	4055.002	0.001773	
	50	a_i	0.0045	0.004374	171.2500	170.5123	0.737748	
16	100	b_i	1.4000	1.435546	275.0000	275.0915	0.091535	1.118103
	150	c_i	90.000	87.80097	401.2500	401.5388	0.288821	
	0	a_i	0.0025	0.002468	10.00000	10.00000	0.000000	
17	50	b_i	0.8500	0.843902	58.75000	58.36492	0.385076	1.315574
	100	c_i	10.000	10.00000	120.0000	119.0695	0.930498	
	50	a_i	0.0045	0.004420	116.2500	115.6299	0.620105	
18	150	b_i	1.6000	1.631116	366.2500	367.1468	0.896771	1.698896
	300	c_i	25.000	23.02344	910.0000	910.1820	0.182020	
	100	a_i	0.0080	0.007983	720.0000	718.5346	1.465383	
19	300	b_i	5.5000	5.514826	2460.000	2460.146	0.146243	1.637241
	600	c_i	90.000	87.22119	6270.000	6270.026	0.025615	

Statistical Analysis

The reliability of the algorithm in finding the best solution can be viewed by performing statistical analysis. The convergence property and the robustness characteristics are focused primly for the statistical analysis. The standard statistical parameters of the objective functions such as best, worst and average values in addition to standard deviation and solution iter are also worked out and analyzed and are given in Table 12. The standard deviation value is small which indicates the accuracy of estimate.

Convergence: The TLBO algorithm is implemented on different scale of test systems and for all cases, the convergence characteristics are plotted. The convergence pattern illustrates the searching ability of algorithm.

Figure 2 shows the convergence characteristics of 10 unit multiple fuels system; the proposed approach converged to a good solution within the 50 iterations for the three different fuels. For the IEEE 30 bus and 62 bus Indian utility systems, the convergence graphs are shown in Figure 3.Figure 2 and Figure 3 show that the TLBO has good convergence property, thus resulting in good evaluation value and low error.

Robustness: Many trials with different initial population have been carried out to test the consistency of the TLBO algorithm. The algorithm is executed for 50 trials and the obtained objective function values are presented in Figure 4. The mean value is close to the optimal value for all the cases. This description clears that the TLBO provides great searching ability, higher solution quality and the best estimate.

Observations

- Accurate parameters are estimated for three different test systems involving a practical test system.
- For the first time in the literature, both the parameter estimation and economic dispatch are carried out concurrently.
- The % error estimation of parameters reveals the rate of change in parameters over the years due to physical conditions like aging.
- The results clearly signify the impact of accurate parameters on ED problems on fetching the accurate dispatch results over existing results.

Table 9: Enhanced parameters for Indian Utility 62 Bus system (in p.u)

Bus		Estimated			% Error with respect to given data			%Efficiency	
nl	nr	R	X	B	R_e	X_e	B_e	Base case	TLBO
1	2	0.003040	0.015610	0.014450	0.33	0.26	0.00	99.6854	99.5225
1	4	0.007158	0.104834	0.033970	0.02	0.08	0.00	99.4529	99.4126
1	6	0.004080	0.070416	0.019500	0.73	0.09	0.05	99.7985	99.5990
1	9	0.002280	0.011710	0.010820	0.44	0.26	0.18	99.9574	99.7074
1	10	0.015690	0.015620	0.074427	0.00	0.03	0.00	98.1424	98.0420
1	14	0.005440	0.006240	0.103900	0.73	0.00	0.02	98.8023	98.5559
2	3	0.002875	0.014870	0.013710	0.52	0.00	0.15	99.7694	99.5076
2	6	0.001670	0.008590	0.007950	0.60	0.23	0.00	99.8778	99.6178
3	4	0.003780	0.048259	0.018070	0.79	0.10	0.00	99.8107	99.7705
4	5	0.007120	0.060220	0.033970	0.56	0.08	0.00	99.5645	99.4913
4	14	0.004090	0.055537	0.019510	0.50	0.09	0.00	99.8105	99.8012
4	15	0.004109	0.027387	0.019509	0.03	0.09	0.01	99.5702	99.4160
5	6	0.005747	0.010570	0.003090	0.05	0.06	0.00	99.7869	99.7711
5	8	0.005711	0.095449	0.003090	0.67	0.09	0.00	99.7041	99.7021
6	7	0.000297	0.001557	0.005766	1.00	0.82	0.24	99.9801	99.6785
7	8	0.000485	0.001680	0.086100	1.02	0.00	0.02	99.9619	99.8603
11	10	0.006868	0.020338	0.032514	0.00	0.06	0.02	99.3434	99.3427
11	16	0.014021	0.015620	0.066654	0.28	0.03	0.07	97.7699	97.7640
12	11	0.019010	0.006240	0.090326	0.21	0.03	0.00	98.3704	98.3658
12	13	0.015363	0.079900	0.072920	0.05	0.02	0.00	81.3413	81.3286
12	20	0.019810	0.062595	0.093950	0.00	0.03	0.00	98.1003	98.1001
13	14	0.013110	0.048259	0.062370	0.08	0.04	0.00	97.3945	97.3887
13	17	0.015600	0.060220	0.074125	0.06	0.00	0.03	97.7258	97.7219
14	15	0.005200	0.055537	0.024462	0.00	0.15	0.01	98.8763	98.8754
14	16	0.003920	0.027387	0.018777	1.01	0.20	0.07	99.6615	99.5581
14	18	0.001336	0.010570	0.025577	1.07	0.43	0.01	99.8328	99.6312
14	19	0.007020	0.095449	0.033500	0.71	0.04	0.09	99.0144	99.0070
16	17	0.003400	0.104834	0.065039	0.87	0.11	0.00	99.2834	99.2784
17	21	0.018500	0.070416	0.088153	0.00	0.03	0.01	97.7762	97.7624
20	23	0.020378	0.020338	0.088157	0.21	0.03	0.00	75.0000	74.9171
21	22	0.013707	0.015620	0.065040	0.02	0.02	0.00	98.3203	98.3186
22	23	0.003910	0.006240	0.075120	1.26	0.06	0.05	99.6443	99.4403
23	24	0.003010	0.079900	0.014446	1.31	0.19	0.03	99.5683	99.3626
23	25	0.001253	0.006423	0.005140	0.56	1.18	1.33	99.9154	99.1389
24	41	0.015527	0.048259	0.073710	0.09	0.04	0.00	99.3582	99.3572
24	45	0.012171	0.060220	0.057809	0.16	0.02	0.00	99.4619	99.4618
25	26	0.009390	0.055537	0.044590	0.10	0.04	0.00	98.7969	98.7914
25	27	0.011709	0.027387	0.055650	0.18	0.07	0.00	98.6864	98.6835
25	28	0.010589	0.010570	0.050370	0.29	0.01	0.00	99.3916	99.3901
27	29	0.005280	0.095449	0.025260	0.94	0.01	0.12	99.8378	99.8347
29	30	0.020540	0.104834	0.097580	0.19	0.03	0.05	99.3531	99.3506

30	31	0.009916	0.050900	0.047010	0.04	0.10	0.09	99.5128	99.5060
32	31	0.017830	0.091400	0.084765	0.22	0.44	0.01	98.8139	98.8117
32	34	0.003908	0.020301	0.075147	1.31	0.24	0.02	99.3724	99.1176
32	36	0.003050	0.015650	0.014430	0.13	0.00	0.14	99.9418	99.8413
32	37	0.022000	0.113006	0.104331	0.00	0.00	0.02	99.4598	99.1606
32	46	0.020920	0.107590	0.099370	0.14	0.02	0.00	99.5276	99.3413
33	32	0.016710	0.086090	0.079470	0.30	0.00	0.03	99.3724	99.3573
34	33	0.017370	0.089211	0.082500	0.00	0.01	0.00	99.8829	99.8826
34	35	0.007010	0.035500	0.033234	0.00	1.39	0.02	99.5225	99.0238
34	37	0.019900	0.010215	0.094371	0.00	0.05	0.01	99.4598	99.4168
35	32	0.000354	0.001834	0.006770	1.64	0.33	0.29	99.9683	99.1634
36	46	0.018278	0.093890	0.086720	0.01	0.02	0.00	99.5501	99.5405
37	46	0.001040	0.005354	0.019500	0.00	0.11	1.52	99.7916	99.7816
38	34	0.010720	0.055220	0.051020	0.37	0.05	0.00	99.3325	99.3124
38	37	0.010400	0.053590	0.049500	0.38	0.04	0.00	98.3801	98.3739
39	37	0.002278	0.011710	0.010828	0.52	0.26	0.11	99.7716	99.5176
39	42	0.006810	0.035190	0.032520	0.73	0.09	0.00	99.3313	99.1268
40	30	0.007136	0.036776	0.033940	0.33	0.01	0.09	99.7272	99.6971
40	41	0.006040	0.031000	0.028902	0.82	0.96	0.03	99.6259	99.1229
41	42	0.000756	0.003910	0.014434	0.48	0.00	0.11	99.9392	99.9381
41	45	0.003326	0.017120	0.015876	0.24	0.00	0.15	99.8795	99.8775
42	43	0.009110	0.046960	0.043320	0.11	0.00	0.09	99.7810	99.7710
42	44	0.014120	0.072750	0.067204	0.11	0.04	0.01	99.3497	99.3472
44	59	0.008840	0.045362	0.041910	0.00	0.06	0.00	98.6675	98.6424
46	44	0.016760	0.086064	0.079476	0.00	0.03	0.02	94.3014	94.2190
47	46	0.007920	0.040644	0.037550	0.00	0.14	0.08	98.1599	98.1283
47	48	0.013680	0.070390	0.065040	0.09	0.06	0.00	96.9166	96.7059
48	50	0.000658	0.003353	0.012406	0.30	0.50	0.12	99.8962	99.8812
48	54	0.012511	0.064380	0.059478	0.13	0.05	0.00	99.4306	99.4116
49	48	0.003646	0.018745	0.069340	0.38	0.19	0.06	99.4691	99.4208
49	50	0.006630	0.034390	0.031800	1.04	0.12	0.00	99.5397	99.1185
51	53	0.011825	0.061100	0.056440	0.63	0.03	0.00	97.9189	97.5027
51	54	0.004058	0.020800	0.019253	0.29	0.48	0.24	99.6324	99.5800
51	55	0.014110	0.072740	0.067210	0.14	0.05	0.00	98.9161	98.9033
52	53	0.011288	0.058100	0.053690	0.28	0.09	0.00	99.2357	99.1319
52	61	0.011200	0.057883	0.053650	0.62	0.05	0.07	99.6475	99.2441

55	58	0.006650	0.034390	0.031790	0.75	0.12	0.03	99.8991	99.4928
56	58	0.002572	0.013251	0.012290	0.69	0.37	0.00	99.6806	99.1762
57	56	0.001520	0.007814	0.007225	0.00	0.20	0.07	99.8051	99.6063
57	58	0.001830	0.009360	0.008670	0.00	0.32	0.00	99.4844	99.2851
58	12	0.012088	0.062180	0.057450	0.18	0.06	0.00	96.4591	96.3550
58	60	0.004110	0.021130	0.019505	0.00	0.00	0.02	99.4220	99.1219
58	61	0.003333	0.017184	0.063590	0.50	0.21	0.00	99.3535	99.1403
59	61	0.009180	0.047300	0.043720	0.43	0.11	0.00	98.6172	98.5121
60	12	0.013600	0.070090	0.064747	0.37	0.04	0.00	96.4907	96.3813
60	61	0.002415	0.012492	0.046250	1.04	0.23	0.00	99.7277	99.1250
61	62	0.014930	0.076997	0.071090	0.40	0.02	0.03	98.0600	98.0401
62	25	0.013800	0.071020	0.065620	0.22	0.06	0.00	99.0983	99.0711

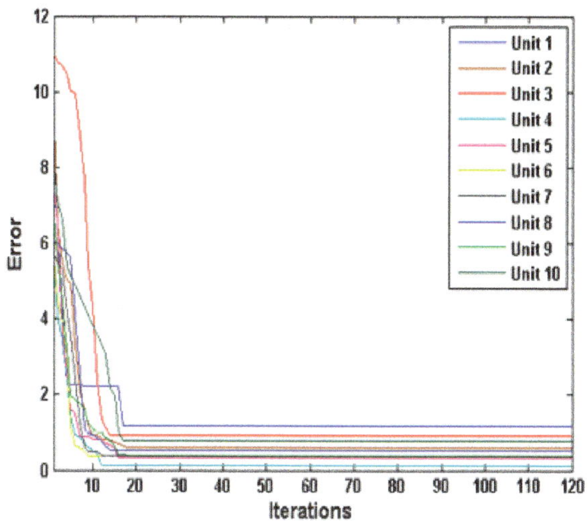

(a)

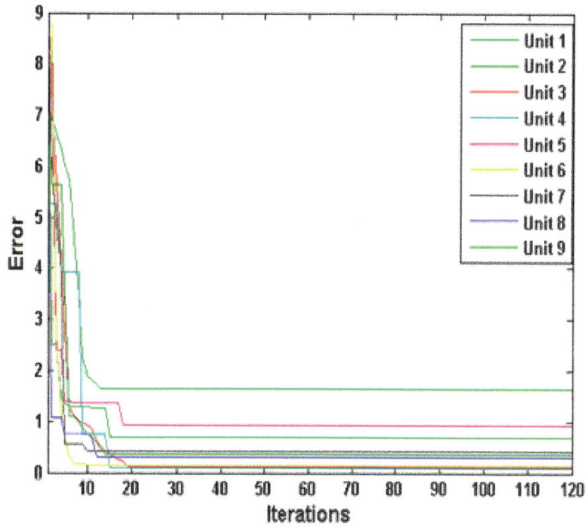

(b)

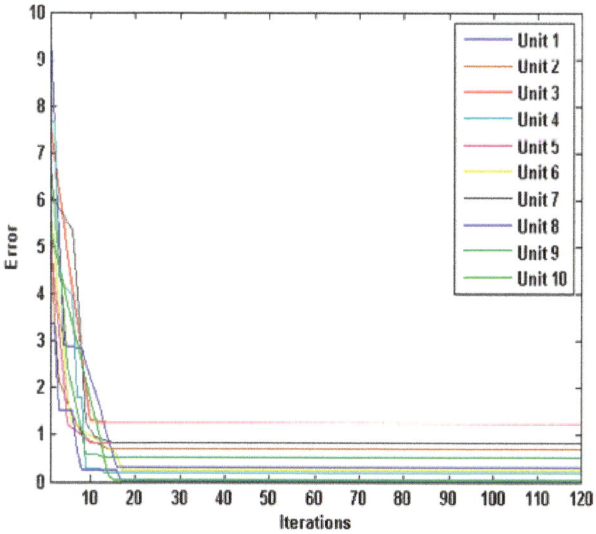

(c)

Figure 2: Cost coefficients convergence curve (a) fuel 1; (b) fuel 2 and (c) fuel 3.

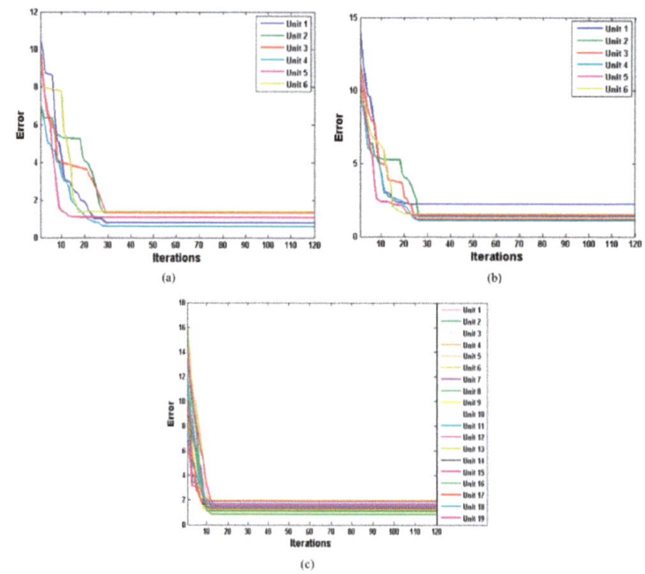

Figure 3: Convergence curve: (a) IEEE 30 bus cost coefficients; (b) IEEE 30 bus emission coefficients; (c) 62 bus cost coefficients.

Table 10: Economic dispatch with improved parameters for Indian utility 62 bus system

Demand (MW)	FC^{Exi} (Rs/h)	FC^{Est} (Rs/h)	%Error	P_L^{Exi} (MW)	P_L^{Est} (MW)	%Error
2967	15651.73	15722.82	0.45420	94.61327	96.25311	1.73321
2900	15522.58	15890.59	2.37080	80.88011	83.48667	3.22275
3200	17493.59	18030.99	3.07198	92.90000	96.90000	4.30571

CONCLUSION

In this paper, the TLBO algorithm is successfully implemented for both the operation of parameter estimations and ED solution. The operational aspects of the generators such as valve-point loading, emission, multiple fuel options and transmission loss are considered. The three different test systems involving a practical test system are selected for this estimation and dispatch problem. The results obtained by satisfying all the constraints for all cases by the proposed algorithm are always comparable or better than the other methods. It is revealed that the

Table 11: Estimated B-Coefficients for 19 unti Indian utility system

	0.00986	0.009897	0.00907	0.009803	0.007403	0.005591	-0.00375	-0.00357	-0.00712	-0.00739	-0.00685	-0.00524	-0.00402	-0.00361	-0.00332	-0.00245	-0.00345	-0.00078	-0.00209
	0.009897	0.012015	0.009786	0.009822	0.007155	0.005615	-0.00452	-0.00406	-0.0089	-0.00845	-0.0071	-0.00445	-0.00396	-0.00328	-0.00352	-0.00247	-0.00354	-0.00056	-0.00266
	0.00907	0.009786	0.011605	0.009023	0.007611	0.005583	-0.00344	-0.0034	-0.0063	-0.00695	-0.00683	-0.00576	-0.00415	-0.00387	-0.00332	-0.00254	-0.0035	-0.00097	-0.00191
	0.009803	0.009822	0.009023	0.011861	0.007375	0.005549	-0.00378	-0.00361	-0.00712	-0.00741	-0.00689	-0.00531	-0.00407	-0.00365	-0.00336	-0.00249	-0.00349	-0.0008	-0.00213
	0.007403	0.007155	0.007611	0.007375	0.008391	0.005861	-0.00242	-0.0027	-0.00409	-0.00558	-0.0064	-0.00651	-0.00408	-0.00411	-0.00294	-0.00237	-0.00325	-0.00106	-0.00114
	0.005591	0.005615	0.005583	0.005549	0.005861	0.008169	-0.00197	-0.002	-0.00446	-0.00527	-0.00539	-0.00468	-0.0032	-0.00306	-0.00242	-0.00182	-0.00261	-0.00058	-0.00097
	-0.00375	-0.00452	--0.00344	-0.00378	-0.00242	-0.00197	0.011636	0.008303	0.001709	0.001041	0.000789	0.000278	7.96E-5	-8.88 E-5	-0.00054	-0.00056	-0.00036	-0.00049	-5.79 E-5
	-0.00357	-0.00406	-0.0034	-0.00361	-0.0027	-0.002	0.008303	0.007873	0.001208	0.000746	0.000705	0.000677	0.000185	6.13E-5	-0.00035	-0.00037	-0.00021	-0.00039	0.000284
	-0.00712	-0.0089	-0.0063	-0.00712	-0.00409	-0.00446	0.001709	0.001208	0.026718	0.019649	0.012062	0.000424	0.002014	0.000593	0.000351	-0.00091	0.000828	-0.00046	-0.00347
B_{ij}=	-0.00739	-0.00845	-0.00695	-0.00741	-0.00558	-0.00527	0.001041	0.000746	0.019649	0.027091	0.013242	0.003525	0.002457	0.00157	0.000264	-0.00104	0.000842	-0.00107	-0.00264
	-0.00685	-0.0071	-0.00683	-0.00689	-0.0064	-0.00539	0.000789	0.000705	0.012062	0.013242	0.014385	0.006924	0.003204	0.00285	0.000609	-0.00071	0.001243	-0.00138	-0.00095
	-0.00524	-0.00445	-0.00576	-0.00531	-0.00651	-0.00468	0.000278	0.000677	0.000424	0.003525	0.006924	0.014532	0.005329	0.005915	0.001883	0.000655	0.002711	-0.00049	0.001063
	-0.00402	-0.00396	-0.00415	-0.00407	-0.00408	-0.0032	7.96E-5	0.000185	0.002014	0.002457	0.003204	0.005329	0.016443	0.014409	0.007107	0.003341	0.008906	0.001521	0.001778
	-0.00361	-0.00328	-0.00387	-0.00365	-0.00411	-0.00306	-8.88 E-5	6.13 E-5	0.000593	0.00157	0.00285	0.005915	0.014409	0.01593	0.006442	0.003321	0.008457	0.001908	0.000824
	-0.00332	-0.00352	-0.00332	-0.00336	-0.00294	-0.00242	-0.00054	-0.00035	0.000351	0.000264	0.000609	0.001883	0.007107	0.006442	0.011956	0.005508	0.010445	0.002346	0.004795
	-0.00245	-0.00247	-0.00254	-0.00249	-0.00237	-0.00182	-0.00056	-0.00037	-0.00091	-0.00104	-0.00071	0.000655	0.003341	0.003321	0.005508	0.01144	0.004947	0.003536	0.003576
	-0.00345	-0.00354	-0.0035	-0.00349	-0.00325	-0.00261	-0.00036	-0.00021	0.000828	0.000842	0.001243	0.002711	0.008906	0.008457	0.010445	0.004947	0.012607	0.002354	0.003416
	-0.00078	-0.00056	-0.00097	-0.0008	-0.00106	-0.00058	-0.00049	-0.00039	-0.00046	-0.00107	-0.00138	-0.00049	0.001521	0.001908	0.002346	0.003536	0.002354	0.010055	-0.0034
	-0.00209	-0.00266	-0.00191	-0.00213	-0.00114	-0.00097	-5.79E-5	0.000284	-0.00347	-0.00264	-0.00095	0.001063	0.001778	0.000824	0.004795	0.003576	0.003416	-0.0034	0.027605
	0.00986	0.009897	0.00907	0.009803	0.007403	0.005591	-0.00375	-0.00357	-0.00712	-0.00739	-0.00685	-0.00524	-0.00402	-0.00361	-0.00332	-0.00245	-0.00345	-0.00078	-0.00209
B_{i0}=	-0.0138	-0.01803	-0.01178	-0.01403	-0.00669	-0.00763	0.016537	0.011528	0.053561	0.037919	0.014398	-0.01547	-0.00371	-0.01247	-0.00268	-0.00539	-0.00256	-0.00102	-0.01581
B_{00}=	0.384294																		

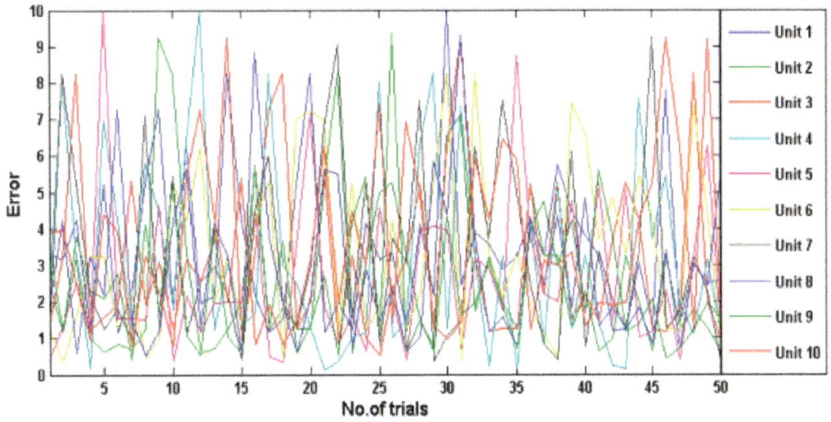
Error
No.of trials
Unit 1
Unit 2
Unit 3
Unit 4
Unit 5
Unit 6
Unit 7
Unit 8
Unit 9
Unit 10

(a)

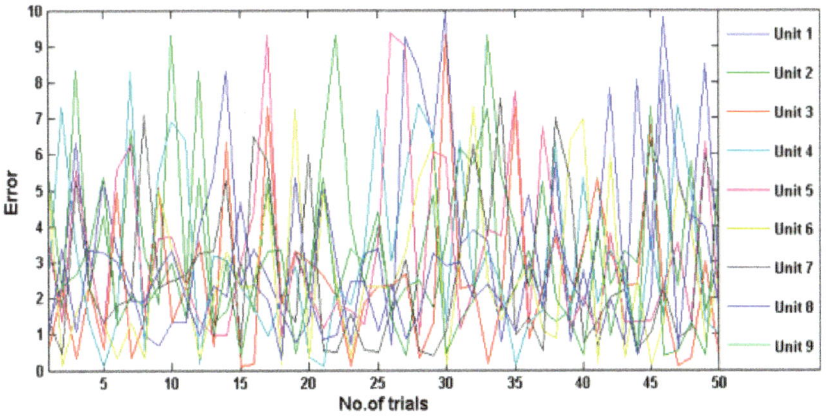
Error
No.of trials
Unit 1
Unit 2
Unit 3
Unit 4
Unit 5
Unit 6
Unit 7
Unit 8
Unit 9

(b)

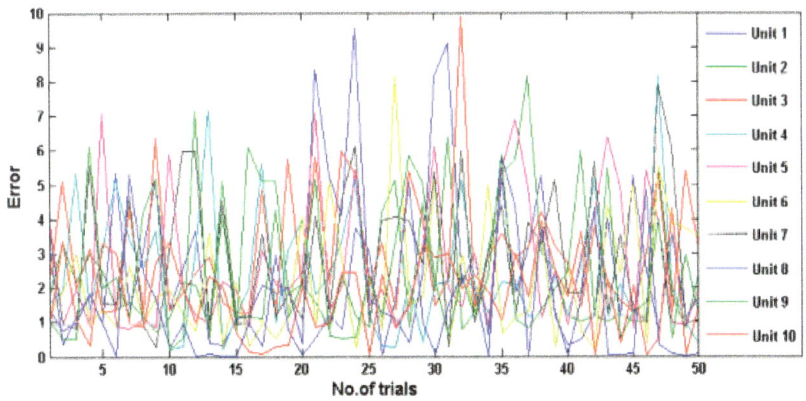
Error
No.of trials
Unit 1
Unit 2
Unit 3
Unit 4
Unit 5
Unit 6
Unit 7
Unit 8
Unit 9
Unit 10

(c)

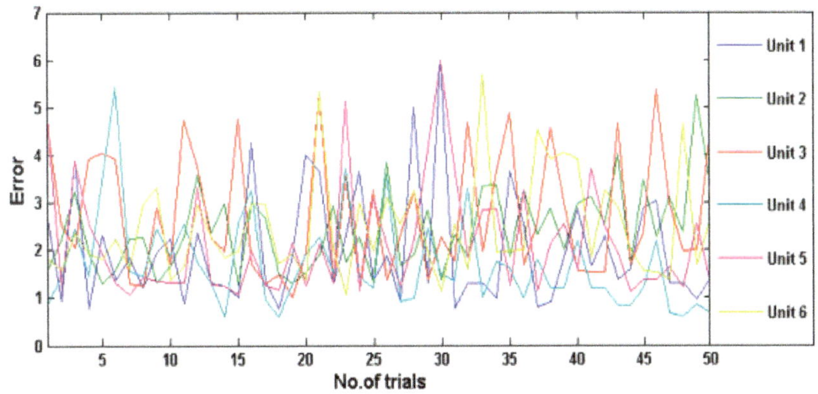
Error
No.of trials
Unit 1
Unit 2
Unit 3
Unit 4
Unit 5
Unit 6

(d)

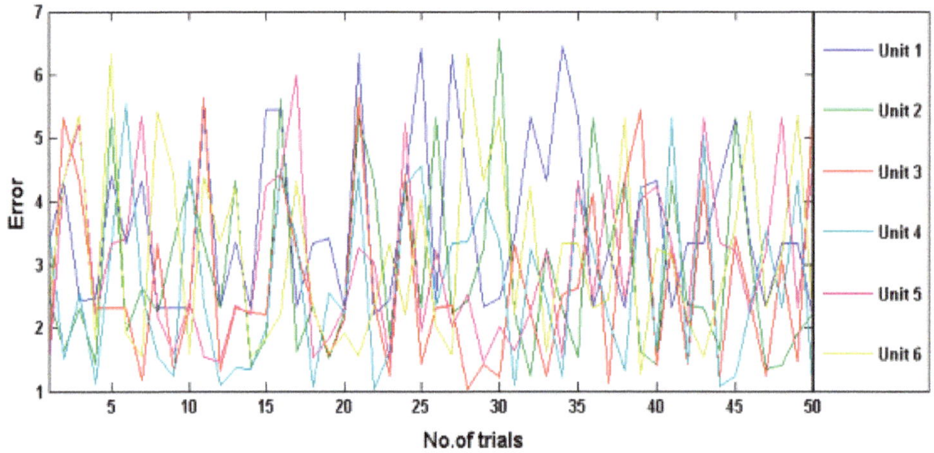

(e)

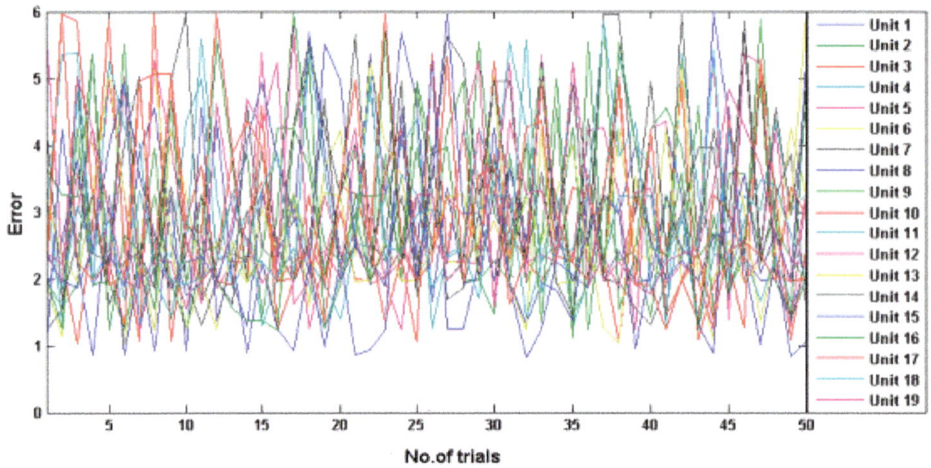

(f)

Figure 4: Robustness characteristics: (a) 10 unit system for fuel 1; (b) 10 unit system for fuel 2; (c) 10 unit system for fuel 3; (d) IEEE 30 bus system for cost function; (e) IEEE 30 bus system for emission function; (f) 62 bus system for cost function.

Table 12: Performance indices by TLBO for various test systems

Test Systems	Units	Best	Worst	Mean	Solution Iter	Standard Deviation
	1	1.168802	9.657846	1.42197	8	1.3247
		0.697156	8.712287	1.51550	17	1.0365
		1.639115	7.88428	1.83870	16	1.9987
	2	0.317203	9.145168	0.18740	9	1.3658
		0.605175	8.510824	1.03250	7	0.6241
		0.936732	8.824971	0.18740	13	0.0874
	3	0.712539	7.533577	2.21470	17	2.1571
10-unit		0.116458	9.139696	0.28540	3	0.1547
		0.122282	8.834306	0.51020	8	0.3251
	4	0.103847	9.809413	1.89980	7	1.6624
		1.261771	5.867897	0.54770	18	0.3198
		0.333505	7.993594	1.69851	5	1.4873
	5	0.936775	9.220367	0.54102	21	0.2163
		0.208988	9.779851	1.80871	15	1.1024
	6	0.371620	9.690017	0.19546	6	0.1687

		0.139008	8.923714	2.10245	14	1.8547
		0.828617	8.869472	2.27845	11	1.9687
		0.383942	9.140136	1.15842	3	0.8554
	7	0.410536	9.799390	1.58751	16	1.3258
		0.251265	7.988590	1.19847	14	0.9962
		0.528055	6.047550	0.89870	17	0.5521
	8	0.313608	9.711690	3.18520	16	2.9385
		0.819262	7.804830	1.84680	11	1.6824
		0.398192	8.165560	1.71870	4	1.2365
	9	0.016320	8.000195	0.95230	6	0.5021
		0.529615	5.556480	1.94770	12	1.6324
		0.781727	6.100437	0.57450	20	0.3219
	10	0.043682	8.128070	1.53210	8	1.3278
		0.368904	6.657846	1.68970	8	1.3014
	1	0.763171	6.001400	1.50080	30	1.2017
	2	1.283524	5.254700	2.41872	28	1.3855
	3	1.286063	5.3606300	2.75281	29	1.3127
6-unit	4	0.586500	5.4471000	1.70420	30	0.9844
	5	1.071758	5.1210000	2.14751	25	1.1795
	6	1.369166	5.6981000	2.47828	20	1.4641
	1	1.292177	5.6987000	2.78649	12	1.2251
	2	1.403032	5.9874000	2.80534	9	1.2379
	3	1.048092	5.6987000	1.51007	10	1.1566
	4	1.290761	4.8746000	1.80493	9	1.3982
	5	1.193407	5.6856000	2.80835	11	1.3069
	6	1.034268	5.2587000	1.51712	12	1.1984
	7	1.302191	5.9655000	2.83551	10	1.4249
	8	0.830749	4.2387000	1.11428	12	1.2307
	9	1.246717	4.5214000	1.77940	8	1.2839
19-unit	10	1.900144	5.8945000	2.48302	10	2.2183
	11	1.396220	5.5854000	1.98359	12	1.4946
	12	1.512366	5.2399000	2.12035	12	1.6254
	13	1.947415	5.9658000	2.78960	11	2.0128
	14	1.618938	5.9574000	2.80584	8	1.7851
	15	1.859955	5.5304000	2.83611	11	1.9962
	16	1.118103	5.8512000	1.92709	10	1.2147
	17	1.315574	5.7415000	2.13565	11	1.4210
	18	1.698896	5.6987000	2.23442	11	1.8745
	19	1.637241	5.5718000	2.77713	13	1.7721

TLBO possesses better convergence characteristics and robustness. The proposed TLBO approach has shown merits such as better results, easy implementation for the accurate parameter estimation for PED problems. It can be concluded that the estimated parameters obtained show that the proposed method can be used as a very accurate tool for estimating the fuel cost, emission and loss coefficients.

ACKNOWLEDGEMENTS

The authors gratefully acknowledge the authorities of Annamalai University, Annamalai Nagar, Tamilnadu, India for the facilities provided to carry out this research work.

Cite This Paper

Selvaraj Durai, Srikrishna Subramanian, Sivarajan Ganesan (2015) Preferred Economic Dispatch of Thermal Power Units. *Journal of Power and Energy Engineering*,**03**,47-69. doi:10.4236/jpee.2015.311005

Appendix Parameter Estimation Problems

The vector equation describing the relationships between the measured values y, the unknown parameters m, the system matrix D and the residual due to the change in values r are as follows,

$$y = Dm + r \tag{A1}$$

The parameter estimation problem for estimating the cost coefficients a, b, c, e, f, emission coefficients e_0, e_1, e_2 and long line transmission parameters R, X, B are formulated as follows: Let, k-Number of measurements, y(k)-Measured value of the kth measurement, D-System matrix

$$m = [a,b,c,e,f]^T, [e_0,e_1,e_2]^T, [R,X,B]^T, \begin{bmatrix} B_{ij} & B_{0i} \\ B_{0i} & B_{00} \end{bmatrix}^T$$

r-Error vector that relates y(k) to m. There are k equations available to represent the parameter estimation that forms measurement matrix y(k). TLBO is used as an estimator to find the unknown values of [m]. These estimates are used to recalculate parameters using consequent equations at each time step. The calculation procedure for the evaluation of parameters such as a, b, c, e, f; e_0, e_1, e_2 and B_{ij}, B_{0i}, B_{00} are detailed in Section 2.2, and the transmission line parameters such as R, X, B are presented in (A.2-A.5).

$$v_s \cos\delta = V_r - B^{Est} X^{Est} V_r + R^{Est} I_r \cos\varphi_r + X^{Est} I_r \sin\varphi_r, \underline{\Delta a}, say \tag{A.2}$$

$$v_s \sin\delta = B^{Est} X^{Est} V_r + X^{Est} I_r \cos\varphi_r - R^{Est} I_r \sin\varphi_r, \underline{\Delta} b, say \tag{A.3}$$

$$I_s(\cos(\delta - \varphi_s)) = -R^{Est} B^{2Est} V_r + I_r \cos\varphi_r - X^{Est} B^{Est} I_r \cos\varphi_r - R^{Est} B^{Est} Ir \sin\varphi_r, \underline{\Delta} c, say \tag{A.4}$$

$$I_s(\sin(\delta - \varphi_s)) = 2B^{Est} V_r - X^{Est} B^{2Est} V_r - I_r \sin\varphi_r + R^{Est} B^{Est} I_r \cos\varphi_r + X^{Est} B^{Est} I_r \sin\varphi_r, \underline{\Delta} d, say \tag{A.5}$$

Next, using the Newton-Raphson (NR) method, the following four non-linear equations of the form F(x) = 0, where $F=(f_1,f_2,f_3,f_4)^T$ and $x = (X, \overline{R}, B)$ are solved for the unknown X, R, B,

$$f_1(x) = -V_s^2 + a^2 + b^2 = 0 \tag{A.6}$$

$$f_2(x) = -\tan\delta + b/a = 0 \tag{A.7}$$

$$f_3(x) = -I_s^2 + c^2 + d^2 = 0 \tag{A.8}$$

$$f_4(x) = -\tan(\delta - \varphi_s) + d/c = 0 \tag{A.9}$$

Transmission Line Loss Parameters The network loss is a function of transmission line parameters and network configurations. Thus accuracy in the transmission line model and B coefficients are necessary. The transmission line model consists of R, X and B parameters and to determine the existing network loss these parameters must be accurate. The estimation of transmission line parameters is mathematically formulated as an optimization problem (Indulkar and Ramalingam, 2008) and the accurate TLP can be determined by using an optimization technique.

$$TLP_i^{Est} = \left[R_i^{Est}, X_i^{Est}, B_i^{Est}\right] \quad i = 1,2,\cdots,TL \tag{A.10}$$

$R_i^{Est}, X_i^{Est}, B_i^{Est}$ are the estimated transmission line parameters

The Kron's coefficients (B_{ij}^{Est}, B_{0i}^{Est} and B_{00}^{Est}) are dependent of network parameters and configuration.

$$KC_i^{Est} = \begin{bmatrix} B_{ij}^{Est} & B_{0i}^{Est} \\ B_{0i}^{Est} & B_{00}^{Est} \end{bmatrix} = f\left(TLP_i^{Est}\right) \tag{A.11}$$

KC_i^{Est} — Estimated loss coefficients

The ideal transmission loss can be determined using KC_i^{Est} by Equation (5).

REFERENCES

1. El-Hawary, M.E. and Mansour, S.Y. (1982) Performance Evaluation of Parameter Estimation Algorithms for Economic Operation of Power Systems. IEEE Transaction on Power Apparatus Systems, 101, 574-582. http://dx.doi.org/10.1109/TPAS.1982.317270.
2. Soliman, S.A., Emam, S.E.A. and Christensen, G.S. (1991) Optimization of the Optimal Coefficients of Non-Monotonically Increasing Incremental Cost Curves.

Electric Power System Research, 21, 99-106. http://dx.doi.org/10.1016/0378-7796(91)90023-G.

3. Liang, Z.X. and Glover, J.D. (1991) Improved Cost Functions for Economic Dispatch Computations. IEEE Transactions of Power System, 6, 821-829. http://dx.doi.org/10.1109/59.76731.
4. Chen, H.Y.K. and Postel, C.E. (1986) On-Line Parameters Identification of Input Output Curves for Thermal Units. IEEE Transaction Power System, 1, 221-224. http://dx.doi.org/10.1109/TPWRS.1986.4334933.
5. Grainger, J. and Stevenson, W. (1994) Power System Analysis. McGraw-Hill., New York.
6. Chan, S.M. (1993) Computing Overhead Line Parameters. Computer Applications and Power, 6, 43-45. http://dx.doi.org/10.1109/67.180436
7. Dommel, H.W. (1985) Overhead Line Parameters from Handbook Formulas and Computer Programs. IEEE Transactions on Power Apparatus and Systems, 4, 366-372. http://dx.doi.org/10.1109/TPAS.1985.319051.
8. Thorp, J.S., Phadke, A.G., Horowitz, S.H. and Begovic, M.M. (1988) Some Applications of Phasor Measurements To Adaptive Protection. IEEE Transaction Power System, 3, 791-798. http://dx.doi.org/10.1109/59.192936.
9. Chen, C.S., Liu, C.W. and Jiang, J.A. (2002) A New Adaptive PMU Based Protection Scheme for Transposed/ Untransposed Parallel Transmission Lines. IEEE Transactions on Power Delivery, 17, 395-404. http://dx.doi.org/10.1109/61.997906.
10. 10. Kim, I.-D., and Aggarwal, R.K. (2006) A Study on the On-Line Measurement of Transmission Line Impedances for Improved Relaying Protection. Electric Power and Energy System, 28, 359-366. http://dx.doi.org/10.1016/j.ijepes.2006.01.002
11. Liao, Y. and Kezunovic, M. (2009) Online Optimal Transmission Line Parameter Estimation for Relaying Applications. IEEE Transactions on Power Delivery, 24, 96-102. http://dx.doi.org/10.1109/TPWRD.2008.2002875
12. 12. Sivanagaraju, G., Chakrabarti, S. and Srivastava, S.C. (2014) Uncertainty in Transmission Line Parameters Estimation and Impact on Line Current Differential Protection. IEEE Transactions on Instrumentation and Measurement, 63, 1496-1504. http://dx.doi.org/10.1109/TIM.2013.2292276.
13. Wagenaars, P., Wouters, P.A.A.F., van der Wielen, P.C.J.M. and Steennis, E.F. (2010) Measurement of Transmission Line Parameters of Three-Core Power Cables with Common Earth Screen. IET Science Measurement and Technology, 4, 146-155. http://dx.doi.org/10.1049/iet-smt.2009.0062.
14. Indulkar, C.S. and Ramalingam, K. (2008) Estimation of Transmission Line Parameters from Measurements. Electric Power and Energy Systems, 30, 337-342. http://dx.doi.org/10.1016/j.ijepes.2007.08.003.
15. Al-Kandari, A.M. and El-Naggar, K.M. (2006) A Genetic-Based Algorithm for Optimal Estimation of Input-Output Curve Parameters of Thermal Power Plants. Electrical Engineering, 89, 585-590. http://dx.doi.org/10.1007/s00202-006-0047-x.
16. El-Naggar, K.M. and Alrashidi, M.R. and Al-Othman, A.K. (2009) Estimating the Input-Output Parameters of Thermal Power Plants Using PSO. Energy Conversion and Management, 50, 1767-1772. http://dx.doi.org/10.1016/j.enconman.2009.03.019.
17. Alrashidi, M.R., El-Naggar, K.M. and Al-Othman, A.K. (2009) Particle Swarm Optimization Based Approach for Estimating the Fuel-Cost Function Parameters of Thermal Power Plants with Valve Loading Effects. Electric Power Components and Systems, 37, 1219-1230. http://dx.doi.org/10.1080/15325000902993589..
18. Sonmez, Y. (2013) Estimation of Fuel Cost Curve Parameters for Thermal Power Plants Using the ABC Algorithm. Turkish Journal of Electrical Engineering and Computer Science, 21, 1827-1841. http://dx.doi.org/10.3906/elk-1203-10.
19. Sinha, A.K. and Mandal, J.K. (1999) Hierarchical Dynamic State Estimator Using ANN-Based Dynamic Load Prediction. IET Generation Transmission and Distribution, 146, 541-549. http://dx.doi.org/10.1049/ip-gtd:19990462.
20. Kumar, D.M.V. and Srivastava, S.C. (1999) Power System State Forecasting Using Artificial Neural Networks. Electric Machine and Power System, 27, 653-664.http://dx.doi.org/10.1080/073135699269091.
21. Lin, J.M., Huang, S.J. and Shih, K.R. (2003) Application of Sliding Surface Enhanced Fuzzy Control for Dynamic State Estimation of a Power System. IEEE Transaction Power System, 18, 570-577. http://dx.doi.org/10.1109/TPWRS.2003.810894.
22. Bhuvaneswari, R., Subramanian, S. and Madhu, A. (2008) A Novel State Estimation Based on Minimum Errors between Measurements Using Ant Colony Optimization Technique. International Journal of Electric Engineering, 15, 457-568.
23. Sakthivel, V.P., Bhuvaneswari, R. and Subramanian, S. (2010) Design Optimization of Three-Phase Energy efficient Induction Motor Using Adaptive Bacterial Foraging Algorithm. International Journal of Computation and Mathematics in Electrical and Electronics Engineering, 29, 699-726.
24. Sakthivel, V.P., Bhuvaneswari, R. and Subramanian, S. (2010) Non-Intrusive Efficiency Estimation Method for Energy Auditing and Management of In-Service Induction Motor Using Bacterial Foraging Algorithm. IET Electric Power Applications, 4, 579-590. http://dx.doi.org/10.1049/iet-epa.2009.0313
25.
26. Sakthivel, V.P., Bhuvaneswari, R. and Subramanian, S. (2011) An Accurate and Economical Approach for induction motor Field Efficiency Estimation Using Bacterial Foraging Algorithm. Measurement, 44, 674-684. http://dx.doi.org/10.1016/j.measurement.2010.12.008.
27. Rao, R.V., Savsani, V.J. and Vakharia, D.P. (2011) Teaching-Learning-Based Optimization: A Novel Method for Mechanical Design Optimization

Problems. Computer Aided Design, 43, 303-315. http://dx.doi.org/10.1016/j.cad.2010.12.015.

28. Rao, R.V. and Waghmare, G.G. (2014) Complex Constrained Design Optimisation Using an Elitist Teaching-Learning-Based Optimisation. International Journal of Metaheuristics, 3, 81-102. http://dx.doi.org/10.1504/IJMHEUR.2014.058863.

29. Rao, R.V., Savsani, V.J. and Vakharia, D.P. (2012) Teaching-Learning-Based Optimization: An Optimization Method for Continuous Non-Linear Large Scale Problems. Information Science, 183, 1-15. http://dx.doi.org/10.1016/j.ins.2011.08.006.

30. Rao, R.V., Savsani, V.J. and Vakharia, D.P. (2011) Teaching-Learning-Based Optimization: A Novel Method for Mechanical Design Optimization Problems. Computer-Aided Design, 43, 303-315. http://dx.doi.org/10.1016/j.cad.2010.12.015.

31. Durai, S., Subramanian, S. and Ganesan, S. (2015) Improved Parameters for Economic Dispatch Problems by Teaching Learning Optimization. Electric Power and Energy System, 67, 11-24. http://dx.doi.org/10.1016/j.ijepes.2014.11.010[Paper reference:2]

32. Chiang, C.-L. (2005) Improved Genetic Algorithm for Power Economic Dispatch of Units with Valve-Point Effects and Multiple Fuels. IEEE Transaction Power System, 20, 1690-1699. http://dx.doi.org/10.1109/TPWRS.2005.857924.

33. Tamilnadu Electricity Board Statistics at a Glance (1999-2000), Compiled by Planning Wing of Tamilnadu Electricity Board, Chennai, India.[Paper reference:2]

CHAPTER 18

A PILOT-SCALE STUDY OF DYNAMIC RESPONSE SCENARIOS FOR THE FLEXIBLE OPERATION OF POST-COMBUSTION CO2 CAPTURE

Paul Tait[a], Bill Buschle[a], Ilja Ausner[b], Prashant Valluri[a], Marc Wehrli[b], Mathieu Lucquiaud[a]

[a]University of Edinburgh, Institute for Materials and Processes, School of Engineering, Sanderson Building, The King's Buildings, Mayfield Road, Edinburgh EH9 3JL, United Kingdom

[b]Sulzer Chemtech Ltd, Sulzer-Allee 48, 8404 Winterthur, Switzerland

ABSTRACT

The ability to operate flexibly is critical for the future implementation of carbon capture and storage (CCS) in thermal power plants. A dynamic test campaign examines the response of a CO_2 absorption/desorption pilot-scale plant to realistic changes in flue gas flow rates and steam supply, representative of the operation of a Natural gas combined cycle (NGCC) plant fitted with post-combustion capture. Five scenarios, demonstrating the operational flexibility that is likely to be encountered in an energy market with significant penetration from intermittent renewables, are presented, with 30% monoethanolamine (MEA) as the absorbing solvent. It complements a wider effort on dynamic modelling of these systems where a lack of dynamic plant data has been reported.

The campaign focuses on analysing critical plant parameters of the response of the pilot plant to a gas turbine shutdown, a gas turbine startup and three enhanced operational flexibility scenarios, including two for power output maximisation and one for frequency response with a rapid increase of steam supply to the reboiler. The campaign also demonstrates the use of continuous in situ solvent lean loading measurement with the use of a novel online continuous liquid sensor.

It confirms that no significant barriers to flexible operation of amine post-combustion capture are found, although there remains scope for the improvement of plant response. Solvent inventory and circulation times are found to have a significant effect on capture rate during certain dynamic operations. A large solvent inventory increases total circulation times, which can result in additional time being required for the plant to return to steady state following a perturbation. The plant is forced to operate with a non-optimal capture rate while the solvent loading at the absorber inlet stabilises is identified as a potential impact.

Citation: Paul Tait, Bill Buschle, Ilja Ausner, Prashant Valluri, Marc Wehrli, Mathieu Lucquiaud, A pilot-scale study of dynamic response scenarios for the flexible operation of post-combustion CO2 capture, International Journal of Greenhouse Gas Control, Volume 48, Part 2, May 2016, Pages 216-233, ISSN 1750-5836, http://dx.doi.org/10.1016/j.ijggc.2015.12.009.

Use of interim solvent storage and continuous online measurement of solvent CO_2loading, combined with comprehensive knowledge of liquid circulation times and potential mixing effects, are suggested as methods for improving plant response to dynamic operation, thereby increasing CCS plant flexibility.

Keywords

CO_2 capture; Post-combustion; NGCC; Flexibility; Dynamic operation; Pilot-plant

INTRODUCTION

In any energy generation portfolio with considerable input from renewable sources such as wind and solar, natural gas or coal-fired thermal power stations will be required to meet variable demand during periods when these intermittent sources are not generating electricity (Pöyry, 2009). As investment in low-carbon renewable technologies continues, the dispatch patterns of thermal power stations will become less regular, with rapid ramping, startup and shutdown operations based on changing weather patterns becoming commonplace (Murray, 2013).

If carbon emission targets are to be met at an acceptable cost to society, thermal power stations must be fitted with carbon capture and storage (CCS) technology. For example, an analysis of possible power sector decarbonisation scenarios for the UK predicts that without CCS, the cost of delivering climate change targets will double from 1% to 2% of GDP (ETI, 2015). The latest IPCC report (IPCC, 2014) states that in 6/10 predicted scenarios, it is not possible to mitigate climate change to within +2 °C with no deployment of CCS and shows significant increase in the costs of mitigation if the technology is not available.

Of the available technologies for CCS in power generation, post-combustion capture (PCC) with amines is a mature and established technology, with an operational full-scale facility on 120 MW coal fired plant at Boundary Dam, Canada, a planned 385 MW project on a natural gas-fired plant at Peterhead, Scotland and numerous pilot-scale facilities worldwide (GCCSI, 2014). However, the majority of research carried out on PCC at the time of writing has assumed steady-state operation with a consistent power plant output and flue gas flow rate/composition. When predicted plant dispatch patterns for electricity generation scenarios which include significant penetration from intermittent renewable sources are considered, it is increasingly likely that dynamic and flexible operation will be highly relevant.

There is a large body of work on the dynamic modelling of PCC; however, there remains a lack of dynamic plant data, leading to many studies being unvalidated or validated using steady-state data, as discussed in an extensive review by Bui et al. (2014). Faber et al. (2011) and Bui et al. (2013) have proposed implementing a series of step-changes in order to obtain dynamic data for model validation. This method has been used successfully to validate the dynamic simulation of PCC for coal (Flø et al., 2014) and NGCC (Flø et al., 2015) flue gas. While step-changes in a single plant variable are useful for the purposes of validation, they are unlikely to be representative of how a real PCC plant will operate under transient conditions. Additionally, a model validated with data from one pilot facility may not agree with one which is validated with data obtained from another, as differences in plant design, liquid inventory and solvent properties can significantly affect dynamic operation (Kvamsdal et al., 2011).

Several modelling studies which investigate realistic scenarios and enhanced plant flexibility are present in the literature. Haines and Davison (2014) have shown the effect of adding flexibility via a rapid "stripper stop" operation, providing a rapid boost of power to the grid by decreasing the flow of steam extracted from the IP/LP crossover. Mac Dowell and Shah (2014) analyse several strategies which could increase profitability for coal plant operators via flexible PCC operation. Ceccarelli et al. (2014) provide an in-depth analysis of load-following, startup and shutdown operations for a PCC retrofit on CCGT plant, with a focus on reducing CO_2 emissions. However, none of these studies are validated with dynamic plant data.

Pilot-scale test campaigns on realistic dynamic operating scenarios are necessary to obtain more data for model validation and provide a more complete understanding of how CCS plants respond to variations in generation plant output. This knowledge allows researchers to devise operating strategies and develop hardware and software capabilities to increase flexibility, allowing PCC to cope with rapidly-changing generation plant output.

In this test campaign the response of a CO_2 absorption/desorption pilot-scale plant to five dynamic operating scenarios is examined. Changes in the flow rate of solvent, flue gas and steam supply are designed to be proportional to, and representative of the operation of a natural gas combined cycle (NGCC) plant fitted with PCC. The focus is on NGCC plants since they offer favourable dynamic operating capability, with short times for startup, ramping and load response, and are likely to play a significant role in providing dispatchable electricity infills for intermittency management in future energy generation mixes. Although current carbon budgets suggest that gas CCS will be important (ETI, 2015) and that a gradual phase-out of coal generation is necessary between now and 2030 in order to meet emissions targets (CCC, 2013), the outcomes could also be applicable to PCC on coal, which is a very similar process to PCC on gas but with less stringent ramping requirements and higher CO_2 concentration in the flue gas.

The pilot-scale facilities of Sulzer Chemtech in Winterthur, Switzerland are configured for atmospheric pressure post-combustion capture and are complemented, in some of the dynamic scenarios, by continuous in situ solvent lean loading measurement with the use of a novel online continuous liquid sensor. An overview of the test facility, measurement techniques and metrics is provided, five relevant dynamic operating scenarios are presented, and the results of the implementation of each scenario during the test campaign are discussed. Full dynamic data obtained from these scenarios are supplied as an attachment to the electronic version of this paper.

CARBON DIOXIDE ABSORPTION/DESORPTION TEST FACILITY

Plant Description

The pilot plant used in this study, represented schematically in Fig. 1, is controlled and monitored using a combination of Labview software and discrete panel-mounted PID microcontrollers. Synthetic flue gas, composed of N_2 and CO_2, is used. Nitrogen is fed to the facility from a tank upstream of a gas blower. After the blower, a mass flow controller determines the rate at which CO_2 is fed to the gas line. For this test campaign, an inlet CO_2 concentration representative of NGCC exhaust (4.3%, v/v) is achieved with a cascade PID control system of the CO2 mass flow controller. The flue gas is heated to an inlet temperature of approx. 45 °C, before entering the absorber column. After exiting the absorber, flue gas is analysed for CO_2 content in order to determine the capture rate, before being cooled and recycled by feeding back to the blower. While not truly representative of real PCC plant operation, this reduces nitrogen usage, and ensures that the majority of the gas stream becomes rapidly saturated, reducing water losses via evaporation in the absorber.

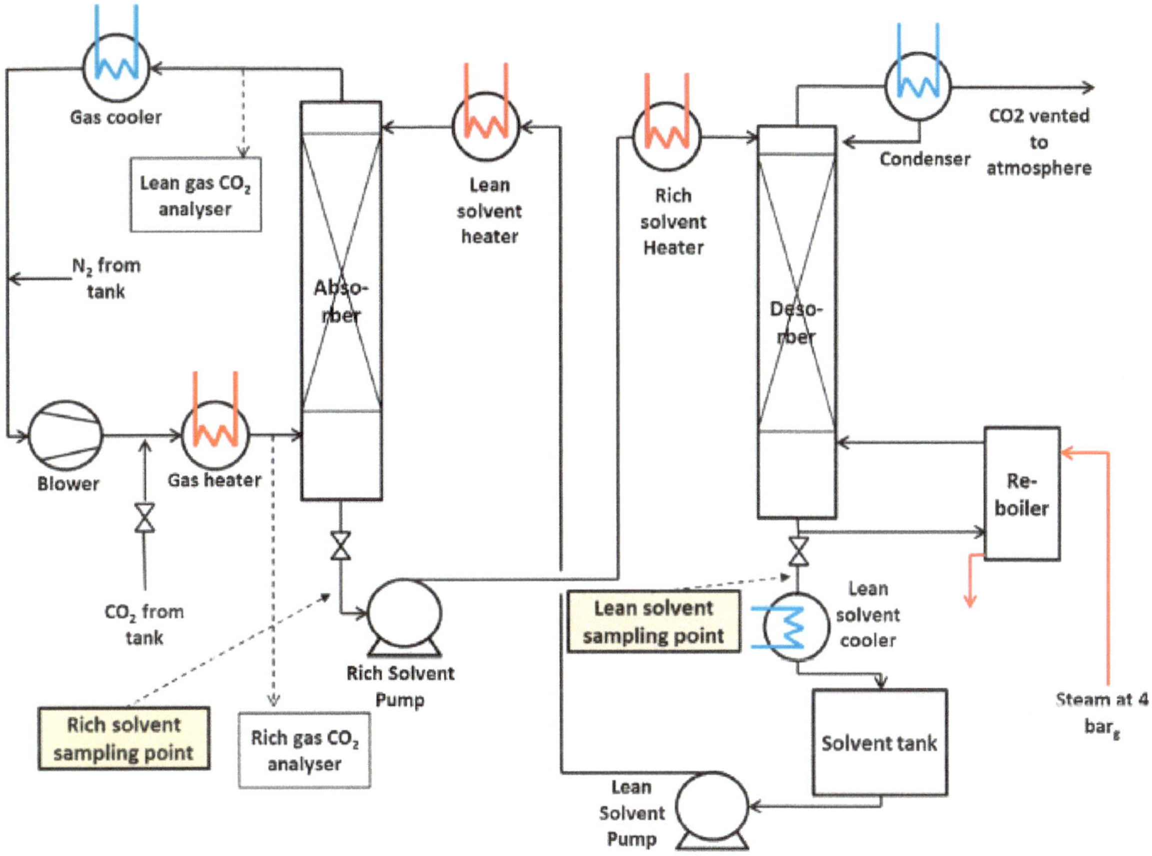

Figure 1: Absorption/desorption test facility process flow diagram.

The solvent tank can hold up to 500 l, and 116.8 kg neat monoethanolamine (MEA) and 270.5 kg deionised H_2O were introduced to produce a solution of 30.16% (w/w) MEA. During plant operation, with solvent circulating in the pipework and vessels, the solvent tank inventory generally remained at around 10% of its maximum level. At the flow rates used in this experimental work, between 40 min and 113 min is required for the entire solvent inventory to circulate once through the plant. The associated effect on plant response is discussed in detail later in this paper.

The main solvent pump circulates liquid to the top of the absorber, where it is heated to approx. 40 °C. The intent is to reproduce the effect of a cross-flow solvent heat exchanger. The absorber sump level is maintained using a Varibell control valve. Upon exiting the absorber, a secondary pump circulates the liquid to the top of the desorber, which is held at a pressure of approx. 1.8 bar.

Steam at a pressure of 5 bar (absolute) is used for solvent regeneration and other heating, for example, mimicking the effect of a cross-flow heat exchanger. CO_2 and water vapour exit the top of the desorber, where the water is condensed and CO_2 vented to atmosphere.

The concentrations of CO_2 at the absorber inlet and outlet are monitored using infra-red (IR) CO_2 analysers. Lean and rich solvent loading are determined by manual sampling at the desorber and absorber outlet, followed by benchtop titration analysis. At standard operating conditions (Table 2) solvent takes approx. 40 min to circulate through the plant, leading to time delays between e.g. a lean sample being taken and this discrete "packet" of solvent reaching the absorber inlet. This effect is exacerbated further at lower flow rates and must be taken into consideration when relating variations in loading to those in capture rate.

The absorber has an inner diameter of 158 mm and contained 6.92 m of Sulzer Mellapak 250.X packing. The desorber has an inner diameter of 350 mm and contained 5.00 m of Sulzer Mellapak 500.X packing. Absorber dimensions are comparable to other pilot plants, as illustrated in Table 1.

Table 1: Comparison of pilot-scale test facilities (Seibert et al., 2011, Kvamsdal et al., 2011, Notz et al., 2012 and Rabensteiner et al., 2014)

Pilot facility	Absorber inner diameter (mm)	Height of packed bed (m)	Packing type	Has cross-heat exchanger Y/N
University of Texas at Austin, Separations Research Program	427	3.07 × 2	Raschig Super-Pak 250	Y
NTNU/SINTEF, VOCC (Validation Of Carbon Capture)	500	5.4	Sulzer Mellapak 2X	Y
University of Stuttgart	125	4.2	Sulzer Mellapak 250.Y	Y
Graz University of Technology, Dürnrohr power station	150	12	Raschig Super-Pak 250	Y
Sulzer Chemtech absorption/desorption test facility, this work	158	6.92	Sulzer Mellapak 250.X	N

The facility in this work is used for a range of applications and is not designed specifically for post-combustion capture at atmospheric pressure. Unlike other pilot plants reported inTable 1, which utilise a cross-heat exchanger to increase the temperature of the rich solvent, the pilot plant used in this study does not have this capability so the effect is simulated using separate heat exchangers. Using a setpoint of 20 °C below the outlet temperature, a rich solvent heater increases the liquid temperature at the desorber inlet. The solvent is cooled to room temperature before entering the main solvent tank, then heated to approx. 40 °C before entering the absorber. A PID control loop with steam supply is used to control the solvent temperature at absorber and desorber inlets. Temperature settings are selected to be representative of the typical temperature pinch of a cross flow heat exchanger (Table 2).

Table 2: Summary of baseload operating conditions

Controlled variable	Value
Gas flow rate at absorber inlet (N m^3/h)	120.5
Gas inlet temperature (°C)	46.14
Inlet gas CO_2 concentration (%, v/v)	4.27
Steam flow rate (kg/h)	19.5
Liquid flow rate (l/h)	344.4
Steam pressure (bar_a)	4.0
Desorber pressure (bar_a)	1.80
Liquid inlet temperature, absorber (°C)	40.05
Liquid inlet temperature, desorber (°C)	104.07

Measured parameter	Value
CO_2 capture rate (%)	89.7
Reboiler duty (GJ/tCO_2)	3.96
L/G ratio (l/m^3)	2.86
Lean solvent loading (mol amine/mol CO_2)	0.232
Rich solvent loading (mol amine/mol CO_2)	0.345

TEST CAMPAIGN METHODOLOGY AND PREPARATION

Determination of Baseload Operating Conditions

The purpose of the test campaign is to implement dynamic post-combustion capture scenarios using a test facility configuration which approximates a PCC pilot plant. A reference "baseload" operation scenario, which all other scenarios are carried out in comparison to, is first determined.

Thirty percent (w/w) MEA solution is used as a liquid absorbent, since it is commonly used as a benchmark for pilot-scale facilities worldwide (Fitzgerald et al., 2014, Hamborg et al., 2014 and Artanto et al., 2012).

A desired baseload capture rate of 90% is selected in order to remain consistent with other pilot-scale studies

(Rabensteiner et al., 2014, Carey et al., 2013 and Mejdell et al., 2011).

Previous work on natural gas flue gas capture (Ystad et al., 2012) indicates that for 30% (w/w) MEA, the liquid to gas flow ratio (*L*/*G*) which results in the lowest energy expenditure per ton CO_2 captured is approx. 1.5l solvent per m^3 gas. Flue gas flow rate is kept constant at 115 m^3/h. With further addition of CO_2 at a rate of approx. 8.1 kg/h and an inlet CO_2 concentration of 4.3% (v/v) (moist), an *f*-factor of 1.85 Pa$^{0.5}$ is obtained. Liquid flow rate is varied to obtain *L*/*G* ratios between 2.4 and 3.6 l/N m^3. For each ratio, at a desorber pressure of 0.8 bar$_g$, the flow rate of steam to the reboiler at 4 bar$_g$ is increased until a CO_2 capture rate of 90% is achieved. A minimum reboiler duty of 3.96 GJ/tCO_2 for this plant is observed at an *L*/*G* ratio of 2.858 l/m^3. Typical reboiler duty values for NGCC base case studies have been reported between 3.4 and 4.04 GJ/tCO_2(Jordal et al., 2012).

It should be noted that the lean loading is low in comparison to what would be expected in a capture plant which is purpose-built and optimised for NGCC flue gas, due to a lack of packing height in the absorber compared to current modelling studies (Biliyok et al., 2014 and Rabensteiner et al., 2014). In order to compensate and achieve 90% capture, a lower lean loading is used to enhance the driving force for CO_2 absorption and the liquid flow rate is increased. This results in a similarly low rich loading, with the effect being exacerbated as the increased liquid flow rate results in a smaller difference in loading between the liquid inlet and outlet. An alternative strategy would be to accept a lower CO_2capture rate while maintaining more representative values of *L*/*G* ratio, lean loading and rich loading, but due to equipment limitations caused by the turndown ratio of the liquid pump and steam inlet valve, further reductions in flow rate would result in considerable flow instability, especially during dynamic scenarios which require additional turndown. Increasing the gas flow rate is also not an option, as the blower has a maximum flow rate of around 130 N m^3/h due to the pressure drop of the packing and liquid holdup.

Additionally, the energy input to the absorber and desorber solvent trim heaters is not accounted for, due to the absence of steam flow rate measurement to each heat exchanger. It is possible that the reboiler inlet liquid temperature is unrealistically high, so less energy is required to provide sensible heat, and if a cross-heat exchanger were installed a higher reboiler duty would be observed. This seems likely, as the figure of 3.96 GJ/tCO_2 is still within the reported range of typical values, with a liquid flow rate which is much higher than optimal.

A full list of conditions is summarised in Table 2.

To allow for comparison with real plant ramping events and future pilot plant campaigns, all flow rates and ramp rates in the dynamic scenarios investigated are subsequently referred to as percentages of this baseload condition.

Due to the low loadings, the solvent retains unused capacity of approximately 0.155 mol CO_2/mol free amine at the bottom of the absorber. This results in a significant driving force in this location during baseload operation. During dynamic scenarios which involve shutdown or reduction of stripping steam, this additional capacity will allow the plant to continue to capture CO_2 for longer periods of time than if the plant were operated with a rich loading which is closer to equilibrium with the inlet gas CO_2 concentration.

Plant Lag Times

The mass inertia of the solvent inventory has an effect on the dynamic response of the plant. Solvent circulation time is obviously dependent on total solvent inventory, which is limited in this study to as little as practically possible. Circulation times at several different liquid flow rates are estimated by inducing a step-change in main pump output, and observing the time required for this to produce a liquid level change of 10% at several points in the liquid line. The solvent pump reaches its new setpoint within approx. 10 s and does not contribute significantly to the overall circulation time. The results are summarised in Table 3. Circulation times between each individual section of the plant, and relevant circulation times at 50% of solvent baseload flow rate are estimated using the measured values and indicated by *. Circulation times through sumps are estimated using (circulation time = liquid holdup/volumetric flow rate) (Fig. 2).

Table 3: Summary of plant solvent circulation times

Liquid flow rate (as percentage of baseload)	Circulation time from main solvent pump to top of absorber sump (min:sec)	Circulation time from top of absorber sump to top of desorber sump (min:sec)	Circulation time from top of desorber sump to top of 500 l liquid tank (min:sec)	Circulation time from top of liquid tank to main solvent pump (min:sec)	Total circulation time required (min:sec)	Circulation time from lean sample point to rich sample point*(min:sec)
100	1:07	6:03	8:35	21:10	36:55	30:17*
75	1:20	7:25	10:36	32:17	51:38	43:38*
50*	1:49*	8:40*	14:25*	48:58*	73:52*	63:11*
30	2:20	12:18	19:40	79:35	113:08	97:50*

Estimated values based on other measurements.

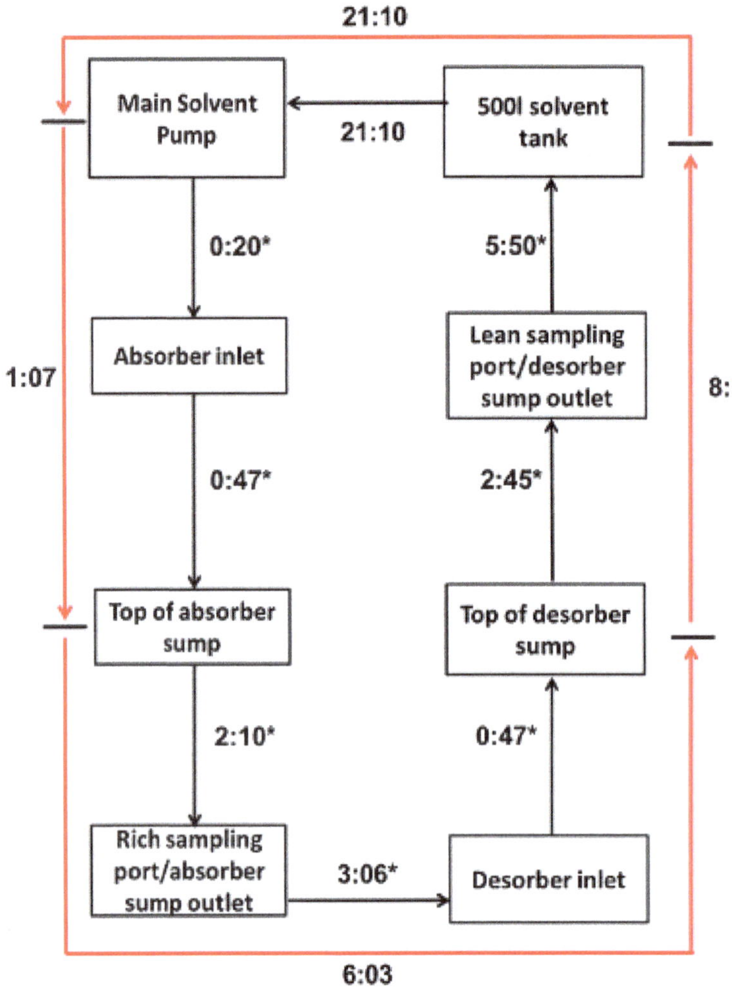

Figure 2: Block diagram of plant solvent circulation times (in min:sec) for baseload solvent flow rate. * indicates estimated value based on other measurements.

Solvent circulation times have a direct impact on the dynamic response of the capture plant and affect the time delay between when a process parameter is changed and when the impact of that change is observed in solvent loading and plant capture rate. For example, after a change in steam flow rate to the reboiler, it takes approximately 30 min (Table 3) at 100% baseload flow rate for the change in lean solvent loading to be observed at the rich solvent sampling point based on solvent circulation times. The amount of solvent in the vessel sumps and solvent tank, and solvent mixing also has an impact on the dynamic response of the plant (Table 4).

Table 4: Estimated circulation times (in min:sec) between relevant plant locations.

	100% of baseload	75% of baseload	50% of baseload	30% of baseload
Main solvent pump – absorber inlet	0:20*	0:27*	0:40*	0:55*
Absorber inlet – top of absorber sump	0:47*	0:53*	1:09*	1:25*
Top of absorber sump – rich sampling port (absorber sump outlet)	2:10*	2:55*	3:24*	5:08*
Rich sampling port (absorber sump outlet) – desorber inlet	3:06*	3:37*	4:07*	5:45*
Desorber inlet – top of desorber sump	0:47*	0:53*	1:09*	1:25*
Top of desorber sump – lean sampling port (desorber sump outlet)	2:45*	3:30*	5:25*	9:03*
Lean sampling port (desorber sump outlet) – top of 500l solvent tank	5:50*	7:06*	9:00*	10:37*
Top of 500 l solvent tank – main solvent pump	21:10*	32:17*	48:58*	79:35*
Total circulation time	36:55	51:38	73:52*	113:53

Rationales for the Choice of Dynamic Scenarios

Five dynamic scenarios are carried out, with the intent of representing to real plant operation as much as practically possible. They are separated into two categories:

Shutdown and Startup Scenarios

1. Gas Turbine Shutdown: The shutdown sequence represents a realistic scenario based on the shutdown sequence of a modern NGCC plant. The gas turbine ramp rates in this and the gas turbine startup scenario are characteristic of a SIEMENS STG5-4000F (Eisfeld and Feldmüller, 2013), and are detailed in Sections 5.1 and 5.2 respectively. The method by which the plant is shut down directly impacts its performance upon restart.Ceccarelli et al. (2014) model two generation/capture plant shutdown cases, one of which is implemented in this work. Liquid and gas flow rates are ramped down simultaneously, maintaining a constant *L*/*G* flow ratio over the course of the shutdown operation until both reach 30% of baseload. Gas flow continues to ramp down to zero, while liquid flow is maintained at 30% to make use of the plant›s cooling duty, resulting in a cooler solvent and promoting CO_2 absorption upon startup. There is, however, some trade-off in implementing this strategy, as the continued circulation causes lean and rich solvent loadings to converge due to mixing effects. This may affect CO_2 absorption performance when the plant is restarted.
2. Gas Turbine Startup: Plant startup is simulated to follow a shutdown event in which the flow of solvent has been maintained at 30% of baseload, causing solvent temperature to be reduced to <35 °C, and lean and rich loading to begin to converge due to mixing. Before the startup sequence is initiated, liquid flow rate is stabilised at 30% of baseload flow rate. Ramp rates for the startup operation are based on an NGCC plant using a state-of-the-art SIEMENS STG5-4000F gas turbine (Eisfeld and Feldmüller, 2013), as with the gas-turbine shutdown scenario, and are provided in Section 5.2. Liquid flow rate is ramped proportionally to that of the gas, maintaining a constant *L*/*G* flow ratio.

Enhanced Operational Flexibility Scenarios

1. Power Output Maximisation by Capture Plant Decoupling

The flow of steam to the reboiler and the flow of flue gas entering the absorber column are rapidly stopped. This represents a bypass of the capture plant, where steam for solvent regeneration is redirected to the combined cycle, the CO_2compressors are shut down and the flue gas is vented directly to atmosphere after the heat recovery steam generator (HRSG). Solvent flow rate is reduced to 50% to reduce the power consumption of the pumps and the extent to which the lean and rich solvent loading will average, due to mixing effects. This mode of operation increases flow rate of steam to the low-pressure (LP) turbine and hence power plant electricity output. It is a useful capability during instances when the selling price of electricity is high, rapidly increasing the amount of energy available for dispatch to the grid (Haines and Davison, 2014). It should be noted that the viability of this scenario depends on carbon cost, electricity selling price and specific regulations regarding emissions costs. Flue gas venting is uneconomical when CO_2 prices are high in comparison to the selling price of electricity.

2. Power Output Maximisation by Reboiler Steam Decoupling Only

This scenario rapidly stops the flow of steam to the reboiler, while flue gas continues to flow through the absorber. This method of operation is similar to capture plant decoupling, but could also be utilised in power plants, where the ability to bypass the absorber has not been implemented or if some level of capture is still desirable. Again, solvent flow rate is reduced to 50% for the reasons stated above.

3. Frequency Response Increase by Rapid Reboiler Steam Flow Increase

In this operation, the flow of steam to the reboiler is rapidly increased to reach 200% of the baseload value. It could be used in power plants as a complementary method of rapidly decreasing plant electricity output where steam extraction from the combined cycle is ramped up as much as practically possible, in response to a requirement to maintain grid frequency within acceptable limits. A doubling of steam extraction from the combined cycle of a gas power plant is consistent with maintaining a minimum steam flow rate at the inlet of the low pressure turbine for blade cooling. With the increase in steam flow rate there exists a risk of accelerating solvent thermal degradation due to the creation of "hot spots" on the heat exchanger, but this will be plant/solvent specific and could be mitigated via plant design or solvent control methods.

For all data presented, the dynamic scenario is initiated at $t = 0$ min. To illustrate that the plant is initially operating at steady state, data was logged for a period of several minutes prior to the first dynamic perturbation. Everything before $t = 0$ is referred to in terms of negative values of time. However, upon analysing the data after the completion of the test campaign, it appears that steady-state operation is not always achieved.

Solvent Circulation Times and Estimation of Real-Time Solvent Working Capacity

To effectively understand the impact of dynamic scenarios on solvent CO_2 loading at the absorber inlet and outlet, time-shifting is utilised in order to estimate the time at which each "packet" of lean or rich solvent taken for analysis will reach the absorber packed bed inlet or outlet, respectively. The continuous online solvent measurement is also time-shifted using the same method. This is illustrated in Fig. 8, Fig. 10, Fig. 12 and Fig. 14. It is possible to estimate when each discrete packet of solvent sampled at the lean inlet reaches the absorber, by estimating the total solvent inventory between the lean sampling port and the absorber inlet and dividing by the flow rate. It is also possible to determine when each rich solvent sample reaches the base of the packed bed by estimating the time required to flow through the absorber sump. The forward time-shifting of the lean loading measurements does not account for potential mixing effects in the main solvent tank, nor does the backwards time-shifting of rich loading measurements account for mixing effects in the absorber sump, as both rely on the assumption of plug flow. This time-shifting method has been applied to lean and rich titration measurements and continuous lean loading measurement values in Fig. 8, Fig. 10, Fig. 12 and Fig. 14, in order to more effectively illustrate how the CO_2 capture rate responds to changes in solvent lean loading. The driving force for CO_2 absorption by the working solvent is dependent on its inlet CO_2 loading, and shifting the lean loading measurements to the absorber inlet shows more clearly how the capture rate and lean loading are related. However, the time-shifting method is unable to account for potential mixing effects in the solvent tank and pipework.

Since the times that the rich and lean samples taken for bench measurements generally do not match, the real-time solvent working capacity (defined as the difference in solvent loading between the absorber inlet and absorber outlet at any given point in time) is estimated using the capture rate and flow rates of CO_2 and MEA. This can be done using two methods:

1. Each lean loading bench measurement is time-shifted forward to the time where it enters the absorber inlet, rich loading at the base of the packing is then estimated using the average capture rate and flow rates of CO_2 and MEA over the following minute

$$\alpha_{\text{rich,calculated}} = \alpha_{\text{lean,measured}} + \left(\frac{\dot{N}_{CO_2} \times \eta_{CO_2}}{\dot{N}_{MEA}} \right)$$

2. Each rich loading bench measurement is time-shifted backwards to the time where it reaches the base of the absorber packing, lean loading at the absorber inlet is then estimated using the average capture rates and flow rates of CO_2 and MEA over the previous minute.

$$\alpha_{\text{lean,calculated}} = \alpha_{\text{rich,measured}} - \left(\frac{\dot{N}_{CO_2} \times \eta_{CO_2}}{\dot{N}_{MEA}} \right)$$

Uncertainty Analysis of Titration Measurements

Based on the analysis of 14 triplicate measurement sets, 42 measurements in total, the expanded repeatability ($k = 2$) is estimated to be ±4.54% of the measured value. The accuracy estimate is ±0.63% of the measured value, based on a single measurement of a 1.110 wt% CO_2 standard. Therefore, the expanded uncertainty estimation for the method is ±5.2% relative (rounded to 2 significant figures) with a coverage factor of $k = 2$. A Bland–Altman plot shows that 40 of 42 measurements have a relative deviation from the mean value which is within 5.2% or lower, and thus the confidence level is approximately 95%. In future test campaigns the uncertainty analysis should be improved, utilising multiple measurements of CO_2 standards which have a range of different CO_2concentrations. Additionally, it is assumed that the nominal amine concentration remains constant throughout the test campaign, when in reality it may vary due to water losses and/or solvent degradation.

Relationship between absorber temperature bulge and real-time solvent working capacity

Calculated real-time solvent capacity measurements are plotted against the temperature at the hottest part of the absorber during baseload operation, at 4.65 m of packing height (Fig. 3).

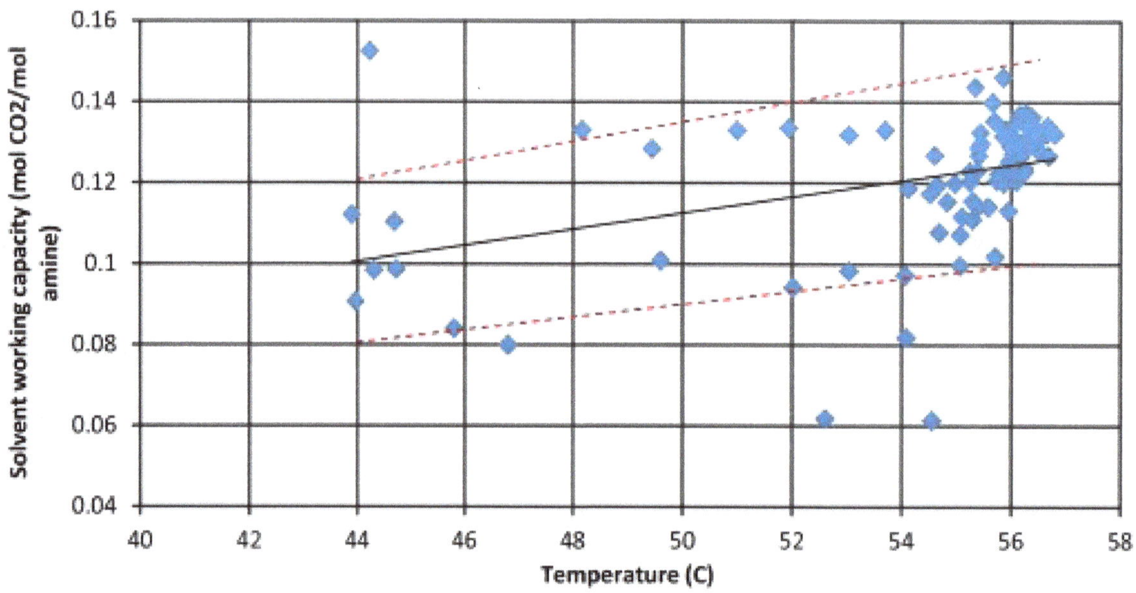

Figure 3: Correlation between calculated solvent working capacity and temperature at packing height of 4.65 m, all scenarios.

The majority of the solvent loading capacity data points fall within 20% of the value predicted by the correlation. It should be noted that solvent capacity values are calculated using the time-shifting estimation discussed in Section 3.4, and mixing effects could have resulted in errors in the calculation. The solvent working capacity may also respond faster to changes in plant operation than the temperature bulge, due to the thermal inertia of the absorber column. This results in a poor correlation, and as such the temperature bulge is not thought to be an effective indicator of solvent working capacity during transient operations.

ONLINE SOLVENT LOADING SENSOR

A novel liquid sensor which differs from previous online measurement techniques (e.g.van Eckeveld et al., 2014) has been developed in-house at Edinburgh University. The sensor is used for continuous measurement of solvent lean loading in three of the dynamic scenarios.

It uses in situ measurements of physical properties to determine solvent loading and amine concentration, using proprietary intellectual property which is under development at the time of writing. Knowledge of the solvent CO_2 loading in real-time can provide CCS plant operators with more frequent solvent analysis measurements, enabling them to react rapidly to perturbations or disturbances and develop automated process control systems for dynamic plant operation. Anecdotal evidence suggests that the established state-of the art for determination of solvent CO_2 loading is accomplished via the use of an online auto-titrator which can take around 20 min or longer. As demonstrated in this work, plant conditions can change significantly within this period of time, therefore a more rapid method of solvent loading determination must be developed in order to incorporate solvent physical properties into future plant control systems.

The online sensor provides a continuous measurement of solvent loading and amine concentration. Further discussion of the operating principle is not possible as the technology is considered commercially sensitive.

Two locations are tested in this study, as illustrated in Fig. 4 and Fig. 5, where the online solvent lean loading sensor takes a slipstream of liquid from the lean solvent line.

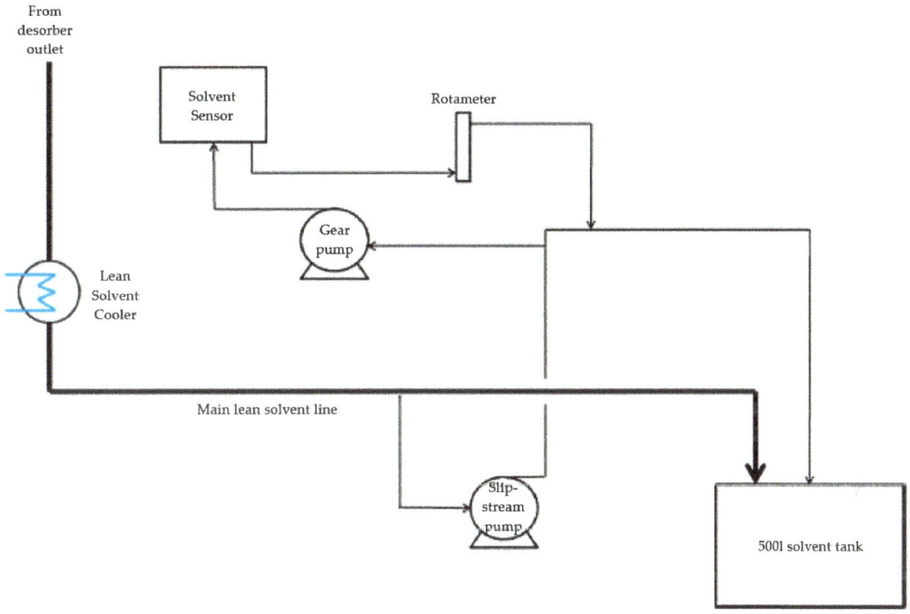

Figure 4: Liquid loop for online solvent sensor – location upstream of solvent tank.

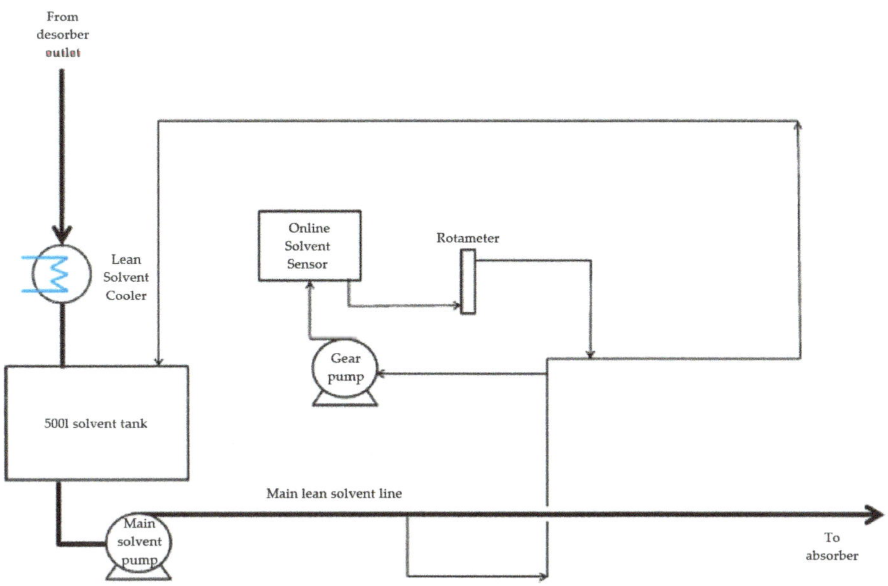

Figure 5: Liquid loop for online solvent sensor – downstream of solvent tank.

For the "gas turbine startup" and "power output maximisation by capture plant decoupling" scenario, the sensor is located in the lean solvent line upstream of the solvent tank, between the desorber sump outlet and the main solvent tank. For the scenarios "power output maximisation by reboiler steam decoupling" and "frequency response" it is located downstream of the solvent tank and the main solvent pump. The sensor output is time-shifted to the point at which each "packet" of lean solvent analysed exits the desorber sump, for comparison with titration measurements. The sensor was unavailable for the "gas-turbine shutdown scenario".

DYNAMIC SCENARIOS – RESULTS AND DISCUSSION

Gas Turbine Shutdown

The shutdown sequence is initiated at t = 0 min as shown in Fig. 6a. Steam flow rate is ramped down to zero over 10 min at a rate of 10% of baseload/min, as is representative of state-of the-art gas turbine operation. Liquid and gas flow rates are ramped down simultaneously at a rate of 2.5%/min for a period of 16 min. At t = 16 min, the shutdown rate of liquid and gas was changed to 7.5% of baseload/min. At t = 20 min, liquid and gas flow rates both reach 30% of baseload. The liquid flow rate is held at this value, while the gas flow is further reduced at a rate of 15%/min until it reaches zero. Gas flow rate reaches zero at t = 22 min, while liquid flow is maintained at 30% of baseload until the absorber temperature bulge has subsided (Fig. 7).

The solvent lean loading at the desorber outlet begins to rise sometime between t = 5 min and t = 10 min due to the reduction in steam flow to the reboiler. Assuming plug flow and no mixing in the solvent tank, the more CO_2-rich "packet" of solvent sampled at t = 10 min does not reach the absorber inlet before the gas flow rate reaches zero, using the estimated solvent circulation times as described in Section 3.4. Lean loading at the desorber outlet increases throughout the shutdown operation, but has no significant effect on capture rate as this "richer" lean solvent is not predicted to reach the absorber inlet before of the flow rate of gas reaches zero.

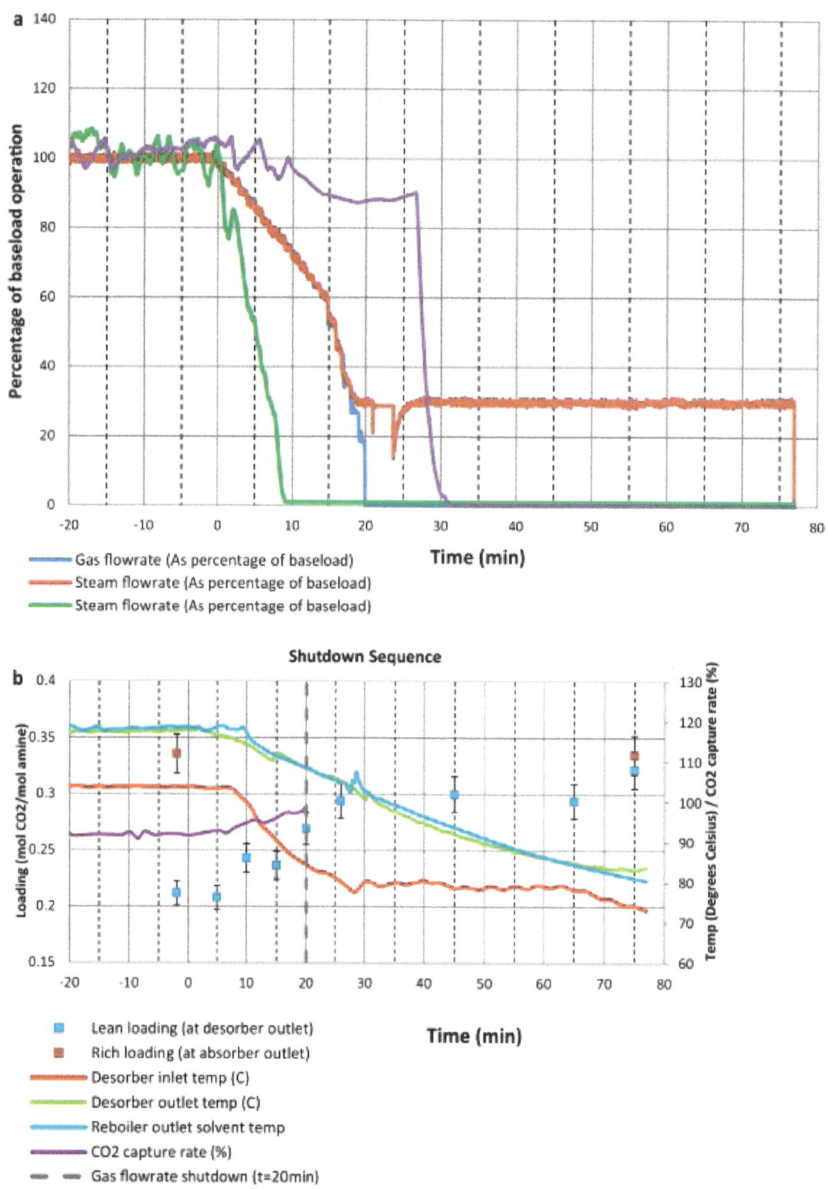

Figure 6: (a) Gas, liquid and steam flow rates as percentage of previously-defined baseload operation, gas turbine shutdown sequence. (b) Rich and lean solvent CO_2 loading, CO_2 capture rate, desorber and reboiler temperatures, gas turbine shutdown sequence.

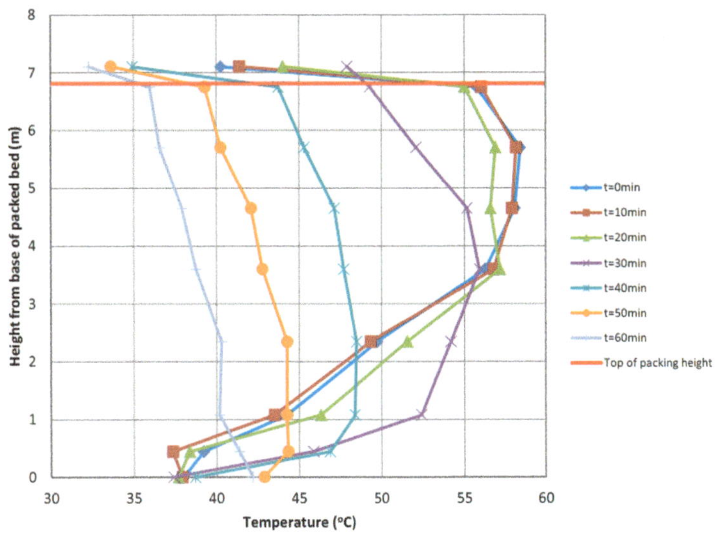

Figure 7: Evolution of absorber temperature profile during gas turbine shutdown operation.

Capture rate is observed to increase from 90% to 97.5% as gas and liquid flow is reduced (Fig. 6b). Since the solvent flow rate is gradually ramped down, the liquid holdup in the absorber requires additional time (compared to baseload) to flow over the packing and into the absorber sump. Although the *L*/*G* flow ratio is controlled to remain constant, the flow rate of gas passing through the absorber decreases more rapidly than the liquid holdup on the packing it comes into contact with, so a greater proportion of the CO_2 in the flue gas may be absorbed as the gas flow rate approaches and eventually drops to zero. Furthermore, the increased residence time of liquid and gas in the absorber column could result in this increased capture rate.

Although capture rate increases, the temperature profile of the absorber (Fig. 7) decreases in magnitude, as the absolute rate of exothermic CO_2 absorption decreases as the gas flow rate approaches zero. After the flow of gas is stopped at $t = 20$ min, the location of the temperature bulge decreases in magnitude and moves down the packing height. Fig. 7 illustrates the evolution of the absorber temperature profile as hot solvent flows down the packed bed and into the absorber sump, with the packing hold up being replaced by slow flowing, semi-stagnant cooler solvent.

Although no exothermic absorption reaction is taking place, residual heat in the metal elements of the absorber column combined with additional insulation to prevent heat loss, result in the absorber requiring some time to cool down.

It should be noted that the values of temperature at a height of 7.1 m in Fig. 7, and all other figures which represent the temperature profile, refer to a temperature measured directly before the absorber inlet. Heat transfer between the gas flow rate and solvent inventory in liquid distributors can, in practice, be neglected to a first order approximation.

It is important to note that the rich and lean loading values, shown in Fig. 6b, are not representative of the solvent loading at the absorber inlet and outlet at the reported time, and instead represent the loadings at the outlet of the desorber (lean) and absorber (rich) sumps, from where solvent samples were manually taken. In other dynamic scenarios with a longer duration, it is possible to estimate the circulation time required for a "packet" of solvent to reach the absorber inlet and outlet from each sampling port and hence, the real-time solvent working capacity, as explained previously in Section 3.4.

The time shifting method predicts that only the first lean solvent "packet" sampled at $t = 0$ will reach the absorber inlet before the end of this scenario. Subsequent lean solvent samples were not given sufficient time to reach the absorber inlet. To observe the increase in solvent loading at the rich loading sampling port, the solvent could have be allowed to circulate for a longer period of time, with lean and rich samples taken at more regular intervals in order to observe the full convergence of lean and rich loading as the solvent continues to circulate.

Gas Turbine Startup

Due to the nature of the shutdown sequence, which involves a continued circulation of solvent at 30% of baseload in order to lower solvent temperature more rapidly, and consequently an equilibration of solvent loadings due to mixing, the startup sequence is initiated with lean loading and rich loading, respectively, higher and lower than baseload values at baseload conditions.

Gas flow is initiated at $t = 0$ and increased at a rate of 5.7% of baseload/min. Once the flow of gas reaches 30%, at approx. $t = 5$ min 15 s, the flow rate of liquid commences ramping at 5.7%/min in order to maintain a constant *L*/*G* flow ratio.

As suggested by the models of Ceccarelli et al. (2014), the CO_2 capture rate (Fig. 8c) is initially higher than that of baseload operation between $t = 0$ and approximately $t = 5$ min, as the gas flow rate is ramped up to reach the operating liquid to gas ratio of the absorber. With the holdup on the packing already established, a cool solvent also results in a higher driving force for CO_2 absorption. Once the target liquid/gas flow ratio is established at $t = 5$ min and the absorber temperature profile increases in magnitude (Fig. 9a), the capture rate begins to drop due to the lean solvent loading becoming gradually higher than at baseload operation, resulting in a lower driving force for the absorption of CO_2 (Fig. 8c).

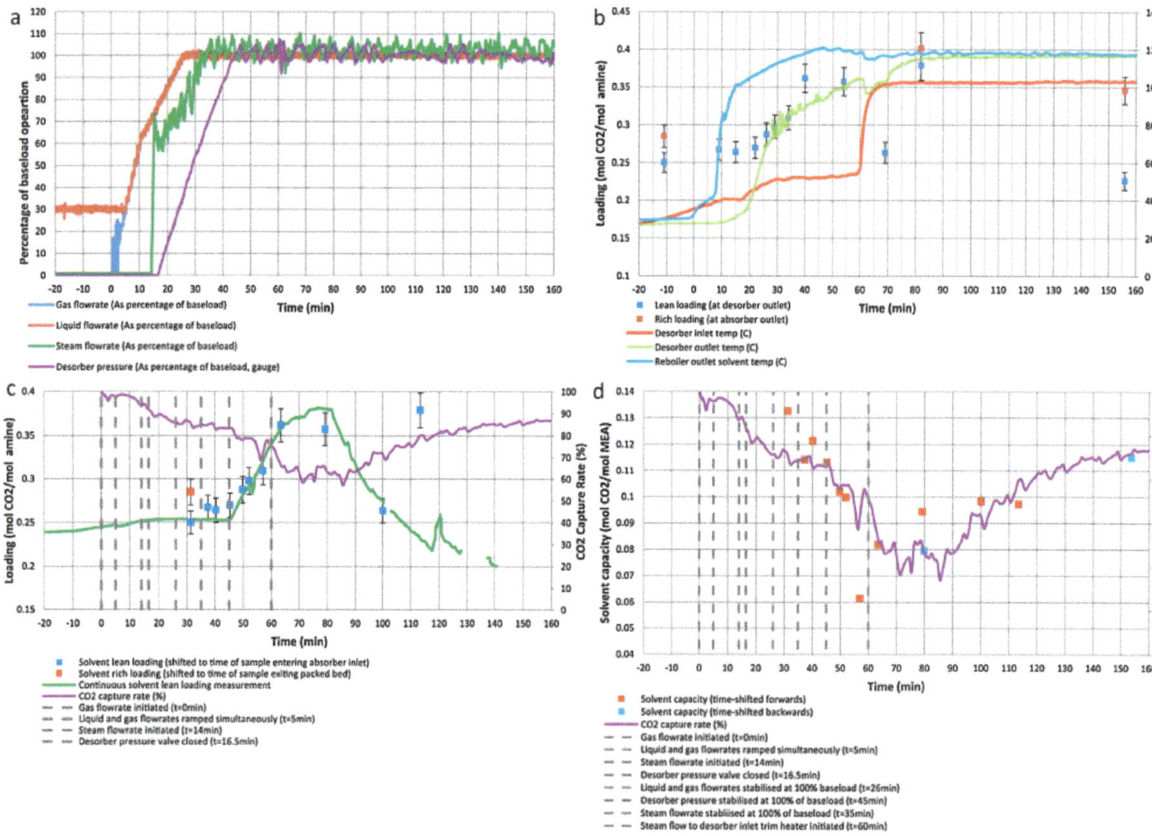

Figure 8: (a) Gas, liquid and steam flow rates as percentage of previously-defined baseload operation, gas turbine startup scenario. (b) Rich and lean solvent CO_2 loading at sampling ports, desorber and reboiler temperatures, gas turbine startup scenario. (c) Lean and rich solvent loading titration measurements, shifted to time of absorber inlet entry and outlet exit respectively, continuous lean loading measurement, CO_2 capture rate, gas turbine startup scenario. (d) Predicted real-time solvent capacity and CO_2 capture rate, gas turbine startup scenario.

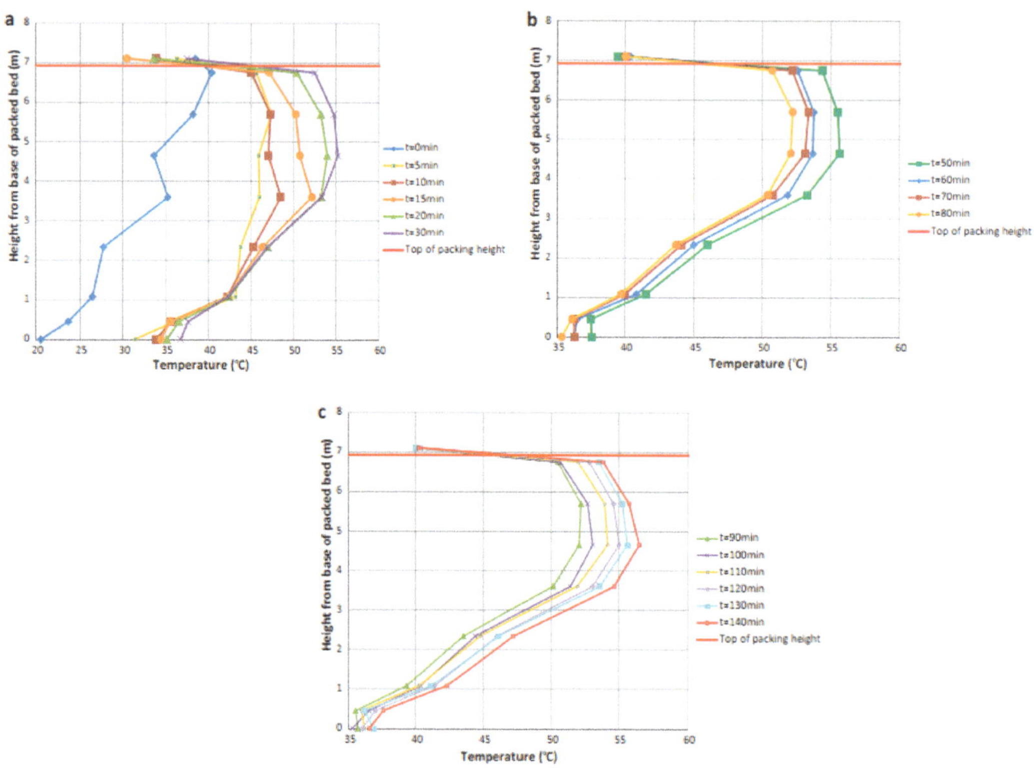

Figure 9: (a) Evolution of absorber temperature profile, $t = 0$ min – $t = 30$ min, gas turbine startup scenario. (b) Evolution of absorber temperature profile, $t = 50$ min – $t = 80$ min, gas turbine startup scenario. (c) Evolution of absorber temperature profile, $t = 90$ min – $t = 140$ min, gas turbine startup scenario.

At t = 11 min, ramp rates of both liquid and gas flow rates are decreased to approx. 2.45%/min and retained at this value until both reach 100% of baseload, at around$t$ = 26 min. Steam flow rate is ramped at a rate of 18.0%/min from t = 15 to t = 18 min, before the ramp rate is decreased to approx. 2.55%/min. The manual valve controlling desorber pressure is closed at t = 17 min. Although the flow of steam to the reboiler is established and stabilised at 100% of baseload by t = 35 min, the steam flow to the rich solvent heater (which is used to simulate the effect of the cross-flow heat exchanger) is not initiated until t = 60 min (Fig. 8b), due to an error by the plant operator. The capture rate continues to decrease until solvent which has been exposed to standard desorber operation circulates through the liquid tank and pipework to the absorber inlet.

This was not identified until after the conclusion of the test campaign, but the data is included since valuable information can still be gained from this scenario.

Until t = 60 min, a significant fraction of the thermal input from the reboiler contributes to increasing the temperature of incoming solvent to the desorber by an additional 50–60 °C compared to baseload operation, and energy which could otherwise be used to strip $CO_2$is lost as sensible heat.

As a consequence, the lean loading gradually increases to around 0.36 mol CO_2/mol MEA which compromises the driving force and solvent capacity in the absorber column, as shown in Fig. 8d. This results in a gradual drop in capture level down to around 65% from t = 65 min to t = 90 min.

Interestingly, this effect is somewhat less pronounced than what would be expected for an absorber configuration optimised for natural gas equivalent CO_2 concentrations, since the pilot plant facility is operated with a lower rich loading. As a result there is additional margin for the rich loading to increase before a pinch in absorption is reached at the bottom of the absorber.

The decrease in absorber temperature profile (Fig. 9b) can be directly linked to this decrease in capture rate, both of which reach a minimum at t = 80 min. Leaner solvent which has been exposed to baseload desorber and reboiler operation is returned to the absorber after t = 80 min (Fig. 8c and d), resulting in a higher driving force for CO_2absorption and a steady increase in capture rate (Fig. 8c and d) and absorber temperature (Fig. 9c). The capture rate, predicted solvent capacity and continuous lean solvent loading measurement all suggest a steady decrease in solvent lean loading at the absorber inlet after t = 80 min, so it is believed that the bench lean loading measurement of 0.3787 at t = 113 min (Fig. 8c) is the result of an abnormal or unrepresentative titration measurement. Triplicate analysis of 14 samples taken over the course of the test campaign resulted in an estimated uncertainty of 5.2%, suggesting that the source of the error most likely lies within the sample, and not the titration method. The sample may be an example of abnormal "noise" and therefore unrepresentative of overall change, or the result of unknown mixing effects in the desorber sump. Several similar measurements are encountered throughout the test campaign, and unfortunately it is not possible to know definitively the source of the abnormal measurement without more regular sampling.

The outcomes from this experiment suggest that a decrease in capture rate following startup is inevitable as plant parameters stabilise, in cases where some solvent flow rate has been maintained between the shutdown and startup to use the cooling capacity available. If maintaining a high capture rate is of critical importance, interim solvent storage has been proposed as a method to increase capture flexibility, by ensuring that a large separate reserve of lean solvent is available until the plant is stabilised at baseload (Lucquiaud et al., 2014).

Power Output Maximisation by Capture Plant Decoupling

Before the capture plant decoupling scenario is initiated, the plant is operating at steady state baseload. The flow rates of steam and gas are reduced to zero at t = 0 min and$t$ = 1 min, respectively, with the capture rate dropping to zero shortly thereafter, at$t$ = 2.5 min, since gas no longer passes through the absorber. The manual valve maintaining pressure in the desorber is released at t = 8 min to prevent creation of vacuum in the desorber. The flow of liquid is turned down in two steps, to 75% of baseload at t = 9 min, then 50% at t = 14 min. Solvent flow is maintained at 50%, reproducing likely operating conditions where power for solvent circulation is reduced to a minimum to maximise electricity power output. Maintaining a minimum solvent flow rate also allows, if necessary, for fast startup of the capture plant than turning off the solvent pumps. This lower solvent flow rate, combined with a constant solvent cooling capacity, ensures that a substantial reserve of solvent capacity is available so that capture rate can rapidly return to normal levels upon reintroduction of flue gas to the absorber column.

Lean solvent loading measured via titration at the desorber outlet appears to rise between t = 1 min and t = 6 min, before dropping once again at t = 11 min. The rise in lean loading around t = 6 min is not observed in the continuous measurement, which indicates that this sample may be unrepresentative of plant trends, as discussed at the end of Section 5.1. The liquid flow rate is still operating at 100% until t = 9 min, and by the time the sample at t = 11 is taken, solvent which has not been exposed to CO_2 in the absorber begins to appear at the desorber outlet. This is consistent with the circulation time between the top of the absorber sump and top of the desorber sump at L = 100% being approx. 8 min in total, allowing 6 min from absorber outlet to desorber, plus an additional 2 min to pass through the desorber sump.

After the flow of gas has been shut down, the absorber temperature bulge decreases in magnitude and moves down the column as hot solvent in the upper region of the absorber flows down the packed bed, being replaced by colder solvent as liquid (Fig. 11a). The absorber is insulated and requires some time to cool down, as heat is transferred from the packing to the incoming cool solvent.

Assuming that the lean titration measurement at t = 6 min is anomalous and unrepresentative of true plant trends, lean loading at the desorber outlet sampling port decreases from 0.202 mol/mol to 0.174 mol/mol from t = 0 min until around t = 50 min (Fig. 10b) as residual heat in the desorber continues to liberate small amounts of CO_2, from solvent not exposed to CO_2 in the absorber.

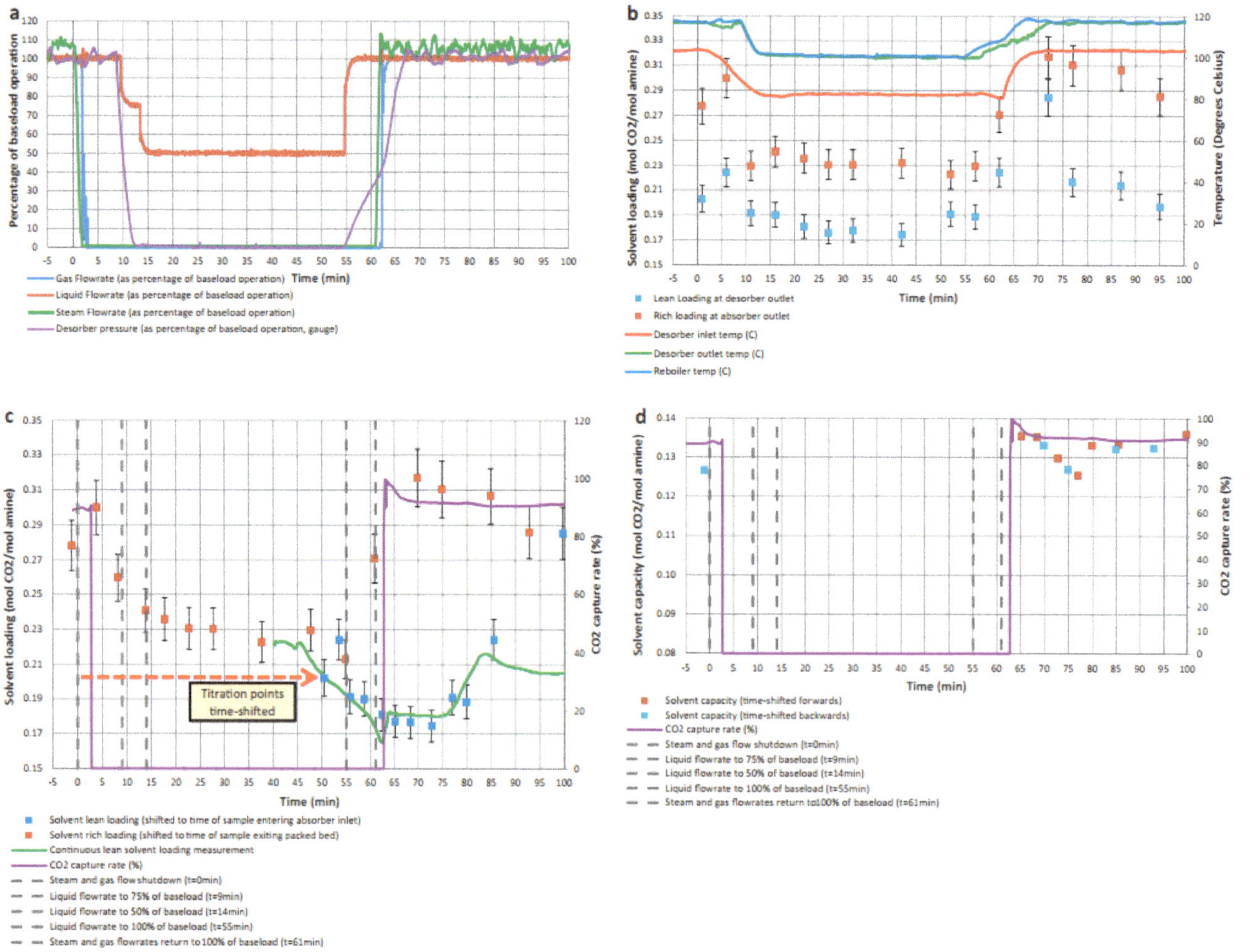

Figure 10: (a) Gas, liquid and steam flow rates as percentage of previously-defined baseload operation, power output maximisation by capture plant decoupling scenario. (b) Rich and lean solvent CO2 loading at sampling ports, desorber and reboiler temperatures, power output maximisation by capture plant decoupling scenario. (c) Lean and rich solvent loading titration measurements, shifted to time of absorber inlet entry and outlet exit respectively, continuous lean loading measurement, CO_2 capture rate, power output maximisation by capture plant decoupling scenario. (d) Predicted real-time solvent capacity and CO_2 capture rate, power output maximisation by capture plant decoupling scenario. (For interpretation of the references to color in the text, the reader is referred to the web version of the article.)

Lean loading titration measurements taken from $t = 0$ min are shifted forward to the time of entry of the corresponding packet of solvent to the absorber inlet (blue markers) in Fig. 10c. The continuous online solvent measurement shows good agreement with the titration data until $t = 85$ min. Based on the liquid flow rate and volume of liquid in the pipework between the lean solvent sampling port and absorber inlet, the lean loading value of 0.204 mol/mol at $t = 1$ min (Fig. 10b), is predicted to reach the absorber inlet at$t = 54$ min (Fig. 10c). This appears very close to the rich loading value of 0.213 mol/mol, corresponding to the time of exit of the absorber outlet at $t = 54$ min. Given that there is no CO_2 flow into the absorber between $t = 1$ min and $t = 54$ min, this indicates that the time-shifting method provides a moderately accurate estimation of solvent circulation time of 53 min. This is also roughly consistent with the estimated circulation time of 63 min at 50% of baseload flow rate shown in Table 3, given than titration measurements are taken every 5 min and that the solvent flow rate does not reach 50% of baseload for the first 15 min.

The temperature profile of the absorber (Fig. 11a) decreases in magnitude after $t = 0$ min, as the exothermic absorption of CO_2 stops after the flow rate reaches zero. The absorber temperature bulge exhibits a similar response to that observed during simulated gas turbine shutdown, decreasing in magnitude and moving down the packing height as hot solvent flows down the packed bed and is replaced by slow flowing incoming cooler solvent. By $t = 15$ min there is no observable bulge, and the temperature across the entirety of the packed bed decreases until flue gas is reintroduced.

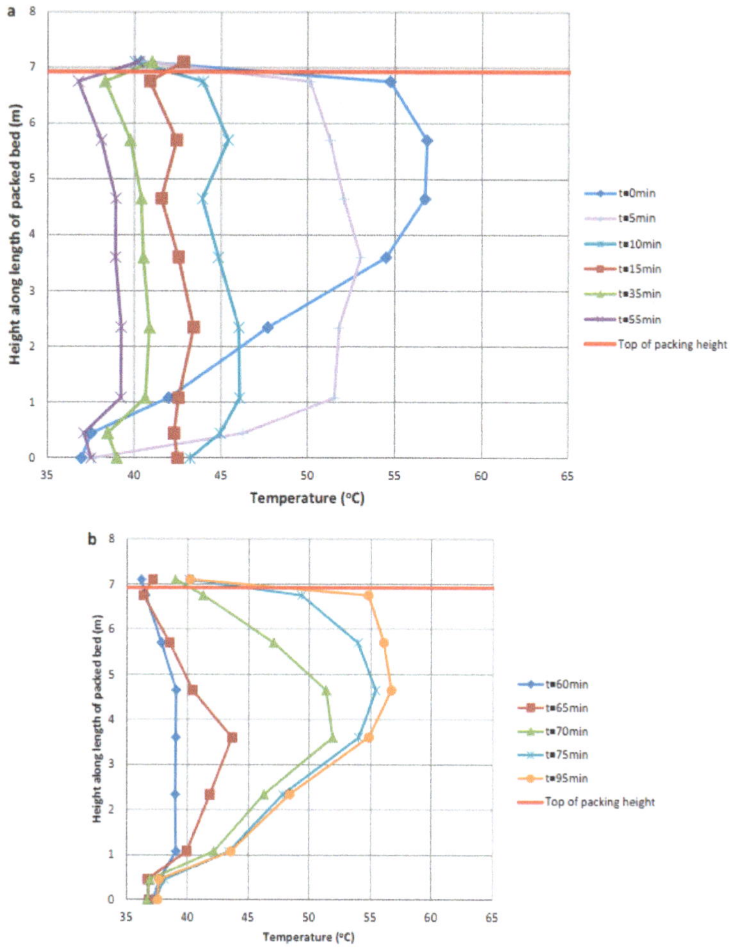

Figure 11: (a) Evolution of absorber temperature profile during power output maximisation by capture plant decoupling scenario – temperature decrease. (b) Evolution of absorber temperature profile during power output maximisation by capture plant decoupling scenario – temperature increase.

The desorber manual release valve is closed and pressure begins to rise at t = 55 min. Att = 60 min, the bypass event ends and the flow rates of gas and steam are ramped back up to 100%.

Rich loading increases rapidly upon reintroduction of CO_2 to the absorber at t = 62 min. The capture rate is initially higher than the baseload value as a batch of cool solvent promotes CO_2 absorption, and this is reflected in the rich loading titration measurements of 0.316, 0.313, 0.306 mol/mol at t = 72, 77 and 87 min (respectively Fig. 10b). The lean loading at the desorber outlet follows a similar trend once the desorber has reached operational temperature, with the circulation time from the absorber sump to the desorber sump being around 6 min at 100% baseload solvent flow rate. A temperature bulge is observed in the absorber at t = 65 min, 3 min after the reintroduction of gas flow, and is fully established by t = 95 min.

Fig. 10c shows the lean and rich solvent measurements time-shifted to the absorber inlet and outlet. It should be noted that the time-shifting method detailed in Section 3.4contains a level of uncertainty due to the hydraulic response of the plant, and in this case predicts an increase in solvent loading at the base of the absorber packing at t = 61 min, 60 s before the flow of gas is reintroduced. A batch of richer lean solvent which is predicted to reach the inlet between t = 85 and t = 99 min appears not to be detected by the online solvent sensor, and has no real effect on the capture rate. It is possible that there is sufficient mixing with leaner solvent between the sampling port and the absorber inlet, that the increase in loading is dampened. It is also possible that the titration measurement at t = 72 min (Fig. 10b)/t = 99 min (Fig. 10c) is unrepresentative of plant conditions, as discussed in Section 5.1.

The time shifting method is used to show the solvent working capacity in the absorber, in mol/mol (Fig. 10d), and how it correlates to capture level. Solvent working capacity is calculated as described in Section 3.4.

In this scenario there is no observable reduction in capture rate upon reintroduction of flue gas. However, it is worth noting that the effect on capture rate upon reintroduction of flue gas will be highly dependent on the length of the shutdown operation, the circulation time, total solvent inventory and extent of mixing in the liquid loop for each individual capture plant.

Potential detrimental effects on capture rate upon restart could be mitigated by using interim solvent storage. This involves holding a batch of lean solvent in reserve so it can be fed to the absorber while the original solvent is being regenerated. The stored lean solvent can also be used to replace the original which is diverted to a rich solvent storage tank for future regeneration when energy selling price is low.

Power Output Maximisation by Reboiler Steam Decoupling Only

Although similar to the previous scenario, a significant difference is that flue gas is continuously fed to the absorber column during reboiler steam shutdown. Steam flow to the reboiler is reduced to zero at t = 0 min, followed by the opening of the desorber pressure release valve at t = 1 min. Solvent flow rate is turned down to 75%, then 50% at$t$ = 8 min and t = 15 min respectively. The capture rate begins to drop at t = 8 min after the initial decrease in solvent flow, but does not decrease sharply until t = 17 min.

Lean loading at the desorber outlet appears to fluctuate between t=6 and t=31 min, based on the titration measurements taken. When the flow rate of solvent is reduced rapidly, the PID control system may fully close the desorber outlet valve for several minutes in an attempt to maintain desorber sump level. Solvent which has not been exposed to steam flow will have ample opportunity to mix with fully regenerated solvent in the desorber sump. This mixing effect makes it very difficult to predict what will happen to the lean loading at the desorber outlet, although based on the results from the online solvent sensor, it is possible that the titration measurements at t = 11 and t = 21 min are abnormal or unrepresentative of the lean solvent loading which will be seen by the solvent sensor, due to mixing effects in the main solvent tank.

In this and the final dynamic scenario (Section 5.5), the online solvent sensor is relocated to just downstream of the main solvent pump. Interestingly, when the continuous measurement output and titration values are shifted to the time at which each packet of solvent would reach the absorber inlet (Fig. 12c), this fluctuating lean loading behaviour is not observed. This may suggest that the fluctuations in lean solvent loading at the desorber are dampened due to mixing in the solvent tank, and will not have a considerable impact on the lean loading at the absorber inlet.

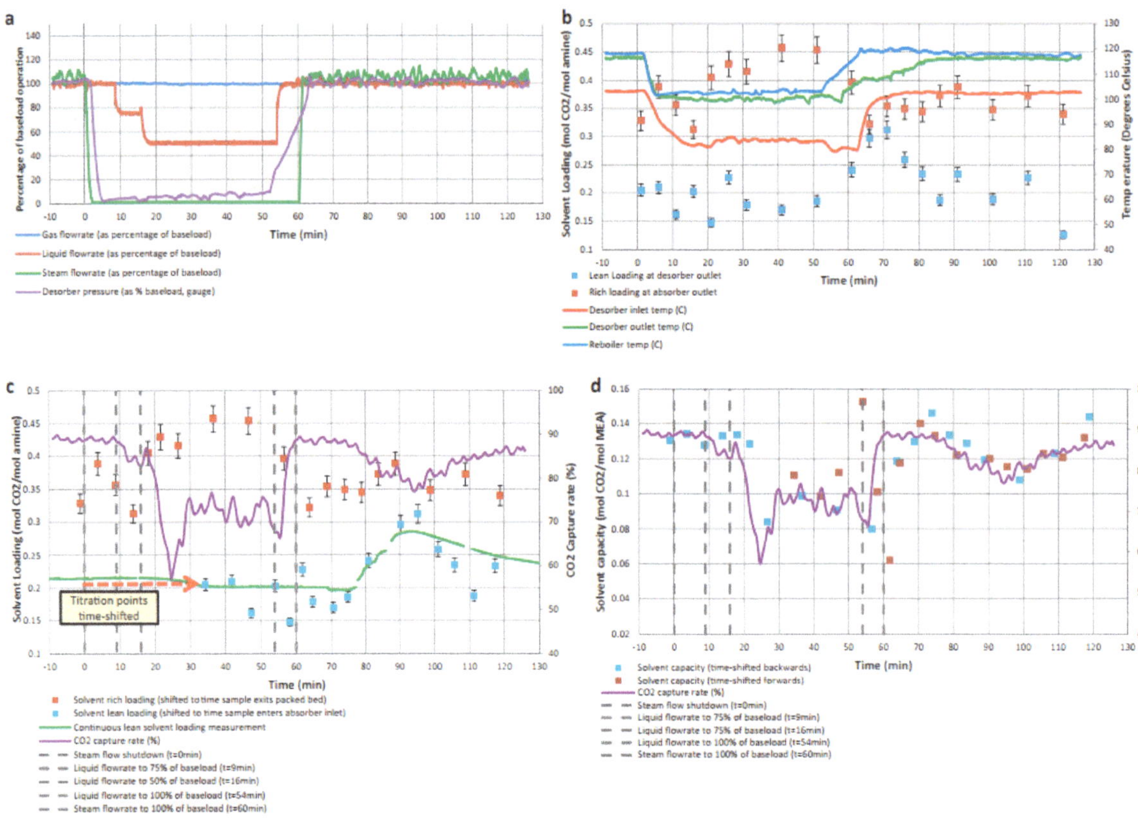

Figure 12: (a) Gas, liquid and steam flow rates as percentage of previously-defined baseload operation, for power output maximisation by reboiler steam decoupling scenario. (b) Rich and lean solvent CO_2 loading at sample points, desorber and reboiler temperatures, power output maximisation by reboiler steam decoupling scenario. (c) Rich and lean solvent loading bench measurements, time-shifted to absorber inlet and outlet respectively, continuous lean loading measurement, CO_2 capture rate, power output maximisation by steam decoupling scenario. (d) Predicted real-time solvent capacity and CO_2 capture rate, power output maximisation by steam decoupling scenario.

The general trend in rich loading at the absorber outlet before the flow rate of liquid is ramped back up to 100% at t = 53 min is an increasing one (Fig. 12b). This is a direct result of the decreased L/G flow ratio, down from 2.86 at baseload to 1.43 from t = 17 min, and the increased liquid residence time on the packing, which results in additional CO_2being absorbed per unit of solvent flowing into the column. Although the rich solvent loading at the bottom of the absorber increases, the liquid flow rate is low enough that the amount of CO_2 captured is reduced overall. The absorber temperature bulge decreases in magnitude (Fig. 13a), consistent with the decrease in capture rate, and moves upwards. The upwards shift may be due to diminished working capacity, capture rate and L/G ratio, which results in higher concentrations of CO_2 in the upper regions of the absorber. The solvent becomes loaded with CO_2 more rapidly upon entry, decreasing the driving force for CO_2 absorption as it flows down the packed bed. Higher CO_2concentration in the upper regions of the packing also results in a higher driving force for CO_2 absorption on the gas-side.

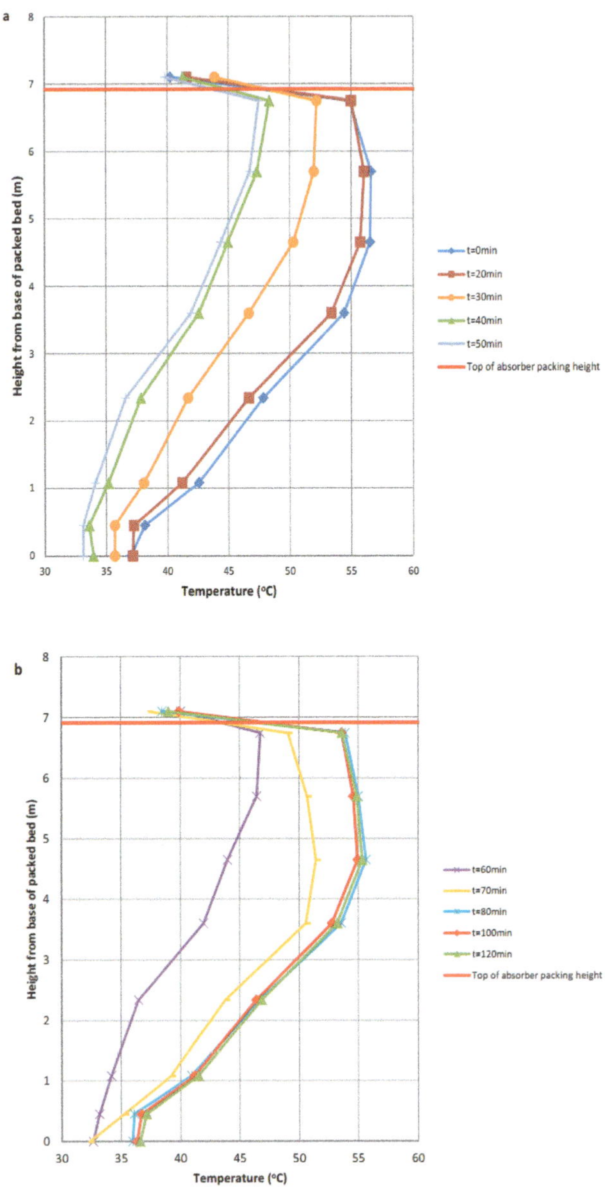

Figure 13: (a) Evolution of absorber temperature profile, power output maximisation by reboiler steam decoupling scenario – temperature decrease. (b) Evolution of absorber temperature profile, power output maximisation by reboiler steam decoupling scenario – temperature increase.

The desorber pressure release valve is closed and liquid flow ramped back up at t = 52 and 53 min, respectively, in anticipation of the plant returning to normal operating conditions. Steam flow is ramped back up to 100% at t = 60 min. In total, the steam shutdown operation lasts 60 min, a plausible duration for a bypass in response to an evening peak in electricity prices. Lean loading at the desorber outlet decreases between$t$ = 71 and t = 76 min, once the desorber reaches operational temperature (Fig. 12b). The estimated real-time solvent loading capacity (Fig. 12d) appears to follow the capture rate for the majority of the scenario, except for two points at t = 54.07 and 61.08. This may be due to the fact that the lean loading titration measurements, which the calculations are based on, are erroneous or that the CO_2 flow rate into the absorber is not entirely stable during this time, with the mass flow controller aiming to maintain a steady CO_2concentration at the absorber inlet while the capture rate and concentration of CO_2 in the recycled flue gas stream is highly variable.

Capture rate is observed to decrease between t = 75 and t = 95 min, corresponding to the time at which solvent not fully regenerated in the desorber is predicted to reach the absorber inlet (Fig. 12c).

There is good agreement around t = 95 min between solvent working capacity and capture rate, yet, in Fig. 12c, the lean loading value of 0.312 mol/mol, time shifted to$t$ = 95 min, is close to the rich loading value of 0.348 mol/mol while a capture rate of >75% is achieved. This is considerably higher than the baseload lean loading value of 0.232 mol MEA/mol CO_2 (Section 3.1) As mentioned in Section 3.4, the time shifting method is not able to account for mixing effects, since it relies on the assumption of plug flow. This illustrates that it is likely that mixing effects in the solvent tank, located between the lean and the rich sampling points, play an important role to maintain solvent working capacity and capture rate around 75–80% for 20–25 min from t = 75 min onwards.

The batch of richer lean solvent passing through the absorber, indicated by the online measurement lean loading readings in Fig. 12c at around t = 95 min, is then followed by a rise in capture rate before the plant returns to steady state baseload operation. The absorber temperature bulge reflects the evolution of the capture rate, increasing in magnitude and moving down the packing height after t = 60 min (Fig. 13b). The temperature profile decreases slightly around t = 100 min, coinciding with the drop in capture rate, before increasing as the plant reaches steady state operation at t = 120 min. This decrease in capture rate observed 20 min after the supply of thermal energy to the reboiler is resumed could be avoided by using interim solvent storage.

A thorough understanding of plant circulation times, combined with continuous monitoring of solvent loading at the liquid inlet and outlet of the absorber would allow operators to anticipate this decrease in capture rate and compensate accordingly by anticipating to adjust solvent working capacity.

Frequency Response by Rapid Reboiler Steam Flow Increase

The flow rate of steam to the reboiler is increased rapidly to 200% at t = 0 and remains at this value for 41 min, while all other plant parameters remain constant. Plant data continue to be obtained for an additional 50 min after steam had been ramped down, in order to observe any further effects on the response of the plant.

It is important to note that the plant may not have fully reached a steady-state operating condition at the start of this scenario, with the initial capture rate at 85% and a lean loading of 0.250 mol/mol.

Lean loading at the desorber outlet is observed to decrease steadily from 0.250 mol/mol after the flow rate of steam is increased at t = 0, eventually stabilising at around 0.2 mol CO_2/mol amine (Fig. 14b). These titration data are shifted to the time at which each sample would reach the absorber inlet in Fig. 14c. The online solvent sensor output displays similar behaviour, but the decrease in lean loading appears less rapid. As the sensor is located after the main solvent tank, any rapid changes in lean loading at the desorber outlet are dampened when that packet of solvent reaches the sensor, due to mixing in the solvent tank, as previously explained. The continuous measurement suggests that the lean loading at the absorber inlet begins to decrease before any solvent which has undergone a greater degree of regeneration due to increased steam flow is expected to reach the absorber inlet. As the plant is not initially at baseload, it is not possible to attribute the observed increase in capture rate and decrease in lean loading at the absorber inlet between t = 0 and t = 25 min to the frequency response operation.

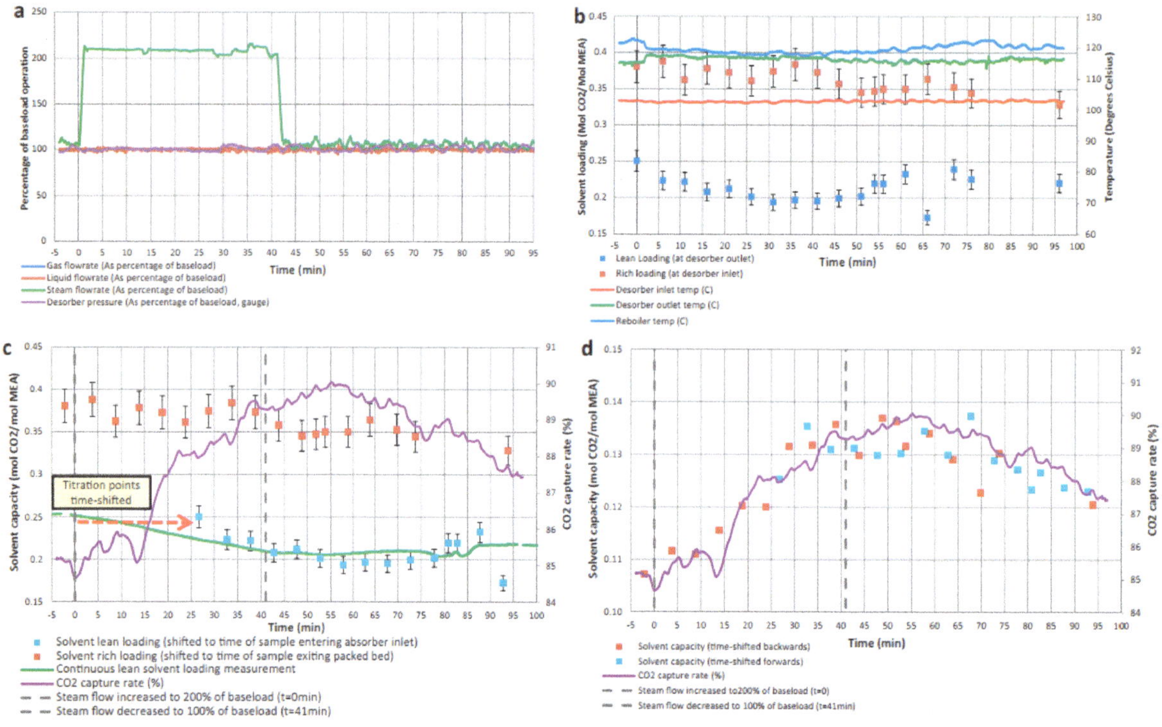

Figure 14: (a) Gas, liquid and steam flow rates as percentage of previously-defined baseload operation, frequency response scenario. (b) Rich and lean solvent CO_2 loading, desorber and reboiler temperatures frequency response scenario. (c) Lean and rich solvent loading bench measurements, time-shifted to absorber inlet and outlet respectively, continuous lean loading measurement, CO_2capture rate, frequency response scenario. (d) Predicted real-time solvent capacity and CO_2 capture rate, frequency response scenario.

Fig. 14c and d shows that, when solvent which has undergone a greater degree of regeneration due to increased steam flow starts entering the absorber from t = 26 min onwards, the increase in solvent working capacity results in an increase in capture rate from 88% to 90%. Steam flow is turned down to 100% of baseload at t = 41 min. The response of the capture rate, starting at t = 70 min, corresponds to an increase in lean solvent loading at the absorber inlet (Fig. 14c).

A titration lean loading measurement at t = 66 min (Fig. 14b)/t = 93 min (Fig. 14c) appears to be anomalous as there is no significant change in the capture rate, and it deviates significantly from the online lean loading measurement at this point in time.

The calculated real-time solvent working capacity (Fig. 14d) follows well the trend of the capture rate over the course of the whole experiment. Accurate monitoring and prediction of real-time working capacity via continuous loading measurements at the absorber inlet and outlet could prove to be a key metric in future advanced control systems for dynamic post-combustion capture.

As there is no significant change in absorber temperature profile, the maximum dT over the course of the experiment being an increase of 1.68 °C at height of 3.6 m, absorber temperature data are not reported for this scenario.

CONCLUSIONS

Five enhanced operational flexibility scenarios of an amine post-combustion capture process treating flue gas from a state of the art combined cycle gas turbine plant are investigated in this work.

No significant barriers to flexible operation were observed for this pilot facility. It is possible that plants operated with higher rich loadings at baseload would have a more limited capacity to absorb CO_2 in the absence of heat input for regeneration, which may result in a more rapid drop-off in capture rate. Solvent inventory and circulation times are observed to have a significant effect on capture rate during certain dynamic operations, with longer circulation times resulting in the plant requiring additional time to return to steady-state following a perturbation. Indeed, if the capture plant is required to operate in continuously changing scenarios, for example load-following, it may never return to steady-state operation.

A reduction of solvent inventory may help to reduce delays due to circulation times, but there are positive effects in having a large solvent inventory – this increases the solvent capacity of the plant, and as observed in one of the two scenarios for power output maximisation, it allows the plant to maintain a CO_2 capture rate greater than 55%, and around 75% for most of the duration of the 60 min whilst steam supply to the reboiler is turned off.

For optimum performance in electricity markets, anecdotal evidence suggests that altering power output every 30 min is desirable, e.g. 30 min is the duration of settlement period in the Balancing and Settlement Code for electricity generation in the UK. Flexible operation can also be of benefit to baseload generation plants equipped with CCS, as they can modify capture rate and electricity output in response to market fluctuations in CO_2 and electricity selling price (Mac Dowell and Shah, 2014).

For the process control loop of an amine post-combustion capture process, this would require solvent loading measurements to be performed continuously if they are to be used as a control variable. An automated control system must perform feedback control loop operations at least 10× faster than the time frame required to change state. Therefore, if the capture plant changes state once every 30 min, the feedback operation must be performed at least once every 3 min. Obtaining a solvent loading measurement within this timeframe is not possible with existing methods – manual sampling followed by loading analysis at regular intervals is time-consuming and labour-intensive, while anecdotal evidence regarding the use of automated titration systems indicates that each loading measurement procedure can take up to 20 min.

The implementation of an online solvent sensor for continuous real-time measurements of lean solvent loading during this test campaign proved a successful step in reducing the time required to obtain solvent loading measurements. Model predictive control has been identified as a method by which PCC could react to perturbations optimally (Mechleri, 2015), and accurate, continuous loading measurements at the absorber inlet and outlet are likely to be key control variables (Seibert et al., 2011 and van der Ham et al., 2014). A further online measurement of loading at the desorber outlet, combined with extensive knowledge of plant circulation times and mixing effects, would allow for greater control over capture rates due to more accurate prediction of solvent loading at the absorber inlet.

Although the sensor performed adequately during this test campaign, there still exists scope for improvement. Current efforts aim to improve sensor stability as rapid changes in solvent loading can result in software errors. Automated data-logging capability will also be added and the sensor will be redesigned for industrial field deployment.

This work indicates that a potential solution to the problem of reduced CO_2 capture as a result of certain dynamic operations is to actively manage solvent inventory. For example, flow rates and/or loadings, combined with forewarning of changes in power plant output and an appropriate control system could allow for the mitigation of detrimental impacts on performance. An option proposed previously is the use of two additional tanks, both separated from the rest of the liquid loop by valves. One tank could contain a reserve of lean solvent to be utilised during steam shutdown situations, while the second tank could act as a receptacle for excess rich solvent, as initially suggested by (Chalmers and Gibbins, 2007) and widely studied since by others, e.g. (Chalmers et al., 2009 and Oates et al., 2014). An alternative proposed by Mac Dowell and Shah (2014) is to vary solvent regeneration levels in anticipation of future power plant output, although this work shows that mixing in the solvent line between the reboiler and absorber inlet will affect lean loading for a short period of time after the level of regeneration is changed.

The added flexibility made available by either method is one benefit of post-combustion capture, and combined with the further development of real-time solvent loading measurement options, is a worthy subject for future investigation at pilot plant scale.

ACKNOWLEDGEMENTS

The authors would like to thank Sulzer ChemTech and the Energy Technology Partnership (ETP) for financial support to Paul Tait. Financial and technical support for the operation of the pilot-scale facility from Sulzer ChemTech is also gratefully acknowledged, and financial support from a UK Carbon Capture Research Centre funded project (UKCCSRC-C2-214), the EPSRC (EP/J020788/1) and the Royal Academy of Engineering for equipment. The online sensor used in this work was developed with funding from Doosan Power Systems and the Energy Technology Partnership (ETP). This research was carried out as part of the UKCCSRC's GAS-FACTS project.

REFERENCES

1. Artanto, Y., Jansen, J., Pearson, P., Do, T., Cottrell, A., Meuleman, E., Feron, P., 2012. Performance of MEA and amine-blends in the CSIRO PCC pilot plant at Loy Yang Power in Australia. Fuel 101, 264–275.
2. Biliyok, C., Mechleri, E.D., Thornhill, N.F., 2014. Dynamic simulation and control of post-combustion CO2 capture with MEA in a gas fired power plant. Comput. Aided Chem. Eng. 33, 619–624.
3. Bui, M., Gunawan, I., Verheyen, V., Artanto, Y., Meuleman, E., Feron, P., 2013. Dynamic modelling and validation of post-combustion CO_2 capture plants in Australian coal-fired power stations. Energy Procedia 37, 2694–2702.
4. Bui, M., Gunawan, I., Verheyen, V., Feron, P., Meuleman, E., Adeloju, S., 2014. Dynamic modelling and optimisation of flexible operation in post-combustion CO2 capture plants – a review. Comput. Chem. Eng. 61, 245–265.
5. Carey, J., Damen, K., Fitzgerald, F.D., Gardiner, R.A., 2013. CCPilot 100+ operating experience and test results. Energy Procedia 37, 6170–6178.
6. Ceccarelli, N., van Leeuwen, M., van Leeuwen, P., Maas, W., Ramos, A., van der Vaart, R., Wolf, T., 2014. Flexibility of low-CO2 gas power plants: integration of the CO2 capture unit with CCGT operation. Energy Procedia 63, 1703–1726.
7. Chalmers, H., Gibbins, J., 2007. Initial evaluation of the impact of post-combustion capture of carbon dioxide on supercritical pulverised coal power plant part load performance. Fuel 86, 2109–2123.
8. Chalmers, H., Lucquiaud, M., Gibbins, J., Leach, M., 2009. Flexible operation of coal fired power plants with postcombustion capture of carbon dioxide. J. Environ. Eng. 135, 449–458.
9. Committee on Climate Change, 2013. 4th Carbon Budget Review Part 2 [.pdf], Available at: https://www.theccc.org.uk/publication/fourth-carbon-budgetreview/ (accessed October 2015).
10. van Eckeveld, A.C., van der Ham, L.V., Geers, L.F.G., van den Broeke, L.J.P., Boersma, B.J., Goetheer, E.L.V., 2014. Online monitoring of the solvent and absorbed acid gas concentration in a CO2 capture process using monoethanolamine. Ind. Eng. Chem. Res. 53, 5515–5523.
11.
12. Eisfeld, T., Feldmüller, A., 2013. Fast cycling and fast start capability of combined cycle power plants with SGT5-4000F gas turbines. In: IMechE Seminar: Demand Response and Flexible Operation in UK Power Stations, 28 November, Nottingham, p. 2013.
13. Energy Technologies Institute, 2015. Building the UK Carbon Capture and Storage Sector – Scenarios and Actions [.pdf], Available at: http://www.eti.co.uk/wpcontent/uploads/2015/03/CCS-Building-the-UK-carbon-capture-and-storagesector-by-2013.pdf (accessed November 2015).
14. Faber, R., Köpcke, M., Biede, O., Knudsen, J.N., Andersen, J., 2011. Open-loop step responses for the MEA post-combustion capture process: experimental results from the Esbjerg pilot plant. Energy Procedia 4, 1427–1434.
15. Fitzgerald, F.D., Hume, S.A., McGough, G., Damen, K., 2014. Ferrybridge CCPilot100+ operating experience and final test results. Energy Procedia 63, 6239–6251.
16. Flø, N.E., Zangrilli, L., Mangiaracina, A., Mejdell, T., Kvamsdal, H.M., Hillestad, M., 2014. Validation of a dynamic model of the Brindisi pilot plant. Energy Procedia 63, 1040–1054.
17. Flø, N.E., Knuutila, H., Kvamsdal, H.M., Hillestad, M., 2015. Dynamic model validation of the post-combustion CO2 absorption process. Int. J. Greenh. Gas Control 41, 127–141.
18. Global CCS Institute, 2014. The Global Status of CCS: 2014, Melbourne, Australia. van der Ham, L.V., van Eckeveld, A.C., Goetheer, E.L.V., 2014. Online monitoring of dissolved CO2 and MEA concentrations: effect of solvent degradation on predictive accuracy. Energy Procedia 63, 1223–1228.
19. Haines, M.R., Davison, J., 2014. Enhancing dynamic response of power plant with post-combustion capture using "Stripper stop". Int. J. Greenh. Gas Control 20, 49–56.
20. Hamborg, E.S., Smith, V., Cents, T., Brigman, N., Falk-Pedersen, O., De Cazenove, T., Chhaganlal, M., Feste, J.K., Ullestad, Ø., Ulvatn, H., Gorset, O., Askestad, I., Gram, L.K., Fostås, B.F., Shah, M.I., Maxson, A., Thimsen, D., 2014. Results from MEA testing at the CO2 Technology Centre Mongstad, part II: verification of baseline results. Energy Procedia 63, 5994–6011.
21. IPCC, 2014. Climate Change 2014: Fifth Assessment Report of the Intergovernmental Panel on Climate Change. Cambridge University Press, Cambridge, UK.
22. Jordal, K., Ystad, P.A.M., Anantharaman, R., Chikukwa, A., Bolland, O., 2012. Design-point and part-load considerations for natural gas combined cycle plants with post combustion capture. Int. J. Greenh. Gas Control 11, 271–282.
23. Kvamsdal, H.M., Chikuwa, A., Hillestad, M., Zakeri, A., Einbu, A., 2011. A comparison of different parameter correlation models and the validation of an MEA-based absorber model. Energy Procedia 4, 1526–1533.
24. Lucquiaud, M., Fernandez, E.S., Chalmers, H., Mac Dowell, M., Gibbins, J., 2014. Enhanced operating flexibility and optimised off-design operation of coal plants with post-combustion capture. Energy Procedia 63, 7495–7507.
25. Mac Dowell, N., Shah, N., 2014. Optimisation of post-combustion CO2 capture for flexible operation. Energy Procedia 63, 1525–1535.
26. Mechleri, E., 2015. Controllability analysis of a post-combustion CO2 capture plant integrated with a coal and natural gas-fired power plan. In: 3rd Post Combustion Capture Conference, 9th September 2015, Regina, Canada.

27. Mejdell, T., Vassbotn, T., Juliussen, O., Tobiesen, A., Einbu, A., Knuutila, H., Hoff, K.A., Andersson, V., Svendsen, H., 2011. Novel full height pilot plant for solvent development and model validation. Energy Procedia 4, 1753–1760.
28. Murray, S., 2013. Why whole portfolio modelling is essential. In: IPA 5th Annual Conference, 12 March 2013, East
29. Kilbride.
30. Oates, D.L., Versteeg, P., Hittinger, E., Jaramillo, P., 2014. Profitability of CCS with flue gas bypass and solvent storage. Int. J. Greenh. Gas Control 27, 279–288.
31. Pöyry, 2009. Impact of Intermittency, How Wind Variability Could Change the Shape of British and Irish Energy Markets [.pdf], Available at: www.poyry.co. uk/sites/www.poyry.uk/files/IntermittencyMethodologyv1 0.pdf (accessed March 2015).
32. Notz, R., Mangalapally, H.P., Hasse, H., 2012. Post combustion CO2 capture by reactive absorption: pilot plant description and results of systematic studies with MEA. Int. J. Greenh. Gas Control 6, 84–112.
33. Rabensteiner, M., Kinger, G., Koller, M., Gronald, G., Hochenhauer, C., 2014. Pilot plant study of ethylenediamine as a solvent for post combustion carbon dioxide capture and comparison to monoethanolamine. Int. J. Greenh. Gas Control 27, 1–14.
34. Seibert, F., Chen, E., Perry, M., Briggs, S., Montgomery, R., Rochelle, G., 2011. UT/SRP CO2 capture pilot plant – operating experience and procedures. Energy Procedia 4, 1616–1623.
35. Ystad, M.P.A., Bolland, O., Hillestad, M., 2012. NGCC and hard-coal power plant with CO2 capture based on absorption. Energy Procedia 23, 33–44.

CHAPTER 19

CONTRIBUTIONS TO THE ANALYSIS AND THE SUPERVISION OF A THERMAL POWER PLANT

Lakhoua M.N, Ben Hamouda M, Glaa R and El Amraoui L

The National Engineering School of Carthage, University of Carthage, Research Unit: Signals & Mechatronic Systems, Tunisia

ABSTRACT

Supervision systems play an important role in industry mainly due to the increasing demand for product quality and high efficiency, and to the growing integration of automatic control systems in technical processes. In fact, the supervision system has a great number of components and interconnections, and it is difficult to describe and understand its behavior. Furthermore, the supervision system in industrial plants, implemented in supervisory control and data acquisition (SCADA) software, must undertake, at least, the following three main tasks: monitoring, control and fault tolerance. So it can be classified as a complex system. The objective of this paper is to show interests of the use of functional analysis techniques such as SADT (Structured Analysis and Design Technique) and SA-RT (Structured Analysis Real Time) for the design of supervisory systems. This is why we present a general model of analysis and supervision of production systems. This model was based on the one hand on the functional analysis (FA) and on the other hand on the SCADA system.

INTRODUCTION

In our days in different industrial areas it become very important to analysis, control and supervise the production systems. In fact, the most used are supervisory control and data acquisition (SCADA) systems; these are used to control systems in different areas, to collect information and to centralized data acquisition. One of the most important concerns in the SCADA design is to assure whether the system could go with the reliability and performance requirements specification. The production systems of increasing complexity present difficulties at the level of their functional analyses. Several types of functional and structural analysis methods exist and are used in many industrial domains since about thirty years. In fact, the functional analysis (FA) permits the synthetic description of fashions of working of a system and the knowledge of functions to guarantee [1]. It establishes of systematic and exhaustive way the functional relations inside and outside of the system. Again, the FA consists in searching for and to characterize functions offered by a system to satisfy its user's needs. The objective of FA methods is to describe and to understand the system. It is gaits that put the accent on what makes the system, to master the complexity and to

Citation: Lakhoua M.N, Glaa R., Ben Hamouda M., El Amraoui L. Contributions to the Analysis and the Supervision of a Thermal Power Plant. International Journal of Advanced Computer Science and Applications. 2016;7(2)93-101 DOI 10.14569/IJACSA.2016.070213

detect defaulting. So the FA serves to model the system while clearing constraints and while specifying the different fashions of working [1].

Many well-known graphical modeling techniques, for example, the Structured Analysis Real Time (SA-RT) method developed by Hatley and Pirbhai [2], provide mechanisms that allow one to: (1) abstract over details best left to later stages of development, (2) model an application along different views, and (3) modularize problems and solutions. It has also been affirmed that the simple and graphical manner of the modeling constructs facilitates the construction of concise and understandable models [3].

This paper can be loosely divided into five parts: first, we summarize the complexity of supervision systems particularly to design SCADA systems and second briefly describe the characteristics of a SCADA system and the problems related to its design. Subsequently, we present the issues involved in the analysis and supervision of the production systems. In section 3, after presenting concepts of SADT and SA-RT, a model of analysis and supervision of production systems is presented. Then, the benefits of using SADT and SA-RT in the design steps are developed. Next, the results of the proposal methodology was used on a case study of a SCADA system of a thermal power plant (TPP) particularly in programming and visualizing new chemical analysis parameters of the water - steam cycle. In the last section, we present a discussion about the advantages and inconveniences of the methodology used in order to describe and understand the behavior of supervision systems.

DESIGN OF SCADA SYSTEMS

With the advances of electronic and software technologies, the SCADA systems are generally used in industrial plant automation. It provides an efficient tool to monitor and control equipment in industrial processes. In fact, SCADA systems can be found in critical infrastructures such as power plants [4] [5] and power grid systems, water, oil and gas distribution systems, building monitoring, production systems and other products. In a typical SCADA system, data acquisition and control are performed by remote terminal units (RTUs) and field devices that include functions for communications and signaling. In fact, the control performed by SCADA is much more superficial and generally applied to correct a fault and the SCADA system monitors and provide control actions based on this monitoring [6].

The whole automation process is done using programmable logic controller (PLC) which has number of unique advantages like speed, reliability, less maintenance cost and reprogram ability [7]. Without reasonably designed system software architectures and hardware structures, it is impossible to handle these tasks efficiently, safely and reliably, with the possibility of online reconfiguration and flexibly embedding applications. Designing, monitoring and controlling such systems is becoming more and more challenging as a consequence of the steady growth of their size, complexity, level of uncertainty, unpredictable behavior, and interactions.

Moreover, SCADA systems have been applied in flexible manufacturing cells for educational purposes in different automation engineering fields [8]. Design and implementation of a low cost compact modular production system controlled by a SCADA system was developed for educational purposes [9]. In this way, students have access to instruments in a lab via the Internet, all by improving their skills related to SCADA systems, used in industry [10]. We present in this part some applications of SCADA systems design that have been presented in various researches. Researchers, Ponsa P. & al. [11], have described how Human-Machine-Interfaces have received appropriate attention in order to improve the design of SCADA applications in industrial domain. It is because of a major concern about aspects related to maintenance, safety, achieve operator awareness, etc has been gained. Even there are in the market software solutions that allow for the design of efficient and complex interaction systems, it is not widespread the use of a rational design of the overall interface system.

The researchers have provided an example of such development also by showing how to include the automation level operational modes into the ergonomic interfacing system. Researchers, Gezer D. & al. [12], have presented a methodology and a case study through which system architecture and dynamic models of related system components are identified in order to design and simulate the SCADA system of a new hydro turbine test laboratory. System architecture model is prepared in System Modeling Language, a system modeling language based on Unified Modeling Language, while the dynamic model of the laboratory is formed in Matlab/Simulink. Some simulations are performed in order to verify the preliminary system design studies and system requirements.

Researchers, Dong W. & al. [13], have presented a remote monitoring system that has been designed by using SCADA. Specifically, the monitor signals of boiler dry spot's worker are been composed first, and a real time database is realized on the basis of the collected information. In summary, these approaches could extend the system function and optimize the system structure. More importantly, it obtains the real-time data of web monitor system, which can draw the real-time curve of the dynamic systems and the curve of history, etc. Furthermore, it could achieve real-time brushless description.

Researchers, Morosan A. & al. [14], have proposed an architecture for the SCADA system, used for making wellorganized production in manufacturing system, in order to can combine all the elements of a flexible manufacturing line. The main characteristic of this SCADA system is to supervise and to control the manufacturing process from a flexible manufacturing line. Researchers, Chun-Lien S. & al. [15], have proposed an approach to take the system performance into account in the reliability analysis of the SCADA system. This approach is based on a model for evaluating data transmission time, which allows us to find the operation time needed to complete SCADA functions. With information on performance, the reliability evaluation technique using fault tree analysis is described and applied for analysis of SCADA system component connectivity to assess the availability of SCADA controls. A sensitivity analysis is also described and used to illustrate the effects of input data uncertainty on the system reliability.

Researchers, Guobing H. & al. [16], have studied an embedded SCADA system in order to manage and control industry objects directly at the production line terminal. A solution is recommended for an embedded SCADA system based on the development of embedded computer and network communication. The design and implementation methods are introduced for technical structure, hardware and software of this system. The first 10 units have been brought into use at power plants and power monitor centers since 2010. The result of site acceptance indicates that measuring precision and reliability follow correlative standards. In the following part, we present the two structured analysis methods.

PRESENTATION OF STRUCTURED ANALYSIS METHODS

This section presents two structured analysis methods used in this research. These methods are: Structured Analysis Design Technique (SADT) and Structured Analysis Real Time (SA-RT). SADT, which was designed by Ross in the 1970s [17], was originally intended for software engineering but quickly other areas of application were found, such as training, operations, manufacturing, finance, etc. Although SADT (Figure 1) does not require any specific supporting tools, a number of computer programs implementing SADT methodology have been developed. Among of them is Design: IDEF, which implements IDEF0 method [18]. Furthermore, SADT/IDEF0 is a confirmed way to model any kind of domain [19].

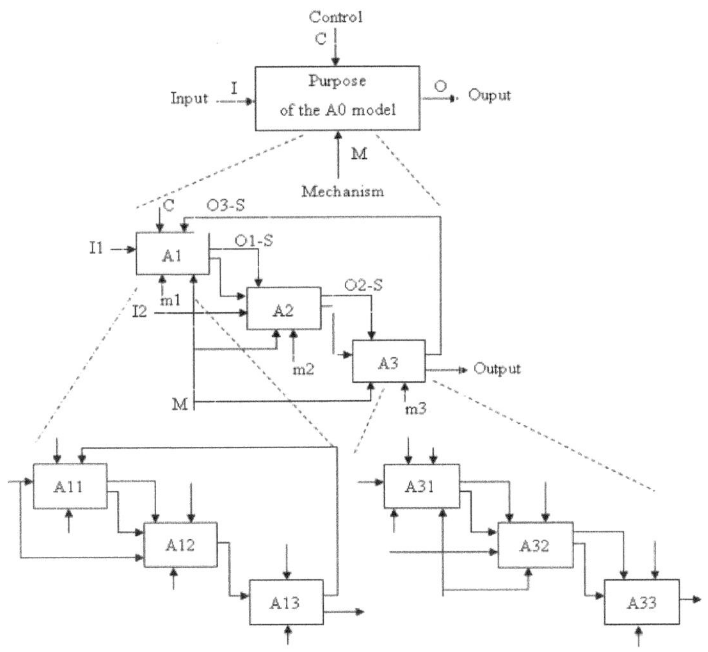

Figure 1: Top-down, modular and hierarchical decomposition of SADT.

Researcher, DeMarco, one of the pioneers of structured analysis (SA), described the need for such a method more than 30 years ago. He suggested that the products of analysis must be maintainable; problems of size must be dealt with using the effective method of partitioning; graphics must be used whenever possible [20]. Then, he proceeded further to establish requirements for the SA method as follows: the method should help us partition our requirements and document that partitioning before specification; it should give us means of keeping track of and evaluating interfaces; it should facilitate the development of new tools to describe logic and policy better than narrative text.

The notation used for creating SA diagrams consists of:

- Data Flow Diagrams (DFDs): these diagrams model the processing of information in terms of data flows; data processing nodes are defined to represent the data processing functions from the requirements of the system.
- Control Flow Diagrams (CFDs): these diagrams model the processing of information in terms of control flows; control nodes are defined to represent the control functions from the requirements of the system under analysis.
- Process specifications: these specifications are used to describe the details of the data processing nodes defined in a DFD. These specifications consist of scripts of pseudo-code or just plain text which explains how the output flows of a particular processing node is generated from its input flows.
- Control specifications: these specifications are used to describe the details of controls nodes in a CFD. These specifications define the behavioral model of the controllers and specify how the output control flows are obtained from the input control flows. They also specify when the data processing nodes are activated or deactivated.
- Data Dictionary: it defines all the information flows and the data and control stores in the system. It contains text that defines each information item and its value range.
- Entity-Relationship Diagrams (ERDs): these diagrams provide an information model for the data items and control signals and the relationships among these data items.

Figure 2 shows three dimensional views of the SA model. The top layer, which is the highest level in the SA-RT model, consists of what is commonly termed as the context diagram (CD). It describes the boundary of the software under analysis as well as the external interfaces and the external entities.

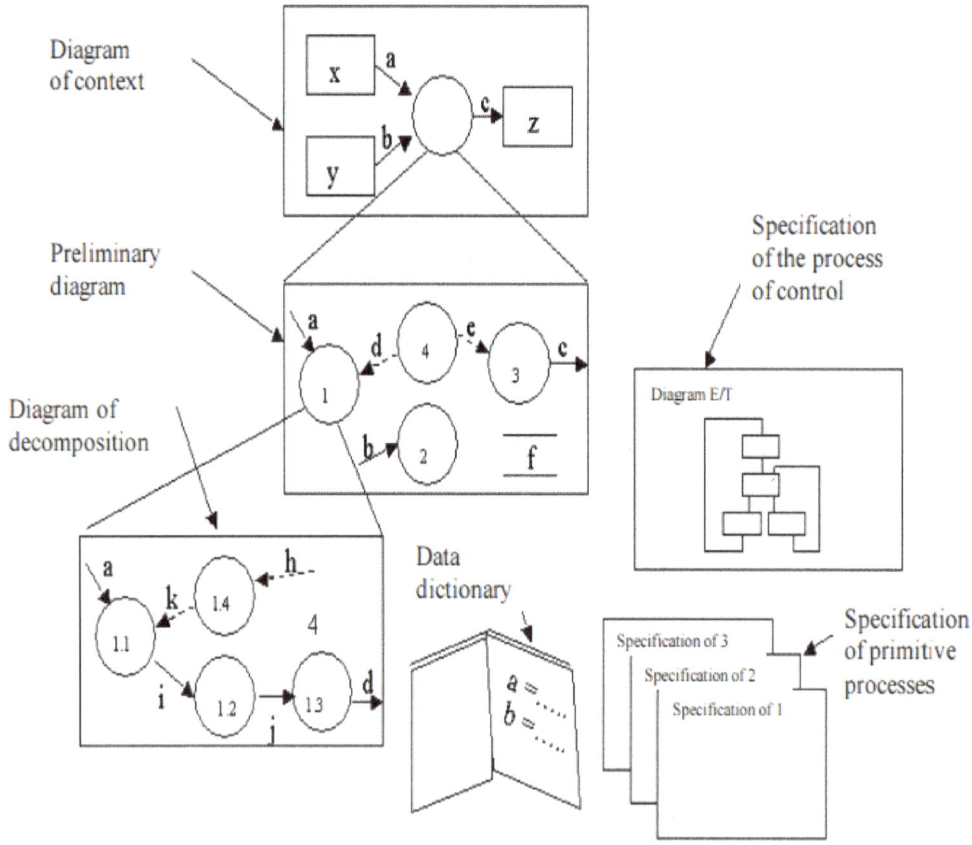

Figure 2: Components of SA-RT and their relationship with each other.

The following level of the hierarchy is a DFD which represents the major functions outlined in the functional requirements. These functions represent a top level decomposition of the software under developed. Consequently the whole DFD0 is viewed as the child of the process representing the system in the top level. Each process in DFD (numbered 1, 2, 3, etc.) will have either its own child DFD or a P-spec sheet describing it in more detail. Also C-spec sheets are used to specify the controllers shown in each DFD. We present in this part some applications of the SADT and SA-RT methods that have been presented in various researches:

Researchers, Benard V. & al. [21], have described the SafeSADT method that enables the explicit formalization of functional interactions, the recognition of the characteristic values affecting the dependability of complex systems, the quantification of the reliability, availability, maintainability, and safety parameters of the system's operational architecture, and its validation in terms of the dependability objectives, as well as constraints set down in the functional requirement specifications.

Researchers, Lauras, M. & al. [22], have presented an approach based on GRAI, SADT/IDEFO that enables the integration of the best practices defined by these methods. In addition the indicators next the results and determinants, three types of indicators are introduced to analyse the performance: the facility viewpoint measures, the appropriateness of the resources available and the determinants of the activity.

Researcher, Marca D.A. [23], has explained how SADT/IDEF0 domain modeling can bring precise and complete context, to today's commonplace disciplines of the UML (Unified Modeling Language), Agile System Development, and Usability Engineering methods. In fact, the power and the rigor of SADT/IDEF0 come from: (1) a synthesis of graphics, natural language, hierarchical decomposition, and relative context coding, (2) distinguishing controls from transformations, (3) function activation rules, and (4) heuristics for managing model complexity.

Researchers, Jimenez F. & al. [24], have developed models and tools for system design and synthesis of MEMS-micro based on SDL (specification description language), SA-RT and PNs. In fact, a main problem concerns the design of these varied circuits because it associates disciplines such as electronics, mechanics, chemistry, etc.

GENERAL MODEL OF ANALYSIS AND SUPERVISION

We present in this part, the development of a general model for the FA and the supervision of a production system in a SCADA environment. The objective of the work is therefore the establishment of a general methodological gait of functional modeling and supervision of the production systems permitting to reach the aimed objectives [35]. The proposed model articulates around three essential phases: functional analysis of production systems; analysis of control-command applications; supervision of the control command applications in a SCADA environment.

Phase 1: Functional Analysis of the Production Systems

The first phase consists in proceeding to a functional modeling of the production systems while using the SADT method. The structure of this method permits to master the complexity of the process thanks to its downward and modular analysis.

The different stages of such a FA by the SADT method of the production systems are:

- To prepare the SADT model: the creation of a SADT model starts with the definition of two concepts that is the goal that stationary objectives of the model and the point of view that establish for what auditorium model it is created. These two determined concepts permit to guide the decomposition of every box;
- To create the diagram A-0 representing the general activity to analyze in an actigram;
- To create the diagram of activity A0: one writes down inside boxes drawn on a new form (3 to 6 boxes) the name of activities and one draws arrows of interfacing between the different boxes;
- From the diagram A0, to select the least complex box and that will give more information to create the diagram by decomposition have ($1 < i < 6$) with corresponding i to the number of the box in A0;
- To redo the same principle of decomposition for the other diagrams of activity;
- To represent all diagrams of the model on the SADT forms.

This first phase of the model enables us the decomposition of the production system under a hierarchized manner to bring back it to the elementary situations and to elaborate a static model describing the activities of the process.

Phase 2: Analysis of the Control-Command Applications

Once functions and activities of every function have been identified, the following stage consists in analyzing control command applications while using the SA-RT method.

The different stages of such a FA by the SA-RT method of control-command applications are:

- To establish the CD while representing the different terminations as well as the incoming and retiring data streams;
- To establish the preliminary diagram (DFD0) that represents the necessary functional process list to the application with the stream of data correspondents.
- To establish the diagram of decomposition partner streams to one of the functional processes identified;
- To establish the DFC representing the controls aspect the SA-RT method;
- To establish the State/Transition diagram representing the behavior aspect or real time of the application;
- Functional process specification (dictionary of data) while leaning on a procedural specification.

This second phase of the model enables us the decomposition of the studied application under a hierarchized manner to bring back it to the elementary situations and to drive to the development of a dynamic model describing the various processes of the application as well as flows of data and control.

Phase 3: Supervision of the Control-Command Application in a SCADA Environnement

After a various process analysis and flows of data and control of the application, different static and dynamic model are elaborated. The last stage of the model of the analysis and the supervision of the production system consists in the establishment of tools and methods of supervision, control and date acquisition using a SCADA system as well as the development of Man - Machine interfacings. In fact, the supervision is the whole of tools and methods that permit to drive some industrial facilities so much in normal working that in presence of failings. It is the tool of reference of the conduct operator but can interact also directly with the control command application. The supervision has for objective to supervise and to control the working of an installation so that it remains in the normal working whatever are the outside disruptions. It permits to detect in real time mistakes from alarms and to identify their reasons. In the following part, we present the results of FA and supervision on an example of a SCADA system of a thermal power plant (TPP) as well as the development of different phases of the general model proposed.

RESULTS OF FUNCTIONAL ANALYSIS AND SUPERVISION

In this section, we present on the one hand an example of a SCADA system in a TPP in Tunisia and on the other hand a control-command application of the water steam cycle of the TPP. By means of a significant example, the objective of this work is to validate the general model of analysis and supervision presented in the previous section [31], [32].

Presentation of the SCADA System and a Case Study of a Water Steam Cycle

In Tunisia, the TPP of Rades (near to Tunis) is composed of thermal power stations that are one of the most important stations of the electric energy production in Tunisia (37% of the national production) [33], [34]. In fact, the SCADA system is used by the power station operators in order to supervise the good working of two production slices of the TPP. This centralized supervision permits to operators, since the control room to control facilities in their domain of exploitation and to treat, in real time, the different types of incidents (Figure 3).

Figure 3: Control room of the TPP.

The electric energy production in the TPP of Rades is based on a set of energies transformations using water as energy support. This water must have a high quality in order to guarantee the security of the installation and to improve the production groups' performances. It is therefore necessary to apply a rigorous treatment of the raw water and a stern control of its quality.

Figure 4 shows the architecture of the SCADA system of a TPP. The pretreatment is constituted of two filtration chains each including a sand filter and an active coal filter. Thereafter, the water passes by the filtration chain then introduced in the inverse osmosis station and thereafter in the demineralization station (Figure 5).

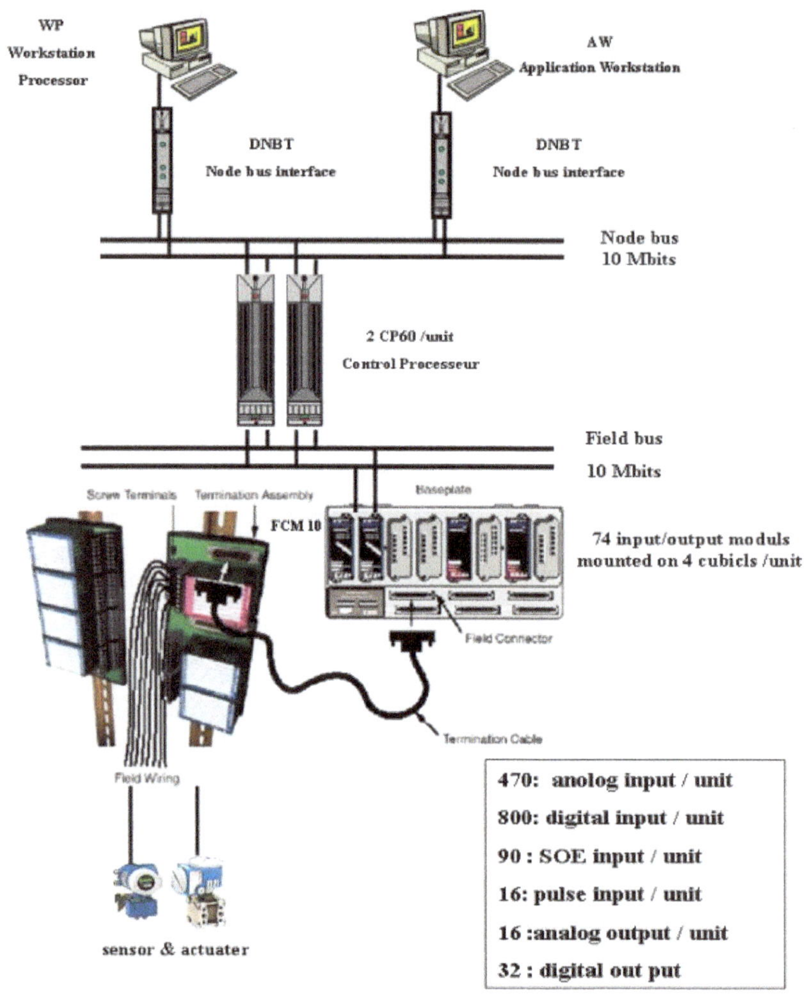

Figure 4: SCADA system of the TPP.

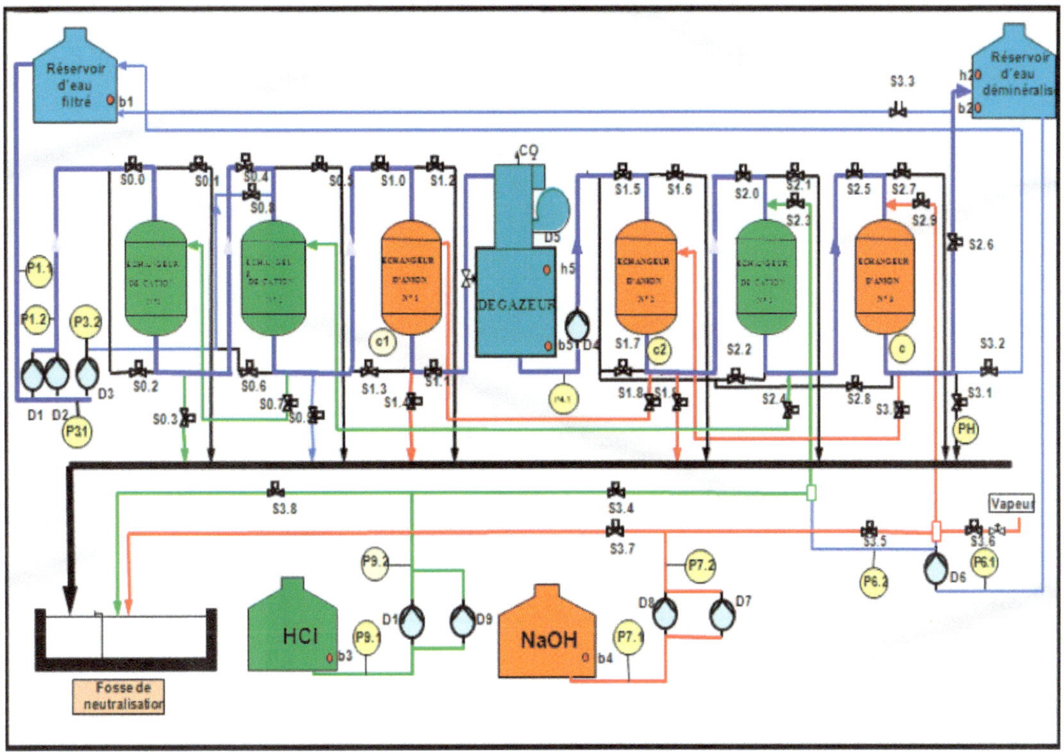

Figure 5: Demineralization station of the TPP.

Results of Analysis

According to the first phase of the methodology established, a FA using the SADT method has been done (Figure 6). This analysis is a very interesting stage because it permits to describe the electric energy process of the TPP. The second phase of the proposed methodology consists on the use of the SA-RT method for the analysis of a control command application of the TPP. Figure 7 shows the CFD of the water-steam station of the TPP.

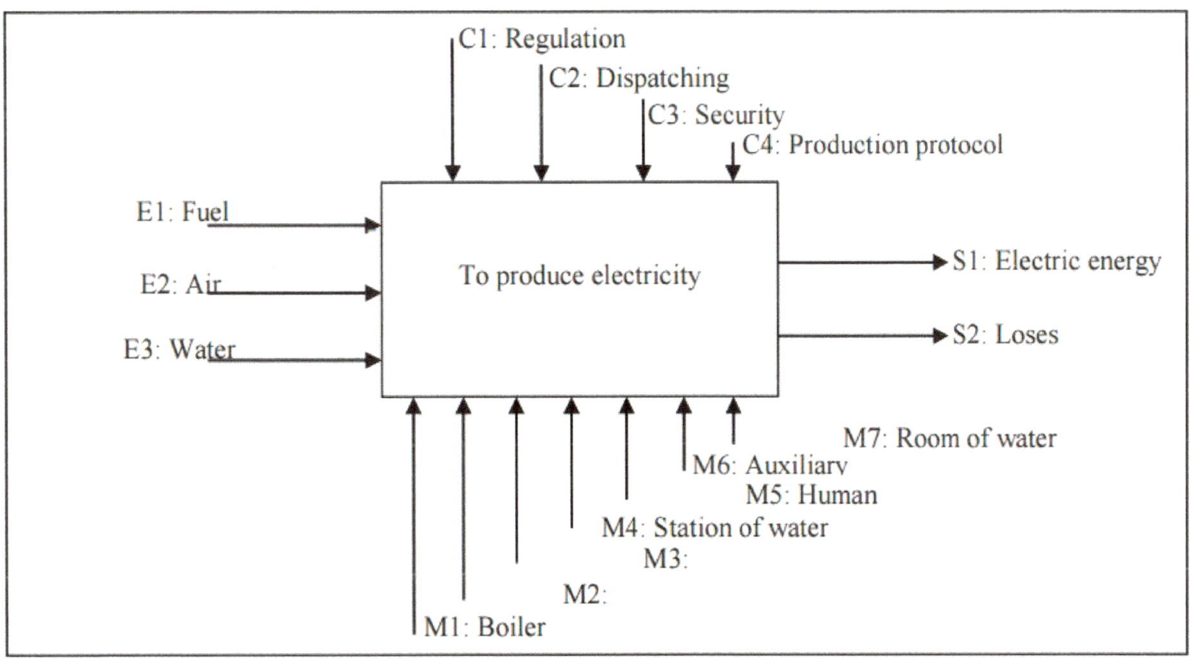

Figure 6: Node A-0 of the SADT model of a TPP.

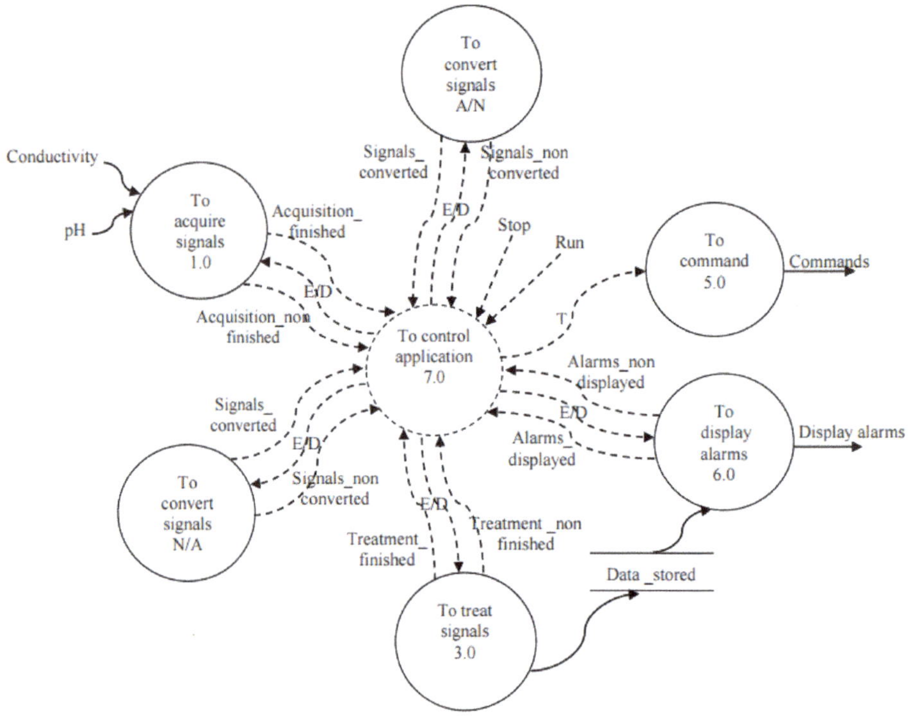

Figure 7: Control Flow Diagram of the SCADA system of the water-steam station.

The CD of the SA-RT model is constituted of one functional process « To pilot the control-command application 0 » and the terminators. It defines perfectly the interface between the designer and the client, that is, to provide or generate data in order to display these data on the tabular of the control-command application. The DFD of the SA-RT model constitutes the first decomposition of the process presented in the CD. Then, we can identify the initial functional processes of the control command application: acquirement process; treatment process; Human-Machine-Interfaces process. The CFD of the SA-RT model includes the control aspect to the DFD elaborated. In fact, the implementation of the monitoring process at the level of preliminary diagram can express the sequence execution of the functional processes. Figure 8 shows the state-transition diagram of the SA-RT model of a control-command application.

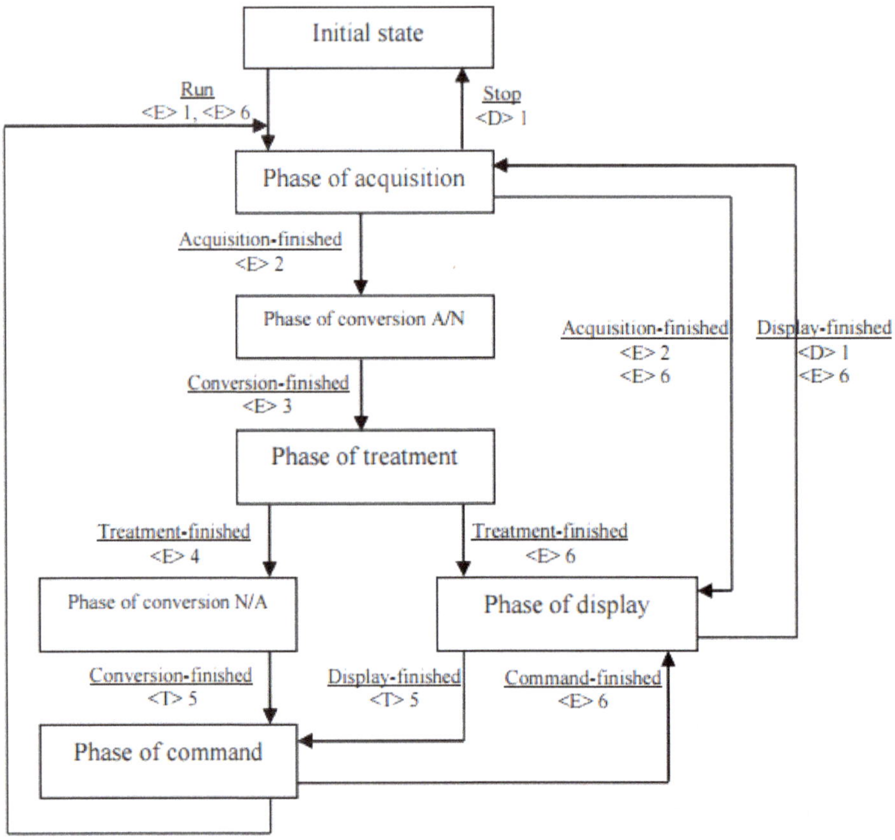

Figure 8. State/Transition diagram of a SCADA system of the water-steam station.

The potential uses for the SA-RT model are the design of the monitoring display and the diagnosis display. For the design of a monitoring display, the preliminary DFD of the SA-RT model presents an overall view of the control command application. Indeed, information relative to each process represented through this level should appear in the monitoring display. For the design of hierarchical diagnosis display, each DFD of the SA-RT model constitutes a vision at a given abstraction level. So, each of these DFDs gives a less or more detailed vision. In function of the objectives defined by the designer for each display, a particular DFD can supply the required information. Finally, this application of the SA-RT method on the SCADA system of a TPP shows briefly the interests of the FA the design of supervisory systems.

Results of Supervision

The last phase of the methodology proposed consists on the supervision of the control-command application. For the example of the water-steam cycle, we study the interfacing of the different signals: pH and conductivity of the ball furnace of the TPP. The application is declined in six stages: 1) choosing the site of the signal (FBM module); 2) programming both AIN and CIN blocks for the supervision of the signals pH (4 to 20 mA) and conductivity (alarm); 3) testing both AIN and CIN blocks by injection of current and by short circuit; 4) passing the cable between the sampling room and the SCADA room; 5) connecting the signals in the two modules 10FBM215 and 10FBM325; 6) conceiving a new tabular for the general vision of the sampling room. Figure 9 shows the display of the sampling room containing the chemical analysis parameters of the water-steam cycle.

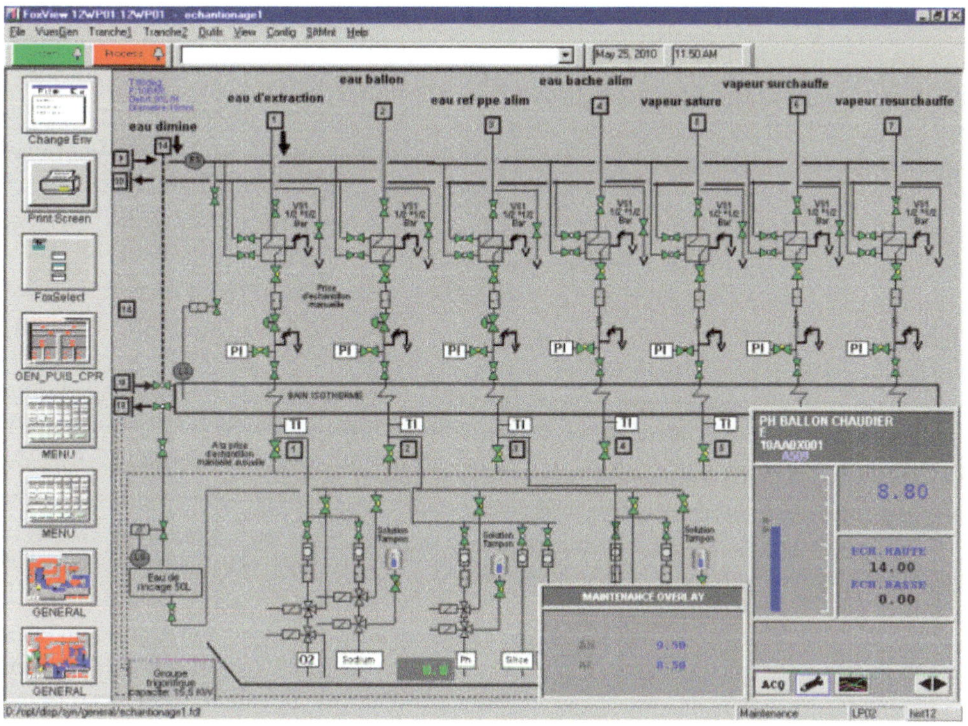

Figure 9: Display of the chemical analysis parameters.

In the first step is used the FoxView software which represents the interfacing operator to visualize the synoptic with a menu bar to activate the main functions of the SCADA system. Then, in the second step is used the ICC (Integrated Control Configurator) to create and to configure programs residing in the CP (Control Process). To access ICC, we need from the menu bar Config choose term Control_Cfg-CIO_STN_Cfg and Config_station_name. It is used to configure the various blocks AIN, CIN, CALC, etc. Indeed, during the programming of a new signal it must be identified: the signal label; the compound and the address signal. For simulation is used FoxSelect software that permitting to reach the various elements of the hierarchy of data base of the CP. The last step of this application is to improve the synoptic the pH meter and the conductivity meter and the design of a new synoptic of the sampling room containing these two data from the Fox Draw software. It provides tools to design displays for monitoring, alarming and process control. To access FoxDraw we can choose function ConfigFoxDraw from the menu bar. Included with FoxDraw is a large library of graphical components ready to be integrated and configured in displays. Thus, we proceeded as follows: creating of pH-meter blocks; operating test of the pH-meter; display configuration of different alarms; configuring of the overleay pH and testing the overleay pH-meter.

CONCLUSION

In order to understand our world we need to think that almost everything can be regarded as a system, where signals must be monitored, controlled and ultimately supervised. In fact, the supervision of the production systems should by nature try to guarantee the observability, the controllability and, most important, the system stability. In this paper, we presented a methodology of analysis and supervision of the production systems. We have suggested three main directions to achieve that methodology: (1) functional analysis of the production systems; (2) analysis of the control-command applications; and (3) supervision of the control-command applications in a SCADA environment. Two interesting functional analysis methods SADT and SA-RT are used on the one hand and a SCADA system for monitoring, control and fault tolerance, on the other hand. In fact, allowing the running of the production equipment to be understood, these techniques permit designers to decide the good information to display through the supervisory interfaces devoted to each type of supervisory task (monitoring, diagnosis, action...). In addition, functional analysis techniques might be a good assist to design support systems such as alarm filtering systems. Staring from this case study of the analysis and supervision of the production systems discussed in this paper, work is in progress to develop a functional analysis and real time for different control-command applications in different production systems.

REFERENCES

1. M. Lambert, B. Riera and G. Martel, Application of functional analysis techniques to supervisory systems, Reliability Engineering and System Safety 64, 1999, pp. 209-224.
2. D.J. Hatley and I.A. Pirbhai, Stratégies de spécification des systèmes temps réel (SA-RT), Masson, Paris, France, 1991.
3. F. Cottet, Systèmes temps réel de contrôle - commande, Dunod, Paris, 2005.
4. M.N. Lakhoua, SCADA applications in thermal power plants, International Journal of the Physical Sciences, Academic Journals, vol.5, N°7, 2010, pp. 1175-1182.
5. M.N. Lakhoua, Surveillance of pumps vibrations using a SCADA, Control Engineering and Applied Informatics, Romanian Society of Control Engineering and Technical Informatics, vol.12, N°1, 2010.
6. Y.L. Kaszubowski, R.S. Rosso, A. Leal, E. Harbs and M.S. Hounsell, Finite Automata as an Information Model for Manufacturing Execution System and Supervisory Control Integration, Proceedings of the 14th IFAC Symposium on Information Control Problems in Manufacturing, Bucharest, Romania, May 23-25, 2012.
7. E.I. Gergely, Dependability Analysis of PLC I/O Systems Used in Critical Industrial Applications, Sudies in Computational Intelligence, 417, Springer, 2013, pp. 201-217.
8. S. Reynard, O. Gomis, F. Bellmunt, A. Sudrià, O. Boix, and I. Benítez, Flexible manufacturing cell SCADA system for educational purposes, Comput Appl Eng Educ 16, 2008, pp. 21-30.
9. R. Yenitepe, Design and implementation of a SCADA controlled MTMPS as a mechatronics training unit. Comput Appl Eng Educ 20, 2012, pp. 247-54.
10. Z. Aydogmus and O. Aydogmus, A web based remote access laboratory using SCADA, IEEE Trans Educ, 52, 2009, pp. 126–132.
11. P. Ponsa, R. Vilanova , A. rez, and B. Andonovski , SCADA design in automation systems, 3rd Conference on Human System Interactions (HSI), 2010, pp. 695 – 700.
12. D. Gezer, H.O. Unver , Y. Tascioglu , K. Celebioglu and S. Aradag, Design and simulation of a SCADA system using SysML and Simulink, International Conference on Renewable Energy Research and Applications (ICRERA), 2013, pp. 1058-1062.
13. W. Dong and S. Xian-li, The boiler design of remote monitoring system based on the SCADA, Chinese Automation Congress (CAC), 2013, pp. 864- 869.
14. A. Morosan and F. Sisak, A SCADA system designed for making more efficient production in flexible manufacturing system, IEEE 13th International Symposium on Computational Intelligence and Informatics (CINTI), 2012, pp. 409- 413.
15. S. Chun-Lien and C. Ya-Chin , A SCADA system reliability evaluation considering performance requirement, International Conference on Power System Technology, 2004, vol.1, pp. 574-579.
16. H. Guobing, M. Guoqiang, Z. Jianmin and D. Jianling, Design for an embedded SCADA system, International Conference on Electrical and Control Engineering (ICECE), 2011, pp. 5819-5821.
17. D.T. Ross, Structured Analysis (SA): A language for communicating ideas, IEEE Transaction on Software Engineering, 3(1), 1977, pp. 16-34.
18. IEEE 1320.1-1998. IEEE Standard for Functional Modeling LanguageSyntax and Semantics for IDEF0, IEEE, 1998.
19. M.N. Lakhoua, Systemic analysis of an industrial system: case study of a grain silo, Arabian Journal for Science and Engineering, ISSN: 1319- 8025, vol.38, 2013, pp. 1243-1254.
20. T. DeMarco, Structured Analysis and System Specification, PrenticeHall Inc., New Jersey, 1979.
21. V. Benard, L. Cauffriez and D. Renaux, The Safe-SADT method for aiding designers to choose and improve dependable architectures for complex automated systems, Reliability Engineering & System Safety, 93(2), 2008, pp. 179-196.
22. M. Lauras, J. Lamothe, and H. Pingaud, Une méthode orientée processus pour le pilotage par la performance des systèmes industriels, Journal Européen des Systèmes Automatisés, 41(1), 2007, pp.71-100.
23. D.A. Marca, SADT/IDEF0 for Augmenting UML, Agile and Usability Engineering Methods, Marca, David., 2012, Software and Data Technologies, pp. 38-55.
24. F. Jimenez, M. Courvoisier, A. Garcia, G. Munoz , N. Harchani, M. AlMohamed and D. Esteve, Tools and models for systems design and synthesis of MEMS based on asynchronous circuits, IEEE International Conference on Industrial Technology, vol.1, 2000, pp. 64- 69.
25. I. Kuuluvainen and V. Ylitolva, The Bubble Operating System: a control extension to data flow diagram, IEEE Transactions on Consumer Electronics, 1991, Vol.37, Issue: 3, pp. 642- 650.
26. A. Flo, M. Kjaernes and A. Skomedal, A bridge from structured analysis (SA/RT) to specification and

description language (SDL), Eighth International Conference on Software Engineering for Telecommunication Systems and Services, 1992, pp. 93-97.

27. N.Sahraoui, Applying specification methods to complex systems, IEEE International Conference on Systems, Man, and Cybernetics, vol.5, 1997, pp. 4488- 4491.
28. L. Urbain and B. Tondu, Robot controller specification using SART approach, Proceedings of the Third IEEE Conference on Control Applications, 1994, pp. 303-308.
29. S. Lihua and J.A. Keane, Algorithmic aspects of hierarchical verification for SA/RT models, IEEE International Conference on Computational Cybernetics and Simulation, vol.3., 1997, pp. 2252- 2257.
30. R.B. France and J. Bruel, Using Integrated Formal and Informal Modeling Techniques to Analyze Software Requirements: A PetriNet/SART Case Study, 1996.
31. M. Ben Hammouda, M.N. Lakhoua and L. El Amraoui, Dependability evaluation and supervision in thermal power plants, International Journal of Electrical and Computer Engineering, vol. 5, N°5: October 2015.
32. M.N. Lakhoua and H.Laadhari, Supervision of the natural gas station using a SCADA System, Journal of Electrical and Electronics Engineering, vol.6, n°1, May 2013.
33. M. Ben Hamouda and M.N. Lakhoua, Methodology of Operating Safety and Supervision of a Production System, CISTEM, IEEE, 3-6 Nov. 2014, Tunisia.
34. R. Glaa and M.N. Lakhoua, Methodology of Analysis and Design of a SCADA System, CISTEM, IEEE, 3-6 Nov. 2014, Tunisia.

CHAPTER 20

HEAT TRANSFER IN FUEL OIL STORAGE TANK AT THERMAL POWER PLANTS WITH LOCAL FUEL HEATING

Svetlana A. Kuznetsova and Vyacheslav I. Maksimov

Institute of Power Engineering, National research Tomsk Polytechnic University, Tomsk 634050, Russia

ABSTRACT

Results of mathematical modeling of the thermal control system in fuel oil storage, in the presence of heat source at the lower boundary of the region, in the framework of models of incompressible viscous fluid are presented. Solved the system of differential equations of non-stationary Navier-Stokes equations, the energy equation and the heat equation with appropriate initial and boundary conditions. Takes into account the processes of heat exchange region considered with the environment. A comparative analysis of the dependence of average temperatures of oil in the volume of the tank on the time calculated by the simplified (balanced) method and obtained as a result of numerical simulation are performed.

INTRODUCTION

The main problems of use (handling, storage, supply) fuel oil at thermal power plants in the colder periods of the year are related with the dependence of the viscosity of this fuel on temperature [1]. In the preparation of fuel oil for burning in the winter, as a rule, expended a lot of energy on its heated to a temperature corresponding to the procedural regimes. Until now the choice of process parameters and schemes of heating such liquid fuel is mainly conducted using relatively simple models [2], do not take into account possible spatial character of the temperature distribution in the conditions of local heating the large volumes of this liquid fuel. At the same time, probably may be reduced the energy consumption for heating oil through the optimal choice of scheme and conditions of the heat supply in fuel oil storage tank. But such choice is difficult on the results of experimental studies. It is expedient to modeling the thermal control system on fuel oil storage tank with the use of systems of equations describing the processes of heat conduction and convection occurring in the system "liquid oil - a source of heat - fuel oil storage wall". In physics under study process the considered problem has much in common with the problem of free convection in a cavity with conducting walls and the local energy source [3, 4]. A feature of heat and mass transfer at movement oil are its high density (compared to air [2]) and the dependence of viscosity on temperature. Within the model of incompressible viscous fluid heating regimes on fuel oil storage tank is practically not been studied until recently. Purpose of work - analysis

Citation: Svetlana A. Kuznetsova and Vyacheslav I. Maksimov. Heat transfer in fuel oil storage tank at thermal power plants with local fuel heating. MATEC Web of Conferences 23, 01047 (2015). DOI: 10.1051/matecconf/20152301047

of non-stationary thermal fuel oil storage mode using a heat transfer model that takes into account the thermal conductivity, natural convection, the local supply of energy and the heat sink through heat conduction and heat storage walls.

STATEMENT OF THE PROBLEM

It has been considered the fuel oil storage tank (Figure 1) of rectangular shape with horizontal and vertical walls of finite thickness and a heat source with a given surface temperature at the bottom edge region. At the initial time the temperature of the liquid fuel is greater than the initial temperature of the tank walls and the environment. It is assumed that at the boundaries of "liquid - solid wall" the condition of perfect contact, at the external borders the conditions of heat dissipation into the environment.

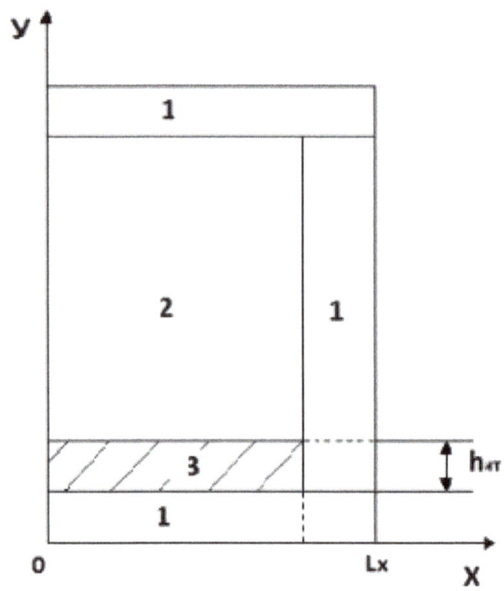

Figure 1: Area of solution the problem.

Formulation of the problem in the spatial coordinates is described by a system of two-dimensional and non-stationary NavierStokes equations and the energy equation for the liquid phase (oil) and heat conduction equation for reinforced concrete wall of the tank, by analogy with [3, 4].

$$\frac{\partial \Omega}{\partial \tau}+U\frac{\partial \Omega}{\partial X}+V\frac{\partial \Omega}{\partial Y}=\frac{1}{\mathrm{Re}}\left(\frac{\partial^2 \Omega}{\partial X^2}+\frac{\partial^2 \Omega}{\partial Y^2}\right)+\frac{Gr}{\mathrm{Re}^2}\left(\frac{\partial \Theta}{\partial X}\right), \quad (1)$$

$$\frac{\partial \Theta}{\partial \tau}+U\frac{\partial \Theta}{\partial X}+V\frac{\partial \Theta}{\partial Y}=\frac{1}{\mathrm{Re}\cdot \mathrm{Pr}}\left(\frac{\partial^2 \Theta}{\partial X^2}+\frac{\partial^2 \Theta}{\partial Y^2}\right), \quad (2)$$

$$\frac{\partial^2 \Psi}{\partial X^2}+\frac{\partial^2 \Psi}{\partial Y^2}=\Omega, \quad (3)$$

$$\frac{1}{Fo}\frac{\partial \Theta}{\partial \tau}=\frac{\partial^2 \Theta}{\partial X^2}+\frac{\partial^2 \Theta}{\partial Y^2}, \quad (4)$$

Here

$Gr=\frac{g\beta L^3\Delta T}{\nu^2}$ - Grashof number;

β– thermal coefficient of volume expansion; g – acceleration created by the mass forces; – coefficient of kinematic viscosity of the fluid; Re - Reynolds number; Pr - Prandtl number; F_0 - Fourier number. The initial conditions for the system of equations:

$$\Psi(X,Y,0)=0,\ \Omega(X,Y,0)=0,\ \Theta(X,Y,0)=0. \quad (5)$$

Boundary conditions:

- on the outer contour of considered area

$$Y=0,\ 0\le X\le\frac{l_1+l_2+l_{um}}{L}; \quad (6)$$

$$X=\frac{l_1+l_2+l_{um}}{L},\ 0\le Y\le\frac{h_1+h_2+H}{L}; \quad (7)$$

$$Y=\frac{h_1+h_2+H}{L},\ 0\le X\le\frac{l_1+l_2+l_{um}}{L}; \quad (8)$$

are set the boundary conditions of the second kind $\frac{\partial \Theta}{\partial n}=Ki$:

- on the axis of symmetry X = 0 conditions of the form

$$\frac{\partial \Theta}{\partial X}=\Omega=\Psi=0; \quad (9)$$

- at the internal borders between the solid and liquid phase, parallel to the axis 0Y accepted:

$$\begin{cases}\Psi=0,\ \frac{\partial \Psi}{\partial X}=0,\\ \begin{cases}\Theta_w=\Theta_f,\\ \frac{\partial \Theta_w}{\partial X}=\lambda_{w.f}\frac{\partial \Theta_f}{\partial X},\end{cases}\end{cases} \quad \text{at } X=\frac{l_1}{L},\ \frac{h_1}{L}\le Y\le\frac{h_1+H}{L}; \quad (10)$$

- at the internal borders between the solid and liquid phase, parallel to the axis 0X accepted:

$$\begin{cases}\Psi=0,\ \frac{\partial \Psi}{\partial Y}=0,\\ \begin{cases}\Theta_w=\Theta_f,\\ \frac{\partial \Theta_w}{\partial Y}=\lambda_{w.f}\frac{\partial \Theta_f}{\partial Y},\end{cases}\end{cases} \quad \text{at } Y=\frac{h_2}{L},\ \frac{l_1}{L}\le X\le\frac{l_1+l_{um}}{L}; \quad (11)$$

- On the local heat source are specified boundary conditions of the second kind:

$$\frac{\partial \Theta}{\partial n}=Ki \quad (12)$$

Here $Ki=\frac{qL}{\lambda_w(T_{it}-T_0)}$ Kirpichev number; λ_w – coefficient of thermal conductivity of the solid phase; λ_f – coefficient of thermal conductivity of the liquid phase; $\lambda_{w,f}=\lambda_w/\lambda_f$ – relative coefficient of thermal conductivity; q – heat flow on the outer boundary solutions.

The boundary value problem (1-4) with appropriate boundary and initial conditions was solved by finite difference method using the algorithm [5, 6] that is designed to the numerical solution of nonlinear problems of heat and mass transfer with inhomogeneous boundary conditions. Equations were solved sequentially; each time step begins with the calculation of the temperature field in the liquid and solid phases, and then solved the Poisson equation (3) for the vector potential. Then was determined the boundary conditions for vector components of the vorticity and solved the equations of motion.

ANALYSIS OF THE NUMERICAL MODELING RESULTS

Numerical studies have been conducted with the following values: T_0=293 K, 253≤T_e≤293 K, 100≤q≤150 kW/m^2. Ranges of variation of size characteristics were chosen from the condition of compliance with range of parameter changes in real-world implementations of such systems. Fig. 2 - 3 shows typical results of numerical investigations as isolines of the stream function and temperature fields.

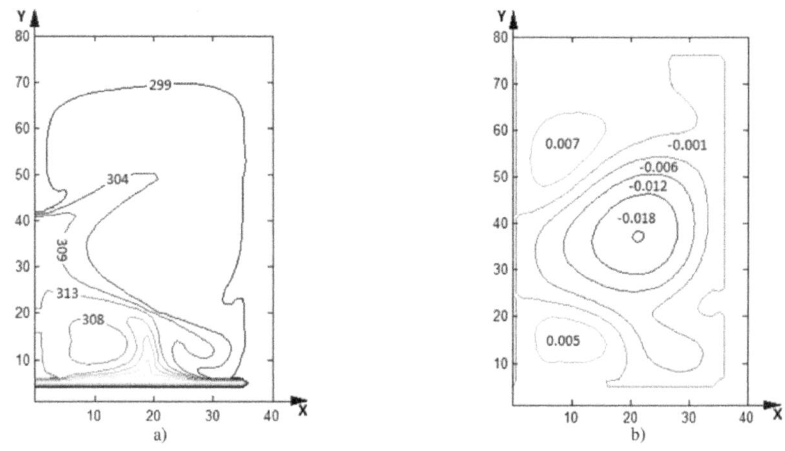

Figure 2: Temperature distribution (a) and contours of the stream function (b) at q=100 kW/m^2.

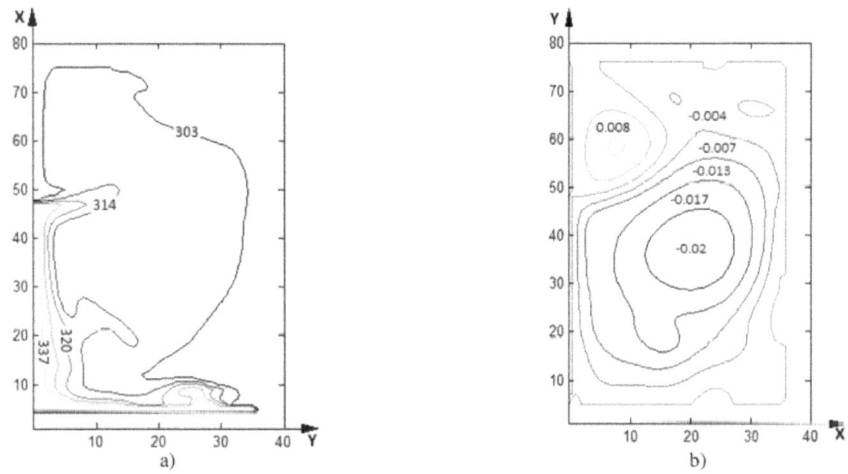

Figure 3: Temperature distribution (a) and contours of the stream function (b) at q=150 kW/m^2.

Figure 2 clearly shows the formation of three vortices in the central part of the tank as a result of natural convection when the heat flux on the heat source q = 100 kW/m^2. This character of vortex formation is due to the heat sink on the vertical boundary of the tank (the problem is solved in an axisymmetric formulation). The bulk of oil during heating process is raised to the top of the tank near its axis of symmetry, and cooled (by heat exchange on the vertical outer boundary of the solution) fuel is moved down. In this case clearly shows that, for example, during the time of warm-up about 3600s oil temperature field is not homogeneous (there are quite significant temperature gradients). Thus the maximum temperature to which the oil has time to warm up under given conditions, it is T_{max} = 320 K, while the minimum temperature is equal to T_{min} = 295 K. Increasing the intensity of the energy supply at the lower boundary (Fig.3.) to 150 kW/m^2 leads to a significant changes in both the hydrodynamics of flow and temperature fields. As a consequence becomes significantly more intense the vortex formation in the central part of the tank and accordingly oil is warmed up to high values of temperature.

Also worth noting the quite obvious asymmetry of the temperature field (Figure 3.), which is due to the heat sink on the outer vertical boundary of the tank. In this case, the maximum temperature of the heated oil during the accounting period of time will be T_{max} = 337 K, and the minimum will be equal to T_{min} = 299 K. In order to justify the advisability of determining the temperature fields on the results of the solution of the problem (1) - (12) were calculated the average temperature of oil in the volume of the tank. Also, average temperatures of oil were calculated by simplified (balanced) method [2]. Comparing the obtained results shown in Fig. 4.

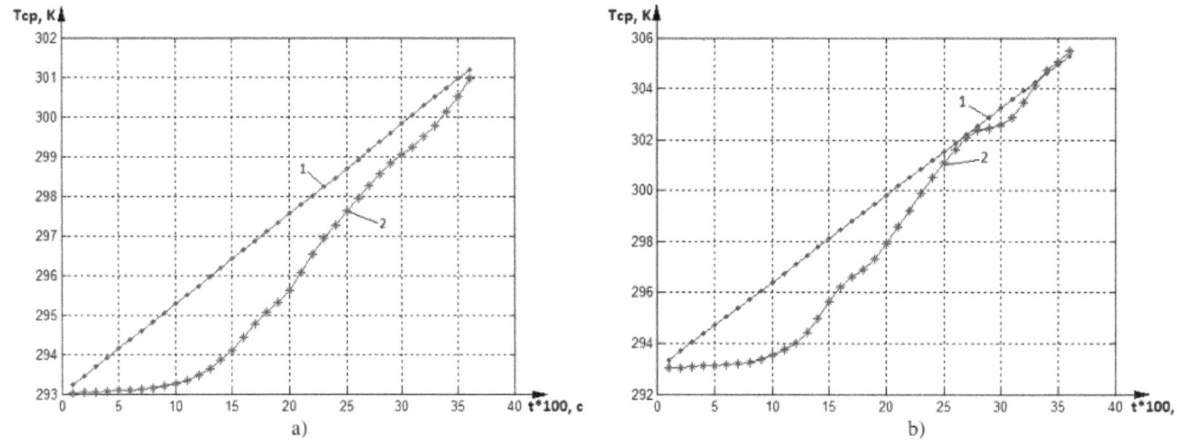

Figure 4: The dependences of the average temperature of oil on the time calculated by the simplified (balanced) method (1) and the obtained from numerical simulation (2). The heat flow is equal accordingly: a) q = 100 kW/m^2; b) q = 150 kW/m^2.

Analyzing the dependence shown in Fig. 4 can be drawing a few conclusions. First, irrespective of magnitude of the heat flux at the lower boundary of the solutions, the largest deviations in the average temperature of fuel in the entire volume of the tank take place at the initial stage of heating. Secondly, it is clear that with increasing of heat supply intensity at the lower boundary of the tank oil temperature the deviation of averages calculated by simplified method and obtained from numerical simulation at the initial stage of heating increases. For example, in the time about 1000s and when the value of heat flux q = 100 kW/m^2, this deviation is 2 K, while for q = 150 kW/m^2 for 3 K. However, it should be noted that by increasing the intensity of heat flux the average temperature value of oil calculated during the numerical simulations at large times slightly different from the average temperature obtained by balanced method. Also, comparison of the minimum temperature of oil in the tank with the mean temperature of the entire volume of fuel at the final time for different values of the heating intensity can be seen that the difference between them varies within 6-6.5 K.

CONCLUSION

It is established that the value deviations of the average temperature of fuel in all storage volume calculated by simplified (balanced) method and obtained by numerical simulation results, can be considered negligible. This is constitute grounds for conclusion about the possibility of investigation of energy efficiency technologies for regulated heating modes of fuel oil storage within the framework of models of incompressible viscous fluid.

REFERENCES

1. ECPs and economy of Russia regions. Reference: V.7. - Moscow: Energy. (2007)
2. Yu.G. Nazmeev, Fuel oil farms at thermal power plant. - Moscow: Publishing House of the Moscow Power Engineering Institute, 2002
3. G.V. Kuznetsov, M.A. Sheremet, Two-dimensional problem of natural convection in a rectangular domain with local heating and heat-conducting boundaries of finite thickness, Fluid Dynamics. V. 41. No. 6. (2006) 881-890
4. G.V. Kuznetsov, M.A. Sheremet, New approach to the mathematical modeling of thermal regimes for electronic equipment, Russian Microelectronics. V. 37. No. 2. (2008) 131-138
5. G.V. Kuznetsov, M.A. Sheremet, Mathematical modelling of complex heat transfer in a rectangular enclosure, Thermophysics and Aeromechanics. V. 16. No. 1. (2009) 119-128
6. V.M. Paskonov, V.I. Polezhaev, L.A. Chudov, Numerical modeling of heat and mass transfer. - Moscow: Nauka, 1984 – 288 p.

CHAPTER 21

DEVELOPMENT OF ENERGY EFFICIENT TECHNOLOGIES FOR BURNING COAL IN MODERN THERMAL POWER PLANTS AND EFFICIENCY ASSESSMENT TOOLS

Vitali Dubrovskiy[1], Marina Zubova[2] , Nikolai Sedelnikov[1] and Anna Dihnova[2]

[1]National Research Tomsk Polytechnic University, 634050 Tomsk, Russia

[2]Institute of Economics and Business Process Management, Department of Economics and Organization of energy and transport sectors, Russia

ABSTRACT

Universal ecological energy-efficient burner was described. The burner allows to burn different types of coal and lignite without the use of fuel oil for kindling the boiler. Efficiency assessment tools of the introduction of the burner for combustion of coal in modern thermal power plants were given.

INTRODUCTION

Modern methods of preparation and combustion of coal in thermal power plants do not completely meet the current requirements of reliability, environmental safety and economic efficiency of the boilers. This fact raises the need to develop high-technology energy use of coal, aimed at a comprehensive solution to the problems of burning coals in modern thermal power plant, namely the exclusion of the use of expensive fuel oil, fuel oil, reducing the nitrogen oxides reduction in slagging of heating surfaces of boiler [1].

Citation: Vitali Dubrovskiy, Marina Zubova, Nikolai Sedelnikov and Anna Dihnova, Development of Energy Efficient Technologies for Burning Coal in Modern Thermal Power Plants and Efficiency Assessment Tools. EPJ Web of Conferences. 2016;11001017. DOI: 10.1051/epjconf/201611001017

ENERGY EFFICIENT TECHNOLOGIES FOR BURNING COAL

Fig. 1 shows a sketch of a universal ecological energy-efficient burner. Universal ecological energy-efficient burner device consists of a retort section which is heated for 15-20 minutes at the start of the boiler, silicon carbide heating elements to a temperature of 550- 650 °C. Then, coal dust is supplied in the burner device; it is gasified at the excess air coefficient which is significantly less than unity. After establishing a stable regime of the heating system it turns off automatically. A mixture of combustible gases and the coke particles is ignited and fed to combustion in the combustion chamber of the boiler at the outlet from the burner it is mixed with the secondary air. Thus, the start of the boiler takes place without the use of expensive fuel oil, thus saving from 40 to 100 tons of fuel oil per kindling. Department of ecological energy-efficient burners can be successfully used with the boiler operating at low load, because flow of gasified coal dust is produced in the volume of the burner and entering the combustion chamber of the boiler will serve as the backlight of the torch instead of currently used expensive liquid fuel – fuel oil. An important novelty of the proposed incineration technology as opposed to analog is its focus on complex solution of problems of the energy efficient use of coal on the basis of universal burners with pre-flow gasification of coal dust in a muffle of the burner. The introduction of universal burners will significantly improve the efficiency of coal-fired thermal power plant units through integrated solutions of coal combustion in boiler units. Similar Russian and foreign developments usually solve maximum of two problems of burning coals.

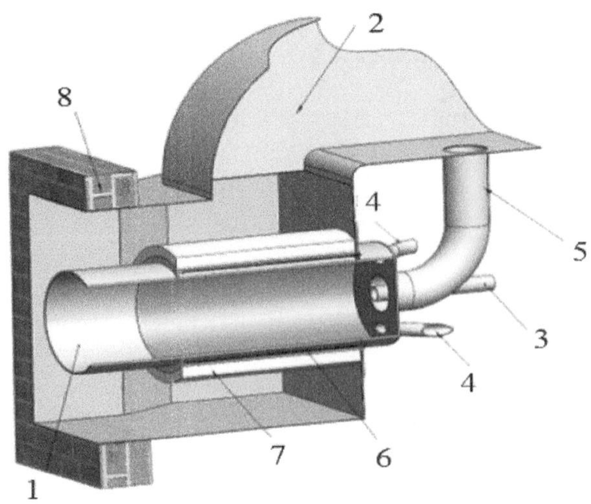

Figure 1: Sketch of the universal energy-efficient burner: 1 – part of the muffle; 2 – secondary air conduit; 3 – device for supplying high concentration of dust; 4 – tangential primary air inlet; 5 – axial feed conduit of primary air; 6 – heating system; 7 – thermal insulation; 8 – boiler loophole.

The technology is designed to burn all types of highly reactive and brown coals, including Kansk-Achinsk and Kuznetsk Basin high-volatile coals, and can work with all types of boilers using coal fuel combustion chamber. Experience of development of the universal burner industrial design installed on the boiler BKZ-420-140 Krasnoyarsk RHPP-2 demonstrated the technical feasibility and a real opportunity to achieve the goals. With full equipment of universal boilers burning devices a real opportunity to provide not only the Plasma-fuel firing the boiler will be achieved, as well: to reduce 2-3 times the emissions of nitrogen oxides; to stabilize the operation of the boiler at low loads; increase the utilization of installed capacity thermal power plants by reducing pollution and slagging of heating surfaces of boilers; provide a payback period for about three years.

EFFICIENCY ASSESSMENT TOOLS OF ECO-LOGICAL ENERGY SAVING TECHNOLOGIES

Efficient energy use of coal from the Kansk-Achinsk basin due to a number of problems and in order to coals of Kansk-Achinsk basin could compete with other types of fuel (including gas), it is necessary to solve a number of major challenges in the development of new methods for the preparation and combustion ensuring high economic and environmental performance of energy equipment. It is necessary to develop more energy-efficient coal combustion technologies. It is extremely important that these technologies can be implemented in existing power plants with little capital investment. In this regard, relevant is the implementation of a new science-intensive technologies for burning coal, aimed at introducing ecological and energy efficient burners, allowing to organize the burning of coal in the existing boiler units so that solve the problems of environment and energy, stabilizes the boiler unit in a given mode [2].

To assess the effectiveness of the implementation of ecological and energy-saving technology to a power station it is necessary to consider projects simultaneously from two perspectives: local (independent investment projects to ensure the recoverability of the costs of their implementation by reducing the fuel component) and system-wide (elements of the system of fuel supply the country's Energy industry, the creation of which allows to reduce damage to the environment and improve the efficiency of energy use of fuel). To assess social efficiency and to determine the prospects of replication technologies it should be modified to use Russia's energy model. This will provide new results on a system-wide definition of the effect of the implementation of energy efficient technologies, to justify the rational scale of development and to identify high-priority technology diffusion regions of Russia in the long term. An important feature of the innovative projects financing is the probabilistic nature of income. As a result of feasibility studies for such projects should be risk-investment analysis. In this regard, assessment of project risks mastering and replication of eco-energy saving technologies in modern thermal power plant should be carried out by a complex procedure consisting of identification and systematization of the risks of development and replication eco-energy saving technologies; methods of sensitivity analysis for the comparative assessment of the effect of individual factors on the final indicators of economic efficiency of the project; probabilistic method to determine expected, the most likely value of the integral economic effect; methods of scenario approach to determine the status of the project break-even point.

REFERENCES

V. A. Dubrovsky, Improving the efficiency of energy use of Kansk-Achinsk coal basin (CPI KSTU, Krasnoyarsk, 2004) [in Russian]

V. A. Dubrovsky, M. V. Zubova, Energy-efficient lighting and kindling the flame combustion chambers of boilers (Teplotechnik, Moscow, 2012) [in Russian]

CHAPTER 22

IMPROVEMENT OF ENVIRONMENTAL ASPECTS OF THERMAL POWER PLANT OPERATION BY ADVANCED CONTROL CONCEPTS

Robert A. Mikulandric[a] , Dražen M. Lon^Ar[a] , Dejan B. Cvetinovi[b]
Gabriel Spiridon[c], and Daniel R. Schneider[a]

[a]Faculty of Mechanical Engineering and Naval Architecture, University of Zagreb, Zagreb, Croatia

[b]Vin~a Institute of Nuclear Science, University of Belgrade, Belgrade, Serbia

[c]R&D Institute for Scientific Research and Technological Engineering for Automation - IPA SA, Bucharest, Romania

ABSTRACT

The necessity of the reduction of greenhouse gas emissions, as formulated in the Kyoto Protocol, imposes the need for improving environmental aspects of existing thermal power plants operation. Improvements can be reached either by efficiency increment or by implementation of emission reduction measures. Investments in refurbishment of existing plant components or in plant upgrading by flue gas desulphurization, by primary and secondary measures of nitrogen oxides reduction, or by biomass co-firing, are usually accompanied by modernisation of thermal power plant instrumentation and control system including sensors, equipment diagnostics and advanced controls. Impact of advanced control solutions implementation depends on technical characteristics and status of existing instrumentation and control systems as well as on design characteristics and actual conditions of installed plant components. Evaluation of adequacy of implementation of advanced control concepts is especially important in Western Balkan region where thermal power plants portfolio is rather diversified in terms of size, type and commissioning year and where generally poor maintenance and lack of investments in power generation sector resulted in high greenhouse gases emissions and low efficiency of plants in operation. This paper is intended to present possibilities of implementation of advanced control concepts, and particularly those based on artificial intelligence, in selected thermal power plants in order to increase plant efficiency and to lower pollutants emissions and to comply with environmental quality standards prescribed in large combustion plant directive.

INTRODUCTION

West Balkan countries (WBC) are heading towards EU integration. Due to EU membership, WBC will be forced to comply with current EU legislative related to thermal power plant efficiency and emission control. All EU member countries have ratified the Kyoto protocol which requires energy consumption and emission pollution decreasing. EU parliament has promoted numerous obligations, laws and Directives to meet goals from the Kyoto protocol and to reduce environment pollution. Most of EU counties have started the process of shutting down old and inefficient power plants which are mostly coal and heavy oil fired. Additional investments into these thermal power plants (TPP) and their

Citation: Mikulandrić Robert, Lončar Dražen, Cvetinović Dejan, Spiridon Gabriel, Schneider Daniel R. Improvement of environmental aspects of thermal power plant operation by advanced control concepts. Thermal Science. 2012;16(3)759-772. DOI: 10.2298/TSCI120510134M refurbishment become unnecessary. Construction of new coal or heavy oil fired power plants is removed from large number of national energy strategies and replaced with construction of renewable energy based facilities in order to meet goals from Kyoto protocol and EU Directives. One solution for emission mitigation problem is implementation of secondary measures technology in TPP. Some of secondary measures technologies for NO_x and SO_x reduction are selective catalytic reduction (SCR), selective non-catalytic reduction (SNCR), ammonia injection, ammonia scrubbers, oxygen-enriched combustion (OEC), and other measures described in [1, 2]. However secondary measures are investment intensive. Another solution is getting primary measures closer to the power plant processes limits using advanced techniques regarding power, temperature, combustion, and frequency control. These measures often prove to be very useful and cost effective [3]. But construction and simulation of such models (especially combustion models) by means of mathematical, physical and chemical analysis (with CFD programs) is very complex process with very long computation time. Because of this reasons, utilization of these models are impracticable for on-line power plant control [4]. Advanced control technologies for improving system operability and environment maintainability based on artificial-intelligence (AI) seems to be promising approach for modelling and controlling large and non-linear power generation processes.

WEST BALKAN COUNTRIES THERMAL POWER PLANT STATUS

The power generated by fossil power plants represents about 59% (52% from coal, 4% from oil, and 3% from gas) of all produced power in the WB region [5]. Most of fossil power plants currently in operation were constructed mainly between 1955 and 1990 during lower emission restrictions. From 1991 till 2008 the technology of power generation and the environmental characteristics of the operating fossil fuel-fired plants in the WBC had not been improved considerably. This is a result of poor maintenance and lack of investments in the energy generation sector. All these factors finally result in high greenhouse gases emissions and low efficiency of WBC thermal power plants. Some of power plants were partially reconstructed (with introduction of ecology friendly technologies) which resulted in lower NO_x, SO_2, and fly ash emissions. Some of the NO_x emissions restricting measures were made as primary measures. On the other hand the secondary measures, which support the inhibition of already created NO_x, were not widely used mostly because of their high investment costs. Currently, the NO_x emissions in most of the fossil fuel power plants vary between 200 and 750 mg/Nm^3 [5], which is a considerable reduction in comparison to the past decades. Some of these emissions however still exceed today's EU limits and most of them exceed the future limits, coming to validity in 2016, setting the NO_x emissions limit to 200 mg/m^3 . Discharge of CO_2 is, however, comparable with EU TPP, which can be explained by the very nature of the system employed for energy transformation.

Domestically produced lignite and natural gas were important fuels for heavy industries during the era of central planning in the 1980s and 1990s. Demand for these fuels had decreased with the closure of most heavy industry. The lignite is of low quality with contents of sulphur (0.3 to 1%), ash (10 to 30%), and moisture (up to 60%) [6, 7]. It also has low calorific value (in the range of 4 500 to 10 000 kJ/kg). Western Balkan energy sector also faces some non-technical problems such as reorganisation of ownership structure, difficult working conditions, social issues, etc. [8].

General Western Balkan energy sector characteristics and key challenges are:

- lack of domestic capacity for thermal electricity generation,
- high energy intensity,
- higher energy consumption in the future due to economical expansion,
- low oil and natural gas reserves, fossil fuel import,
- domestically produced lignite of low quality,
- old TPP technology,
- lack of TPP maintenance,
- low TPPs efficiency,
- frequent TPP outages,
- SO_2 and NO_x emissions above EU limits,
- low investment possibilities in energy sector,
- coal preparing problem (milling, dosing, pre-heating), and
- poor power plant personnel process understanding.

INFORMATION AND COMMUNICATION TECHNOLOGY BASED SYSTEMS

The primary task for power plant operation is to meet the load demand for electric power and to ensure stable, safe and efficient power generation. However, task of establishing optimal power plant operation processes seems to be very demanding. Purely software-oriented approach to optimizing processes has been an ongoing success in the world's power plants for just on a decade now and has made many improvements regarding power plants operability, efficiency, etc. (tab. 1). Besides being widely incorporated into new power plants, process optimization is nowadays a popular choice for power plant upgrades. More and more operators are going beyond simply replacing their old instrumentation and control systems and are taking the opportunity of a scheduled modernization to make their power plants more flexible to current market situation, which means greater flexibility and higher efficiency, better availability and lower emissions. General benefits of information and communication technology (ICT) based systems for high flexibility is more flexible power generation which implies lower process losses during the load changes, start-ups and shut-downs. These loss reduction increases annual profit due to lower operational costs. Main disadvantages of ICT based systems for high flexibility is reduction of overall TPP efficiency and increment of annual emissions. Due to poor electricity market optimisation in WBC and high possibility of load regulation with hydro power plants, installation of high flexibility ICT systems do not offer great opportunities for TPP efficient increment and emission reduction in WB TPP.

Table 1: General classification of ICT based systems

ICT based system group	ICT based system
High flexibility	Power optimization
	Economic load allocation for boilers and turbines
	Combined cycle control tools
	Minimum load reduction
	Maximum load extension
	Fast load increase, fast TPP start-up, low loss TPP start-up
	Frequency control
High availability	Automatic runback control
	Low-stress operation
Low emissions	Emissions control
High efficiency	Advanced combustion optimization
	Temperature optimization

Main advantage of higher TTP availability optimisation is extended service life. Although it deals with WB TPP requirements for extended lifetime (in correlation with lack of domestic capacity for thermal electricity generation), it does not lower emissions or increase efficiency, which is main challenge for WB TPP. Poor power plant personnel process understanding and marginal emission reduction with no efficiency increment are key drawbacks for emission control ICT systems implementation in WB TPP. Generally, implementation of emission control systems in TPP is key requirement to reduce emissions and to comply with large combustion plants (LCP) emission standards.

Table 2: ICT based systems used in WBC TPP [9]

ICT based system group	ICT based system implemented	Thermal power plant	Country
High efficiency	Steam temperature optimization	Coal TPP 210 MW Coal TPP 675 MW	Croatia Macedonia
High flexibility	Fast TPP start-up Frequency control	Coal TPP 210 MW Coal TPP 210 MW	Croatia Croatia
High availability	Automatic runback control	Coal TPP 1650 MW	Serbia
Low emissions	Emissions control	Crude oil TPP 320 MW	Croatia

ICT based systems for high efficiency optimise TPP processes to increase TPP efficiency and due efficiency increment to lower emissions. Mentioned improvements are key challenges of WB TTP. Old emissions monitoring technology in WB TPP encounters with possibilities of ICT systems to improve efficiency. Implementation of ICT based systems for higher efficiency gives great opportunity to increase efficiency and to reduce emissions within primary measures technologies. With the need of power production increment in WB sector, introduction of ICT based systems for high efficiency into TPP control system as one of primary measure seems to be promising action to meet this goal.

Most of ICT based systems are implemented in coal TPP, mainly to increase efficiency and flexibility and therefore reduce emissions and operational costs. Croatia is most prominent in ICT based systems implementation (tab. 2).

Comparative analysis of ICT based systems showed that the strongest way to compile with EU Directives and to minimize emissions and increase TPP efficiency is to introduce combustion or/and steam temperature optimization into WBC TPP. But only combustion optimization directly successfully encounters thermal power plant efficiency and emission mitigation at the same time [10].

POSSIBILITIES FOR ARTIFICIAL INTELLIGENCE SYSTEMS IMPLEMENTATION IN THERMAL POWER PLANTS

Artificial intelligence systems (AIS) are widely accepted as a technology offering an alternative way to tackle complex and undefined problems. They can learn from examples, they are fault tolerant in the sense that they are able to handle noisy and incomplete data, they are able to deal with non-linear problems, and once trained can perform prediction and generalization at high speed. They have been used in diverse applications in control, robotics, pattern recognition, forecasting, medicine, power systems, manufacturing, optimization, signal processing, and social/psychological sciences. They are particularly useful in system modelling such as in implementing complex mappings and system identification [11].

As mentioned in the section Information and communication technology based systems, the strongest way to reduce emissions and to improve efficiency is combustion process optimisation. Combustion optimisation is mostly conducted through fuel and air flow regulation. Conventional air flow control for combustion processes (see fig. 1) in coal TPP consists of four proportional-integral (PI) controllers. Based on required boiler thermal load, measured O_2 in flue gases on the boiler exit and firing requirements, first PI controller partially controls total air flow to minimize share of O_2 in flue gases. Secondary PI controller, based on boiler thermal load, measured total air flow and regulated air flow for O_2 minimization controls total air flow in combustion process. Total air flow is then divided into secondary air and tertiary air (air for additionally burning). Secondary air is subdivided for coal feeders where it is controlled separately.

Based on coal feeder load, total secondary air for coal feeder is calculated. Secondary air for burners (inside same coal feeder) is then controlled separately, based on measured air temperature and calculated secondary air for coal feeder. Total tertiary air is calculated from boiler thermal load and total air flow. After that, tertiary air for every burner is controlled separately based on measured air temperature and calculated total tertiary air flow. Parameters for conventional PI controller in air flow control must be carefully tuned to meet control demands in various operating condition. One approach to tackle changing process states is gain scheduling approach, where parameters for linear PI controller can be adjusted regarding different operating point of the system.

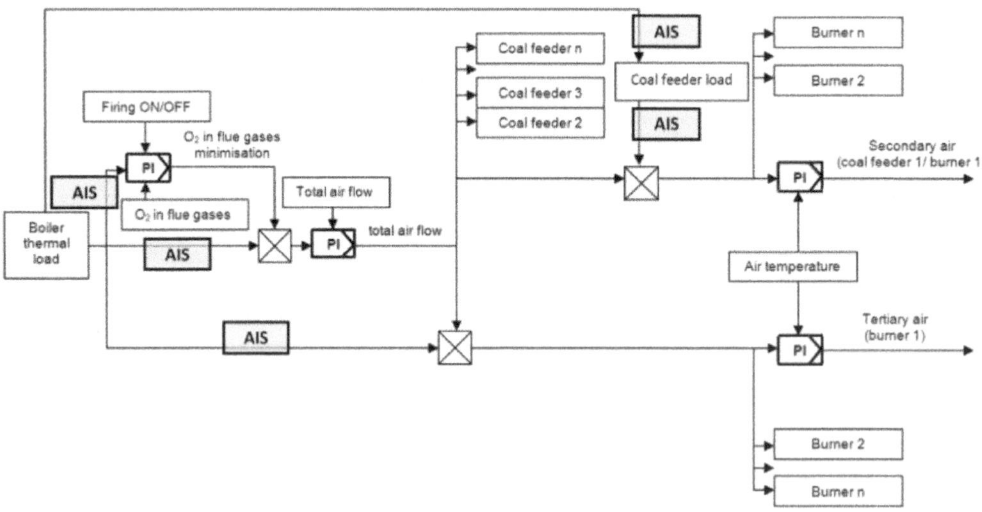

Figure 1: Appropriate places for AIS introduction to conventional air flow control.

Calculation of total, secondary, and tertiary air flows based on boiler thermal load, coal feeder load and air temperature are purely linear. This calculation gives linear approximation of required air flows regarding input parameters. Processes behind these linear approximations are very complex and highly non-linear, so by introducing linear approximations into process control, we introduce a certain error between calculated approximations and real process values. Neuro-fuzzy control has possibility to "learn" from input parameters to approximate parameter values from these processes. With better process models, the calculated approximations are closer to real process values. That gives more realistic input parameters for process control, and improve overall process control. Places for Artificial Intelligence Systems introduction to conventional air flow control is shown in fig. 1.

Artificial neural network and neuro-fuzzy combustion modelling and optimization methods can be divided into two stages. In the first stage, the relation between NO_x emissions and various operational parameters of the boiler is modelled. After that, operational parameters for low NO_x combustion are optimized. Optimization is based on previously constructed NO_x emissions model. NO_x emissions are often multi-dimensionally and highly non-linearly correlated to boiler operational parameters, so it is difficult to establish a perfect NO_x emissions predicting model. Due to high non-linearity, establishing NO_x emission predicting model depends on acquired (measured) emission data. Artificial neural-networks are well-known tools among artificial intelligence techniques, which are able to reproduce the relationships existing between input and output variables of highly non-linear systems [12]. In the second stage of combustion modification, some optimization algorithms are used to manipulate the inputs of the model in order to minimize the emissions output. Coal combustion and NOx emission modelling can be conducted through computational fluid dynamic (CFD) models [13, 14], but their very long computation time imposes need for different model that will be practical for on-line power plant control. Artificial neural networks and neuro-fuzzy models are currently the most researched approaches to and NO_x emission modelling [15]. They have proved their effectiveness on emissions prediction and control [16, 17]. For coal combustion modelling, support vector regression approach [12] and generic algorithm approach [12] can be used, where unknown parameters act as random variables with a known aprior probability distribution. The process identification then shifts into process observation (measured data).

Data Processing and Selection

For utilizing neuro-fuzzy learning algorithm, the probability model has to learn from observed/measured data. Data for learning algorithm were extracted from minute based field measurements on the 650 MWe TPP „Nikola Tesla B" (TENT B), Obrenovac, Serbia. Parameters, with resolution of 1 minute, selected for the purposes of neuro-fuzzy learning algorithm are: power output [MWe], coal/heavy oil flow rate [%, m3 h–1], total air quantity [$Nm^3 h^{-1}$], oxygen content in furnace/flue gasses [%], feeder loads [%], secondary and tertiary air flow [$Nm^3 h^{-1}$], fuel (coal and heavy oil) composition and air temperature [°C]. Measurements of some of these parameters are shown in fig. 2. Measured data have often some erroneous data due to human errors, faulty sensors etc. Because of that, before process prediction modelling and neuro-fuzzy learning algorithm utilization (neuro-fuzzy training), training data preprocessing is needed. Erroneous data (called "outliers") have to be indentified and removed from neuro-fuzzy training data. Erroneous data is often very difficult to indentify due to large data sets. They can be identified and removed either by observing data sets or by implementation of user defined rule system for erroneous data identification. For initial erroneous data identification, standard deviation rule system (±3σ) has been used.

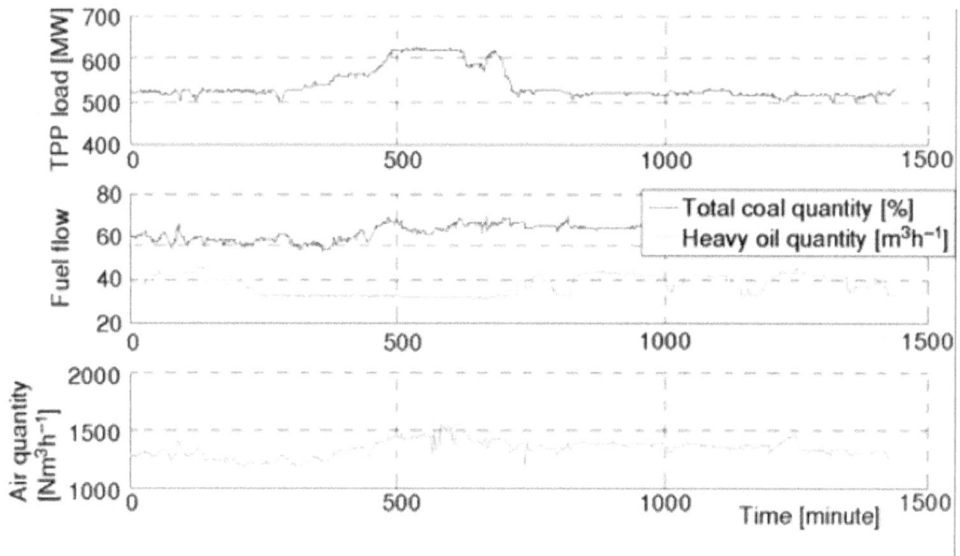

Figure 2: Measured data from 650 MWe coal fired/heavy oil TPP unit.

Some of secondary air flow measured values exceed upper or lower standard deviation limits (fig. 3). Due to this, they are removed from training data and replaced by interpolated value between values that are 3 minute before and 3 minute after erroneous data. Data values for interpolation are also checked to meet standard deviation rules before their usage.

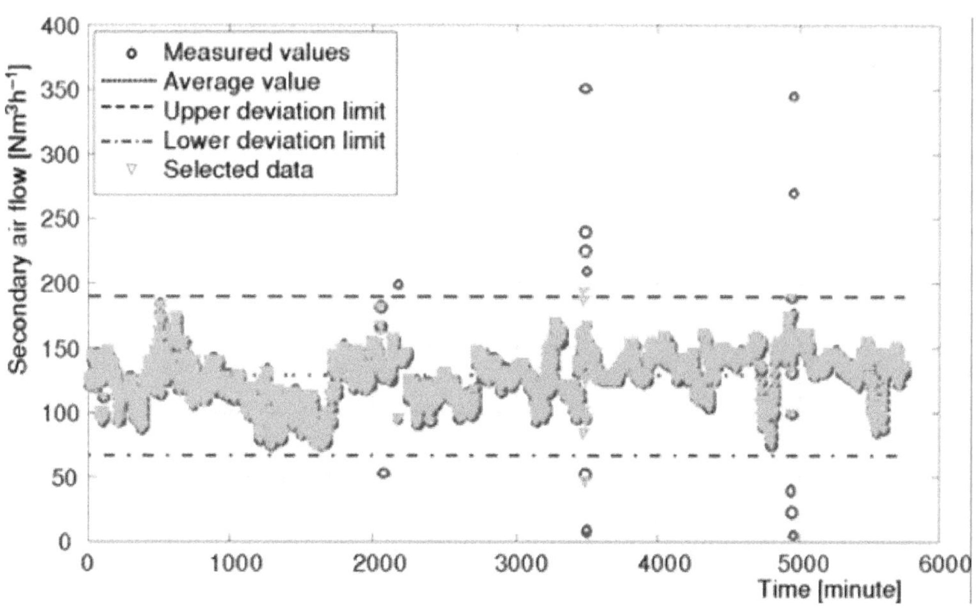

Figure 3: Data selection process after initial erroneous data identification.

In some cases measured data can exceed deviation limits for a longer time. This is not because of human measuring error or sensors fault, but due to power plant operation changes. In this case, for shorter time, power plant operator had to increase heavy oil flow (fig. 4). Heavy oil flow values exceed upper deviation limits, but they are not erroneous data, so by that, they cannot be excluded from training data. For identification of such process data, user defined data identification rule system has been devised. It takes into consideration measured values that are 5 minutes before and 5 minutes after current measured value. If the measured value that is 5 minute before or 5 minute after current value is also over (or below) deviation limit, current value for identification is considered valid for training data.

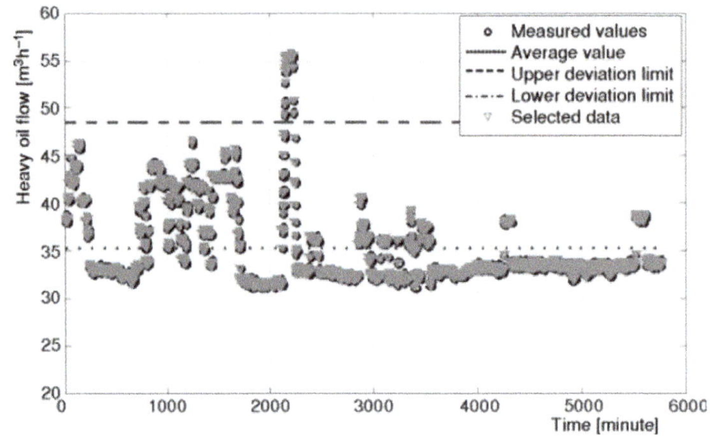

Figure 4: Data selection process after user defined data identification.

After data selection, ANFIS total air flow prediction test model has been devised in MATLAB® programming environment [18]. Average error between measured and calculated values without data selection is 4.97%, while average error with data selection is 2.95% (fig.5).

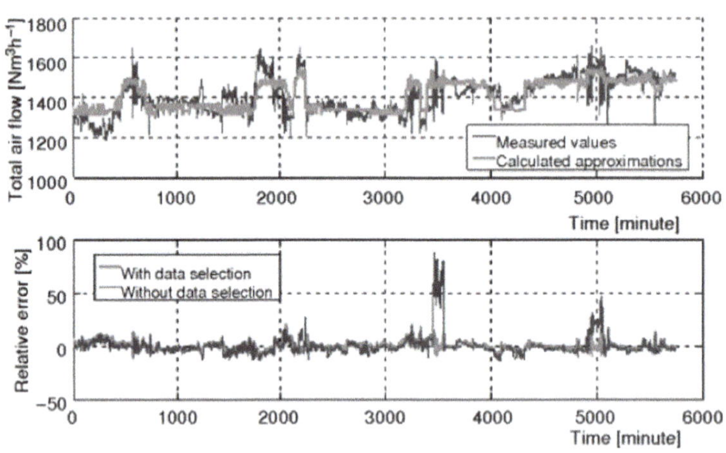

Figure 5: Selected data for artificial neural network training.

Process Parameter and Emission Prediction Modelling

Measured process data such as NO_x and SO_x emission values are necessary for overall process evaluation. With NO_x and SO_x emission values database, process performance can be evaluated and eventually improved regarding emission mitigation. In combustion process there are three primary sources of NO_x formation:

- NO_x formation due to bound nitrogen in fuel (fuel NO_x).
- Formation of NO_x due to high-temeprature combustion and residence time of nitrogen molecules at that temperatures (thermal NO_x).
- NO_x formation due to reaction of atmospheric nitrogen (promt NO_x).

For NO_x emission calculation, Thompson et al. model has been used [18]. The NO_x formation rate is primary function of combustion process temperature. The higher the temperature, the NO_x formation rate is faster. In combustion process, fuel and air mixing imperfections effects NO_x formation. Consequently, due to fuel and air mixing imperfections, NO_x formation rate becomes affected by combustion temperature and air distribution. When the fuel flow in burner increases, the turbulence in combustion area becomes greater and it improves mixing process. Simultaneously, the temperature is raised due to additional fuel. With adding more primary air to the burners, turbulence becomes grater, but the combustion temperature falls due to higher combustion losses:

$$\frac{dNO}{dt} = \alpha_0 W_f^r \left(1 + \alpha_1 \frac{\xi(t) - 55}{90}\right)\sqrt{\lambda_{st} - \lambda} \quad (1)$$

where α_0 and α_1 are reactions coefficients regarding fuel flow and burner tilt positions, W_f^r – fuel mass flow rate, $\xi(t)$ – burner tilt position (in percentage), λ_{st} and λ – stoichiometric and real fuel to air ratio. The more primary air is added to burner for combustion process, the higher oxygen concentration in flue gases is occurred:

$$\lambda_{st} - \lambda = \frac{v_a}{\beta}(O_2) \quad (2)$$

From eq. (2), eq. (1) becomes:

$$\frac{dNO}{dt} = \alpha_0 W_f^r \left(1 + \alpha_1 \frac{\xi(t) - 55}{90}\right)\sqrt{\frac{v_a}{\beta}(O_2)} \tag{3}$$

where v_α is specific volume of air ($v_\alpha = 0.7767$), β – theoretical oxygen volume percentage for combustion process, and O_2 – oxygen concentration. Theoretical oxygen volume percentage for combustion processes can be calculated from fuel composition that is defined by:

$$\beta = 1.87C + 0.70S + 5.6H \tag{4}$$

where C, S, and H are mass content of carbon, sulphur and, hydrogen in fuel. Substituting eq. 4 into eq. 3 NO_x formation rate becomes:

$$\frac{dNO}{dt} = \alpha_0 W_f^r \left(1 + \alpha_1 \frac{\xi(t) - 55}{90}\right)\sqrt{\frac{v_a}{1.87C + 0.70S + 5.6H}(O_2)} \tag{5}$$

In this case, all the secondary damper positions are fixed during operation. Burner tilt positions which as percentage is ranging from 10% (lowest position) to 100% (highest position), setting 55% as middle position, are also fixed at 60%. After measurement of burner fuel and air flows on 650 MWe coal fired TPP, parameters α_0, α_1 and r were chosen as 23.77, 0.438 and 0.25. NO_x formation model has been used for NO_x values calculation on 650 MWe coal fired TPP. Burner load rate, burner primary air flow and O2 concentration, both with coal composition are considered in equation.

RESULTS

After processing all the data from 650 MW_e coal fired/heavy oil TPP unit, ANFIS model for total, secondary and tertiary air values approximation has been devised in MATLAB® programming environment. For devising ANFIS model, Mamdani model with Gauss membership functions has been used. In dependence of unit load and O_2 content in flue gases total air values have been calculated. Steam boilers of the TENT B are designed for the domestic lignite from the coal mine „Kolubara" as the main fuel. As start-up fuel heavy fuel oil is used. Fuel heating value has been taken as constant (based on main guarantied design parameters of the coal). The lower heat capacity value of the fuel is 6.699 MJ/kg, the moisture content is 47.8%, the ash content is 19%, and the content of sulphur is 0.5%. In the real power plant operation this will not be the case. Fuel with different composure (heating value, moisture composition etc.) is used for combustion process. Lack of coal quality homogenization can cause problems from the point of combustion optimisation. Only small divergence of the coal quality parameters (within the range of ±5%) enables the optimization of the combustion process in the furnaces. However, due to the lack of homogenization, i. e. the equalization of the coal quality, the lower heat capacity of the coal supplied to the unit is in the range from 5 to 9 MJ/kg. Very often, during the winter season when energy demands are the highest, coal supplied to the TPP is extremely low quality with very high moisture content. In this situation, combustion support with heavy oil burners is necessary to reach nominal boiler load. Co-combustion of low quality lignite and heavy oil in the boiler furnace has to be carefully led from the unit operators to achieve optimal combustion. That is the additional reason for implementation of artificial system for combustion optimization.

ANFIS approximation model shows good corelation with measured values for total air flow calculation (fig. 6). The error between measured and calculated values is mostly between ±10% which represents good aproximation of total air values in dependance of unit load and O2 content in flue gases. Approxiation errors of secondary and tertiary air are mostly between ±20% but in some marginal cases they can reach ±50%. Measurement errors should also be considered.

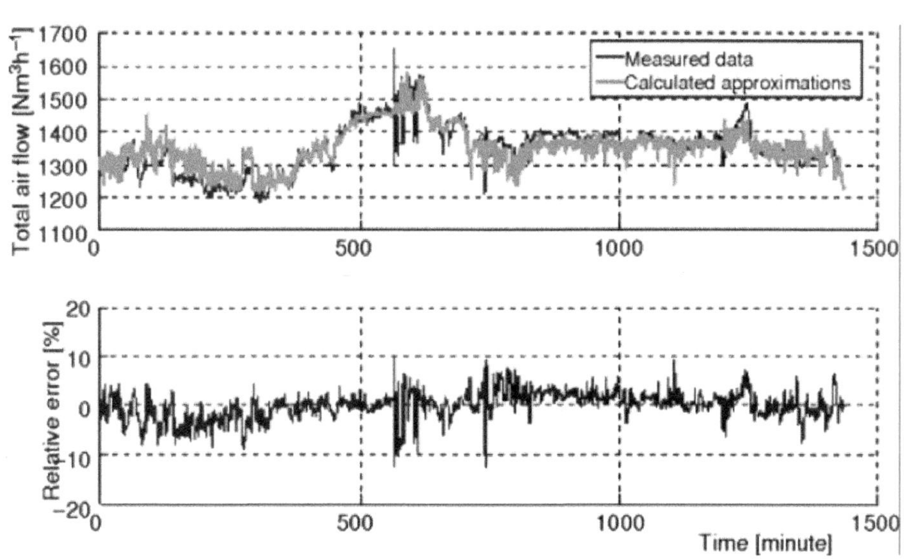

Figure 6: Calculated approximations of total air flow in comparison with measured values.

Figure 7 shows calculated approximations of total air flow in dependence of unit load and O_2 content in flue gasses. It is obvious that total air flow dependence is highly nonlinear and that values of total air flow in power plant unit have a "saddle" around unit load of 540 MW. By increasing or decreasing unit load, current process total air flow increases. On constant unit load there is no obvious correlation between O_2 content and total air flow values.

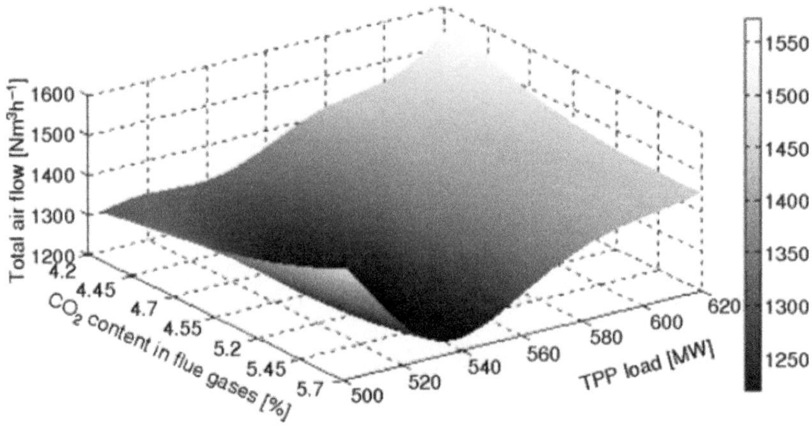

Figure 7: Calculated approximations of total air flow in dependence of unit load and O_2 content in flue gasses.

Figure 8 presents dependence of secondary air flow (on burner 1) on coal feeder load and total air flow while fig. 9 presents dependence of secondary air flow on total air flow and hot air temperature. Similar like in previous case, there is no obvious or linear defined correlation between secondary air flow, coal feeder load and total air flow.

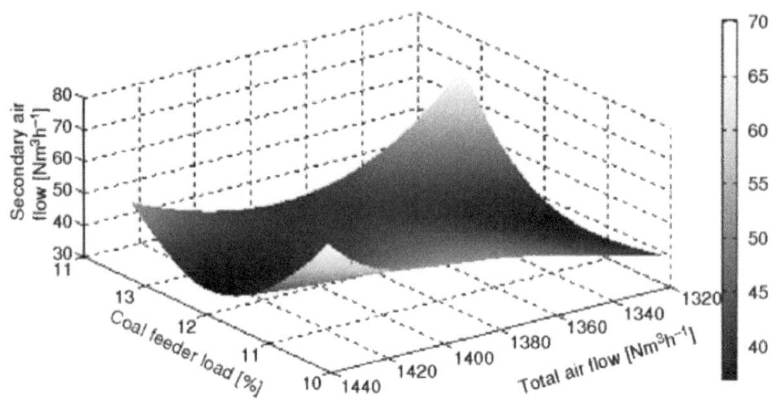

Figure 8: Secondary air flow calculations in dependence of coal feeder load and total air flow.

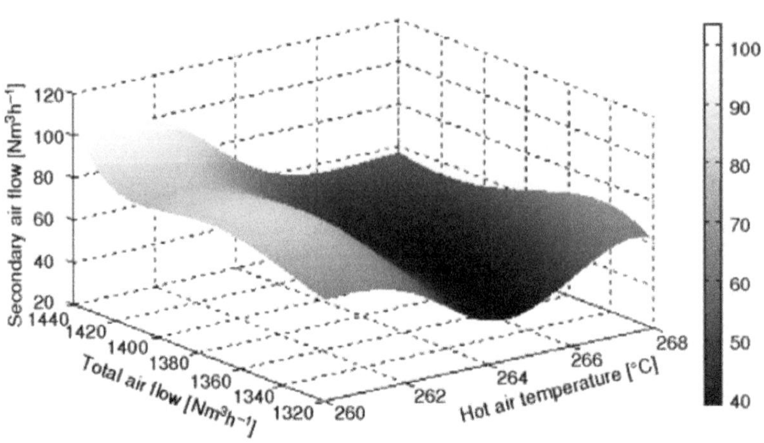

Figure 9: Secondary air flow calculations in dependence of total air flow and hot air temperature.

Figure 10 shows NO_x values during 24 hours power plant operation. NO_x values are ranging between 335 and 395 ppm, while power load is ranging between 500 and 625 MW_e. NO_x emissions are generally higher on lower power plant loads.

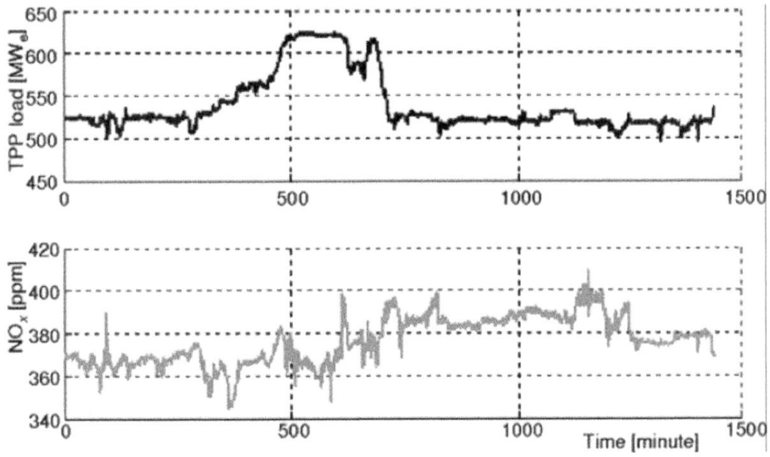

Figure 10: NOx emission calculation in 650 MWe coal fired unit.

Generally, with unit load increment, NO_x emission values declines. On the same TPP output, with air flow increment, the NO_x emission rises (fig. 11). This is result of higher nitrogen input (derived from air) for combustion process. In this case, the temperature of combustion process is decreased which results in lower NO_x formation rate, but additional nitrogen that is introduced through air flow increases NO_x formation rate. General proposition for NO_x minimisation is to keep real fuel to air ratio close to its recommended values that ensures complete combustion process.

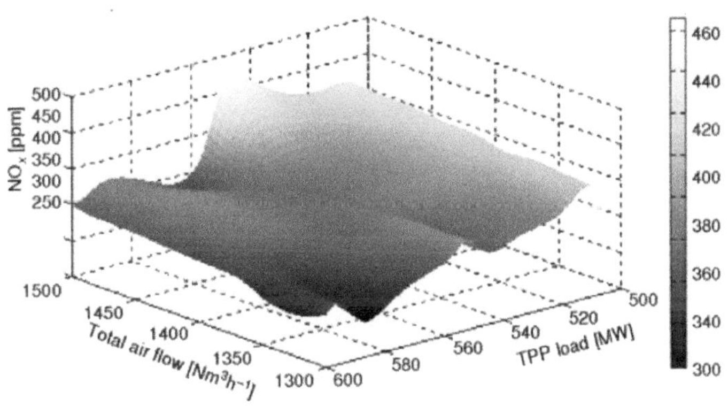

Figure 11: NO_x emission calculations in dependence of total air flow and unit load.

CONCLUSIONS

This paper has analysed the possibilities of implementation of advanced control concepts, particularly those based on artificial intelligence, in WBC TPP in order to optimize combustion, increase plant efficiency and to lower pollutants emissions. General Western Balkan energy sector characteristic is old and inefficient TPP technology and coal dependence with very low investment possibilities. Lack of investment possibilities imposes necessity of easily implementable, low cost solutions that encounters emissions and raises efficiency. Combustion optimisation is the strongest way to directly influences emission mitigation and efficiency increment. Adaptive neuro-fuzzy algorithms can be used for combustion process parameter and emission prediction modelling. ANFIS approximation model for total, secondary and tertiary air values calculation both with fuel flow calculations shows good correlation with measured values and by this they can be used for process parameter and emission values prediction. ANFIS approximation model can be further used for combustion optimisation and emission minimisation purposes. Lack of some detailed process parameter data for process parameter prediction modelling could lead to misleading results and conclusions regarding further TPP combustion process interpretation and control and thus it should be analysed.

ACKNOWLEDGMENTS

This paper has been created within WBalkICT-Supporting Common RTD actions in WBC for developing Low Cost and Low Risk ICT based solutions for TPP Energy Efficiency increasing, SEE-ERA.NET plus project in co-operation among partners from IPA SA-Romania, University of Zagreb, Croatia and Vin~a Institute of Nuclear Sciences Serbia. The project has initiated a strong scientific co-operation, with innovative approaches, and high scientific level, in order to correlate ICT last generation solutions, procedures and techniques from fossil fuelled burning processes thermodynamics, mathematical modelling, modern methods of flue gases analysis, combustion control and AIS systems with focus on Expert Systems category in an optimal form.

REFERENCES

1. Barrett, M., Atmospheric Emissions from Large Point Sources in Europe, Swedish NGO Secretariat on Acid Rain, Goteborg, Sweden, 2004
2. Chatel-Pelage, F., et al., Applications of Oxigen for NOx Control and CO2 Capture in Coal-Fired Power Plants, Thermal Science, 10 (2006), 3, pp. 119-142
3. Kaminski, J., Technologies and Costs of SO2-Emissions Reduction for the Energy Sector, Applied Energy, 75 (2003), 3-4, pp. 165-172
4. Zheng, L., Shuijun, Y., Minaggo, Y., Monitoring NOx Emissions from Coal-Fired Boilers Using Generalized Regression Neural Network, Proceedings, International Conference on Bioinformatics and Biomedical Engineering, Shanghai, China, 2008, Vol. 1, pp. 1916-1919
5. ***, Energy in the Western Balkans, http://www.iea.org
6. Kazagić, A., Smajević, I., Duić, N., Selection of Sustainable Technologies for Combustion of Bosnian Coals, Thermal Science, 14 (2010), 3, pp. 715-727
7. Đurić, S., et al., Qualitative Analysis of Coal Combusted in Boilers of Thermal Power Plants in Bosnia and Herzegovina, Thermal Science, 16 (2012), 2, pp. 605-612
8. Gligorić, Z., et al., Simulation Model – Support to Investment Decision-Making in the Coal Industry, Thermal Science, 14 (2010), 3, pp. 835-844
9. ***, Instrumentation & Controls – P3000 solution compendium, http://www.energy.siemens.com
10. Li, S., et al., NOx Emission and Thermal Efficiency of a 300 MWe Utility Boiler Retrofitted by Air Staging, Applied Energy, 21 (2009), 4, pp. 269-360
11. Kalogirou, S., Applications of Artificial Neural-Networks for Energy Systems, Applied Energy, 67 (2000), 1-2, pp. 17-35
12. Zheng, L., et al., A Comparative Study of Optimization Algorithms for Low NOx Combustion Modification at a Coal-Fired Utility Boiler, Expert Systems with Application, 36 (2009), 2, pp. 2780- 2793
13. Sijerčić, M., Belošević, S., Stefanović, P., Modeling of Pulverized Coal Combustion Stabilization by Means of Plasma Tourches, Thermal Science, 9 (2005), 2, pp. 57-72
14. Filkoski, R., Petrovski, I., Karas, P., Optimisation of Pulverised Coal Combustion by Means of CFD/CTA Modelling, Thermal Science, 10 (2006), 3, pp. 161-179
15. Kalogirou, S., Artificial Intelligence for the Modeling and Control of Combustion Processes: A Review, Renewable and Sustainable Energy Reviews, 5 (2001), 4, pp. 373-401
16. Yap, W. K., Karri, V., ANN Virtual Sensors for Emissions Prediction and Control, Applied Energy, 88 (2011), 12, pp. 4505-4519
17. Ćojbašić, Ž., et al., Computationally Intelligent Modelling and Control of Fluidized Bed Combustion Process, Thermal Science, 15 (2011), 2, pp. 321-338
18. Li, N., Thompson, S., Modelling of Combustion and Emissions for a Cyclic Power Plant, Proceedings, 13th Triennial World Congress, San Francisco, Cal., USA, 1996, Vol. 1, pp. 37-42

INDEX

G

H

I

L

M

N

O

P

Q

R

S

T

U

V

W

Y